Biobased Products and Industries

Biobased Products and Industries

Edited by

Charis M. Galanakis
Research & Innovation Department
Galanakis Laboratories
Chania, Greece

Food Waste Recovery Group
ISEKI Food Association
Vienna, Austria

ELSEVIER

Elsevier
Radarweg 29, PO Box 211, 1000 AE Amsterdam, Netherlands
The Boulevard, Langford Lane, Kidlington, Oxford OX5 1GB, United Kingdom
50 Hampshire Street, 5th Floor, Cambridge, MA 02139, United States

Library of Congress Cataloging-in-Publication Data
A catalog record for this book is available from the Library of Congress

British Library Cataloguing-in-Publication Data
A catalogue record for this book is available from the British Library

ISBN: 978-0-12-818493-6

For information on all Elsevier publications visit our website at https://www.elsevier.com/books-and-journals

Publisher: Susan Dennis
Acquisition Editor: Susan Dennis
Editorial Project Manager: Charlotte Rowley
Production Project Manager: Kamesh Ramajogi
Cover Designer: Victoria Pearson

Typeset by TNQ Technologies

Transferred to Digital Printing in 2020

Contents

9. Bio-based electric devices

V. Hoffmann, M.P. Olszewski, K.M. Swiatek, B. Musa, P. J. Arauzo Gimeno, C. Rodriguez Correa and A. Kruse

10. Bio-based textile coatings and composites

David De Smet, Frederik Goethals, Brecht Demedts, Willem Uyttendaele and Myriam Vanneste

Contributors

Oscar Aguilar-Juárez, CIATEJ, Guadalajara, Jalisco, Mexico

Florent Allais, URD Agro-Biotechnologies Industrielles, CEBB, AgroParisTech, Pomacle, France

Luis Arellano-García, CONACYT-CIATEJ, Guadalajara, Jalisco, México

Neydeli Ayala-Mendivil, CIATEJ, Guadalajara, Jalisco, Mexico

Iliana Barrera-Martínez, CONACYT-CIATEJ, Guadalajara, Jalisco, México

Gustavo Cabrera-Barjas, University of Concepcion, Technological Development Unit, Concepcion, Chile

Leticia Casas-Godoy, CONACYT-CIATEJ, Guadalajara, Jalisco, México

Patricia Castaño, University of Concepcion, Technological Development Unit, Concepcion, Chile

Catalina Castillo, University of Concepcion, Technological Development Unit, Concepcion, Chile

C. Rodriguez Correa, University of Hohenheim (UHOH), Conversion technologies of biobased resources (440f), Stuttgart-Hohenheim, Baden-Württemberg, Germany

Brecht Demedts, Centexbel, Textile Competence Centre, Ghent, Belgium

David De Smet, Centexbel, Textile Competence Centre, Ghent, Belgium

Fatima Charrier - El Bouhtoury, CNRS/Univ. PAU & PAYS ADOUR/ E25 UPPA, Institute of Analytical Sciences and Physico-Chemistry for Environment and Materials (IPREM), UMR 5254, IUT des Pays de l'Adour, Mont de Marsan, France

P. J. Arauzo Gimeno, University of Hohenheim (UHOH), Conversion technologies of biobased resources (440f), Stuttgart-Hohenheim, Baden-Württemberg, Germany

David J. Glass, D. Glass Associates, Inc., Needham, MA, United States

Frederik Goethals, Centexbel, Textile Competence Centre, Ghent, Belgium

V. Hoffmann, University of Hohenheim (UHOH), Conversion technologies of biobased resources (440f), Stuttgart-Hohenheim, Baden-Württemberg, Germany

Hamed Issaoui, CNRS/Univ. PAU & PAYS ADOUR/ E25 UPPA, Institute of Analytical Sciences and Physico-Chemistry for Environment and Materials (IPREM), UMR 5254, IUT des Pays de l'Adour, Mont de Marsan, France

A. Kruse, University of Hohenheim (UHOH), Conversion technologies of biobased resources (440f), Stuttgart-Hohenheim, Baden-Württemberg, Germany

J.W.A. Langeveld, Biomass Research, Wageningen, the Netherlands

Andrés Méndez-Zamora, CIATEJ, Guadalajara, Jalisco, Mexico

Antonio D. Moreno, CIEMAT, Department of Energy, Biofuels Unit, Madrid, Spain

Louis M.M. Mouterde, URD Agro-Biotechnologies Industrielles, CEBB, AgroParisTech, Pomacle, France

B. Musa, University of Hohenheim (UHOH), Conversion technologies of biobased resources (440f), Stuttgart-Hohenheim, Baden-Württemberg, Germany

María José Negro, CIEMAT, Department of Energy, Biofuels Unit, Madrid, Spain

Aleksandra Nesic, University of Belgrade, Vinca Institute for Nuclear Sciences, Belgrade, Serbia; University of Concepcion, Technological Development Unit, Concepcion, Chile

José Miguel Oliva, CIEMAT, Department of Energy, Biofuels Unit, Madrid, Spain

M.P. Olszewski, University of Hohenheim (UHOH), Conversion technologies of biobased resources (440f), Stuttgart-Hohenheim, Baden-Württemberg, Germany

Ana Laura Reyes, CIATEJ, Guadalajara, Jalisco, Mexico

J.P.M. Sanders, Wageningen University, Department of Biobased Chemistry and Technology, Wageningen, the Netherlands

Georgina Sandoval, CIATEJ, Guadalajara, Jalisco, Mexico

Jesus Serrano, Cipa Chile - Advanced Polymer Research Center, Concepcion, Chile

Ana Susmozas, CIEMAT, Department of Energy, Biofuels Unit, Madrid, Spain

K.M. Swiatek, University of Hohenheim (UHOH), Conversion technologies of biobased resources (440f), Stuttgart-Hohenheim, Baden-Württemberg, Germany

Michail Syrpas, Kaunas University of Technology, Department of Food Science & Technology, Radvilėnu pl. 19, Kaunas, Lithuania

Willem Uyttendaele, Centexbel, Textile Competence Centre, Ghent, Belgium

Myriam Vanneste, Centexbel, Textile Competence Centre, Ghent, Belgium

Petras Rimantas Venskutonis, Kaunas University of Technology, Department of Food Science & Technology, Radvilėnu pl. 19, Kaunas, Lithuania

Foreword

I know Charis Galanakis as an authoritative expert from the bioeconomy community, well recognized for his expertise in food and environmental science and technology, innovation, and sustainability. He has a profound knowledge of the bioeconomy in all its dimensions, which he constantly shares with a wide range of audiences, as he does in this book.

It is not simple to explain the bioeconomy to a wide public, due in particular to its still fragmented value chains and geographic areas. This latest work of Charis Galanakis and his coauthors is a successful step in this direction and will represent a much needed major source of information and inspiration for all the professionals, researchers, policy makers, and entrepreneurs who aim to develop industrial applications of biobased products and bring them to the market.

After an exhaustive general overview of the biobased industries in Europe and of the development perspectives for the biobased economy, the book provides a far-reaching and thorough analysis of the biobased products and their respective industries, focusing on biofuels and chemicals, biobased products from algae, from wood materials, as well as biobased packaging materials, electric devices, textile coatings, and composites. Advanced production methods, new technological approaches, current challenges, and future prospects are discussed in depth for each sector.

Through this work, readers will appreciate how the bioeconomy provides society with the way forward to better resource efficiency. What emerges from the book is a clear picture of how we can expand the use of renewable biological resources as possible substitutes for fossil ones, in a context where decreasing the dependency of the European economy on fossil resources is of utmost urgency, taking into account their increasing depletion and their impact on climate change.

In the European Union, the biobased industries already account for 3.7 million jobs and provide €700 billion in turnover. According to the Bio-based Industry Consortium, the sector is expected to generate by 2030 hundreds of thousands of skilled and nonskilled jobs, 80% of which in rural areas. There is potential to regenerate underdeveloped and marginalized regions, contributing to increase and diversify the income of primary producers. Moreover, this will make it possible for the European Union to diminish its dependency on the import of strategic raw material—such as fossil resources and proteins for animal feed. By 2030, by pursuing the appropriate kind of development,

30% of fossil-based products could be replaced by biobased alternatives, resulting in a reduction in GHG emissions of 50%.

As Executive Director of the Bio-based Industries Joint Undertaking, an ambitious €3.7 billion joint initiative between the Bio-based Industry Consortium and the European Union, I am delighted to see Charis Galanakis and his coauthors sharing their extensive knowledge on the biobased economy with the new generation of engineers, scientists, entrepreneurs, and innovators.

BBI JU was part of the bioeconomy strategy approved by the European Parliament in 2012. This book comes at a strategic moment as the updated bioeconomy strategy for Europe was approved in October 2018. Most of the objectives of 2012 are still relevant, but the updated strategy emphasizes further the need to improve communication and better understand the challenges of the bioeconomy. Therefore, education at all levels is key to increase consumers' knowledge.

Whether you are an engineer, scientist, decision-maker, or citizen, this book provides all readers with an overview of the current technological, socioeconomic, and environmental challenges contributing to the current bioeconomy revolution, which is all about preparing Europe for the postpetroleum era. I gladly welcome the initiative of Charis Galanakis and his coauthors and highly recommend this book to anyone as a fascinating review of the current state of the art for a bioeconomy for Europe.

Philippe Mengal
Executive Director
Bio-based Industries Joint Undertaking
Brussels, Belgium

Preface

The increasing population, rapid depletion of nonrenewable resources, and global warming are major driving forces to radically change our production and consumption approaches. Current chemicals are typically based on fossil fuels. On the other hand, the use of renewable biomass as a source could be an excellent alternative, taking account future shortage of petrochemicals. Biobased industries use as many renewable raw materials as possible to generate products such as chemicals, materials, and energy. This is vital if further depletion of nonrenewable raw materials and an increase in CO_2 emissions are to be avoided. Biobased products provide additional product functionalities, less resource-intensive production, and efficient use of all natural resources. They could reduce dependency on fossil fuels, developing a more sustainable economy.

Food Waste Recovery Group (www.foodwasterecovery.group of ISEKI Food Association) has prepared several books dealing with food waste recovery technologies, saving food actions, valorization of different food processing by-products (e.g. from olive, grape, cereals, coffee, meat etc.), sustainable food systems, innovations strategies in the food and environmental science, innovation in traditional foods, nutraceuticals and nonthermal processing, shelf life and food quality, and personalized nutrition. The group has also prepared books dealing with food components such as polyphenols, proteins, carotenoids, glucosinolates, dietary fiber, and lipids, as well alternative food products and nonalcoholic drinks.

Nowadays, there is a need for a new reference highlighting principally the biobased products and industries beyond the biochemical processes. Following these efforts and considerations, the current book aims to cover all kind of biobased products and respective industries. Its ultimate goal is to support the scientific community, professionals, and enterprises who aspire to develop industrial and commercialized applications of biobased products.

The book consists of 10 chapters. Chapter 1 provides a general overview of prominent biobased industries working on in Europe and globally and lists some of the most relevant projects that are leading the transition to a nonfossil era. Focusing at processing, storage, recycling, and disposal of biological resources, biobased industries are leading the transition towards a more sustainable bioeconomy worldwide. These industries use starch-based, sugar-based, lignocellulose, algal biomass, and waste-derived feedstocks to produce a wide range of biofuels, polymers, and other significant products within a biorefinery perspective. Nowadays, there is an aggregating demand of existing and new biobased products. However, in the modern bioeconomy environment, companies, researchers, and consumers are confronted with numerous uncertainties that may limit the development of new products and technologies into full-scale industrial applications. Therefore, Chapter 2 discusses development perspectives for the biobased economy.

In spite of their expected ecological benefits, biobased chemicals will be subject to laws and regulations requiring governmental review of new chemicals, while biologically produced fuels may require certification for commercial or military use. Companies may face certain regulatory obligations to qualify for governmental programs to promote the use of renewable or sustainably produced products for the benefits of such programs. In addition, the use of genetically modified microorganisms or plants might subject biobased processes to additional levels of government regulation under national biotechnology laws or regulatory programs. Chapter 3 explains in details these regulatory programs and their requirements.

Biofuels have been playing a key role in biobased economy, besides the environmental benefits derived from its use, which sometimes are controversial. But a closer look at the recent application data of biofuels and latest technologies on advanced biofuels makes evident the great contribution of biofuels to the renewable energetic matrix and their potential for sustainable industrial development in the near future. In Chapter 4, the latest statistics of use, targets, main biomass feedstocks, and production technologies for conventional and advanced liquid (bioalcohols, biodiesel, HVO, and aviation), solids, and gaseous (biomethane and biohydrogen) biofuels are presented. Chapter 5 focuses on drop-in fuel that can be used directly in current engines and obtained through metabolic engineered microorganisms. Many efforts have been made in this field, and research teams were capable to produce biohydrocarbons with concentrations up to c.500 mg/L. Yet, these accumulation rates are still insufficient for a profitable process at the industrial scale. The chapter also discusses the bottlenecks and solutions that could allow solving current limitations.

Over the last years, there has been a significant amount of research focusing on proper strain selection, cultivation conditions and systems, and downstream processing of algae. Apart from a few commercially cultivated strains, large-scale production of algal biomass still faces a number of issues. Moreover, it is becoming evident that biomass production for biofuel-only purposes is not feasible from a technoeconomic perspective. A potential solution to this issue could be the biorefinery concept, where besides biofuels, algae biomass will serve as a feedstock for a number of biobased products. Chapter 6 provides an overview of algal biomass production methods and biobased products and discusses current challenges and future prospects in the nascent algae-based sector.

Wood is an important material for the circular economy, and the actual desire of our society to get out of our dependence on oil for ecological and economic reasons pushes scientists and industrialists to become interested in valorizing the wood industry by-products and especially wood polymers and extractives. Among them, lignins and tannins, which are abundant, represent a main source of phenolic compounds. Chapter 7 refers to the valuation of lignins and tannins. It gives an overview of lignins and tannins, structures, and extraction methods and summarizes the progress in using these compounds in composites manufacturing, adhesives, foams, adsorbent, and as carbon fibers precursor.

Chapter 8 summarizes the current state of biobased materials in food packaging sector. A range of main biobased polymers targeted for food packaging industry are presented focusing on structural properties of biobased polymers, general processing techniques, main limitations of these materials, and current state on market. Moreover, the chapter examines the new technological approaches in upgrading the properties of biobased packaging materials to a commercial application, together with potential solutions, as well as discusses the major industry players who are bringing these materials to the market.

Decoupling of our economy from the fossil resource base by developing a biobased one can only be successful, if advanced materials are produced from renewable resources. There is a considerable potential of substitution in the field of electrochemical energy storage technologies (supercapacitors or batteries) and energy conversion systems (fuel cells), which are of central importance in the context of renewable energies and e-mobility. Chapter 9 gives an overview about the operation principles and state of the art in the field of energy storage and conversion systems to provide a deeper insight into the application potential of biobased carbon materials.

Because of their characteristic (flexible, lightweight, and strong) properties, textiles know a wide range of applications: clothing, interior textiles, and technical textiles (e.g. medical textiles, automotive textiles, etc.). Properties of textiles will vary according to the type of fibers they are composed of. For all applications, textiles need to meet stringent performance requirements. Textiles composites are a combination of (biobased) polymer matrix and (biobased) textile reinforcement. Changes in legislation and consumers mindset result in an increasing use of biobased materials in textile coatings and composites. Chapter 10 presents an overview of implementing biobased materials in textile-reinforced composites and textile coating and finishing.

Conclusively, the book addresses researchers, practitioners, and scientists engaged in the study of biorefineries, professionals working in the area of bioresource technology, and those who are interested in the development of innovative products from biomass. It could be used by university libraries and institutes all around the world to be used as a textbook and/or ancillary reading in undergraduate- and postgraduate-level multidiscipline courses dealing with chemical engineering, bioresource technology, and agricultural engineering and technology.

At this point, I would like thank and acknowledge all authors for their collaboration and efforts in bringing together different topics of biobased products and biobased industries. Accepting my invitation, following editorial guidelines and respecting timeline schedule are highly appreciated. I consider myself fortunate to have had the opportunity to work together with different experts from Belgium, Chile, Germany, France, Lithuania, Mexico, Serbia, Spain, the Netherlands, and the United States. I would also like to thank the acquisition editor Susan Dennis, the book managers Katerina Zaliva and Charlotte Rowley, and Elsevier's publication team for their help during editing and production of this book. Last but not least, a message for the readers: those collaborative editing projects of hundreds thousands words may contain errors or gaps. Any kind of comments or even criticism is always welcome, so please do not hesitate to contact me to discuss any relevant issue.

Charis M. Galanakis
Galanakis Laboratories
Chania, Greece
cgalanakis@chemlab.gr
Food Waste Recovery Group, ISEKI Food Association,
Vienna, Austria
foodwasterecoverygroup@gmail.com
Research & Innovation Department

Chapter 1

Overview of bio-based industries

Antonio D. Moreno, Ana Susmozas, José Miguel Oliva, María José Negro
CIEMAT, Department of Energy, Biofuels Unit, Madrid, Spain

1.1 Introduction

Sustainable development is among the most important issues for both global research and political agenda today. This is due to the serious challenges that our society is facing, including climate change, resource depletion, and environmental degradation. The implementation of the so-called bioeconomy will promote the use of renewable biological resources to produce food, materials, and energy, while balancing, for instance, waste production and/or greenhouse gas emission.

Targeting at reducing our dependence on fossil-based products and meeting the global sustainability objectives, a strong bio-based industrial sector is needed to lead such transition. Bio-based industries are therefore in charge of using renewable biomass (i.e., any biological resource that can be used as raw material) to deliver and place in the market a full pallet of products with application in different sectors, ultimately creating new jobs opportunities and economic growth. Moreover, within the concept of a *circular economy* where the value of products, materials, and resources should be extended in the economy for as long as possible, the bio-based industry contributes to a better management of current biological resources, thus minimizing waste generation.

Bioindustry mainly uses animal fats, vegetable oils (rapeseed oil, palm oil, soybean oil, etc.), sugar, and/or starch crops (maize, wheat, sugar beet, etc.), lignocellulosic and algal biomass (wood, straw, sugarcane bagasse, corn stover, seaweed, etc.), and waste-derived feedstocks (brewer's spent grain, the organic fraction of municipal solid waste, residues from the paper and pulp industry, etc.) as bio-based raw materials. Intermediate compounds and other coproducts such as glycerol and ethanol are also considered as platform chemicals by these companies. Special attention must be paid to the use of

Biobased Products and Industries. https://doi.org/10.1016/B978-0-12-818493-6.00001-4

food-derived residues such as raw material, since the Food and Agriculture Organization of the United Nations has estimated the production of about 1.3 billion tons/year worldwide (FAO, 2011). Food waste can either be of animal and plant origin and may be converted into a wide range of products such as antioxidants, dietary fibers, essential oils, carotenoids, oils rich in omega-3 fatty acids, or chitosan polymers for biomedical applications (Galanakis, 2012; Bastidas-Oyanedel et al., 2016).

Bio-based products traditionally include wood, cork, natural rubber, paper, textiles, and/or wooden construction materials. Notwithstanding, bio-based chemicals, bioplastics, and biofuels are also relevant products, thus covering a long list of **organic acids** (e.g., lactic acid, succinic acid, acetic acid), **alcohols** (e.g., ethanol, lauryl alcohol, furfuryl alcohol), **polymers** (e.g., polyethylene terephthalate, polyhydroxyalkanoate, starch-based plastics), **surfactants** (e.g., glycolipids, sophorolipids, carboxymethyl starch), **solvents** (e.g., isobutanol, ethyl acetate, acetone), **adhesives** (e.g., methacrylates, epoxy resins, tall oil rosin), **cosmetics** (e.g., limonene, xanthan, vanillin), **lubricants** (e.g., isoalkanes, fatty acid methyl esters), etc (Spekreijse et al., 2019). Moreover, bio-based products can be classified into "drop-in" alternatives, when they are homologous to those obtained from fossil resources, and novel products, when having new functionalities and potential markets (European Commission, 2018).

The most important drivers for the development of the bio-based industry are the economic impact and process sustainability (Nattrass et al., 2016). These drivers include the profitability of the company and the environmental performance of the products. Policy may also be considered a significant driver with a secondary role, although it may become more important in the future. Although it is an important driver, the economic impact exhibits a dual role by representing one of the most prominent constrains at the same time. Major economic constrains include production costs, the availability of funds to invest in production capacity, and the variable feedstock prices. In this context, several countries have launched different national investment programs and public-private partnerships to promote research and innovation as well as cooperation between both academy and industry, such as the European Horizon 2020 program (https://ec.europa.eu/programmes/horizon2020/en), the European Bio-Based Industries Joint Undertaking (BBI-JU) action (https://www.bbi-europe.eu), the US BioPreferred program (https://www.biopreferred.gov/BioPreferred), the US BETO program (https://www.energy.gov/eere/bioenergy), the Brazilian BIOEN-FAPESP program (http://bioenfapesp.org), the Canadian BIOTECanada (http://www.biotech.ca), the Indian BIRAC (http://www.birac.nic.in), the Malaysian Bioeconomy Transformation Program (http://www.bioeconomycorporation.my), the Bio-industry to expand action of the Chinese 13th Five-Year Plan (http://en.ndrc.gov.cn), or the Argentinian PROBIOMASA Program (http://www.probiomasa.gob.ar/sitio/es) among others.

Besides Europe (EU-28), USA, Brazil, Canada, China, and Malaysia are the global leading countries of the bio-based sector according to the existing production capacity, the planned production capacity, the industrial innovation, and the availability of feedstock (Natrass et al., 2016). The present chapter is intended to provide a general overview of the bio-based sector, highlighting major industries and the most relevant projects working on in Europe and globally.

1.2 Bio-based industries from a bioeconomy perspective

Bioeconomy can be defined as the use of biomass feedstock sources for the production of food and animal feed, bio-based products, and bioenergy, thus promoting the replacement of fossil resources. In spite of this general meaning, the strategies and the main actions taken to foster bioeconomy very much depend on each country's agenda. For instance, the European bioeconomy strategy seeks ensuring food and nutrition security, managing natural resources sustainably, reducing dependence on nonrenewable, unsustainable resources (independently of having a domestic origin or sourced from abroad), mitigating and adapting to climate change, and strengthening European competitiveness and jobs opportunities (European Commission, 2018). The US bioeconomy supports the implementation of novel technologies to enable increasing domestic bioenergy production, creating new jobs, boosting economic growth, and encouraging national investment, advancing US competitiveness in global energy and bioproduct markets, maximizing the use of biomass resources, and improving the quality of life of American citizens (BETO, 2016). The Chinese bioeconomy promotes the expansion of bio-based industry by supporting the application of biotechnological tools (e.g., genomics, biocatalysts) and the progress on developing new medicines, biological breeding, and new-generation biotechnology products and services (Chu, 2016). Overall, 41 states worldwide have expressed the intention of pursuing explicit political strategies to develop and stimulate their bioeconomies (Dietz et al., 2018). These countries are identified in Fig. 1.1.

In addition to these 41 states, several international governments are also in the process of developing policy initiatives toward implementing a dedicated bioeconomy strategy or have other related strategies at national level. In this context, bioeconomy promoting countries can be classified into four categories according to their implemented strategies (European Commission, 2018; Rodríguez et al., 2017):

(1) Dedicated bioeconomy strategy at national level (e.g., Germany, Spain, United States of America, Malaysia, South Africa, Argentina)
(2) Dedicated bioeconomy strategy at national level under development (e.g., Austria, United Kingdom, Lithuania)

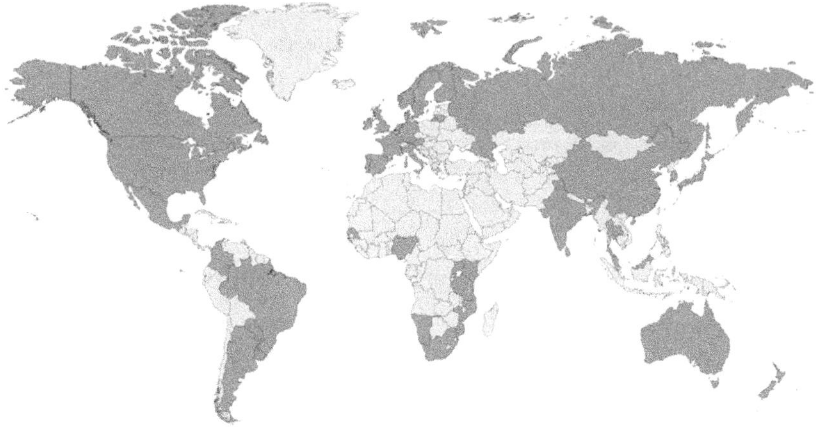

FIGURE 1.1 World map showing the states with explicit political strategies for the development and stimulation of bioeconomy. These countries have been identified by Dietz et al. (2018). *Map created with mapchart.net (https://mapchart.net/world.html).*

(3) Other policy initiatives dedicated to the bioeconomy (e.g., Sweden, Canada, Brazil, China, India, Kenya, Senegal)
(4) Other related strategies at national level (e.g., Greece, Portugal, Cyprus)

The development of innovative biomass conversion processes aimed at guaranteeing environmental protection and sustainable growth while ensuring people's health and improving agricultural productivity is therefore of utmost importance toward building a solid bio-based industry and reaching global bioeconomy goals. In this context, efforts to improve and optimize biomass-related processes have increased exponentially worldwide. According to the European Commission (2018), the EU bio-based industry created a turnover of €2.3 trillion in 2015. Fifty percent of the turnover was generated by the food, beverages, and tobacco industry; 17% by agriculture; and 8% by the paper industry. Also, it created €621 billion of value added and employed 18 million people. In 2014, the bio-based industry contributed to the US economy with $393 billion and 4.2 million jobs (USDA, 2016). Similarly, the Chinese bio-based industry created a total value added of about ¥4 trillion (equivalent to about €560 billion) in 2015 (Wang et al., 2018). These figures are representative of the large number of bio-based products (either as novel "drop-in" alternatives or as new products) that are already placed in the market today, simultaneously encouraging industrial revitalization. Furthermore, a bunch of novel products are expected to be commercialized in the near future (European Commission, 2018; Kovacs, 2015). The pulp, paper, and board sector have developed novel cellulose-based applications with the aim of replacing fossil-based textiles and plastics; boosting the use of nanofibril applications in bio-based adhesives, 3D printing, and flexible electronics; using foldable cardboard for the packaging business; or converting side streams into biofuel and/or other fine and commodity chemicals, healthcare, automotive, consumer goods, construction, etc. Similarly, the textile sector has provided more sustainable products from wood-derived textile fibers, by only considering mechanical processing and avoiding the use of harmful chemicals. Other sectors, such as construction, are also replacing energy-intensive nonrenewable building materials with alternative wood-related products, thus contributing to reduce greenhouse gas emissions from this sector, which emitted 1.45 Gt CO_2 in 2016 (Andrew, 2018; FAO, 2016).

The participation and impact of bio-based products on current market are estimated according to present classification systems, which group these compounds into several sectors or categories depending on different criteria. Both general and detailed lists can be found worldwide. For instance, Chinese bioindustry classifies bio-based products into seven major clusters including biomedicine, biomedical engineering, bioagriculture, bio-based manufacturing, bioenergy, bio-based environmental protection, and biotechnology services (Wang et al., 2018). On the other hand, the Statistical classification of economic activities in the European Community, NACE, classifies the economic activities related to the production and manufacturing of biomass into the sectors listed in Table 1.1 (EUROSTAT, 2008). Similar to NACE classification, the North American Industry Classification System

TABLE 1.1 NACE sectors integrating the economic activities related to the production and manufacturing of biomass.

Bioeconomy sector[a]	NACE codes
Agriculture, forestry, fishing and aquaculture	A01, A02, A03
Manufacture of food, beverages, and tobacco	C10, C11, C12
Manufacture of textiles (textiles, wearing apparel, leather)	C13*, C14*, C15
Manufacture of wood products and furniture	C16, C31*
Manufacture of paper	C17
Manufacture of bio-based chemicals, pharmaceuticals, plastics, and rubber	C20*, C21*, C22*
Manufacture of liquid biofuels (bioethanol and biodiesel)	C20.14*, C20.59*
Production of bioelectricity, heat, and power	D35.11*, D35.30*
Manufacture of biogas	D35.21*

[a]*NACE classification does not differentiate bio-based and nonbio-based activities. Therefore, some economy sectors' (those with an asterisk) use of biomass and other kinds of feedstocks.*

(NAICS) codes products according to the general economic activity and is used by the government and business in Canada, Mexico, and USA (United States, 2017). These lists (NACE and NAICS) do not however differentiate between bio-based and nonbio-based activities, which represents an important limiting factor when determining the contribution of the bio-based product to each sector.

In order to identify bio-based products and label them, different international certification bodies audit commercial or industrial goods and determine the bio-based content of each compound according to international laws (IEA Bioenergy, 2018). Hence, Europe has launched the EU Ecolabel Product Catalogue, which covers a wide range of product groups including lubricants, sanitary products, food disposables, and office materials (http://ec.europa.eu/ecat). The US Department of Agriculture has also released the BioPreferred Program Catalog (USDA, 2018), which groups all different product categories according to its function and showed the minimal bio-based content on each product category. This bio-based content is estimated by using the ratio of "new" organic carbon (i.e., plant or agricultural-based component) to total organic carbon ("new" and petroleum-based carbon) (https://www.biopreferred.gov). The calculation of bio-based content excludes inorganic carbon (e.g., carbonate) and noncarbon molecules. Table 1.2 shows the product categories with higher bio-based content from each functional group of the BioPreferred Program Catalog (USDA, 2018).

1.3 The bio-based industry in Europe

Considering the implementation of regulation mechanisms, governmental development of positive incentives, government support for private standards and certifications, and international cooperation, European states have developed the most advanced sustainable bioeconomy strategies when compared to any other country (Dietz et al., 2018). The European bio-based sectors related to primary production, including agriculture, forestry, and aquaculture, are well covered in terms of statistical databases and mainly include figures for biomass production. In average, Europe produces about 1466 Mt in dry matter, mainly coming from agriculture (956 Mt) and forestry sectors (510 Mt) (Camia et al., 2018). However, due to soil maintenance practices (i.e., part of the biomass is left in the field to maintain the carbon sink) and other ecosystem services, only 806.03 Mt in dry matter were harvested (578 Mt agriculture, 227 Mt forestry, 1.5 Mt fisheries and aquaculture, 0.03 Mt algae). From about 1 billion tons of biomass, EU-28 uses 60% of them in the feed and food sector, 19.1% in the bioenergy sector, and 18.8% in the biomaterial sector (Camia et al., 2018).

After excluding the food, beverages, and tobacco products and the primary biomass production/extraction Piotrowski et al. (2018) have estimated an annual turnover of €695 billion (out of €2.3 trillion) for the European

TABLE 1.2 USDA BioPreferred Program product categories by functional area showing high bio-based content. Data obtained from USDA (2018).

Product category	Minimum bio-based content
Construction/Renovation:	
Blast media	94%
Composite panels: Structural wall panels	94%
Floor coverings (noncarpet)	91%
Fleet/Transportation:	
Bioremediation materials	86%
Diesel fuel additives	90%
Gasoline fuel additives	92%
Facilities (operations and maintenance):	
Fluid-filled transformers: Vegetable oil-based	95%
Multipurpose lubricants	88%
Office/Shipping:	
Electronic components cleaners	91%
Films: Nondurable	85%
Thermal shipping containers: Nondurable	85%
Food service:	
Disposable containers	72%
Oven and grill cleaners	66%
Groundskeeping/Road and parking lots:	
Compost activators & accelerators	95%
General purpose deicers	93%
Mulch and compost materials	95%
Water clarifying agents	92%
Household:	
Air fresheners and deodorizers	97%
Candles and wax melts	88%
Industrial:	
Expanded polystyrene foam (EPS) recycling products	90%

Continued

TABLE 1.2 USDA BioPreferred Program product categories by functional area showing high bio-based content. Data obtained from USDA (2018).—cont'd

Product category	Minimum bio-based content
Heat transfer fluids	89%
Sorbents	89%
Turbine drip oils	87%
Personal care:	
Cuts, burns, and abrasions ointments	84%
Foot care products	83%
Lip care products	82%
Shaving products	92%
Janitorial/Custodial:	
Bathroom and spa cleaners	74%
Floor cleaners and protectors	77%
Floor strippers	78%

bio-based industries in 2015. The sectors with largest shares are the paper and the forest-based industries (wood products and furniture), accounting for roughly 27% and 25%, respectively. The industry related to biomass manufacturing also contributed with 3.6 million jobs, with the forest-based sector, the sector related to the paper and paper products, and the textile industry being the most prominent categories. The chemical industry is another important sector for the European economy. With an annual turnover of €542 billion (2018) and 3.3 million people directly employed, Europe is the second largest chemical producer after China (CEFIC, 2018). Since 2008, this sector has also increased its relative content of bio-based products by 27% (from 11% in 2008 to 14% in 2015) and has created the highest number of jobs (DataM, 2018; Piotrowski et al., 2018), highlighting the commitment of this industry for the realization of a European bioeconomy. The European Chemical Industry Council (CEFIC) has estimated the use of 77.7 Mt of organic raw materials by the chemical industry, of which 7.8 Mt (10%) came from renewable sources (European Commission, 2018; CEFIC, 2018). However, due to the large product diversity of this sector, the bio-based contribution to the EU chemical market is highly different depending on each product category. For instance, solvents and platform chemicals include a very small percentage of biocompounds (1.5% and 0.3%, respectively), while surfactants and cosmetics both have a bio-based contribution of 50% and 44%, respectively (Spekreijse et al., 2019). In general, between 2009 and 2015 the overall growth rate of the value added of bio-based sectors has been 23% for the pulp, paper, and cardboard; 17% for the bio-based chemicals, pharmaceuticals, plastics and rubber; and 14% for the manufacture of wood and wooden products (DataM, 2018).

In contrast to the primary production systems, the industrial sectors related to biomass manufacturing are poorly defined due to the vast number of products and their complex supply chains (Dammer et al., 2014). This might lead to underrepresentation of the contribution of bio-based compounds to each product category, which may increase after proper data collection. As an effort to better classify and fill the missing data from the bio-based sectors, Parisi (2018) has identified 803 biorefineries in Europe (Fig. 1.2). By definition, biorefineries are processing plants where biomass feedstocks are

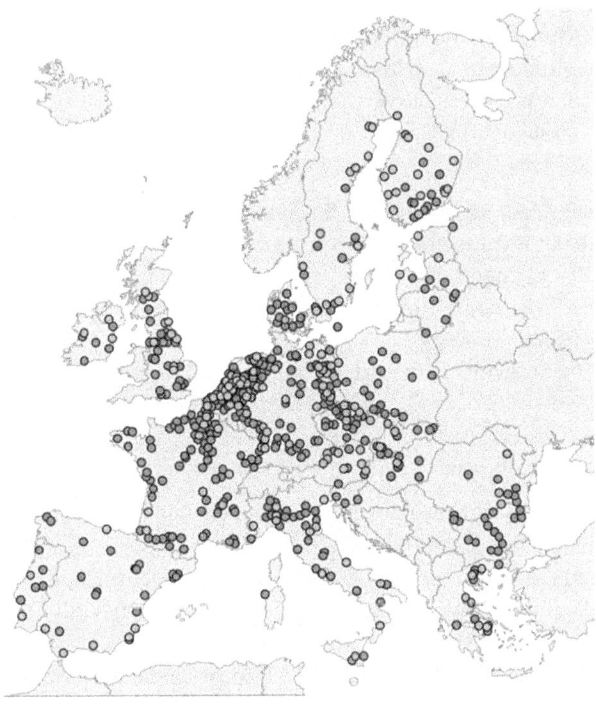

FIGURE 1.2 European map showing biorefineries producing bio-based chemicals, liquid biofuels, and composites and fibers. Orange dots (light gray in print version) (177) highlight those biorefineries involving the integrated production of bio-based products (chemicals and/or composites) and bioenergy (biofuels and/or other biomass related energy). Map adapted from Parisi (2018).

converted and extracted into a range of valuable products. Among the identified biorefineries, 507 facilities produce bio-based chemicals, 363 produce liquid biofuels, and 141 produce bio-based composites and fibers (multi-product facilities are counted more than once). Furthermore, 177 facilities can be considered as integrated biorefineries, which are specific plants targeting at producing both bio-based products and bioenergy. With regards to the feedstock sources, 781 facilities use agriculture-derived materials (e.g., sugar- and starch-based feedstocks, oil- and fat-based feedstocks, intermediate products derived from agriculture-based feedstock such as ethanol or lactic acid, or vegetable fibers), 210 facilities use forestry-derived materials (e.g., wood, forestry intermediate products), and 195 facilities use other bio-based feedstocks (e.g., grasses, waste, insects-derived feedstocks).

The map depicted in Fig. 1.2 shows certain degree of correspondence between biorefinery placing and the locations of ports and chemical clusters in the EU (Parisi, 2018). In this context, biorefineries are mainly localized in Belgium, The Netherlands, and the industrialized regions of Germany, France, and Italy. In contrast, Eastern European countries show lower density of biorefineries and therefore offer higher development potential (European Commission, 2018). To face this gap, the Visegrad Group Countries (Czech Republic, Hungary, Poland, Slovakia, and joined by Bulgaria, Croatia, Latvia, Lithuania, Republic of Estonia, Romania, Slovenia) started the BIOEAST Initiative that provides active support mechanisms to develop dedicated Bioeconomy strategies in Central and Eastern European countries (http://www.bioeast.eu). This initiative particularly focuses on the following objectives:

(1) Initiate cooperation and the development of knowledge-based policies at European level to facilitate joint actions
(2) Identify common challenges and validate common research topics
(3) Initiate strategies for the development of a national circular and bioeconomy strategy
(4) Provide an evidence base to promote continuous, long-term observation
(5) Improve skills of dedicated multi-stakeholder actors
(6) Initiate development of synergies between countries
(7) Increase visibility by involving society and promoting public awareness

With a strong industry, research, and renewable resources, Europe has, in general, excellent potential to develop new biorefining technologies and increase the number of bio-based products in the upcoming years. The development and deployment of new biorefineries greatly depends on the profit margins of bio-based products and the successful commercialization of novel technologies; the availability of local and/or regional feedstocks at competitive prices; suitable infrastructure (e.g., logistics); skilled personnel and private and public support services; and the implementation of supportive policy and regulatory enabling environment (European Commission, 2018). The identification of new potential feedstocks and products is also crucial to increase the final value during biomass processing chains. In this context, the use of materials currently considered as residues or wastes may offer huge potential to obtain alternative and novel bio-based products. This is the case, for instance, of the lignin and/or lignosulfonate streams derived from the pulp and paper industry, food-derived waste, industrial residues (e.g., brewer's spent grain, fruit pomace, potato peels, etc.), or the organic fraction from the municipal solid waste, which can be used as energy sources or to obtain lightweight materials, phenol-based aromatic compounds, biofertilizers, biotextiles, or bioplastics (Galanakis, 2012; Beres et al., 2017; Cao et al., 2018; Pathak et al., 2018). Prior to the implementation of new biorefinery plants, lifecycle and technoeconomic assessments are also required to guarantee the environmental standards and profitability of the developed processes (Susmozas et al. 2016, 2019).

The European Union is an active partner when promoting bioeconomic transformations and has implemented different action programs to strengthen the European Bioeconomy Network by funding projects and initiatives, as well as to support all related sectors at the national, regional, and local level. Horizon 2020 is an outstanding funding program that supports research projects at all developmental stages through different investment calls (https://ec.europa.eu/programmes/horizon2020). For instance, the BESTF3 call (Bioenergy Sustaining the Future) provides funding to proposals based on precommercial bioenergy projects, with especial emphasis on those projects demonstrating collaboration, innovation, and industry focus (https://www.era-learn.eu/network-information/networks/bestf3). WASTE2BIO ("*Valorization of urban WASTEs to new generation of BIOethanol*") is one of the BESTF3-funded projects (http://www.waste2bio.com). WASTE2BIO is a 3-year project with a 4-partner consortium (from two different countries) that focuses on the valorization of urban wastes to new generation of bioethanol. Furthermore, it subsequently processes the residual feedstock obtained to produce biogas and biofertilizers. Both bioethanol (considered as a petrol substitute and/or additive) and biogas (a renewable fuel that can also be used for electricity and heat production) have several economic and environmental benefits as they are clean, renewable, and efficient biofuels. On the other hand, biofertilizers are free from pollution hazards and increase soil fertility. This project also seeks to strengthen the competitiveness and growth of the European biofuels industry by validating a cost-effective innovation technology that meets the needs of the world market in terms of fuel and energy quality. Another example of BESTF3 funded projccts is **SEGRABIO** (https://energiforskning.dk/da/project/segrabio; https://www.swecris.se/betasearch/details/project/P426741Energi). Similar to WASTE2BIO, SEGRABIO ("*Second grade biomass for biofuels*") is a 3-year project with a 4-partner consortium (from two different countries) intended to develop and demonstrate the production of bioethanol and biogas from second-grade and low-cost biomass feedstocks (animal bedding and second-grade straw). This project also aims at verifying the synergies from integrated biogas and bioethanol production in terms of energy efficiency, and improving the critical components during biomass processing (pressurized acid hydrolysis).

Besides the BESTF3 call, the BBI-JU action can be also found under Horizon 2020. The BBI-JU action is a public-private partnership that promotes the development of a sustainable bio-based industry in Europe through different investment calls (https://www.bbi-europe.eu). This EU body has therefore a central role to increase the industrial competitiveness of Europe by funding key research projects toward achieving a robust bio-based sector. Particularly, the BBI-JU action focuses on contributing to more efficient resource utilization as well as to increase economic growth and employment (paying special attention to rural areas), using sustainable low-carbon technologies. The projects funded through BBI-JU aim at:

(1) Demonstrating technologies to enable the production of new chemical building blocks and novel products from European biomass feedstocks
(2) Developing new business models varieties within the value chain, such as creating new interconnections between cross-sectors
(3) Promoting the setup of flagship biorefinery plants with novel technologies and business models showing their competitiveness as alternatives to the fossil-based ones

The BBI-JU action has funded several ambitious projects targeting different bio-based processes and sectors. Some of the granted projects in 2016 and 2017 are listed in Table 1.3.

As examples of projects funded through the BBI-JU action, a brief description of WoodZymes, BioBarr, URBIOFIN, BIOMOTIVE, and POLYBIOSKIN projects can be found below:

- **WoodZymes** (https://www.woodzymes.eu). The WoodZymes project ("*Extremozymes for wood-based building blocks: from pulp mill to board and insulation products*") aims at developing novel extremophilic enzymes with capacity of transforming wood feedstocks to obtain different sugars and phenolic compounds as chemical building blocks for the subsequent production of medium-density fiberboard, polyurethane insulation foam, and bleached paper. The underutilized lignin and the hemicellulose fractions extracted from the pulping processing will be used as raw material for producing the aforementioned compounds. The WoodZymes project has a total duration of 3 years, and the project consortium involves up to 11 institutions from four different countries, including world's leading companies of the pulp and paper sector, the fiberboard manufacture and insulation materials sectors, a small biotech company with expertise in extremophilic enzymes, and diverse research institutes and technology centers.

TABLE 1.3 Projects funded through the Bio-Based Industries Joint Undertaking (BBI JU) action (2016 and 2017).

Acronym[a]	Title	Products[b]	Feedstock	Coordinator	Partners	Chain Value	Date	Type
iFermented	Conversion of forestry sugar residual streams to antimicrobial proteins by intelligent fermentation	High value galactose (food/ feed preservative)	Residual sugars from biorefineries	Norges Teknisknaturvitenskapelige Universitet (Norway)	11 (5 countries)	Lignocellulose	05/ 2018–04/ 2022	Research & Innovation Action
SWEET-WOODS	Production and deploying of high purity lignin and affordable platform chemicals through wood-based sugars	High-purity lignin and depolymerized lignin compounds (elastomer foams, rigid polyurethane foam panels, and polymer compounds for insulation and injection molding), high-purity glucose, fructose, xylose, and glucosone (bio-IBN, xylitol, and lactic acid)	Wood-processing residues	Graanul Biotech OÜ (Estonia)	9 (6 countries)	Lignocellulose	06/ 2018–05/ 2022	Innovation Action Flagship
UNRAVEL	UNique refinery approach to valorize European Lignocellulosics	High-value lignin Sugars (fermented into chemical building blocks) Lignin-based polyols (polyurethane and polyisocyanurate foams)	Mixtures of hardwood wood chips, forest residues, bark, and straw	Fraunhofer-Gesellschaft zur Förderung der angewandten Forschung e. V. (Germany)	10 (6 countries)	Lignocellulose	06/ 2018–05/ 2022	Research & Innovation Action

Continued

TABLE 1.3 Projects funded through the Bio-Based Industries Joint Undertaking (BBI JU) action (2016 and 2017).—cont'd

Acronym[a]	Title	Products[b]	Feedstock	Coordinator	Partners	Chain Value	Date	Type
WoodZymes	Extremozymes for wood-based building blocks: From pulp mill to board and insulation products	Hemicelluloses-derived sugars (papermaking additives) Lignin-based resin and phenols (medium-density fiberboards)	Pulp and wood residues	Consejo Superior de Investigaciones Científicas (Spain)	11 (4 countries)	Lignocellulose	06/2018–05/2021	Research & Innovation Action
Pro-Enrich	Development of novel functional proteins and bioactive ingredients from rapeseed, olive, tomato, and citrus fruit side streams for applications in food, cosmetics, pet food, and adhesives	Protein (food, adhesives, pet feed) Bioactives molecules (cosmetics)	Agricultural residues from rapeseed meal, olives, tomatoes, and citrus fruit industries	Danish Technological Institute (Denmark)	16 (7 countries)	Agro-based	05/2018–04/2021	Research & Innovation Action
SUSFERT	Sustainable multifunctional fertilizer—combining bio-coatings, probiotics, and struvite for phosphorus and iron supply	Bio-coatings, probiotics, and struvite (phosphorus and iron fertilizers)		RTDS Association (Verein zur förderung der Kommunikation und Vermittlung von forschung, Technologie and Innovation) (Austria)	11 (7 countries)	Agro-based	05/2018–04/2023	Innovation Action - Demonstration

EXCornsEED	Separation, fractionation, and isolation of biologically active natural substances from corn oil and other side streams	Concentrate proteins Bioactive compounds	Side streams and coproducts of two biotech processes (bioethanol and biodiesel production)	Università degli Studi di Roma La Sapienza (Italy)	13 (8 countries)	Organic waste	06/2018–11/2021	Research & Innovation Action
Prolific	Integrated cascades of PROcesses for the extraction and valorization of proteins and bioactive molecules from Legumes, fungi, and Coffee agro-industrial side streams	Proteins Bioactive molecules	Agro-industrial residues of legumes, fungi, and coffee	Fachhochschule Nordwestschweiz (Switzerland)	17 (8 countries)	Organic waste	09/2018–08/2022	Research & Innovation Action
SUSBIND	Development and pilot production of SUStainable bio BINDer systems for wood based panels	Bio-based binders	Feedstock from existing European starch-based and vegetable oil-based biorefineries	RTDS – Verein Zur forderung der Kommunikation und Vermittlung Von forschung, Technologie Und Innovation (Austria)	10 (5 countries)	Organic waste	05/2018–04/2022	Research & Innovation Action

Continued

TABLE 1.3 Projects funded through the Bio-Based Industries Joint Undertaking (BBI JU) action (2016 and 2017).—cont'd

Acronym[a]	Title	Products[b]	Feedstock	Coordinator	Partners	Chain Value	Date	Type
EFFECTIVE	Advanced Eco-designed Fibers and Films for large consumer products from biobased polyamides and polyesters in a circular EConomy perspecTIVE	Bio-based polyamides (garment, carpet, and packaging) Polyesters (packaging)	Sustainable biomaterials	Aquafilslo Proizvodnja poliamidnih filamentov in granulatov d.o.o. (Slovenia)	12 (6 countries)	Across VCs	06/ 2018–05/ 2022	Innovation Action - Demonstration
NEWPACK	Development of new competitive and sustainable Bio-Based plastics	PHB PLA Blends of PHB/PLA Cellulose and chitin nanowhiskers (plastics)	Agro-food waste, including potato peelings and crab shells	Oulun Yliopisto (Finland)	13 (7 countries)	Across VCs	06/ 2018–05/ 2021	Research & Innovation Action
ReInvent	Novel products for construction and automotive Industries Based on Bio materials and natural fibers	Bio-based rigid and soft foams NFRP (construction and automotive industries)		Centro Ricerche FIAT (Italy)	19 (8 countries)	Across VCs	06/ 2018–05/ 2022	Innovation Action - Demonstration

VIPRISCAR	Validation of an industrial process to manufacture isosorbide bis(methyl carbonate) at pilot level	IBMC polyurethane dispersions (PUDs) based on isosorbide bis(methyl carbonate)-derived materials Antibacterial and antithrombotic compounds from isosorbide bis(methyl carbonate)-based NIPU	Forestry and timber industries	Fundación Tecnalia (Spain)	9 (4 countries)	Across VCs	06/2018–05/2021	Research & Innovation Action
SHERPAC	Innovative structured polysaccharides-based materials for recyclable and biodegradable flexible packaging	Microfibrillated cellulose (flexible paper-based packaging material)	Wood- and cereal-based feedstocks	Centre Technique de l'Industrie des Papiers, Cartons et Celluloses (France)	6 (5 countries)	Across VCs	05/2017–11/2020	Research & Innovation Action
EUCALIVA	EUCAlyptus LIgnin VAlorization for advanced materials and carbon fibers	Lignin (carbon fibers and other carbon-based materials)	Black liquors from Kraft pulping	Contactica S.L. (Spain)	6 (4 countries)	Lignocellulose	09/2017–02/2021	Innovation Action—Demonstration
LigniOx	Lignin oxidation technology for versatile lignin dispersants	Lignin	Biomass Side stream lignins from the pulp and paper industry and otherbiorefineries	Teknologian tutkimuskeskus VTT Oy (Finland)	10 (6 countries)	Lignocellulose	05/2017–04/2021	Innovation Action—Demonstration

Continued

TABLE 1.3 Projects funded through the Bio-Based Industries Joint Undertaking (BBI JU) action (2016 and 2017).—cont'd

Acronym[a]	Title	Products[b]	Feedstock	Coordinator	Partners	Chain Value	Date	Type
PEFerence	From bio-based feedstocks via di-acids to multiple advanced bio-based materials with a preference for polyethylene furanoate	Furan dicarboxylic acid PEF,PBF and polyurethanes		Synvina CV (The Netherlands)	11 (6 countries)	Lignocellulose	09/2017–08/2022	Innovation Action—Flagship
SSUCHY	Sustainable structural and multifunctional bio-composites from hybrid natural fibers and bio-based polymers	Aliphatic polyesters (thermoplastics) and Thermosetting polymers (epoxy) Composite materials	Lignocellulosic feed-stocks	Université de franche-Comté (France)	17 (6 countries)	Lignocellulose	09/2017–08/2021	Research & Innovation Action
BioBarr	New bio-based food packaging materials with enhanced barrier properties	PHA PHA-PLA	Agro-industry side-products and wastes	Tecnoalimenti S.C.p.A. (Italy)	7 (4 countries)	Agro-based	06/2017–05/2021	Research & Innovation Action

AgriChem-Whey	An integrated biorefinery for the conversion of dairy side streams to high value bio-based chemicals	L-Lactic acid Polylactic acid Minerals (human nutrition) Bio-based fertiliser	Side streams from dairy industry	Glanbia Ingredients (Ireland)	11 (5 countries)	Organic waste	01/2018–12/2021	Innovation Action—Flagship
BARBARA	Biopolymers with advanced functionalities for building and automotive parts processed through additive manufacturing	Pigments Fragrances Reinforcing agents Biocide compounds	Food waste and agricultural byproducts	Fundación AITIIP (Spain)	10 (4 countries)	Organic waste	05/2017–04/2020	Research & Innovation Action
EMBRACED	Establishing a Multi-purpose Biorefinery for the Recycling of the organic content of AHP waste in a Circular Economy Domain	Fermentable sugars Bio-based building blocks Polyesters	Absorbent hygiene products waste	Fater Spa (Italy)	13 (7 countries)	Organic waste	06/2017–05/2022	Innovation Action Demonstration
PERCAL	Chemical building blocks from versatile MSW biorefinery	Lactic acid (hot-melt adhesives) Ethyl lactate Succinic acid (polyols, polyurethane industry) Biosurfactants	Municipal solid waste	Industrias Mecánicas Alcudia SA (Spain)	12 (8 countries)	Organic waste	07/2017–06/2020	Research & Innovation Action

Continued

TABLE 1.3 Projects funded through the Bio-Based Industries Joint Undertaking (BBI JU) action (2016 and 2017).—cont'd

Acronym[a]	Title	Products[b]	Feedstock	Coordinator	Partners	Chain Value	Date	Type
URBIOFIN	Demonstration of an integrated innovative biorefinery for the transformation of municipal solid Waste into new bio-based products	Bioethanol Volatile fatty acids Biogas Polyhydroxyalkanoates Bioethylene Biofertilizers	Organic fraction of municipal solid waste	Industrias Mecánicas Alcudia SA (Spain)	16 (8 countries)	Organic waste	06/2017–05/2021	Innovation Action—Demonstration
BIOMOTIVE	Advanced BIObased polyurethanes and fibers for the autoMOTIVE industry with increased environmental sustainability	Bio-based polyesters-polyols Thermoplastic polyurethanes Thermosetting polymers foams Regenerated natural fibers	**Wood pulp**	Selena Labs Spółka Z Ograniczoną Odpowiedzialnością (Poland)	16 (8 countries)	Across VCs	06/2017–05/2021	Innovation Action—Demonstration
BIOSMART	Bio-based smart packaging for enhanced preservation of food quality	Bio-based packaging compounds	Natural organic raw materials	IK4-Tekniker (Spain)	11 (8 countries)	Across VCs	05/2017–04/2021	Research & Innovation Action
ECOXY	Bio-based recyclable, reshapable and repairable (3R) fibre-reinforced EpOXY composites for automotive and construction sectors.	Bio-based epoxy resins fiber reinforcements (composites) Polylactic acid		Fundación CIDETEC (Spain)	13 (8 countries)	Across VCs	06/2017–11/2020	Research & Innovation Action

OPTISO-CHEM	OPTimized conversion of residual wheat straw to bio-ISObutene for bio based CHEMicals	Bio-isobutene oligomers (DIB, TIB, TeIB) Polyisobutylenes (PIBs)	Wheat straw	Global Bioenergies (France)	6 (5 countries)	Across VCs	06/2017–05/2021	Innovation ActionDemonstration
POLYBIO-SKIN	High performance functional bio-based polymers for skin-contact products in biomedical, cosmetic and sanitary industry	Polyhydroxyalkanoates (PHAs) Biopolyesters Polysaccharides (cellulose, starch, and chitin/chitosan)	Bio-based raw materials.	Innovacio i Recerca Industrial i Sostenible SL (Spain)	12 (7 countries)	Across VCs	06/2017–05/2020	Research & Innovation Action
REFUCOAT	Full recyclable food package with enhanced gas barrier properties and new functionalities by the use of high performance coatings	PHA Glycolic acid Olyglycolic acid		Asociación de Investigación de Materiales Plásticos y Conexas (Spain)	12 (4 countries)	Across VCs	06/2017–05/2010	Research & Innovation Action
SYLFEED	From forest to feed: enable the wood industry to bridge the protein gap	Single-cell protein	Wood residues	Biométhodes SA (France)	8 (5 countries)	Across VCs	09/2017–08/2021	Innovation Action - Demonstration

[a] *The project link can be found in the reference list.*
[b] *Information about the compound uses is indicated in brackets.*

- **BioBarr** (http://www.biobarr.eu). The BioBarr project ("*New bio-based food packaging materials with enhanced barrier properties*") is intended to respond to the industrial and technological challenges of developing new biodegradable food packaging materials, overcoming the barriers that have limited the applications in food packaging of 100% biodegradable biopolymers. In this context, the project will combine advanced technological elements including new coating treatments applied to polyhydroxyalkanoates (PHA) and the development and application of a completely biodegradable bio-ink for printing on the packaging. The BioBarr consortium comprises seven partners (4 countries), providing expertise from the whole relevant value chain.
- **URBIOFIN** (https://www.urbiofin.eu). URBIOFIN ("*Demonstration of an integrated innovative biorefinery for the transformation of Municipal Solid Waste (MSW) into new BioBased products*") is a 4-year project focusing on demonstrating the viability of an integrated innovative biorefinery to transform the organic fraction of municipal solid waste into several bio-based products, including bioethanol, volatile fatty acids, biogas, and biopolymers. This project will therefore valorize at the semi-industrial scale the organic fraction of urban wastes into new bioproducts, validating the entire value chain of the process (i.e., from waste management authorities to the validation of products selected by end users). The URBIOFIN consortium represents eight countries in total and lists 16 well recognized institutions (including 10 industries, five research organizations, and 1 university) from the waste management, bioprocessing technologies, bioproducts and biomaterials, and biofuels sectors.
- **BIOMOTIVE** (https://biomotive.info). The BIOMOTIVE project ("*Advanced BIObased polyurethanes and fibers for the autoMOTIVE industry with increased environmental sustainability*") aims at replacing the fossil-based, nonbiodegradable automotive interior parts by more sustainable alternatives to improve the environmental profile and economic competitiveness of automotive sector. The project intends to reduce the greenhouse gas emissions of the sector by 58% by demonstrating the production of bio-based raw materials and building blocks to formulate bio-based polyesters-polyols and thermoplastic TPUs (Thermoplastic PolyUrethane), and the production of novel cellulose-based regenerated fibers from paper pulp. The project consortium includes 16 partners from eight different countries, covering both academia and industry institutions.
- **POLYBIOSKIN** (http://polybioskin.eu/). POLYBIOSKIN ("*High performance functional bio-based polymers for skin-contact products in biomedical, cosmetic and sanitary industry*") is an interdisciplinary 3-year project directed to broaden the use of biopolymers in biomedical, cosmetic, and sanitary skin-contact applications. Instead of using traditional fossil resources, POLYBIOSKIN seeks the use of bio-based raw materials (including primary and food waste biomass) to obtain antimicrobial and antioxidant

compounds, absorbents, and skin compatible biopolymers to incorporate them in price-competitive everyday products such as baby diapers, facial beauty masks, etc. The project consortium of POLYBIOSKIN consists of 12 institutions from seven countries, having deep expertise on biopolymer development that include a range of processing technologies such as electrospinning, surface modification, and impregnation of films and fibers.

Overall, the projects funded through these competitive calls will definitely benefit the European Bioeconomy Strategy by contributing to increasing the potential of biomass feedstocks, especially agricultural and forestry residues and industrial wastes; diversifying and growing farmers' incomes; replacing oil-based resources and products with bio-based and biodegradable alternatives; creating a competitive bio-based infrastructure in Europe and boosting job creation; and reducing CO_2 emissions when compared to the fossil-based production processes (https://www.bbi-europe.eu).

1.3.1 The bio-based industry in non-EU states

In addition to European states, several countries are developing or implementing a bioeconomy-related strategy in order to increase the national shares of bio-based sectors. USA, Brazil, Canada, India, Malaysia, China, and Argentina are some of these countries. The following paragraphs summarize their main bioeconomy strategies as examples of the global bio-based industry.

As a leading biotechnology nation, the US federal government views bioeconomy as an opportunity to enhance the US security and economic growth through transformative science, technology innovation, and market solutions. Since 2006, the US bioeconomy has been expanding at an annual rate of about 10%, generating almost \$400 billion in 2014 (Carlson, 2016; USDA, 2016). Still, challenges such as sustainable production of biomass feedstocks, innovative biotechnology development, proficient workforce preparation, additional biorefining and manufacturing capacity, and a robust market for biofuels and bioproducts are important aspects to address toward reaching bioeconomy goals (BRDB, 2016). Among the strategies to increase the commercialization and use of bio-based products, the US BioPreferred program (www.biopreferred.gov) aims at reducing the US dependency on petrochemistry industry by creating new jobs and establishing new market options for farm commodities. This program was launched in 2002 under the Department of Agriculture and has a heterogeneous representation of companies (from small businesses to large companies) and bio-based products (e.g., intermediate products, platform chemicals, and final products). The program includes two main actions: the mandatory purchasing requirements for federal agencies and their contractors and a voluntary labeling initiative for bio-based products. The first action is planned to enforce by law the purchase by the federal agencies and contractors of bio-based products from the identified 109 categories (e.g., cleaners, carpet, lubricants, or paints). On the other hand, the second action encourages vendors and manufacturers to apply for certification of their bio-based products to be evaluated and tested (by a third-party laboratory) for their eligibility to display the USDA certified bio-based product label (www.biopreferred.gov). In spite of promoting the bio-based industry, the BioPreferred program offers no financial support for its participants. However, the US government has other investment programs including loans and grants available within rural development for businesses. The BETO Strategic Plan from the US Bioenergy Technologies Office is one of these funding opportunities (BETO, 2016). Similar to the European BBI-JU action, the BETO plan is intended to sponsor public-private partnerships to support targeted research, development, and demonstration (RD&D) projects and market transformation activities through an integrated supply chain approach (i.e., addressing feedstock supply, conversion, distribution, and end use). BETO's investment approach supports those transformative and revolutionary bioenergy technologies where the private sector or nongovernmental stakeholders have identified a noninvestment scenario due to risk, scale, or the time frame required for commercialization. The main objectives covered by BETO plan include (BETO, 2016):

(1) The development and demonstration of innovative and integrated value chains for biofuels, bioproducts, and biopower
(2) The reduction of delivered cost and risks associated with feedstock quality and volume
(3) To meet early-adoption market demands and catalyze new markets
(4) To grow an informed community of public and private stakeholders

Production of lignocellulosic biofuels is one of the main targets from BETO plan. Chemical intermediates that can be converted into high-value bioproducts such as bioplastics, bio-based chemicals, cosmetics, or food ingredients are also relevant compounds for this plan, since they can be placed in most cases in the current market as alternative to the oil-derived ones. Several strategic cutting-edge projects have received funding through BETO program. Table 1.4 lists few examples of projects funded by BETO. All these projects have been already reviewed by each expert review panel in 2017, obtaining a "Weighted Project Score" above 7.5 (BETO, 2017). "*Algae production CO_2 absorber with immobilized carbonic anhydrase*" is one of the granted projects. This project targets at improving the CO_2 capture efficiency (up to 80%) and the carbon utilization efficiency (up to 90%) during open raceway algae cultivation. The system uses and advances CO_2 supply method (the CO_2 has been previously captured, stored, and distributed from a power plant flue gas) that incorporates an absorber and a carbonate shuttle to ensure an ample supply of CO_2 for photosynthesis. In order to improve the efficiency of this system, the project studies the use of immobilized carbonic anhydrase to speed up the absorption of CO_2 into the media and reduce the final cost of algae biofuel. "*Upgrading lignin-containing biorefinery residues for bioplastics*" is another example of

TABLE 1.4 Projects intended to promote the US bio-based industry through the BETO Strategic Plan.

Company/Institution	Project Title	Date
Antares Group Inc.	Enabling sustainable landscape design for continual improvement of operating bioenergy supply systems	04/2016–03/2021
Southern Research Institute	Biomass conversion to acrylonitrile monomer-precursor for production of carbon fibers	01/2015–04/2018
Oak Ridge National Laboratory	Biomass-derived pyrolysis oil corrosion studies	10/2016–09/2019
Global Algae Innovations Inc. (GAI)	Algae production CO_2 absorber with immobilized carbonic anhydrase	07/2016–07/2019
Texas A&M University	Upgrading lignin-containing biorefinery residues for bioplastics	07/2016–06/2019
Ohio University	Biomass electrochemical reactor for upgrading biorefinery waste to industrial chemicals and hydrogen—BCU ALT	04/2016–03/2019
University of California, San Diego	Integrated pest management for early detection algal crop production	10/2015–03/2019
National Renewable Energy Laboratory (NREL)	Pretreatment and process hydrolysis—pretreatment	01/2003–09/2018
National Renewable Energy Laboratory (NREL)	Biogas to liquid fuels and chemicals using a methanotrophic microorganism	10/2013–09/2017
Sapphire Energy Inc.	Biomass productivity technology advancement toward a commercially viable, integrated algal biomass production unit	10/2013–08/2016
LanzaTech Inc.	A hybrid catalytic route to fuels from biomass syngas	10/2011–06/2016
FDC Enterprises Inc. (FDCE)	Design and demonstration of an advanced agricultural feedstock supply system for lignocellulosic bioenergy production	09/2010–06/2015
University of Minnesota	Pathways toward sustainable bioenergy feedstock production in the Mississippi river watershed	09/2010–12/2015

Continued

TABLE 1.4 Projects intended to promote the US bio-based industry through the BETO Strategic Plan.—cont'd

Company/Institution	Project Title	Date
Virent Energy Systems Inc.	Catalytic upgrading of thermochemical intermediates to hydrocarbons: Conversion of lignocellulosic feedstocks	10/2011–10/2015

More information about these projects and other granted projects can be found in "The 2017 Peer Review" document from US Department of Energy – Bioenergy Technologies Office (https://www.energy.gov/sites/prod/files/2018/02/f48/2017_project_peer_review_report.pdf) (BETO, 2017).

granted project by BETO program. This project investigates the upgrading of biorefinery intermediates for the production of bioplastics. In particular, the project has three main goals:

(1) Engineering and optimizing microorganisms capable of converting biorefinery waste streams into PHA for the production of bioplastics
(2) Characterizing biorefinery residues, optimizing lignin treatment and fermentation, and designing the novel bioprocess
(3) Studying process integration and optimization paths for the lignin-to-PHA upgrading process by conducting biorefinery onsite scale-up and assessing technoeconomic and lifecycle analyses (TEA, LCA)

Overall, the project aims to achieve a PHA concentration of 8 g/L with 50% lignin utilization, which is an important step forward in the development of modern biorefinery approaches. With a similar concept, the project "*Biomass conversion to acrylonitrile monomer-precursor for production of carbon fibers*" is intended to convert nonfood, biomass-derived sugars into acrylonitrile monomers (the major precursor worldwide for carbon fibers), currently produced from petroleum-based feedstocks (e.g., propylene). This conversion process is performed at mild conditions and includes multiple catalytic reaction steps (including hydrocracking, dehydration, and ammoxidation), yielding 35%–40% and 60%–80% of product recovery and carbon recovery, respectively. Compared to the conventional processes for producing acrylonitrile monomers, TEA and LCA analyses have estimated a cost reduction of 15%–20% and 37% reduction in greenhouse gas emissions by using this approach.

Brazil is another important country in terms of bioeconomy and has a strong bio-based sector due to the extensive industry involving sugarcane. It is among the top five users of genetically modified crops, and bioethanol represents about a quarter of the total fuel consumption. However, Brazil has no dedicated bioeconomy strategy yet, even though the government has encouraged bioeconomy with several policy strategies and laws since the oil crisis in the 1970s (Bioökonomierat, 2018). Since 2011, the National Confederation of Industry (CNI) has also a specific agenda to stimulate life sciences, biotechnology, and biodiversity as key factors for success in developing a bio-based industry (CNI, 2018). This strategy seeks to provide novel solutions when facing the main societal challenges (e.g., water; food and energy security; urbanization; climate change) (CNI, 2018; Bioökonomierat, 2018). Brazil has also created the Brazilian Industrial Biotechnology Association, ABBI, to foster the organization of events and discussion forums in order to (http://www.abbi.org.br/pt/home_pt_br):

(1) Stimulate better regulation policies
(2) Upgrade Brazilian infrastructures
(3) Direct the development of novel biotechnology and technology tools

ABBI is made up of 13 international leading companies from Brazil, Europe, and the USA (e.g., Raizen, Novozymes, or Dupont). Although the associated companies mainly focus on biofuels, biochemical, and biomass-related process, ABBI is open to any other bio-based sector, such as the paper and pulp industry or the textile-related sector. Similar to the BioPreferred Program, ABBI has no investment calls for funding projects. Nevertheless, the Brazilian bio-based industry has been promoted with programs such as PADIQ (http://www.finep.gov.br/apoio-e-financiamento-externa/programas-e-linhas/programas-inova/padiq), PAISS (http://www.finep.gov.br/apoio-e-financiamento-externa/programas-e-linhas/programas-inova/paiss), RENOVABIO (http://www.mme.gov.br/web/guest/secretarias/petroleo-gas-natural-e-combustiveis-renovaveis/programas/renovabio/principal), or BIOEN-FAPESP (http://bioenfapesp.org). For instance, the BIOEN-FAPESP program supports research and development activities by public-private partnerships, paying marked attention to those projects, focusing at comprehensive research on sugarcane and other bio-based resources. In particular, this program focuses on five main points:

(1) Biomass research with focus on sugarcane, including plant improvement and farming
(2) Biorefinery technologies and alcohol chemistry
(3) Biofuel industrial technologies
(4) Ethanol applications for motor vehicles: otto cycle
(5) Research on sustainability and impacts: social, economic, and environmental studies

Table 1.5 lists few projects that have received funding through the BIOEN-FAPESP program. For instance, the project "*Bioproduction of enzymes for cellulose refining*" focuses in the development of customized enzymes (specifically recombinant xylanases and endoglucanases) for the refining of virgin cellulose fiber and cellulose from recycled paper (https://bv.fapesp.br/en/auxilios/87015/bioproduction-of-enzymes-for-cellulose-refining). These

TABLE 1.5 Projects intended to promote the Brazilian bio-based industry through PIPE (*Pesquisa Inovativa em Pequenas Empresas*) call under the BIOEN-FAPESP program.

Company	Project title	Date
Qfir Eireli Probiotic Juices (EPP)	Standardization of a new drink with probiotic potential based on kefir and fruit juice: Evaluation of microbial population dynamics during fermentation and refrigeration storage	02/2019–10/2019
Bioprocess Improvement Consulting and Research in Bioprocess Ltda	Virtual prototyping as a platform for the development of industrial processes: Application of selective centrifugation in alcoholic fermentation	02/2019–10/2019
Bioinfood Solucoes Em Biotecnologia Ltda	Biotechnology platforms for the food and beverage industry	12/2018–08/2019
Metamorphosis Biotechnology Pesquisas Científicas Eireli	Development of mineral protein functional supplement of animal origin (insects) for animal feed	11/2018–07/2019
Biocelere Agroindustrial Ltda	Transgenic cane-energy for high biomass production	07/2017–05/2019
Algae Biotecnologia Ltda (ALGAE)	Astaxantin production from fixed cultivation of *Haematococcus pluvialis*	02/2018–10/2018
Itatijuca Biotech Ltda	Development of a broad spectrum herbicide produced by microorganisms	09/2017–07/2018
Algae Biotecnologia Ltda (ALGAE)	Production of microalgae protein biomass from brewery wastewater for use as feed formulation for aquaculture	08/2017–04/2018
Itatijuca Biotech Ltda	New biological devulcanization processes	07/2016–05/2017
Verdartis Desenvolvimento Biotecnológico Ltda	Bioproduction of enzymes for cellulose refining	09/2014–04/2017
Accert Pesquisa e Desenvolvimento em Química e Biotecnologia Ltda (ACCERT)	Development of fermentation process for obtaining *Saccharomyces boulardii*	02/2016–10/2016
Bio & Green Industria de Produtos Biodegradáveis Ltda	Production of biodegradable germination and planting tubes from industrial and urban waste from renewable sources	07/2013–03/2016

TABLE 1.5 Projects intended to promote the Brazilian bio-based industry through PIPE (*Pesquisa Inovativa em Pequenas Empresas*) call under the BIOEN-FAPESP program.—cont'd

Company	Project title	Date
Biocelere Agroindustrial Ltda	Cane: Transgenic energy for high biomass production	05/2015–01/2016
Accert Pesquisa e Desenvolvimento em Química e Biotecnologia Ltda (ACCERT)	Drug recycling: An approach involving green chemistry and biotechnology	06/2013–11/2015
Fermentec Assistência Técnica em fermentação Alcoólica Ltda	Valuation of selected industrial yeast for high fermentative performance and consolidation of a new business model	04/2013–11/2014

More information about these projects can be found in the following web: https://bv.fapesp.br/pt/empresas.

customized enzymes will be obtained by using PERSOZYME technology (http://verdartis.com.br/Sites_Ingles/services.html), which enables directed evolution of target enzymes to increase their productivity under specific industrial process conditions (pH, retention time, and temperature). On reaching the productivity target at bench scale (600,000 IU/L), the process will be scaled up for the commercialization of such developed enzymes. Furthermore, the use of resulting enzymes will be subsequently evaluated to estimate the potential energy consumption savings during the refining stage. "*Standardization of a new drink with probiotic potential based on Kefir and fruit juice: evaluation of microbial population dynamics during fermentation and refrigeration storage*" is another example of project granted by BIOEN-FAPESP program (https://bv.fapesp.br/en/auxilios/102651/standardization-of-a-new-drink-with-probiotic-potential-based-on-kefir-and-fruit-juice-evaluation-o). This project studies the population dynamics of the microbial cultures used in fermentation and during the shelf life of the Qfir beverage, a water kefir-based drink containing fruit juice (developed by QFIR Sucos Probióticos Eireli: https://lojaqfir.com.br), to understand the behavior of the microorganisms and their potential probiotic uses and health benefits.

In Canada, bioeconomy is mainly stimulated by regional approaches (Birch, 2016). Nonetheless, in 2017 the Canadian Council of Forest Ministers launched the first federal policy: "*A Forest Bioeconomy Framework for Canada*" to promote the use of biomass as raw material for the production of bio-based chemicals, plastics, and composites (CCFM, 2017). In this context, the production of industrial relevant bioproducts is seen as an opportunity to (Rancourt et al., 2017):

(1) Open new agricultural markets
(2) Provide add value to commodity production

(3) Diversify operations and increase revenues
(4) Offer new ways for the management of resources
(5) Serve as sustainable alternative to nonrenewable petroleum-based resources
(6) Provide economic opportunities for rural communities

Therefore, this federal policy seeks to increase the competitiveness of the Canadian forestry sector by promoting new businesses, new technologies, or better inventory information to promote the use of forest biomass in the Canadian economy. In addition to policy and regulation strategies, Canada has implemented several investment programs to support novel concepts and breakthrough technologies associated with the invention, development, production, and use of bio-based products and processes within the health, energy, agriculture, chemicals, manufacturing, automotive, aerospace, and materials industries (Bioökonomierat, 2015a, 2018). BIOTECanada is the national industry association (having over 200 members) committed to lead the advancement of all Canadian bio-based sectors (http://www.biotech.ca). The major BIOTECanada's missions are:

(1) The development of a constructive working relationship between policy-makers and regulators to support the industry/government objectives
(2) To increase investment in Canadian biotechnology innovation, research, and commercialization
(3) To grow the capacity of Canada to attract and develop C-suite entrepreneurial talent and biotechnology leadership

Bioindustrial Innovation Canada is a not-for-profit business accelerator (based in Sarnia, Ontario) which invests in early stage companies who develop clean, green and sustainable technologies, and simultaneously provides business development and job benefits at both local and global levels (https://www.bincanada.ca). So far, Bioindustrial Innovation Canada has funded over 20 projects to key companies from the chemical sector. For instance, the company Advance Chemical Technologies Inc. has received funding to improve the process performance and mitigate the key risks during the production of green methanol (http://advancedchemicaltech.com). This product can be directly used as a low carbon energy source or as raw material for the production of derivative products such as adhesives. The process also utilizes the CO_2 emitted by other industries, thus reducing the related carbon footprint. The company Mirexus Biotechnologies Inc. has also received funding from Bioindustrial Innovation Canada to validate and optimize the use of its self-developed technology, PhytoSpherix (a natural form of glycogen with humectant and film-forming properties due to its unique structure), to enhance virus yields during vaccine manufacturing (https://mirexus.com). This technology has been already implemented and is being commercialized in different skin care products to improve skin feel, hydration, calming, or skin redness. The project will therefore broaden the application scope of the PhytoSpherix-related products to other bio-based sectors such as health. In addition to BIOTECanada and Bioindustrial Innovation Canada, the Canadian Agricultural Partnership (http://www.agr.gc.ca/eng/about-us/key-departmental-initiatives/canadian-agricultural-partnership/?id=1461767369849) and FPInnovations (https://fpinnovations.ca) are also important programs to foster bio-based industry in Canada.

In India, the bioeconomy strategy covers all sectors with potential applications of biotechnology, aiming at translating life science knowledge into socially relevant eco-friendly and competitive products (Bioökonomierat, 2015b). Among the priority areas, the Indian bioeconomy strategy promotes pharmaceutical biotechnology, bioenergy, nanobiotechnology, and other bio-based related technologies. Nutrition and food security is an important pillar for this strategy. In this context, modern tools such as genetic engineering are used to improve the productivity of the food sector through innovative research (DBT, 2015). However, novel genetically modified crops and/or food products have to be tested prior to their commercialization, according to the Good Manufacturing Practices (GMP) standards (https://ispe.org/initiatives/regulatory-resources/gmp). To implement this strategy and meet the industry requirements in terms of funds and nurturing the innovation ecosystem, the Indian Department of Biotechnology created the "Biotechnology Industry Research Assistant Council" (BIRAC). BIRAC is a not-for-profit agency aimed at strengthening and empowering the different biotechnology sectors by supporting strategic research and innovation to fulfill the national needs (http://www.birac.nic.in). This agency particularly stimulates, fosters, and enhances the research and innovation capabilities of startups and SMEs as major agents of change toward building an Indian bioeconomy. The BIRAC's public-private partnership program is an industry-academia interface that implements a wide range of impact initiatives to provide targeted funding to risk capital, technology transfer, IP management, and handholding schemes, thus contributing to make them competitive (DBT, 2015). Major investments programs and partnerships include a list of potential funding options to support both early and late stage research, translation research, and social innovation. Biotech ignition grant scheme (BIG), bioincubators nurturing entrepreneurship for scaling technologies (BioNEST), small business innovation research initiative (SBIRI), biotech industry partnership program (BIPP), or Grand challenge—India (GCI) are some of these investment programs. These programs have provided funding to numerous industries and research groups to foster research on bio-based products. A full list of funded projects under BIRAC's programs can be found at BIRAC's web page (http://www.birac.nic.in/desc_new.php?id=145). For instance, Avinash Dental Laboratories & Research Institute Pvt. Ltd. has received funding from BIPP program to design and develop silicone-based cartilage implants (with ear, nose, and eye shape) for craniofacial reconstruction. BIPP program has also invested on Aegis Agro Chemicals to investigate a novel micro RNA-induced gene silencing strategy for the development of pest and disease-tolerant cotton plants. Jubeln lifesciences has received funding from BIG to study a novel drug eluting biofilm platform for oral and topical delivery with applications for the nutraceutical, pharmaceutical, and cosmetics industries.

Malaysia has been the first country in South-East Asia to establish its own national bioeconomy initiative. The Malaysian bioeconomy is very much related to industrial upgrading and the application of biotechnology approaches, having its main pillars at sustainability and circular economy. In 2015, bioeconomy (including all sectors of economy that could possibly benefit from application of bio-based technologies) contributed to 11.3% of the total Malaysian GDP (RM131 billion, equivalent to about €28 billion), with palm oil, fruit and vegetables, livestock, and fishing industries as major contributing sectors (BTP, 2016). In addition, the total employment in bioeconomy represented 2.34 million and the overall investment accounted for RM24.7 billion (ca. €5 billion), spending 42% in agriculture and 58% in manufacturing-related sectors. In 2012, Malaysia introduced the Bioeconomy Transformation Programme (BTP) to support the so-called "trigger projects" and maximize the commercial opportunities of the bio-based industries. In its last report (BTP, 2016), the Malaysian Ministry of Science, Technology and Innovation informed that a total of 61 Trigger Projects has been funded through BTP program in the AgBiotech (hih-value bioingredients; high-value food varieties; and bio-based farm inputs), BioIndustrial (industrial bioproducts; bio-based chemicals; bioremediation; and biomaterials), and BioMedical (molecular screening and diagnostic; biopharmaceuticals; drug discovery and preclinical services; stem cells and regenerative medicine; and bio-based materials) group areas. Among these funded projects, seven of them have been identified as high impact projects based on the gross national income, investment, and social contributions. These high impacts projects are:

(1) Setting up of 10 MW palm oil biomass power generation plant in Mukah, Sarawak (Olive Energy Sdn. Bhd.)
(2) Setting up of a biogas power generation plant at Rompin palm oil mill in Pahang based on the feed-in-tariff model (GLT Renewable Sdn. Bhd.)
(3) Setting up of a biogas power generation plant at Setia Kawan Kilang Kelapa Sawit in Kedah based on the feed-in-tariff model (GLT Eco Sdn. Bhd.)
(4) Setting up of a 9.95 MW rice husk and woodchip biomass power generation plant in Naka, Kedah (Majunaka Eco Energy Sdn. Bhd.)
(5) Setting up of a biogas power generation plant at Sri Jelutong palm oil mill in Pahang based on the feed-in-tariff model (Metro Havana Sdn. Bhd.)
(6) Production and commercialization of high-value agarwood products (Asia Plantation Capital, Berhad)

(7) Development, production, and commercialization of biodegradable products (Telic Paper Sdn. Bhd.).

Under the BTP program, Malaysia has also established the Bioeconomy Community Development Program (BCDP) to encourage the enlistment of rural farmers and settlers to cultivate raw materials as potential inputs for the companies being funded by BTP program (BTP, 2016). Developing a secure, local, high quality source of raw materials will subsequently produce a ripple effect throughout the entire industry value chain, enabling capacity expansion for downstream processes and improving market penetration. So far, the BCDP program has funded 500 farmers within 30 projects involving bee farming, stevia farming, high-value herbs farming, shrimp aquaculture, fish aquaculture, oyster farming, dairy farming, mushroom farming, seeds production, and seaweed aquaculture as major strategic subsectors.

In China, the bioeconomy strategy is strongly linked to biotechnology development, since it provides innovative solutions to guarantee environmental protection and sustainable development, as well as to ensure people's health (Wang et al., 2018). In this context, the Chinese Bioindustry covers agriculture, bioenergy, manufacturing, and the health sector (including biomedicine and biopharma), and it is clustered in three main regions that have most of the capital, specialized personnel, universities and factories: Beijing-Tianjin, Yangtze River Delta (Shanghai), and Pearl River Delta (Guangzhou-Shenzhen-Hong Kong). China is still a developing country and despite the massive resources and increasing utilization, bioenergy and bio-based products need further development (Ellis, 2018; Wang et al., 2018). Among the main challenges to address for a rapid bioindustry development include (Wang et al., 2018):

(1) To improve the quality and level of original research and technological innovation still
(2) To advance the cross- or multi-disciplinary research in biological big data and gene manipulation
(3) To improve the research capacity for manufacturing new vaccines, antibodies and other bio-based chemicals
(4) To increase the number of innovation-driven companies

The bio-based industry in China is promoted by the "*Bio-industry to expand*" action of "*The 13*th *Five-Year Plan for Economic and Social Development of the People's Republic of China,*" which highlights the progress on biotechnology as an engine for sustainable development (13th FYP, 2016). Within this plan, the main strategies intended to promote the Chinese bio-based industry are:

(1) To move faster to facilitate the wide application of genomics and other biotechnologies
(2) To create demonstrations of network-based biotech applications

(3) To stimulate the large-scale development of personalized medical treatment, new drugs, bio-breeding, and other next generation biotech products and services
(4) To promote the creation of basic platforms such as gene and cell banks

In addition to the 13th Five-Year Plan, China has other approaches and programs to foster biotechnology (ISDP, 2018; GS-RhG, 2019; Nibbi et al., 2019). For instance, Made in China (2025) is an important government strategy to promote high-tech and high-value industries, including new energy vehicles, biopharmaceuticals, and biomedical technologies (ISDP, 2018). This strategy also emphasizes on using international resources to further contribute to their own industries. Thus, China highlights the concepts such as "opening-up," "going-out," and "bringing-in" as major ways to utilize international expertise for its benefit. As an example of international collaboration, the project BBChina aims at helping People's Republic of China to exploit its biotechnological potential through a Master Program on bio-based circular economy (Nibbi et al., 2019). This project is supported by the Education, Audiovisual and Culture Executive Agency of the European Commission under the ERASMUS+ program and takes place in collaboration with six higher education institutions among China and Europe: Tongji University (Shanghai), East China University of Science and Technology (Shanghai), Sichuan University (Chengdu), University of Florence (Italy), University of Rostock (Germany), and Mälardalen University (Sweden). The Master Program includes different education and training activities (e.g., technical laboratory, international mobility of teachers and students, seminaries) as well as other actions to trigger and encourage young entrepreneurship and innovation. Within a similar context, China has also begun to attract talent back, especially in science and engineering fields, under the "*Thousand Talents Program,*" "*Hundred Talents Program,*" and the "*National Science Fund for Distinguished Young Scholars*" (GS-RhG, 2019). Another key strategy that China is implemented to advance its bioindustry is building biotechnology parks to place high-tech companies with a common theme, such as biopharmaceuticals or nanotechnology (GS-RhG, 2019).

Argentina has tremendous possibilities for bioeconomy development due to biomass availability, well-developed capacities in the scientific and technological disciplines, and long standing entrepreneurial and institutional structures related to the agricultural sector (Trigo et al., 2015). Moreover, Argentina is among the top five countries leading in bioenergy production and use of genetically modified crops (Bioökonomierat, 2018). In 2017, the Argentinian Ministry of Agro-Industry launched "*Bioeconomía Argentina: Visión desde Agroindustria*" to foster bioeconomy and bio-based industries (Trigo et al., 2017). This document identifies as main priority areas the production and use of local bioresources, improving the efficiency of the supply chain to ensure food security, the development of new feedstock resources (e.g., nonfood crops, residues from forestry and agriculture, and side-streams from the food industry) for a sustainable bioenergy supply, and the development of large-scale biorefineries to produce higher value bioproducts. Thus, the strategy focuses on promoting innovation by strengthening public R&D and novel technologies (e.g., bioinformatics, "omics" techniques), supporting international cooperation with centers of excellence abroad, intensifying public-private partnerships, and investing at local and regional levels for the development of new infrastructure (giving priority to the development of large-scale biorefineries). To meet the great investment efforts needed to support the transition toward novel productive systems, Argentina has several public and private funding programs and plans. FONCYT ("*Fondo Nacional de Ciencia y Tecnología*"; https://www.argentina.gob.ar/ciencia/agencia/fondo-para-la-investigacion-cientifica-y-tecnologica-foncyt) and FONTAR ("*Fondo Tecnologico Argentino*"; https://www.argentina.gob.ar/ciencia/agencia/fondo-tecnologico-argentino-fontar) programs are specific actions of the Ministry of Science, Technology and Productive Innovation (MINCYT) designed to fund R&D and innovation activities in the private sector (Trigo et al., 2015). MINCYT also encourages public-private partnerships through FONARSEC ("*Fondo Argentino Sectorial*"; https://www.argentina.gob.ar/ciencia/agencia/fondo-argentino-sectorial-fonarsec) fund. This fund has a central role for the implementation of new policies and strengthens the cooperation between the science and technology and productive sectors. The National Research Council (CONICET; https://www.conicet.gov.ar) also fosters and promotes innovation projects to enhance the national economy through a specific strategic plan that provides subsidies to high impact research projects (https://www.conicet.gov.ar/?lan=en). In the 2017/2019 call, CONICET has funded several projects intended to promote the bio-based industry. The full list of funded projects can be found in the CONICET web page (https://convocatorias.conicet.gov.ar/investigacion-y-desarrollo). For instance, three of such bio-based related projects are:

(1) Blue-light bacterial photoreceptors for biotechnological applications (Instituto de Bionanotecnología del NOA)
(2) Processing of zirconia bioceramics for dental applications (Centro de Tecnología de Recursos Minerales y Cerámica)
(3) Use of seaweed as bioadsorbent to prevent mycotoxicosis in birds (Universidad Nacional de Río Cuarto)

1.4 Concluding remarks

Several countries are developing and implementing different strategies toward establishing a sustainable bioconomy. EU-28 is considered to house one of the most advanced bio-based industries, but other countries such as USA, Brazil, Canada, or Malaysia also have a strong bio-based sector. These countries pursue the general use of national biomass feedstocks to obtain the food, feed, energy, materials, and products necessary to sustain the increasing world population. Furthermore, they have launched specific investment programs to support both public and private organizations, such as the BESTF3, BBI-JU, BETO, and BIOEN-FAPESP actions. These funding programs are a key aspect to grant ambitious, crucial projects and there is absolutely no doubt about the importance of these investment strategies to success in the optimal development of the technology required to definitively deploy a sustainable bioeconomy worldwide.

Acknowledgments

Authors gratefully thank Dr. María Cristina Canela for her valuable contribution in providing funding information with regards to Brazilian bioeconomy. Authors also acknowledge the European Commission and the Spanish Ministry of Science, Innovations and Universities for partially funding this work through URBIOFIN (Grant agreement ID: 745785), WASTE2-BIO (PCIN-2016-098) and ACMIBIO (ENE2017-86864-C2-1-R, AEI/FEDER, UE) projects.

References

AgriChemWhey project, 2019. Web source: https://www.agrichemwhey.com/.

Andrew, R.M., 2018. Global CO_2 emissions from cement production. Earth Syst. Sci. Data 10, 195–217. https://doi.org/10.5194/essd-10-195-2018.

BARBARA project, 2019. Web source: http://www.barbaraproject.eu/.

Bastidas-Oyanedel, J.-R., Fang, C., Almardeai, S., Javid, U., Yousuf, A., Schmidt, J.E., 2016. Waste biorefinery in arid/semi-arid regions. Bioresour. Technol. 215, 21–28.

Beres, C., Costa, G.N.S., Cabezudo, I., da Silva-James, N.K., Teles, A.S.C., Crus, A.P.G., Mellinger-Silva, C., Tonon, R.V., Cabral, L.M.C., Freitas, S.P., 2017. Towards integral utilization of grape pomace from winemaking process: a review. Waste Manag. 68, 581–594.

BETO, 2016. The Strategic Plan for a Thriving and Sustainable Bioeconomy. Bioenergy Technologies Office. Available at: https://www.energy.gov/sites/prod/files/2016/12/f34/beto_strategic_plan_december_2016_0.pdf.

BETO, 2017. The 2017 Project Peer Review. Bioenergy Technologies Office. Available at: https://www.energy.gov/sites/prod/files/2018/02/f48/2017_project_peer_review_report.pdf.

BioBarr project, 2019. http://www.biobarr.eu/.

BIOMO-TIVE project, 2019. Web source: https://biomotive.info/.

Bioökonomierat, 2015a. Bioeconomy Policy (Part I). Synopsis and Analysis of Strategies in the G7. A Report from the German Bioeconomy Council. Available at: https://biooekonomierat.de/fileadmin/Publikationen/berichte/BOER_Laenderstudie_1_.pdf.

Bioökonomierat, 2015b. Bioeconomy Policy (Part II): Synopsis of National Strategies Around the World. A Report from the German Bioeconomy Council. Available at: https://biooekonomierat.de/fileadmin/international/Bioeconomy-Policy_Part-II.pdf.

Bioökonomierat, 2018. Bioeconomy Policy (Part III). Update Report of National Strategies Around the World. A Report from the German Bioeconomy Council. Available at: https://biooekonomierat.de/fileadmin/Publikationen/berichte/GBS_2018_Bioeconomy-Strategies-around-the_World_Part-III.pdf.

BIOSMART project, 2019. Web source: http://biosmart-project.eu/.

Birch, K., 2016. Emergent imaginaries and fragmented policy frameworks in the Canadian Bioeconomy. Sustainability 8, 1007.

BRDB, 2016. Federal Activities Report on the Bioeconomy. The Biomass Research & Development Board, Washington (DC). Available at: https://www.biomassboard.gov/pdfs/farb_2_18_16.pdf.

BTP, 2016. Enriching the Nation, Securing the Future. Bioeconomy Transformation Programme: Annual Report. BioEconomy Malaysia. Available at: http://www.bioeconomycorporation.my/wp-content/uploads/2011/11/publications/BTP_AnnualReport2016.pdf.

Camia, A., Robert, N., Jonsson, R., Pilli, R., García-Condado, S., López-Lozano, R., van der Velde, M., Ronzon, T., Gurría, P., M'Barek, R., Tamosiunas, S., Fiore, G., Araujo, R., Hoepffner, N., Marelli, L., Giuntoli, J., 2018. Biomass Production, Supply, Uses and Flows in the European Union. First Results from an Integrated Assessment. Publications Office of the European Union, Luxembourg, ISBN 978-92-79-77237-5. Available at: http://publications.jrc.ec.europa.eu/repository/bitstream/JRC109869/jrc109869_biomass_report_final2pdf2.pdf.

Cao, L., Yu, I.K.M., Liu, Y., Ruan, X., Tsang, D.C.W., Hunt, A.J., Ok, Y.S., Song, H., Zhang, S., 2018. Lignin valorization for the production of renewable chemicals: state-of-the-art review and future prospects. Bioresour. Technol. 269, 465–475.

Carlson, R., 2016. Estimating the biotech sector's contribution to the US economy. Nat. Biotechnol. 34, 247–255.

CCFM, 2017. A Forest Bioeconomy Framework for Canada. Canadian Council of Forest Ministers, ISBN 978-0-660-09391-8. Available at: http://cfs.nrcan.gc.ca/pubwarehouse/pdfs/39162.pdf.

CEFIC, 2018. Facts and Figures of the European Chemical Industry. The European Chemical Industry Council. Available at: https://cefic.org/app/uploads/2018/12/Cefic_FactsAnd_Figures_2018_Industrial_BROCHURE_TRADE.pdf.

Chu, W., 2016. China's 13th Five-Year Plan: The Challenges and Opportunities of Made in China 2025. HKTDC Research. Available at: https://hkmb.hktdc.com/en/1X0A6918/hktdc-research/China%E2%80%99s-13th-Five-Year-Plan-The-Challenges-and-Opportunities-of-Made-in-China-2025.

CNI, 2018. Mapa estratégico da indústria 2018–2022. Confederação Nacional da Indústria – CNI [Portuguese]. Available at: https://bucket-gw-cni-static-cms-si.s3.amazonaws.com/media/filer_public/ee/50/ee50ea49-2d62-42f6-a304-1972c32623d4/mapa_final_ajustado_leve_out_2018.pdf.

Dammer, L., Piotrowski, S., Carus, M., 2014. Study on: Methodology Framework for the Bioeconomy Observatory. Project: Bioeconomy Information System and Observatory Project (BISO) – Set up of the Bioeconomy Observatory. FP7 Grant Agreement No. 341300.

DataM, 2018. Jobs and Wealth in the European Union Bioeconomy. Data Portal of Agro-Economics Modelling. Web source: https://datam.jrc.ec.europa.eu/datam/mashup/BIOECONOMICS/index.html.

DBT, 2015. National Biotechnology Development Strategy. Department of Biotechnology (DBT), Government of India. Available at: http://www.dbtindia.nic.in/wp-content/uploads/DBT_Book-_29 december_2015.pdf.

Dietz, T., Börner, J., Förster, J.J., von Braun, J., 2018. Governance of the bioeconomy: a global comparative study of national bioeconomy strategies. Sustainability 10, 3190. https://doi.org/10.3390/su10093190.

ECOXY project, 2019. Web source: http://www.ecoxy.eu/.

EFFECTIVE project, 2019. Web source: http://www.effective-project.eu/.

Ellis, S., 2018. Biotech booms in China. Nature 553 (7688), S19–S22.

EMBRACED project, 2019. Web source: https://www.embraced.eu/.

EUCALIVA project, 2019. Web source: https://eucaliva.eu/.

European Commission, 2018. A Sustainable Bioeconomy for Europe: Strengthening the Connection between Economy, Society and the Environment. Publications Office of the European Union, Luxembourg, ISBN 978-92-79-94144-3. Available at: https://ec.europa.eu/research/bioeconomy/pdf/ec_bioeconomy_strategy_2018.pdf.

EUROSTAT, 2008. NACE Rev.2. Statistical Classification of Economic Activites in the European Community. Office for Official Publications of the European Communities, Luxembourg, ISBN 978-92-79-04741-1. Available at: https://ec.europa.eu/eurostat/documents/3859598/5902521/KS-RA-07-015-EN.pdf.

EXCornsEED project, 2019. Web source: https://www.excornseed.eu/.

13th FYP, 2016. The 13th Five-Year Plan for Economic and Social Development of the People's Republic of China. National Development and Reform Commission (NDRC) People's Republic of China. Available at: http://en.ndrc.gov.cn/policyrelease/201612/P020161207645766966662.pdf.

FAO, 2011. Global food losses and food waste – extent, causes and prevention. Rome, Italy. Available at: http://www.fao.org/3/mb060e/mb060e.pdf.

FAO, 2016. Forestry for a Low-Carbon Future. Integrating Forests and Wood Products in Climate Change Strategies. Food and Agriculture Organization of the United Nations, Rome, ISBN 978-92-5-109312-2. Available at: http://www.fao.org/3/a-i5857e.pdf.

Galanakis, C.M., 2012. Recovery of high added-value components from food wastes: conventional, emerging technologies and commercialized applications. Trends Food Sci. Technol. 26 (2), 68–87.

GS-RhG, 2019. China's Biotechnology Development: The Role of US and Other Foreign Engagement. A Report Prepared for the U.S.-China Economic and Security Review Commission. Gryphon Scientific, LLC. and Rhodium Group, LLC. Available at: https://www.uscc.gov/sites/default/files/Research/US-China%20Biotech%20Report.pdf.

IEA Bioenergy, 2018. Standards and Labels Related to Biobased Products. Developments in the 2016–2018 Triennium, ISBN 978-1-910154-51-9. Available at: https://www.ieabioenergy.com/wp-content/uploads/2018/10/Standards-and-Labels-related-to-Biobased-Products-2016-to-2018.pdf.

iFermented project, 2019. Web source: https://ifermenter.eu/.

ISDP, 2018. Made in China 2025. Institute for Security and Development Policy. Available at: http://isdp.eu/content/uploads/2018/06/Made-in-China-Backgrounder.pdf.

Kovacs, B., 2015. Sustainable Agriculture, Forestry and Fisheries in the Bioeconomy – A Challenge for Europe. 4th SCAR Foresight Exercise. Publications Office of the European Union, Luxembourg, ISBN 978-92-79-47538-2. Available at: https://ec.europa.eu/research/scar/pdf/ki-01-15-295-enn.pdf.

LigniOx project, 2019. Web source: http://www.ligniox.eu/.

Nattrass, L., Biggs, C., Bauen, A., Parisi, C., Rodríguez-Cerezo, E., Gómez-Barbero, M., 2016. The EU Bio-Based Industry: Results from a Survey. Joint Research Centre, ISBN 978-92-79-54969-4. Available at: http://publications.jrc.ec.europa.eu/repository/bitstream/JRC100357/jrc100357.pdf.

NEWPACK project, 2019. Web source: http://newpack-h2020.eu/.

Nibbi, L., Chiaramonti, D., Palchetti, E., 2019. Project BBChina: a new master program in three Chinese universities on bio-based circular economy; from fields to bioenergy, biofuel and bioproducts. 10th international conference on applied energy (ICAE2018), 22–25 August 2018, Hong Kong, China. Energy Procedia 158, 1261–1266.

OPTISOCHEM project, 2019. Web source: https://ecoxy.eu/.

Parisi, C., 2018. Research Brief: Biorefineries Distribution in the EU. European Commission – Joint Research Centre, ISBN 978-92-79-94882-4. Available at: http://publications.jrc.ec.europa.eu/repository/bitstream/JRC113216/online_biorefineries_research_brief.pdf.

Pathak, P.D., Mandavgane, S.A., Puranik, N.M., Jambhulkar, S.J., Kulkarni, B.D., 2018. Valorization of potato peel: a biorefinery approach. Crit. Rev. Biotechnol. 38 (2), 218–230.

PEFerence project, 2019. Web source: http://peference.eu/.

PERCAL project, 2019. Web source: http://www.percal-project.eu/.

Piotrowski, S., Carus, M., Carrez, D., 2018. European Bioeconomy in Figures 2008–2015. Bio-based Industries Consortium. Available at: https://biconsortium.eu/sites/biconsortium.eu/files/documents/European_Bioeconomy_in_Figures_2008-2015_06042018.pdf.

POLYBIOSKIN project, 2019. Web source: http://polybioskin.eu/.

Pro-Enrich project, 2019. Web source: https://www.pro-enrich.eu/.

Prolific project, 2019. Web source: http://www.prolific-project.eu/.

Rancourt, Y., Neumeyer, C., Zou, N., 2017. Results from the 2015 Bioproducts Production and Development Survey. Statistics Canada – Ministry of Industry. Available at: https://www150.statcan.gc.ca/n1/pub/18-001-x/18-001-x2017001-eng.pdf.

REFUCOAT project, 2019. Web source: https://www.refucoat.eu/.

ReInvent project, 2019. Web source: https://www.bbi-europe.eu/projects/reinvent.

Rodríguez, A.G., Mondaini, A.O., Hitschfeld, M.A., 2017. Bioeconomy in Latin America and the Caribbean: Global and Regional Context and Perspectives. United Nations – ECLAC [Spanish]. Available at: https://www.cepal.org/es/publicaciones/42427-bioeconomia-america-latina-caribe-contexto-global-regional-perspectivas.

SHERPAC project, 2019. Web source: http://www.sherpack.eu/.

Spekreijse, J., Lammens, T., Parisi, C., Ronzón, T., Vis, M., 2019. Insights into the European Market for Bio-Based Chemicals. Publications Office of the European Union, Luxembourg, ISBN 978-92-79-98420-4. Available at: https://ec.europa.eu/jrc/en/publication/eur-scientific-and-technical-research-reports/insights-european-market-bio-based-chemicals.

SSUCHY project, 2019. Web source: https://www.ssuchy.eu/.

SUSBIND project, 2019. Web source: https://susbind.eu/.

SUSFERT project, 2019. Web source: https://www.susfert.eu/.

Susmozas, A., Iribarren, D., Zapp, P., Linβen, J., Dufour, J., 2016. Life-cycle performance of hydrogen production via indirect biomass gasification with CO_2 capture. Int. J. Hydrogen Energy 41 (42), 19484–19491.

Susmozas, A., Moreno, A.D., Romero-García, J.M., Manzanares, P., Ballesteros, M., 2019. Designing an olive tree pruning biorefinery for the production of bioethanol, xylitol and antioxidants: a techno-economic assessment. Holzforschung 73 (1), 15–23.

SWEETWOODS project, 2019. Web source: https://sweetwoods.eu/.

SYLFEED project, 2019. Web source: http://www.sylfeed.eu.

Trigo, E., Regúnaga, M., Costa, R., Wiemy, M., Coremberg, A., 2015. The Argentinean Bioeconomy: Scope, Present State and Opportunities for its Sustainable Development. Bolsa de Cereales, ISBN 978-987-97337-7-6. Available at: http://grupobioeconomia.com.ar/wp content/uploads/2017/02/The_Argentinean_Bioeconomy_Scope_present_state_and_opportunities_for_its_sustainable_development.pdf.

Trigo, E., Morales, E.V., Grassi, L., Losada, J., Dellisanti, J.P., Molinari, M.E., Murmis, M.R., Almada, M., Molina, S., 2017. Bioeconomía Argentina: Visión desde Agroindustria. Ministerio de Agroindustria – Presidencia de la Nación [Spanish]. Available at: https://www.agroindustria.gob.ar/sitio/areas/bioeconomia/_archivos//000000_Bioeconomia%20Argentina.pdf.

United States, 2017. North American Industry Classification System. Executive Office of the President Office of Management and Budget. Available at: https://www.census.gov/eos/www/naics/2017NAICS/2017_NAICS_Manual.pdf.

UNRAVEL project, 2019. Web source: http://unravel-bbi.eu/.

URBIOFIN project, 2019. Web source: www.urbiofin.eu/.

USDA, 2016. An Economic Impact Analysis of the US Biobased Products Industry. United States Department of Agriculture. Available at: https://www.biopreferred.gov/BPResources/files/BiobasedProductsEconomicAnalysis2016.pdf.

USDA, 2018. The BioPreferred Catalog. Product Categories by Functional Area. United States Department of Agriculture. Available at: https://www.biopreferred.gov/BioPreferred/faces/pages/ProductCategories.xhtml.

VIPRISCAR project, 2019. Web source: https://vipriscar.eu/.

Wang, R., Cao, Q., Zhao, Q., Li, Y., 2018. Bioindustry in China: an overview and perspective. N. Biotech. 40, 46–51. https://doi.org/10.1016/j.nbt.2017.08.002.

WoodZymes project, 2019. Web source: https://www.woodzymes.eu/.

Chapter 2

Development perspectives for the bio-based economy

J.P.M. Sanders[1], J.W.A. Langeveld[2]
[1]*Wageningen University, Department of Biobased Chemistry and Technology, Wageningen, the Netherlands;* [2]*Biomass Research, Wageningen, the Netherlands*

2.1 Introduction

World population is expected to exceed 9 billion in 2050. People on average will be more prosperous than today. We will need much more food, thus increasing pressure on the planetary boundaries as described by Rockström et al. (2009). Already, four of nine planetary boundaries have been crossed as a result of human activities: climate change, biosphere integrity, land-system change, and biogeochemical cycles (phosphorus and nitrogen). Other boundaries include ozone depletion, ocean acidification, and freshwater use. Climate change and biosphere integrity are considered "core boundaries"; significantly altering any of these "core boundaries" would irrevocably change the character of the Earth System (Steffen et al., 2015).

Several authors address the supply of food and the consequences for the planetary boundaries. Tilman et al. (2011) describes the growing needs in more detail, presenting strategies to generate the necessary calorie intake and the amount of protein. Development of better agricultural technologies and/or the transfer toward countries that still deprive these technologies will be essential. In order to provide the food that is needed, we will require much more nitrogen fertilizers, and/or agricultural land. Both developments will increase greenhouse gas emissions.

Today, global use of nitrogen fertilizer is 110 million tonnes (Mtonne). In addition, 60 Mtonnes nitrogen is fixated by plants (Conijn et al., 2018) which contribute in the same way to ammonia emissions but less to N_2O volatilization and nitrate leaching or runoff. If we have to double global protein production from an estimated 550 Mtonnes in 2010 to about 1300 Mtonnes, needed to feed 9 billion people in 2050 with a European-style diet, innovations are essential. By applying the same technologies we are using today, we will need 260 Mtonnes of nitrogenous fertilizer while the planetary boundary has been defined as about 80 Mtonnes by Rockström et al. (2009).

Biobased Products and Industries. https://doi.org/10.1016/B978-0-12-818493-6.00002-6

Conijn et al. (2018) presented different scenarios how our global food system can be sustained within planetary boundaries; diet change, 50% reduction of animal products in the food consumption, 25% improvement of animal feed conversion rates, and 50% increase in crop yields including grass provide the best perspectives to prevent major increase of land use and nitrogen flows. Nitrogen flows can only be limited to 80M tonnes, if all of the scenarios are implemented together.

Springmann et al. (2018) analyzed options for reducing environmental effects of the food system including dietary changes (more plant-based), technology and managerial improvements, and reduction of food losses. It is concluded that no single measure alone is sufficient to keep the effects within all planetary boundaries.

None of the authors cited above considered the need for raw materials for the bio-based economy, an economy in which at least 30% of fossil resources used are substituted by biomass. Biorefinery technologies as a means to improve resource use efficiency of biomass and create more added value also were not included. In this chapter, approaches are presented that address environmental issues by integrating the bio-based economy and food production in the so-called bioeconomy. This will be done by focusing on technology development that enables increased resource use efficiency such as (renewable and nonrenewable) energy, biomass, and land use in general, protein and nitrogen in particular, and of course the other planetary boundaries as defined by Rockström (2009).

Bio-based technologies have great potential, as they facilitate the production of animal proteins and other products and services (e.g., energy) in a more efficient way. In this way, we can stay within the planet's boundaries for sustainable production. In order to realize this we suggest six development rules ("principles") to be implemented in the design and development of sustainable bio-based production chains (Table 2.1).

The ecopyramid presented in Fig. 2.1 demonstrates how high-value biomass endproducts, such as pharmaceuticals and food ingredients, will be profitable. The amount of fossil resources substituted by these products will however be limited. Large volumes of fossil resources can be substituted by biomass, reducing major CO_2 emissions, but this will not be economical because these applications like the production of electricity and heat have to compete with (cheap) fossil alternatives, especially coal[1]

Products in the middle of the pyramid combine large market volumes with high added value; this is where Planet and Profit meet each other. Three groups providing large market volumes at a high value are (i) nutritious proteins for animal feed, (ii) chemical building blocks, and (iii) materials substituting

1. Recent development suggests that renewable alternatives are now becoming cheaper than any fossil energy source, including coal. While this currently is a regional development (i.e., not applying in all regions of the globe), it is expected to become a generic situation.

TABLE 2.1 Development rules for a sustainable bioeconomy.

Every project should follow the principles of *People, Planet, Profit*
The aim is to improve overall energy efficiency of the entire system
Increase field yield but keep plant components that are required for soil fertility on the field
Use all biomass components and choose the right raw material
Use each component at its highest value: (molecular) structure is much better than caloric
Reduce capital cost to speed up innovation and to benefit from small scale without the disadvantages

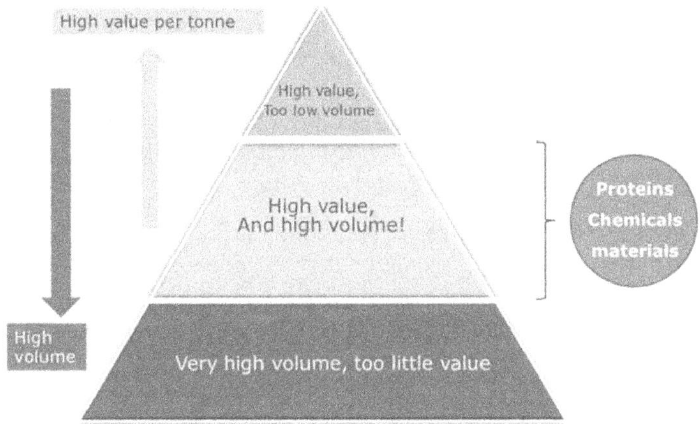

FIGURE 2.1 Ecopyramid.

synthetic polymers. After obtaining these components, residues will be available for high-volume energetic applications.

2.2 Principles of bio-based production

2.2.1 Six development rules

Six development rules that were introduced in Table 2.1 are elaborated in this section.

2.2.2 People, planet, profit

In order to reach sustainable results, projects should cope with People, Planet, and Profit conditions simultaneously. If only two of them are addressed, situations may be expected that may put the project results at risk. Focusing on two conditions, assuming that the third can be addressed later will cause ineffective investments to be done and money and peace being lost. It is challenging to address Planet and People issues while earning money. Generally, this cannot be done without development of new technologies or concepts. The main drawback of such an approach is that a vision needs to be developed how to cope with challenges, while funds are needed to develop the concepts and time required often exceeds 6 or 10 years.

If quicker results are needed, focus should be on technologies that are already developed but not available in the open domain. In this case one can cope with Planet and People issues already with less than say 6 years, but there will not be much profit or loss. Shorter implementations times, say 2–3 years, require technologies in the open domain that address People and Planet issues at cost that has to be paid annually. One can discriminate five different approaches to cope with environmental issues:

- Curing or end-of-pipe solutions at very high annual cost;
- Dilution by, for example, a chimney or spreading manure over the field will cost much less;
- Recycling materials will generally cost even less;
- Preventing will have no costs by definition but the fifth approach;
- Add value will prevent and at the same time enable to earn money.

In general, adding value and at the same time solving People and Planet issues is not possible. If this would be the case, solutions would already have been applied. On the other side, development of new technologies for end-of-pipe solutions would be a waste of time and resources that could be spent more effectively.

2.2.3 Improve efficiency of material and energy use

Even when renewable energy is available, costs have to be made to generate them. Preventing and/or reducing energy use is always to be preferred. The same applies to material use. If we express our daily food in energy terms at 2500 kcal per person, at a European level about 1800 PJ would be needed on an annual basis (Sanders, 2014; Wirsenius, 2000). When we compare this with the amount of energy used in food chains, this equals to 20,000 PJ biomass (Fig. 2.2). About the same amount of fossil energy is required by tractors on the field, for heating greenhouses, for transport of the harvested materials, for processing, for refrigeration and food storage, and for cooking. Consequently, we need more than 20 times more energy compared to the amount of digestible energy that is generated.

Over the past 50 years, there was not much pressure to produce food in a more efficient way: European farmers were paid to limit production, and fossil energy was amply available at low prices. Two paradigms, however, have

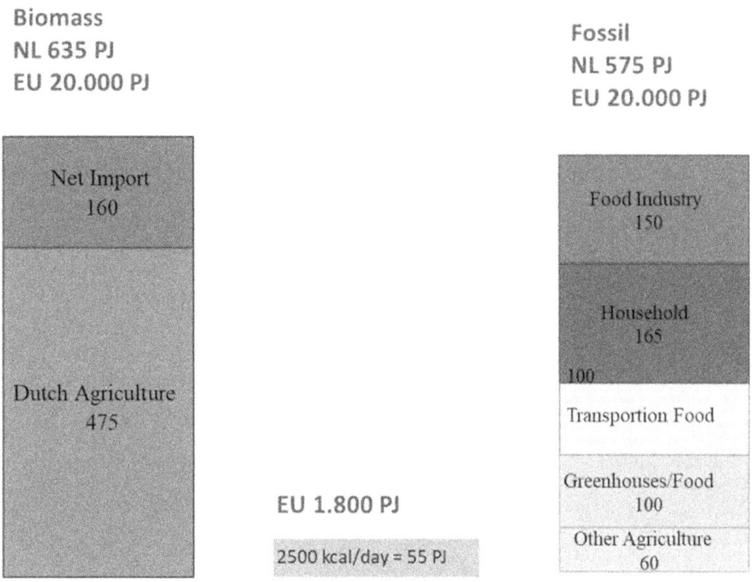

FIGURE 2.2 Indirect energy use food production in the Netherlands.

changed drastically. We need to increase efficiency of input use which requires a drastic change in the way production is organized. By halving energy use in food production, indirect energy use is only 10 times higher than food energy output. Still, 20,000 PJ of energy, or more than 20% of the total European energy consumption, is saved.

Further, the use of reactive nitrogen needs to change in a similar way (Leip et al., 2017). The top of Fig. 2.3 shows how every kg of nitrogen in European food (mainly protein) requires an input of 6 kg of nitrogen. This nitrogen comes from chemical fertilizers, imported animal feed, or even from deposition—mainly ammonia that originally was emitted. Excess nitrogen from animal manure is recycled to agricultural fields and meadows. All 5 kg of nitrogen that was in the input and does not end in food will be emitted as ammonia, nitrous oxide (N_2O), or as N_2 to the air or as nitrate to the groundwater. Nitrogen use efficiency will not only lead to lower costs (saving on nitrogen use), but will also reduce nutrients that often have become environmental burdens.

2.2.4 Increase crop yields but leave field residues for soils

Plants need a variety of components such as minerals including the spore elements; also organic components are needed in soils as these provide structure and contribute to water and nutrient buffering. As plants have taken up a considerable portion of the nutrients, it is better not to remove residues from the field. Instead, these should be recycled as early as possible and

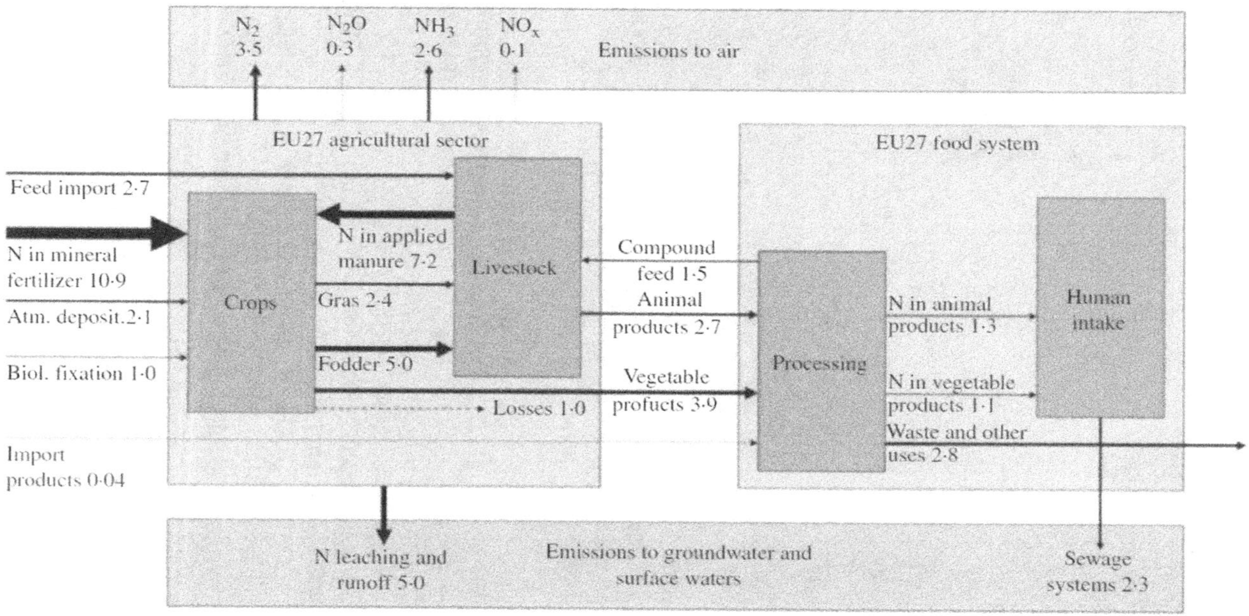

FIGURE 2.3 Nitrogen flows associated with EU food production. *Source: Nitrogen flows in EU27 2005 Leip, A.F., Weiss W., J.P. Lesschen, H. Westhoek., 2017. The nitrogen footprint of food products in the European Union. J. Agric. Sci. p. 1–14. doi:10.1017/S0021859613000786.*

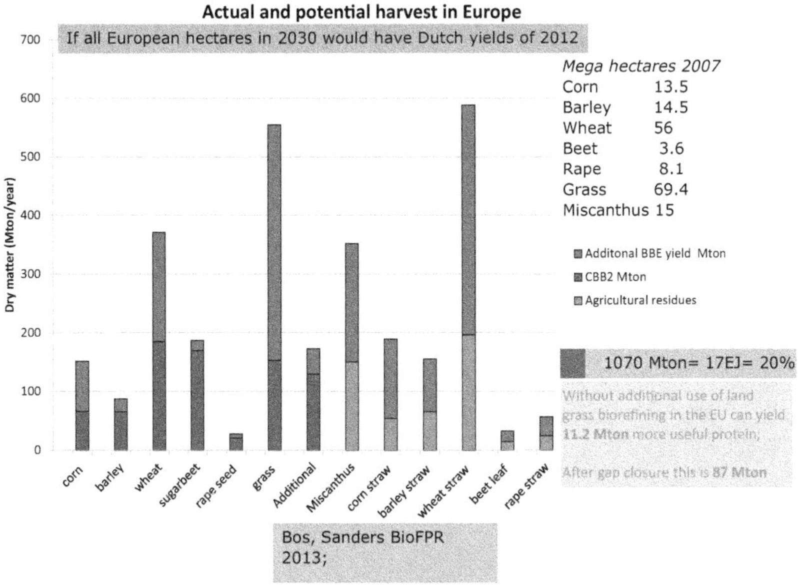

FIGURE 2.4 Actual and potential harvest in Europe. *Source: Bos, H.L., Sanders J.P.M., 2013. Raw material demand and sourcing options for the development of a bio-based chemical industry in Europe. Biofpr 7(3), 246–259.*

close to the field. This will not only save the minerals that otherwise would have ended in a waste water treatment system or even worse in a river that ends in the sea, but it will also save cost of transporting especially moist crops and also will save costs that are required to bring these minerals back to agricultural fields including the costs for concentrating and storage.

Fig. 2.4 depicts total dry matter (DM) production for the six most important crops in Europe (*red bars* [dark gray in print version]). The acreage of these crops is given as well. If annual yield would be increased to levels that currently are obtained in the Netherlands, an additional amount of biomass would be produced indicated by the *green bars* (gray in print version). If this would be used just for energy purposes, this could substitute about 20% of our European energy that mainly originates from fossil origin (Table 2.2).

Increasing yields to prevailing Dutch levels will increase protein production in Europe from 76 to 148 Mtonne. The largest contribution comes from grassland, mainly because the yield gap between the European average and that of the Netherlands is large. It should be mentioned that closing this yield gap will require additional use of fertilizers. Considering the fact that the EU imports about 35 Mtonnes of proteins, the relevance of closure of the mentioned yield gap becomes clear.

TABLE 2.2 Protein production in Europe (Mtonne of dry matter per year).

Crop	Mha	Actual yield	2050 potential	Protein (% of dry matter)	Actual protein yield	Potential protein yield
Maize	13.5	61	145	10	6.1	14.5
Barley	14.5	61	89	10	6.1	8.9
Wheat	56	170	340	10	17	34
Sugar beet	3.6	160	180	2	3.2	3.6
Rape seed	8.1	22	28	25	5.5	7
Grass	69.4	140	500	16	22.4	80

2.2.5 Use all biomass components

Often plant materials consist of a combination of components, such as protein, fat, lignin, cellulose, minerals, pectin, etc. If these components are not separated, the application of the harvested products benefits often from only one of these components while the other elements provide not a burden (at best). Often, however, they cause waste problems or contain potentially toxic components for human or animal food. The separation of biomass into several components often is called biorefining; it is performed in biorefineries (Task 42 IEA Bioenergy, 2009).

Fig. 2.5 presents a process overview for a traditional wheat biorefinery. Plant material is separated into three fractions of which the grain is milled and separated further into flour, germ oil, shorts, and middlings that are used in food, and bran residue that ends in animal feed. The flour can further be separated into gluten and starch. Another biorefinery process example is to use the grain without further separation and just turn the starch into ethanol, while all the residues will end in the distiller's grain, a low-value raw material for the animal feed industry. The combined value of gluten, starch, germ oil, shortlings and bran is much higher than the value of the alcohol and the distiller's grain.

2.2.6 Apply biomass components at highest value

Once the components are available in a pure form, one can freely choose the application with the highest economical and/or environmental and/or social value, of with a combined income higher than the original value including the additional costs for the biorefining process.

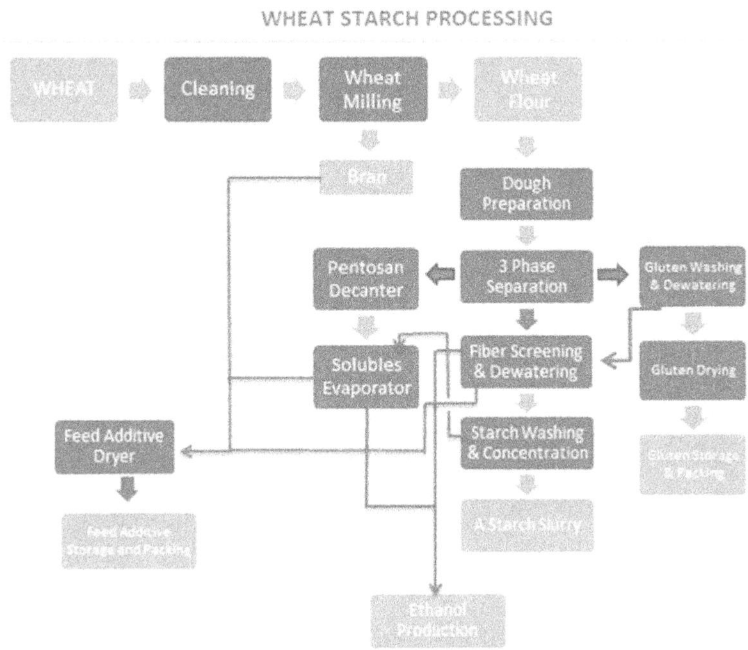

FIGURE 2.5 Process scheme traditional biorefinery. *Source: Sigma process technology, Istanbul, Turkey Sigma, 2019. Wheat Proccessing. https://sigma-process.com/wheat-processing.*

The ecopyramid presented above already pointed out that one should strive for a combination of high value and high volume. The F stairs (Table 2.3) helps to identify high-value options in cases where the value of an actual application is not high enough. The indicative values the applications biomass can get are ranked from high to low per tonne of biomass. "Fire" is the most popular at the moment, in developing countries on large scale to heat and to cook, but also in Europe where coal is substituted by wood pellets to produce electricity. Since coal has to be substituted, the value of these wood pellets would be equal in energy terms to coal, being about 50€ per tonne of biomass.

The cost of growing wood, harvesting, and the logistics in Europe often are higher than this. Wood pellets arriving in the port of Rotterdam cost around 120€. Therefore, a subsidy is required to use wood pellets to substitute coal. Burning methane produced during manure storage to reduce the emission of this potent greenhouse gas helps to reach our environmental goals and it represents some value replacing fossils fuels. When biomass contains toxic components such as heavy metal, this biomass should be discarded in landfills which will cost about 300€/tonne of biomass.

There are many applications with higher value than just the calorific value as "Fire." The biomass that is suitable to produce transportation fuels often has a higher value than just its calorific value, since transportation fuels are liquid

TABLE 2.3 F-stairs: Applications of plant components and their estimated market value.

Application	Value (€/tonne)
Farma (pharma)	High
Fun	High
Food ingredients	5–20,000
Food nutritional	100–500
Feed protein nutritional	500–800
Feed pigs	100–300
Feed cattle	50–250
Functional chemical	500–800
Fiber	500
Fermentation	150–400
Fermentation bulk	100–300
Fuel	100–300
Fertilizer	–/– 200 – 100
Fire	50–150
Flare	0
Fill	–/– 300

and should have a high energy density. Certainly not all biomass types that can be burned are suitable for conversion to biofuels. In several countries where subsidies are provided to enable electricity production, these subsidies prevent more valuable use of raw materials. As indicated, "Fiber" represents materials such as construction wood and other biomass-derived materials that can substitute fossil-derived polymers. "Functionalised chemicals" represent the chemical building blocks that contain apart from carbon and hydrogen one or more atoms of oxygen or nitrogen.

Examples of functionalized chemicals and feed protein uses in food are presented in Section 2.4. Residues obtained after the high-value applications of several fractions of the biorefinery have been elaborated can of course serve as the raw materials for electricity production or heat. We should be happy that these large outlets exist and that no waste is accumulated. If possible these raw materials for electricity production can be optimized by, for example, reducing the presence of mineral salts that will lead to slag formation.

One might ask: "what is the reason that this component cannot be applied at a higher value?" Often, values presented in Table 2.3 are not realized if the biomass fraction is not pure enough. This illustrates an important reason for fractionating biomass, and it will depend on the nature of biomass residues or primary crops.

2.2.7 Reduce capital costs

In academic research of the bioeconomy, raw materials and the related CO_2 reduction often play the dominant role but there are financial boundary conditions, certainly including the capital cost of processes. In the forthcoming decades, we will require many innovative processes with the "new" bio-based raw materials. Hence, not only the capital cost per unit product will play a role but also the absolute amount of money that should be invested for a "first of a kind, new process."

Many companies shy away from an investment of 60 M€ or 100 M€ required for a commercial plant when a "first of a kind" process has to be introduced, even when the process developed up to the pilot scale promises a favorable return on investment. We need to lower the investments required in absolute terms for new processes, since we need a significant number of new processes in the recently initiated bioeconomy. Reducing capital intensiveness will often also enable implementation in small-scale factories. These will also increase the rate of transition because they can even benefit when the market outlets still are limited and/or raw materials are still scarce.

Current chemical production routes include many steps often requiring frequent heating and cooling to provide conditions required for chemical reactions. Each of these steps generally involves a lot of energy, thus making production processes capital-intensive. Consequently, economics of scale dictate that production units are large which means a lot of materials have to be transported over long distances to feed them. Saving on capital and energy costs provides more opportunities for smaller units requiring less transport and facilitating short-cycle circulation of residues. This can be done by making sure that feedstock use is matched by the (chemical) composition of the end-product.

Fig. 2.6 depicts a large number of commodity chemicals that are produced from fossil resources (Scott et al., 2013). Production costs mainly are built up from the capital cost and the raw materials per tonne of end product. Furthermore, there costs are related to personnel, waste treatment, etc., but as these often contribute less than 15% these costs are not shown. The sum of capital and the raw material cost is given by the diagonals. In principle, there would be no difference in total cost in case 1000 € comes from capital cost having no raw material cost, or having 1000 € raw material cost combined with no capital costs. All the points between these imaginary combinations will give a total cost of 1000 €.

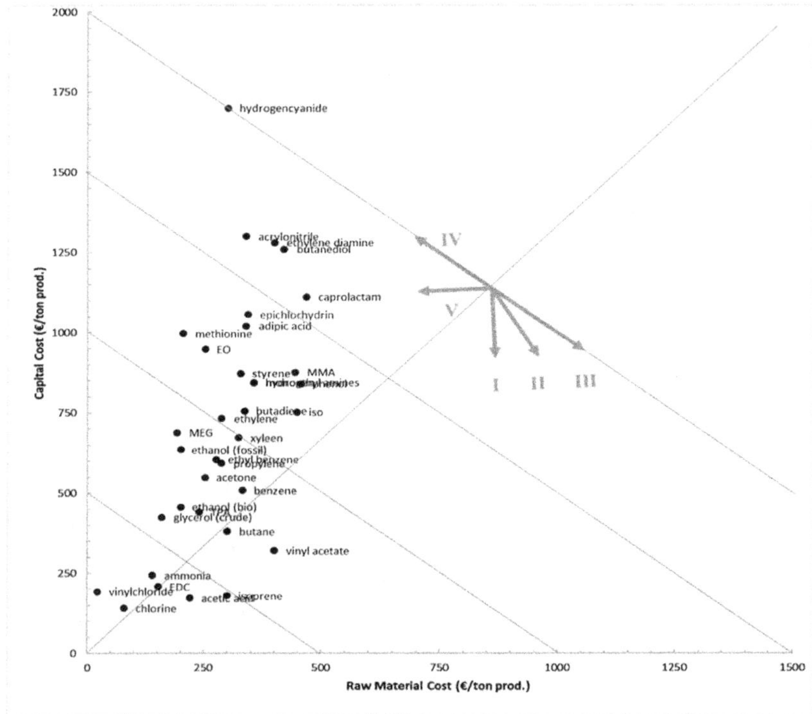

FIGURE 2.6 Capital and raw material costs of major bulk chemicals.

In the figure, we see a bias for chemicals for which the capital cost is (much) higher than the raw material costs. For these chemicals there will be chances to lower the cost price by increasing the capacity of the factory and benefit from economies of scale if the market volume will allow this. This reduction of capital cost is indicated with strategy I. Strategy V benefits from lower raw material cost price which occurs when oil price or energy prices will drop, but this more often cannot be influenced by the chemical company, and prices can go up as well.

A lower capital cost allows producers to afford higher raw material costs and still have a lower overall cost (strategy II), or have the original cost price as in strategy III in order to absorb even higher raw materials costs. These latter two strategies give biomass an advantage over fossil oil under a number of specific conditions that will be explained below. A company can afford to pay higher raw material costs if the capital cost can be reduced as compared to the fossil production route.

2.2.8 Circumvent economies of scale

Lange (2001), on analyzing performance of a large number of chemical factories as well as power plants, concluded that the fundamental reason for

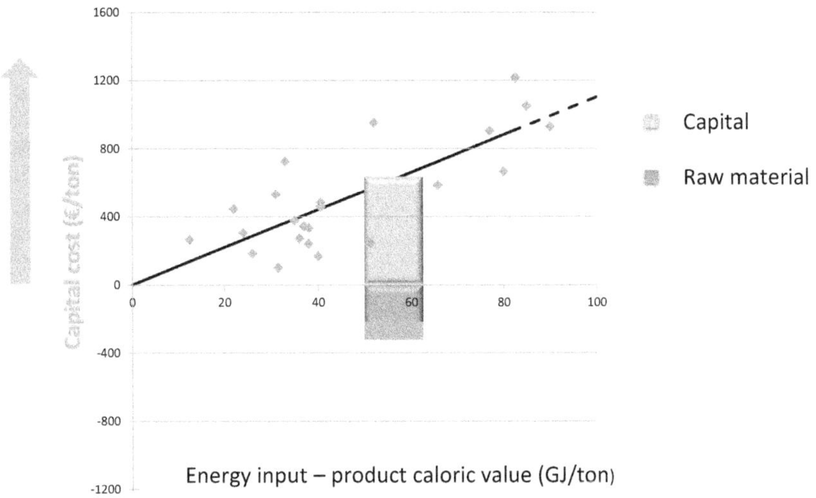

FIGURE 2.7 Capital costs and heat loss over the entire bulk chemicals production chain.

the economy of scale in these types of processes is the need to generate heat. Therefore, it is possible that there could be some reduction in capital cost if the biomass power plant can be built larger than the existing fossil ones. There is a limit, however, in the size of a power plant because large power plants require investments in excess of one billion euros. Logistics and cooling could also become a problem, while the room for increased raw material cost will not be much greater than 5%–10%.

For the production of bulk chemicals, however, there might be some hope as is depicted in Fig. 2.7. A large number of bulk chemicals were analyzed for their (cumulative) capital cost per tonne of end product. We ascertained a correlation between the capital costs for each of these chemicals with the amount of energy that is lost in the entire product chain to synthesize this chemical (Scott et al., 2013).

Fig. 2.7 depicts capital cost for the average chemical product with the *orange bar* (light gray in print version) (the *gray bar* representing the average raw material costs) while the costs are expressed as €/tonne of product. The blue dots (light gray dots in print version) indicate the calculated capital costs for approximately 40 bulk chemicals. Lifecycle analyses of all of these chemicals have become available within the past 10 years which allows the calculation of the energy losses over the entire production chain: energy losses = INPUT (raw materials + energy) minus OUTPUT—being the calorific value of the bulk chemical. The capital costs have been calculated by subtracting the raw material cost, labor costs, and estimated margin from the market price.

Although still a simplification of the actual process, we can reason that capital costs can be considerably less if we begin with biomass raw materials that do not require processing causing extensive energy losses to provide the final product. Biomass components are very appropriate for this because they already have existing functional groups.

As a consequence, higher biomass raw materials can be afforded if capital cost can be reduced. The chemical industry is the only sector that can afford to pay more for its feedstocks, as this can be compensated by a significant reduction in capital costs. Secondly, because of the reduced capital costs per unit of product, scale of operation will become less dominant in process design. It is anticipated that factories producing 10,000 tons of product per year can become as competitive as the large petrochemical factories with annual capacities of 200,000 and even 500,000 tons.

Finally, labor requirements can be expected to increase compared to fossil-based petrochemical production routes since biomass production and biorefining are relatively labor-intensive. The costs for labor can be covered from reduced capital requirements. The dominant competitive factor will become how to obtain the most appropriate raw material/conversion process combination.

2.2.9 Choosing the best biomass raw materials

Fig. 2.8 gives a schematic representation of the chemical industry in Rotterdam. On the left, the import of approximately 140 million tons of oil is represented. The oil is cracked and refined into fractions such as diesel, kerosene, bitumen, and naphtha which are used to produce all fossil-based chemicals. The first step is to produce base chemicals like ethylene, propylene, benzene, toluene, and xylene, all from naphtha. Ethylene and propylene can be polymerized to obtain polyethylene and polypropylene. The base chemicals are also employed to obtain other building blocks that are more functionalized by the introduction of oxygen and/or other atoms.

After this, functionalized, products are generated. These are located to the right in the scheme. The higher the added value per tonne of product, the lower, in most cases, the market volume. Each square represents a factory and each color represents the owner of the factory, that is, the more yellow, the more Shell. In the left part of the scheme, chemicals are found with high energy content per tonne of product; on the right side, energy content is lower. It is obvious that, in moving from left to right, energy is lost in the value chain, and this leads to the capital costs as indicated above.

In a biorefinery, it is best to limit horizontal transformations as these require considerable exchange of heat and, consequently, high capital investments. In principle, one should not oxidize or reduce materials as these processes are energy-intensive. Instead, it is recommended to identify and use raw materials with similar energy contents as the desired endproduct.

Oxidizing materials (moving from left to right) leads to energy loss and added costs for heat exchange. Reduction (moving to the left) leads to mass loss and/or energy loss at the cost of heat exchange. While the petrochemical industry has no alternative feedstock other than oil or gas which both have a

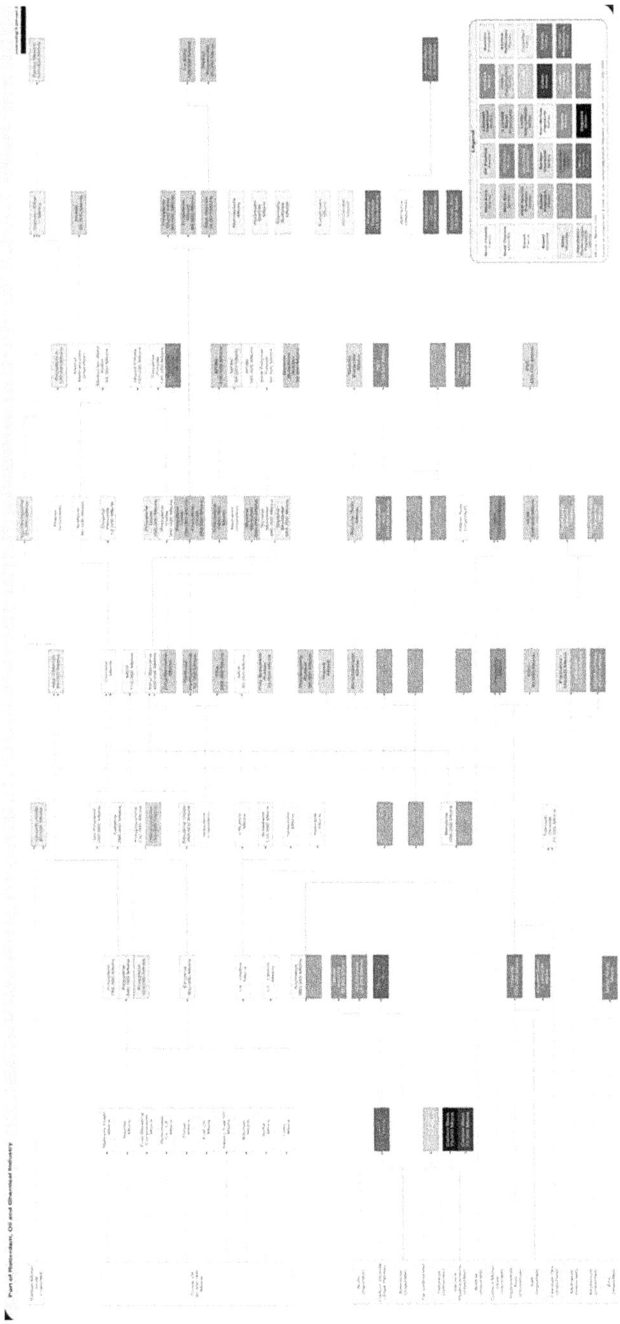

FIGURE 2.8 The chemical industry in the port of Rotterdam.

high state of reduction, in order to make functionalized chemicals, oxidation is required at the cost of energy loss and the need for heat exchange which add to the capital cost. With biomass there is a choice because there are biomass molecules very much reduced as plant oils but also very much oxidized as some organic acids and many forms in between such as sugars.

Terephthalic acid, for example, is a chemical with an annual production volume of around 8 Mtonnes. It is used to make the plastic bottles for a significant number of popular (soft)drinks such as Coca Cola. A common production route in the USA is to convert sugar into butanol that is further converted into xylene (Peters et al., 2011), not considering losses to byproducts. Xylene then is oxidized to terephthalic acid. A shorter route, however, might be developed from glucose directly to terephthalic acid.

If one wants to produce more reduced chemicals such as ethylene from biomass, one should start with biomass that is much more reduced than sugar. Fatty acids, waxes, and the like could become the raw material of choice if these can be sourced on large scale in a sustainable way.

2.2.10 Sustainability principles

More principles, related to sustainable production, are identified in this chapter. They refer to sustainable production along dimensions of social ("people"), ecological ("planet"), and economic ("profit") sustainability, and include application of biorefinery technologies, short circularity/circular economy, and application of the value pyramid.

2.3 Circular economy

We are used to a linear economy in which a company purchases raw materials that are converted to a product that is sold on the market and eventually waste streams are discarded. In a circular economy there will be no waste streams. Optimally these components are reused. The Ellen McArthur Foundation that has introduced the term *Circular Economy* makes a distinction between the "Technical" World and the "Biological" world. This is important especially for materials that are scarce like several metal elements and will be exhausted if consumption continues with today's volumes.

In the biological world when fertilizers are given, there will not be exhaustion since the used biomass can be cultivated again. In principle the technical and the biological materials are the same: we should strive for the highest Resource Use Efficiency. This efficient use can be quantified and it is a wrathful goal to strive for. Circular is not a goal but just a means to obtain Resource Use Efficiency. Often it is preferred to make materials that do not degrade because they can be reused over a longer period, which is called "life extension." Even burning these materials after long periods of time can give a better footprint than the energy and capital consuming recycling of a polymer that is degraded after its first use. In the processing of crops it can also be exemplified that not all circular processes lead to optimal footprints.

The upper panel of Fig. 2.9 exhibits the traditional value chain of sugar beets: they are harvested and placed at the side of the field where they are collected by a lorry that brings them to the factory that produces crystalline sugar. As a residue, molasses is produced and shipped, for example, to a baker's yeast factory such as the one where I began my professional career, Gist brocades in Delft. The sugar in the molasses is an inexpensive substrate, and the other components of the molasses that are not used by the yeast were discarded into the North Sea until our government directed us to build a waste water treatment plant.

The technology was working satisfactorily. While it was not effective in discarding the minerals, this was not regarded as a problem: Gist brocades purchased an evaporator and energy was provided at low cost by a Central Heat and Power plant (CHP). Then the problem was not solved at all because the vinasse containing the concentrated minerals that were originally present in the beets that were cultivated on 50,000 ha had to be distributed again over an agricultural surface of approximately the same area, posing major logistic problems.

The lower panel shows if we could process the beets close to the fields and recycle most of the water from the beets containing the minerals, then much less water will need to be transported over the roads. The minerals do not need to be concentrated because, if the size of operation is around 500 ha, the distance to recycle is 1 km to the right and 1 km to the left. Furthermore, the farmer would incur lower costs for his minerals. In addition, if we can produce an intermediate product that can be stored over the entire year with simple process steps, we would facilitate the more difficult and more capital-intensive steps to be performed during 12 months per year and not just for the 4–5 months of the traditional beet-campaign in the central factory (Bruins and Sanders, 2012).

All of this would reduce a significant amount of capital cost, the amount of energy required to concentrate the minerals from the molasses, and transport costs, and, at the first step in the chain, there would be additional work to perform which makes the farmer more independent as an entrepreneur. Both examples are circular processes but the small-scale one that recycles the minerals very close to the fields of production and prevents lots of transport cost and energy and capital to concentrate diluted streams is of course the preferred one.

2.3.1 Small-scale biorefinery processes

Intuitively, we know that the larger a factory, the lower the cost per unit product as many processes follow the rules of Economies of Scale. If we have a specific process design in mind, this is very true but only within the limits of

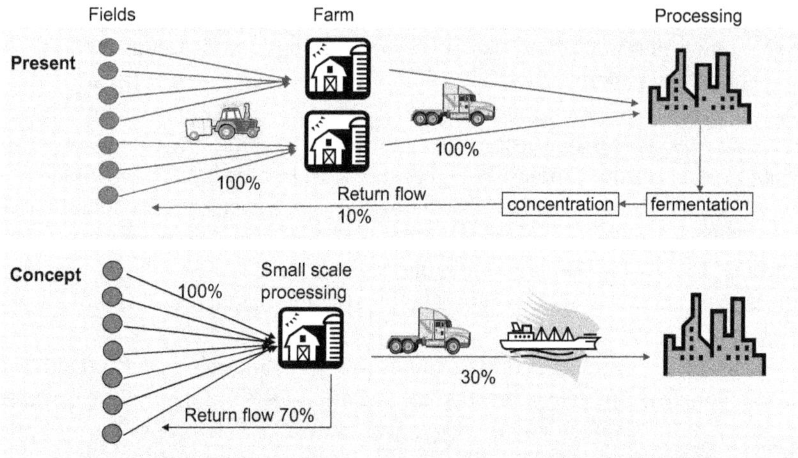

FIGURE 2.9 Small-scale biorefinery reduces transport cost and seasonality.

the process itself, certainly when the raw materials like oil are delivered at a specific point. But for agricultural crops it will be understood that the larger the factory, the longer the lines of collecting the raw materials to the factory. Less obvious is the cost of recycling as has been explained above. There will be other advantages (Bruins and Sanders, 2012):

- One can modify the process in order that on a small scale only low capital investments are required and that and intermediate product is made that does not have the loads of water to be transported and that can be stored over the whole year. In a more capital-intensive factory the second part of the process is performed 12 months of the year. This gives another advantage especially when the crop can be stored only for a short time after harvest as is the case with many crops with high moisture content.
- Water and the minerals stay on the field
- Less waste treatment is the central factor
- Gradual development of the market in case a new product is launched
- Gradual development of processed raw materials, in case that raw materials are not abundantly available or are costly to contract on large scale.
- Will create more income to the farmer and makes a farmer less dependent of downstream markets

2.4 Developing bio-based production chains

Implementation of the development and sustainability principles will be discussed following a few themes. First, we further explain how development principles can be used to design production chains in practice. Next, we demonstrate how rules can be developed for production chains.

2.4.1 Using design rules

In order to fulfill People, Planet, Profit most of the design rules given should be applied at the same time. One can start from different points: Is there a demand for bio-based products? We call them product-driven value chains; one can start from the raw material which obviously can be an existing commodity like sugar, but often the side products of the biofuel industry because of attractive cost price since biofuels are nowadays by far the largest application of biomass. We call these value chains: biofuel-driven. There can be other starting points like the anticipated availability of sustainable raw materials like algae, seaweeds, and grasses. We call them raw material-driven.

After the successful implementation, these indications will not be relevant anymore because in biorefinery concepts it is always the combination of the incomes from the different products that result from such a biorefinery. Another design rule that has not been introduced yet is the message that Ansoff (1957) already gave more than 60 years ago that one should be careful to invest in more than one uncertain dimension. He even mentioned that the combination of entering a new market with customers whose wishes you do not know at the same time that you introduce a new product of which nobody exactly know the performance will lead to suicide.

Later people have elaborated on his warning with other dimensions such as new technology and also new raw materials. To cope with these uncertainties, a strategy would be to collaborate with other partners. But the more partners, the more complex the collaboration will become, certainly when each of these partners have different expectations of how to share risks. These risks can come from outside: the market prices of raw materials or end products. If that would be just one of the multitude of products, who will pay the bill? What would be the value of the contributions to the collaboration such as technology, market knowhow, etc.? The risk can be that the development of a process takes longer and more money. Will all partners share the costs?

Who will take the initiative and which raw materials are attractive? Companies with knowhow of their products in the market are a good starting point because they can estimate the value which a new product can have in the market or an existing product for the production of which the raw material will be switched from fossil to biomass. These companies have invested in their market position and there is not always the drive to innovate in times that they are earning money. Companies with raw materials can choose almost every product that is technologically feasible, and they need support from potential buyers. This is certainly the case for technology companies, and these have to ensure in addition to obtaining raw materials at a good price for a reasonably long period of time.

Table 2.4 provides an overview of recent bio-based production chain developments. Who took the lead? With which partners, right from the beginning or at the moment of a milestone in the technology development? Was it merely

TABLE 2.4 Overview of recent bio-based production chain developments.

	Feedstock		Technology supplier		Market application	
Product	Residue	Primary	Core	Other	Existing	New
Epichlorhydrin	Biodiesel		Solvay		Solvay	
1,3 propanediol		Tate & Lyle	Genencor		Dupont	
Succinic acid	Roquette		DSM			DSM
1,4 Butanediol		Cargill	Genomatica	Cargill	Novamont	DOW
Polylactic acid		Cargill	Natureworks			DOW
Foamed PLA		PLA	Wageningen UR		Synbra	
FDCA		Tereos	Avantium Corbion			BASF
		Sugar Union	Bird Engineering			
3-hydroxy-propanoic		ADM		ADM		
Ethylacetate		Suiker Unie	Wageningen UR		AKZO Nobel	
Fatty amines			Arkema			Arkema
Methionine			Cheil Jedang	Arkema	Cheil Jedang	

Bold, commercial.

the transition to a better footprint for the same chemical, a so-called drop in, with a bio-based raw material, residue, or commodity? Or was it to introduce a totally new product?

The drop-in epichlorhydrin was an existing product of Solvay, who in the past had developed technology to produce glycerol using the epichlorhydrin chemistry. Now that the fast growing biodiesel industry would produce large amounts of glycerol as a residue from the transesterification, they invested with their own technology in the sustainable production of this chemical and applied for a good patent protection to keep competitors that also had similar technology available, but started to slow. An essential step was that the glycerol needed to be purified to fulfill the high purity requirements of chemical processes.

The drop in 1,3 propanediol technology was developed by a technology supplier, a biotechnology company avant la letter, with roots in Genentech and Gist brocades, a Dutch traditional fermentation company. Once the development of a fermentation process that took more than 100 full time equivalent was well underway, they started a collaboration with Dupont as the market party and Tate and Lyle the starch company to supply the raw material, corn starch, for the process.

DSM started the development of a totally new building block, succinic acid, that chemically had similar properties as their adipic acid that they apply in their polyesters, also by a fermentation process; and they lined up with the French raw material supplier Roquette Freres. Succinic acid as an intermediate in the citric acid cycle in the central metabolism of microorganisms was an obvious candidate to produce by fermentation.

Several other initiatives to produce succinic acid by a fermentation process were started by the French technology company with their own supply of raw materials sourced from the very active Reims area, but without a market partner; by the Dutch fermentation company Corbion after that they acquired the Bird Engineering company, but without a specific raw material supplier. Other players include AVA Biochem, SynbiaS, Tokyo Chemical Industry, and V & V Pharma Industries.

A very early initiative was the DOW/Cargill collaboration on the development of polylactic acid. The technology was more or less available in-house, but the market introduction of a new polymer proved to be very labor-intensive and took at least 10–15 years. Properties that gave the advantage over existing chemicals that were being manufactured at huge scale, benefitting from economies of scale, were mainly related to food packaging. Worries about low concentrations of monomers from existing polymeric products such as polystyrene gave PLA an advantage, as the building block for lactic acid already is a component of many food products.

Only a few developments have been initiated from the availability of raw materials. ADM started with the production of 3 hydroxypropanoic acid, a chemical with technical similarities as lactic acid.

For each of these fermentation approaches, the companies benefitted from the fermentation knowhow and from the fact that in fermentations no pure raw materials are required as is the case in the chemical industry. The challenge is to purify the fermentation products up to the high standards, however.

2.4.2 Ethanol production driven routes

Worldwide ethanol is the largest bio-based product. Major residue products remain such as bagasse in case ethanol is derived from sugarcane and Dried distillers Grain and Solubles (DDGS) when the ethanol is derived from corn, wheat, or sorghum. The volume of DDGS in weight is almost as much as the ethanol that is produced from the cereals. The value as animal feed is however modest because the digestible energy is already converted to ethanol and the residual protein does not contain high fraction of essential amino acids and as a consequence of the drying step of the DDGS, part of the essential acid is damaged. Furthermore the DDGS contains a high amount of minerals such as potassium and phosphate that cause either problems in mono gastric animals or environmental issues because of eutrophication.

If one would consider refining the DDGS before drying, and one would use components such as glutamic acid, which is present as 10% of the dry matter as a raw material for the production of chemicals, one could benefit from the structure of glutamic acid. This approach has been shown for a number of chemicals (Lammens et al., 2010) such as N-vinylpyrrolydone, acrylonitril, diaminobutane but also N-methyl- pyrrolydone, a solvent that has been used frequently and that can be replaced by its analog N-ethyl pyrrolydone which does not have severe toxic properties that NMP has.

Fig. 2.10 exhibits how utilizing the structure of glutamic acid can help to simplify the production of N-methyl pyrrolidone as compared to the traditional production route using natural gas and nitrogen from the air. The first step is an enzymatic decarboxylation to GABA. The second step, at modest reaction conditions, provides the desired endproduct (Lammens et al., 2010).

The petrochemical route involves eight reaction steps, each with their respective losses and capital requirements. The synthesis of NMP or better NEP exemplifies the advantage of using biomass raw materials because much less capital is required and a high yield can be obtained because only two reactions are required instead of eight.

As a general consequence, using only 10% of the dry matter of the DDGS that is produced in a 400,000 tonne/year ethanol factory in Rotterdam, a value of 58 M€/y can be created from a residue worth only 46 M€. If a second component can be upgraded, or when some adverse properties of DDGS such as the over excess of minerals can be alleviated, the combined value of these residue-derived products can be in the order of the 200 M€ currently generated by its main product (ethanol). This will provide an important improvement for the profitability of the corn or wheat biorefinery. It shows,

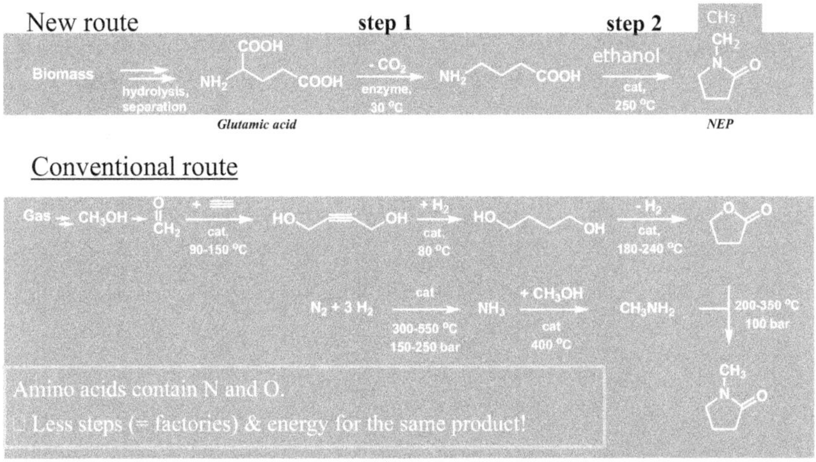

FIGURE 2.10 Using glutamic acid as a suitable building block for the synthesis of N-methyl pyrrolidone. The new production route is simpler than the original fossil-based route.

once again, that chemical products can benefit more from biomass as compared to just using biomass for its calorific value.

2.4.3 Protein production routes

As indicated by the ecopyramid, chemical building-blocks and materials are linking value with volume. This explains the success of the approaches presented above. Proteins, especially for animal nutrition, are the third group of products that combine high volume with high value. Proteins do not need to be generated by chemical processes, as they already are constituents of biomass feedstocks, often as a small fraction and as a consequence of that with little value. Fig. 2.11 provides an overview of the composition and value of major biomass residue components.

If we would rank all agricultural residues (and primary crops) by the fraction of protein content and we would fractionate the components, the value of the protein would be enhanced as it can be employed in a more valuable application compared to the original residue.

The first group of feedstocks presented in Fig. 2.11, with only minimal protein such as wheat straw, would best not be addressed, as fractionation will leave little return (lignocellulosic residues only can be used as raw material for biofuel production). The fifth group with high protein contents (e.g., soy meal) should also not be addressed because it already has a high value.

However, groups in between are candidates for biorefining and the lignocellulose even for the production of biofuels since many costs that have to be made for the protein extraction and the biofuel can be shared. On examining the fourth group with a protein content of approximately 35% such as rape meal, it was found that this group is just good enough for a pig

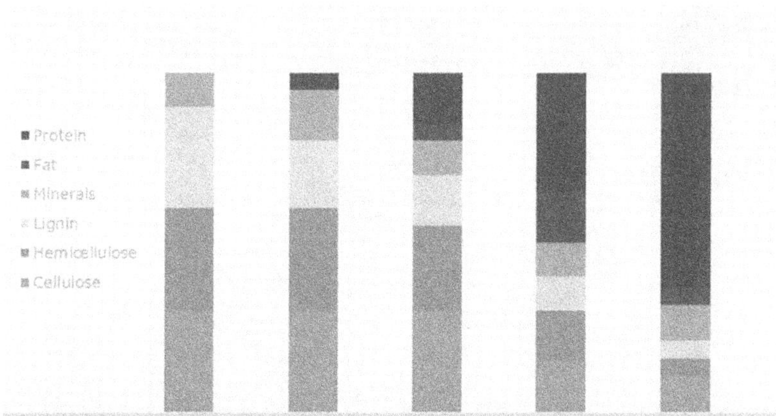

Group	1	2	3	4	5
Protein content	0	5 %	15 %	35 %	60%
Examples	Wheat straw	Cocoa hulls Corncobs Sugarcane leaf	Coffee pulp Rapeseed straw Beet leaf	Rapeseed leaf	Soy meal
Cost (€/ton)	50 - 80	50 - 110	100 - 140	150 - 180	300 - 350

FIGURE 2.11 Biorefining of agricultural residues.

feed component. The third group is only suitable as cattle feed because it contains some protein and many fibrous materials, while the second group is not currently valuable enough to be collected from the fields.

We were challenged by an electricity company to design a biorefinery process that would increase product values as well as provide low cost feedstocks that can compete with coal in power generation. Currently, because of the mandate and subsidies, the cost of wood pellets that are imported from Canada is at least two times higher than coal (indicated with the *dotted line* in Fig. 2.12).

Straw on the field is inexpensive, but costs are increased by collecting and washing needed to obtain the quality that is required for power generation. In the end, it is as expensive as wood in Europe. In considering the rape seed meal, which is more expensive, the cost will increase even more after fractionation, but this will be rewarded as the protein component receives a greater value per tonne, amino acids are made available at modest prices, and lactate can be produced from the cellulose and the hemicellulose fractions. Lactate, of course, can be utilized for the production of polylactate, a novel biopolymer that is currently used for packaging and insulation.

Lactate can also be employed as a substrate for the production of bulk chemicals in a more improved manner than other second-generation

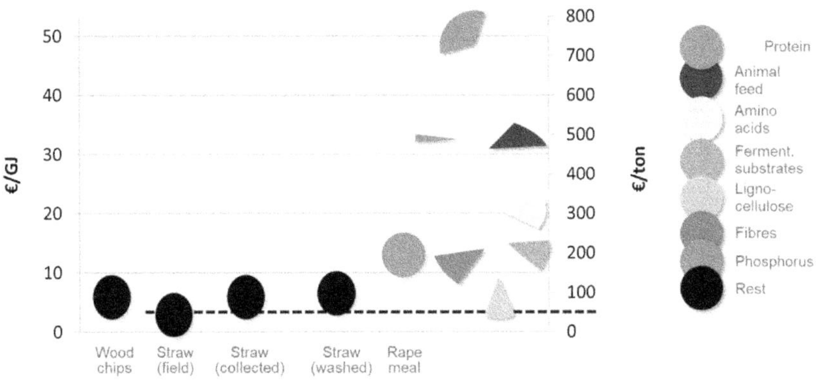

FIGURE 2.12 Rapeseed meal can be a beneficial substrate for high value proteins, chemical building blocks, animal feed components, as well as raw material for power generation.

fermentation substrates because we predict that the purity and the concentrations are much better than the sugars that are obtained from second-generation processes in the manner that they are developed by the US ethanol producers. Maybe somewhat surprising is the fact that lactate can substitute starch in animal feed. By doing so illustrates how the "food versus fuel" controversy is an artificial debate—inedible cellulose being converted into a starch substitute in pig and poultry feed. Finally, the undigested cellulose and lignin become available at the price of coal for the power companies.

2.4.4 Second-generation bioethanol

Many hope that second-generation raw materials will provide the solution to climate change because they claim that, if waste streams are exploited, sufficient inexpensive biofuels can be generated which of course is not the case. Second-generation biofuels require large production plants and, consequently, require substantial raw materials that often are not available at low prices.

In accordance with this line of thinking, it can be reasoned that wood would not be attractive as a feedstock because the production process is complicated. It seems to make more sense to use forest, agricultural, or industrial residues such as straw that is abundantly available and which is the resource of choice in the USA for the production of second-generation ethanol.

Fig. 2.13 depicts wheat straw being treated to open the structure for the enzymes that will enable the production of ethanol (Maas et al., 2008). The maximum straw concentration in water that still can be processed is very low. With straw at 15% dry weight concentration, the concentration of ethanol produced will be limited and, again, there is a need for significant heat transfer to distil the ethanol from a low concentration of only 5%.

Perspectives for second-generation technology (in Europe) is challenged by low market value and capital-intensive production routes which means it

FIGURE 2.13 Wheat straw, $Ca(OH)_2$ pretreated wheat straw, and pretreated and cellulose-treated wheat straw.

can only be operated on a large scale. This requires enormous quantities of raw material, which has its implications on availability. Furthermore, prices for this type of biomass often are affected by mandated use in power generation.

Below, a comparison is made for the process economy in the USA and in Europe. There is one second-generation factory running in the USA in Iowa (POET/DSM) with a turnover of about 50 M€ and a capital cost calculated at least as 10% of 250M€ investment, leaving 22M€ room for raw materials and all other costs. Raw material cost in the USA is about 14 M€ leaving 8 M€ for labor, insurance, maintenance, etc. While in Europe the cost of capital might be higher because of more strict environmental regulations, the cost of raw material could well be between 20 and 25M€ which would leave a negative margin for labor, maintenance, etc., costs (Fig. 2.14).

For the nutritional value of proteins several properties should be mentioned: amino acid composition and the digestibility of the proteins. Since monogastric animals like pigs and poultry cannot live without enough essential amino acids in their feed, a balanced composition of these amino acids is important. Nowadays it is common practice to add the four most limiting amino acids such as methionine, lysine, threonine, and tryptophan to the feed, because in so doing, the amount of protein can be reduced. In general terms oilseed proteins such as soymeal and leaves such as grass are high in essential amino acids while cereal proteins like corn or wheat-derived proteins are low in essential amino acids.

Glutamic acid and glutamine are generally the most abundant amino acids. Apart from their presence the amino acids should also become available in the small intestines (the ileum). Often some 20% of the protein is not digested to free amino acids when the feed arrives in the ileum. Then these proteins will travel to the large intestines where they are hydrolyzed and stimulate bacterial growth if enough energy is available for these microorganisms, but the amino acids are lost with feces.

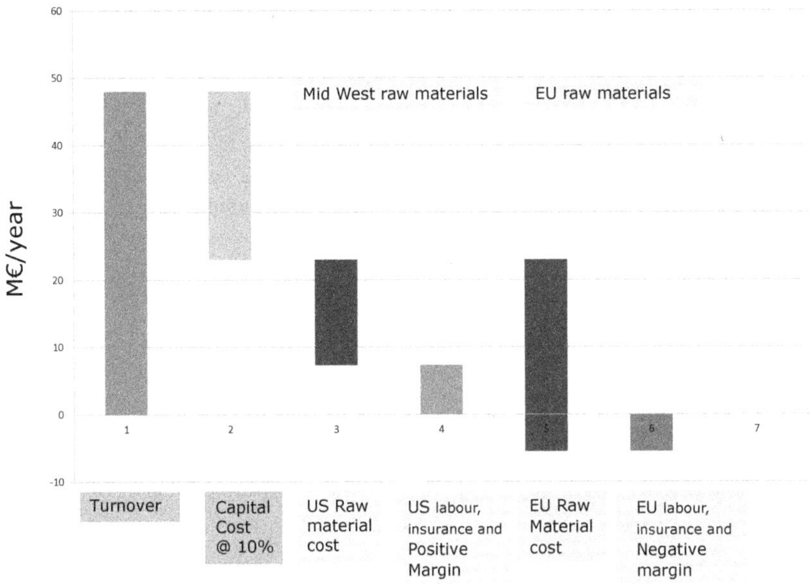

FIGURE 2.14 Heat exchange requires high capital cost, reducing room for high raw material prices.

Digestion by cattle means that protein should be degraded in the rumen and serve as building blocks for microorganisms that form new protein structures. In this case, it is not important whether the primary proteins did contain many essential amino acids. The microbial protein is of high quality. But degrading proteins in the rumen and resynthesized proteins cost a lot of energy often in the form of carbohydrates.

Resistant proteins that are not degraded in the rumen but in the second stomach or ileum are much more attractive as cattle feed, since it reduces energy requirements. When proteins with high share of essential amino acids are given, the cow and the farmer will benefit. Since in grass part of the protein is "rumen bypass" and the other part is not, it is worthwhile to separate these two forms of protein. With the Grassa technology, a mobile biorefinery unit that will be explained below, this fractionation is enabled.

2.4.5 Fermentation sugar-driven developments

Microorganisms can convert sugar into a great variety of components such as ethanol and lactic acid, processes of which our society already has benefitted for thousands of years. Since mid-1800 (Pasteur) in Europe, other products have been produced by microorganisms in fermenters including baker's yeast, acetone/butanol, enzymes, and antibiotics. Since the introduction of micro-organisms, genetic modification in the late 70s of the last century, academia and industries increased the number of products that can be made by microorganisms and also commodity chemicals.

Fermentation processes require sugars which can be derived from beet or cane sugar or converted from corn, wheat, and cassava starch. All these raw materials are available in large volumes. Most fermentation processes are aerobic which means that microorganisms produce their products in the presence of oxygen from the air. As a consequence, they oxidize available sugars, requiring a lot of biological energy to grow cells and production of heat.

Sugars that are used for producing heat and growing cells cannot be used for making the desired products. If the production of the chemicals could be done under anaerobic conditions, there will not be a lot of energy available; there will not be much cell growth and there will not be a lot of heat produced. It means a double advantage: a much higher yield and a lot less capital is required since there is little heat to cool away as has been already explained in Fig. 2.9A here above but also less capital is required because no air compressor is required and not much stirring is required.

The cost price of the fermentation products will go down typically from 1500 €/tonne to some 1000 €/tonne. Fig. 2.15 depicts more products to be generated from fermentation of sugar.

For some chemicals, cost price will not be low enough because the synthesis from fossil resources is cheap because the raw materials are of low cost and/or the synthesis benefits the economies of scale of the large factories. Anaerobic fermentations, since they do not produce much heat, can be performed at reasonably small scale: 50,000 tonnes/year per factory. Further reduction of cost price can be obtained by using a sugar source that is just good enough for fermentation purposes, which means that purity and also concentration is much less of an issue. In this way a further cost reduction in the order of 100 € can be obtained.

Further cost reductions can be obtained as has been shown for the fermentative production of acetic acid, a commodity chemical with a market price of about 400 €/tonne. In traditional acetic acid fermentation processes,

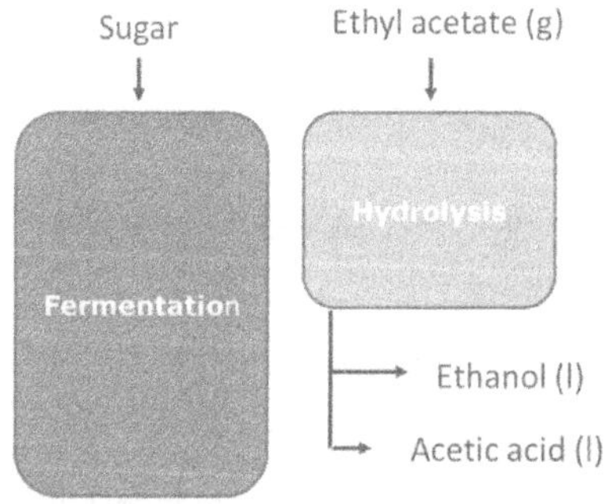

FIGURE 2.15 Acetic acid and ethanol via ethyl acetate.

the produced acid will reduce the pH requiring base applications in order to keep the pH constant. Upon further accumulation of the acetate the fermentation rate slows down because of product inhibition. The concentration of acetate in water is still low and the product can be recovered, for example, by ion exchange technology. Again, chemical need to be used at high cost.

Wageningen University developed a process to produce ethyl acetic ester. This compound has a boiling point of 72°C. When a thermophilic microorganism is used, the ester can be evaporated from the fermenter into a compound from which glacial acetic acid is obtained, allowing harvest of ethanol in high purity and concentration. In this way no chemicals are required to neutralize the acid formed, no product inhibition occurs, and an expensive purification step can be prevented. As a side product ethanol results, which enjoys a large market volume. Initially the ethyl acetate can be marketed as a solvent with a higher market price than ethanol and acetic acid, but the volume of this product is limited.

2.5 Examples

2.5.1 Using biomass' molecular structures to synthesize chemicals

Fig. 2.16 exhibits an early example of the principle to benefit from the molecular structure of residues. The Belgian company Solvay has begun to establish its third factory producing epichlorhydrin from glycerol (Solvay, 2014), which is a waste stream from the biodiesel industry. Dow Chemical Corporation and Samsung specialty chemicals have followed. Epichlorhydrin is used for resin productions as this is use for swimming pools or the blades of modern windmills. In the petrochemical process, chlorine (Cl_2) is required to functionalize propylene.

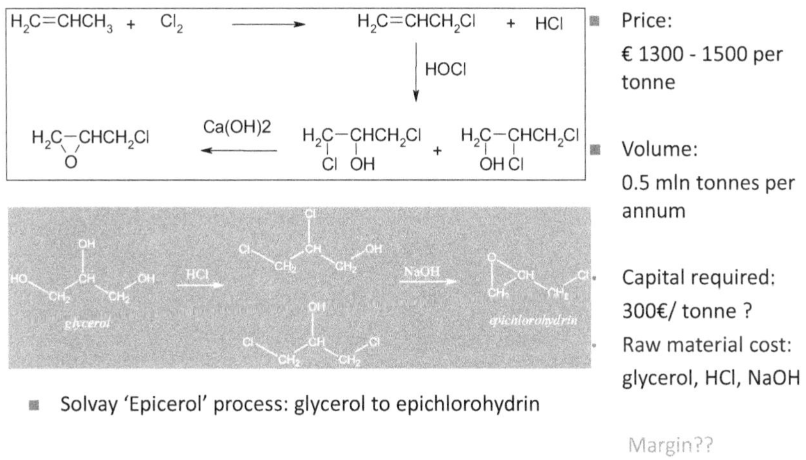

FIGURE 2.16 Epichlorohydrin from glycerol requires little heat exchange.

The large amount of energy accumulated in chlorine must be dissipated from the process which expends heat exchange capacity and, in this case, under quite severe conditions of the corrosivity of chlorine. As the bio-based process does not require chlorine, electricity normally consumed for the synthesis of chlorine no longer is needed. This process requires less capital while, in this specific case, the raw materials are not more expensive than the fossil alternatives. Epichlorhydrin is sold at 1300 to 1500 Euro per tonne. At glycerol prices of 350 Euro per tonne, production margins are 40%−50% which is very interesting.

2.5.2 1,3 propanediol

Another early example of a successful biorefinery approach is production of 1,3 propanediol (PDO) by Dupont for the synthesis of Sorona, a polyester built on terephthalic acid and 1,3 propanediol. 1,3 PDO is produced naturally by microorganisms such as *Clostridium* and *Lactobacillus*, but there is no microorganism that directly synthesizes 1,3 PDO from glucose. Genencor developed an *E. coli* production system modified with genes from a yeast strain. This system can produce glycerol from glucose. In 2007 the production was started up in a 100M$ factory with a capacity of 50,000 tonnes of 1,3 PDO. The *E. coli* produced titers of 135 g/L at a volumetric productivity of 3.5 kg/m^3.hour and a yield on glucose of 0.50 kg/kg. Dupont teamed up with Tate&Lyle for the supply of corn starch that serves as the substrate for 1,3 PDO (Fig. 2.17).

2.5.3 Succinic acid

Succinic acid is usually manufactured from petroleum-based feedstock, maleic anhydride. It can be used in a wide range of applications, including polymers, fibers, solvents, pharmaceuticals, surfactants, food, and pigments. Global demand for succinic acid is estimated at 40,000t/y−45,000 t/y.

FIGURE 2.17 Propanediol factory.

Reverdia's Biosuccinium Technology was developed to produce bio-based materials. It is offered under license to partners and coproducers. It uses a special low-pH yeast which requires few basic chemicals for neutralization during the fermentation process. The pH is stabilized at 3, while more than 50% of the succinic moieties are present in the acid form (Jansen and Verwaal, 2010). Effective April 1, 2019, the joint venture Reverdia will be dissolved and the partners will transfer the rights and obligations related to Reverdia's Biosuccinium plant in Cassano, Italy, to Roquette. Under a nonexclusive license from DSM, Roquette will operate the plant and continue serving Biosuccinium customers. Customer service, order processing, and marketing and sales will be integrated into Roquette's existing business to ensure a smooth transition.

A competitor, the French Amber group, started their production using a bacterial system, but later acquired a yeast strain from Cargill. Still later, a license of the Reverdia yeast strain was taken. Obviously, the low pH has a big advantage. Amber has built a factory in Canada in Ontario. They started a collaboration with Evonik to expand the succinic market in the direction of butanediol; market size of this chemical is several million tonnes while succinic acid market is limited to some 50,000 tonnes. A third group, Myriant, started a 13,500 tonnes factory in Lake Providence, Louisiana, United States, in 2013. Construction costs amounted to 80M$.

For a reasonably small market of 50,000 tonnes, even more than the three mentioned groups stepped in. The aim was to produce 1,4 butane diol, a chemical with a market volume about 2 million tonnes annually. 1,4 BDO is used for the manufacturing of PBT (polybutylene terephthalate) mostly used in cell phones and automotives, furthermore in the production of poly urethanes.

1,4 BDO can be produced from succinic acid by reduction with hydrogen. The energy contained in the required amount of hydrogen is about equal to the energy in the succinic acid and a capital-intensive process is needed for this reduction as has been explained in Section 12.2 (capital intensive processes and how to prevent economies of scale). The reduction step can be circumvented by producing 1,4 BDO using anaerobic fermentation straight on.

2.5.4 1,4 butanediol

Mater-Biotech, a 100% of the Italian Novamont teamed up with technology supplier Genomatica and started a 1,4 BDO factory with a capacity of 30,000 tonnes annually in 2016. An *E. coli* strain is used that produces the product straight from glucose without the intervening succinic acid as is described above.

2.5.5 Benefitting from polymer structures

Synbra is a Dutch food packaging company that used expanded polystyrene foams to keep fish at low temperatures. Together with Wageningen University they were able to make expanded foams from polylactic acid and obtained similar isolation properties with which they could substitute their polystyrene for fish packaging but also make isolation plates for the isolation of houses. This is a nice example of how biomass can come to value as a structured material.

2.5.6 Production of starch from fresh cassava roots in a mobile factory

Fig. 2.18 illustrates a small-scale biorefinery designed in 2003: a mobile factory that can be transported to the fields where it can process cassava roots to a crude starch meal within a short time following their harvest. This will strengthen the position of the farmer because the roots deteriorate within 24 h after harvest if they are not processed. It prevents factories using this fact to take advantage of the vulnerable position of farmers.

The farmer can either attempt to locate another factory 100 km further at the end of his working day while the roads become dangerous and probably hear the same story at the other central factory, or he can begin negotiating, while both men know that, every hour, the transfer price will decrease. While processing nearer to the fields, the minerals from the root that would have been discarded into the river by traditional factories can now be recycled without any problems. These mobile factories, of which there are now more than 12 in operation in Africa, are a very positive example of where the advantages of small-scale processing can be obtained without having the disadvantages.

2.5.7 Benefitting from proteins

The Dutch company Grassa developed a mobile biorefinery system for which all the equipment is mounted on a trailer of a lorry as can be seen in Fig. 2.19.

FIGURE 2.10 Production of starch from fresh cassava roots in a mobile factory.

FIGURE 2.19 Grassa mobile biorefinery system DEMO.

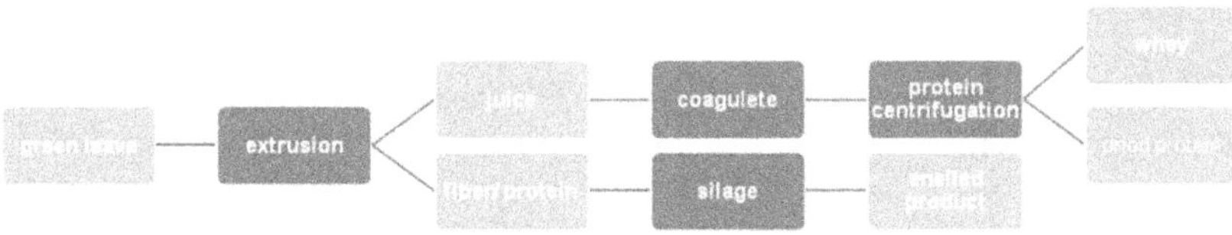

FIGURE 2.20 The Grassa process.

After mowing, the grass is immediately loaded in the machine and washed with water. Then a grinder/extruder opens up the cell structure. This also has been tested for other leafy feedstocks. Fig. 2.20 presents thesimplified process scheme. A press collects the press cake in which fibers as well as rumen bypass protein is enriched, while the juice containing all the soluble components is collected for further processing. The press cake is wrapped in plastic to obtain a sort of silage that can be stored over long periods before feeding the press cake to cattle. On a 100% substitution basis, the silage has a 15% improved feed conversion rate and show a 40% improve nitrogen and phosphate use efficiency in cattle feed.

The juice is heated to 90°C in order to allow proteins to coagulate and be collected by centrifugation or filtering. The protein can be fed to pig or poultry or dried in order to transport and/or store it for long periods. The protein-product is at least equivalent to soymeal containing about 45% protein on a dry matter basis, with a very similar (essential) amino acid pattern. Furthermore, it contains no antinutritional factors and much less phosphate and potassium as compared to soymeal or other oil meal-protein products.

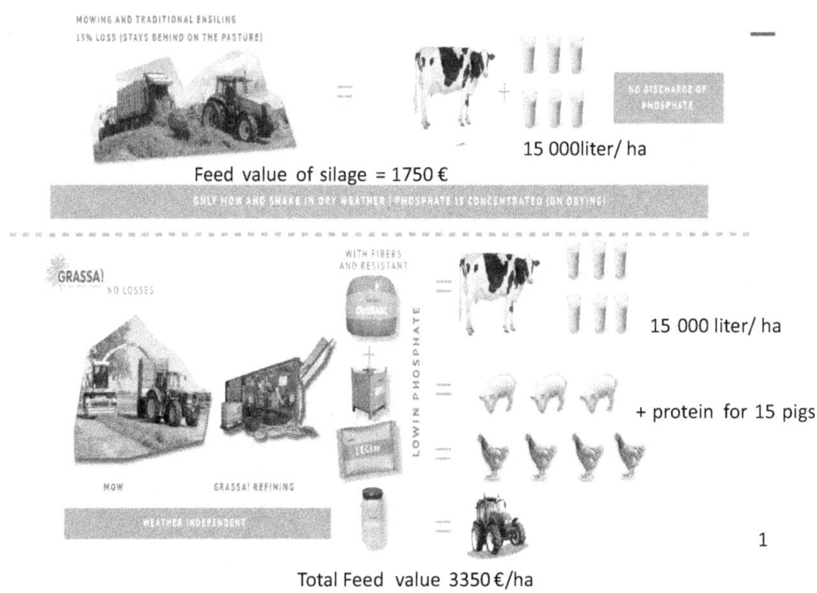

FIGURE 2.21 Realizing a 50% increase in animal protein production per ha.

The filtrate or the supernatant of the centrifugation, referred to as "whey," contains the mineral fraction but also sugars such as fructose oligosaccharides that can be isolated and applied in animal feed as a prebiotic to improve the health of the large intestine. The minerals can be recycled to the field immediately, or used in a concentrated form as soluble fertilizers that can be sprayed in green house fertilization-systems.

As this approach reduces the amount of proteins in cow feed, and as several other losses can be prevented, the net amount of animal protein production per hectare will increase by about 50% (see Fig. 2.21 for details). For Europe, It would mean that less proteins would have to be imported. Income from grassland would increase by sale of protein streams to pig and poultry farmers. Furthermore, in areas with excess of nitrogen and phosphate, the manure problem and costs will be reduced.

Fig. 2.22 compares nitrogen flows (expressed in kg/ha/year) in traditional grass feeding systems in the Netherlands, where 75% of the grass is ensiled each year, with the new biorefinery feeding system that includes pig or poultry production. The lower panel depicts that 40% of the nitrogen administered to the field from recycled manure and 175 kg fresh fertilizer is lost; only 60% ends up in the grass mainly as protein. After the biorefinery, 60% of the protein is given to cows and the other 40% to pigs. Since the fiber/protein output increases, cows will produce 53 kg of N in the milk, which is very similar to the cows in the traditional feeding that got all the grass. The pigs generate as much as 34 kg of nitrogen in meat proteins.

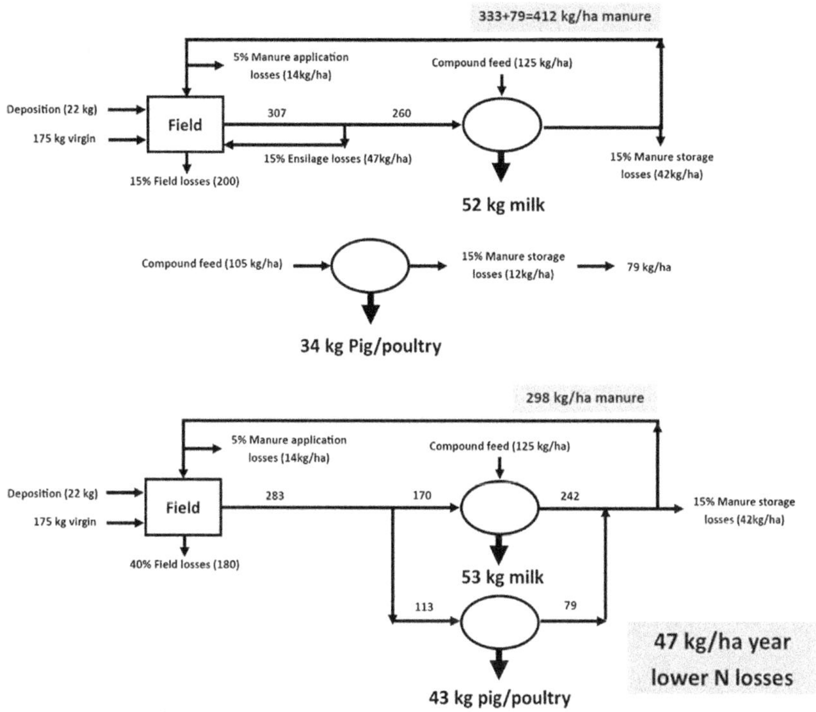

FIGURE 2.22 Nitrogen flows in kg per ha per year.

In a traditional production system, 105 kg of nitrogen in compound feed per hectare would have been imported (Fig. 2.22). This would be provide a load of 79 kg of nitrogen in pig manure which would, together with cow manure, bring the total load to 412 kg on an annual basis to be recycled back to the field. In the biorefinery system, "only" 298 kg ends up in the manure to be recycled. In the two processes there are several losses to be considered: losses during manure storage in the houses and in the silos during winter time and losses during recycling the manure to the field.

In the traditional system, 15% of the grass is lost during ensiling while the biorefinery has no losses, since this processes the grass immediately after mowing (even when it is raining). The overall loss in the traditional feeding system is 115 kg/ha per year while in the biorefinery system this is limited to 31 kg. The losses of phosphate and methane will be lower and also the volume of the manure which makes it cheaper to store and to spread over the fields.

2.6 Discussion and conclusion

By exploiting the most appropriate biorefinery technologies in combination with the appropriate type of biomass, raw materials can be supplied for the chemical industry, for the production of electricity and transportation fuels, and simultaneously for animal feed production. In order to benefit best from our limited resources and in addition to this technological development we need to collaborate between sectors of our society that did not collaborate much until now. This not only means that crop growers and animal farmers can benefit from more integrating their activities, one produces the feed for his neighbor while this neighbor ships back the manure for fertilization, but also the chemical industry can benefit from collaboration with the compound feed industry.

Food processors can improve their resource efficiency by upgrading their "waste" streams into valuable food products by very much the same biorefinery technologies that are developed for animal feed resources. Since we will feed animals in the future with less components that do not contribute to the nutritional value, there will be less of these components in the manure leading to reduction of ammonia emissions and probably methane emissions from the rumen and from manure storage. We will need less land and less reactive nitrogen to feed an increased world population.

With the grass example given we can make progress from different angles at the same time: we can increase field yield by mixed culturing with legumes and herbs (Fig. 2.23), the increased production after biorefining the different components from grass can not only feed the cows but also large volumes of pig and poultry. Since grassland is about 40% of the European agricultural area, these combined developments can change the feed industry in a short time.

We are also afforded an opportunity to substitute capital costs by labor and change the direction of the development of requiring less and less labor inputs

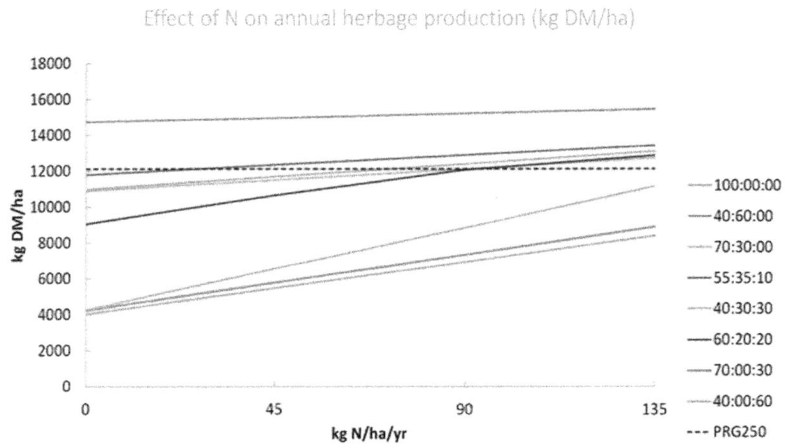

FIGURE 2.23 The effect of a functional group and nitrogen input on grass yields: yield (kg DM/ha) dependent on fertilizer application and sward composition (Grace, 2018).

per economic output that began with the onset of the steam engine. We should therefore choose the right raw material for each product in order to develop the conversion process that produces only little heat and therefore will not become capital-intensive. The direct fermentative production of 1,4 butanediol is an example for this although the reduction of succinic acid to 1,4 BDO is a well-developed process.

Many politicians would support development of additional employment, but this fundamental change of opportunities largely remained unnoticed. This may be caused by the misconception that employment depends on today's large companies with their large-scale capital investment needs. In this way, politicians have been captured in an incorrect trust in second-generation transportation fuels even while these, at least in Europe, are far too expensive. Many do not foresee that the production of bio-based chemicals could be initiated by a new group of companies, for example, the agricultural companies that have the raw materials and the biorefinery expertise. These new factories do not need to be built in big ports or large chemical complexes, but they are likely to thrive on locations close to the agricultural raw materials, which might even be in countries that lack tradition in chemical production.

The transition to a bio-based economy has started but up till now not at a very high pace. We need to accept that for People, Planet, Profit solutions we need time to define options that cannot be realized with knowledge that is already available to everybody because in that case these profitable opportunities would not have existed yet. We need to develop new technologies and that takes time and money. This is also caused by the cost of changing toward a totally new resource base, with novel production and supply chains, with novel processes and challenging new networks of collaboration. The chemical industry with its high capital investments based on oil as the resource is worried that changing to a new raw material will lead to fast losses of their investment base. The Paris Agreement of 2015 and the followed realization that 2050 is coming closer is creating the required sense of urgency in our society in order to make progress in the next century.

For many challenges there are opportunities that fulfill People, Planet, and Profit at the same time if we give ourselves the time to design our efforts right from the beginning.

References

Ansoff, H.I., 1957. Strategies for diversification. Harv. Bus. Rev. 113–124. September–October.

Bos, H.L., Sanders, J.P.M., 2013. Raw material demand and sourcing options for the development of a bio-based chemical industry in Europe. Biofpr 7 (3), 246–259.

Bruins, M.E., Sanders, J.P.M., 2012. Small-scale processing of biomass for biorefinery. Biofuels Bioprod. Biorefining 6 (2), 135–145.

Conijn, J.G., Bindraban, P.S., Schroder, J.J., Jongschaap, R.E.E., 2018. Can our global food system meet food demand within planetary boundaries? Agric. Ecosyst. Environ. 251, 1244–1256.

Grace, C., 2018. The Potential of Multispecies Swards Compared to Perennial Ryegrass Only Swards for Dry Matter Yield, Chemical Composition and Animal Performance. PhD dissertation. School of Agriculture and Food Science, University College Dublin, Ireland.

Jansen, M.L.A., Verwaal, R. Low pH Dicarbonic Acid Production, Patent WO 2010/003728.

Lammens, T.M., Franssen, M.C.R., Scott, E.L., Sanders, J.P.M., 2010. Synthesis of biobased N-methylpyrrolidone by one-pot cyclization and methylation of c-aminobutyric acid. Green Chem. 12 (8), 1430–1436.

Lange, 2001. Fuels and chemicals manufacturing: guidelines for understanding and minimising the production costs. J.P. Cattech. 5, 82–95.

Leip, A.F., Weiss, W., Lesschen, J.P., Westhoek, H., 2017. The nitrogen footprint of food products in the European Union. J. Agric. Sci. https://doi.org/10.1017/S0021859613000786.

Maas, R.H.W., Bakker, R.R.C., Boersma, A.R., Bisschops, I., Pels, J.R., Jong, E. de, Weusthuis, R.A., Reith, H., 2008. Pilot-scale conversion of lime-treated wheat straw into bioethanol: quality assessment of bioethanol and valorisation of side streams by anaerobic digestion and combustion. Biotechnol. Biofuels 1, 1–13.

Peters, M.W., Taylor, J.D., Jenni, M., Manzer, L.E., Henton, D.E., 2011. US20110087000A1 Integrated Process to Selectively Convert Renewable Isobutanol to P-Xylene.

Rockström, et al., 2009. Planet. Bound. Ecol. Soc. 32 (2), 1–33, 14.

Sanders, J.P.M., 2014. Biorefinery. The Bridge between Agriculture Ands Chemistry. Wageningen University, Wageningen. http://edepot.wur.nl/298449.

Scott, E.L., Bruins, M., Sanders, J.P.M., 2013. Rules for the bio-based production of bulk chemicals on a small scale- Can the production of bulk chemicals on small scale be competitive? Rep. BCH 16. https://biobasedeconomy.nl/wp-content/uploads/2011/08/Can-production-of-bulk-chemicals-on-a-small-scale-be-competative-formated-20131126FINAL.pdf.

Sigma, 2019. Wheat Proccessing. In: https://sigma-process.com/wheat-processing.

Solvay, 2014. http://www.ti.kviv.be/11ASCGROEN/public/Klumpe.pdf.

Springmann, et al., 2018. Options for keeping the food system within environmental limits. Nature. https://doi.org/10.1038/s41586-018-0594-0.

Steffen, et al., 2015. Planetary boundaries: guiding human development on a changing planet. Science 347, 6223. https://doi.org/10.1126/science.1259855.

Task 42 IEA Bioenergy, 2009. Biorefineries, Adding Value to the Sustainable Utilization of Biomass. IEA Bioenergy, Paris. https://www.ieabioenergy.com/wp-content/uploads/2013/10/Task-42-Booklet.pdf.

Tilman, D., Balzer, C., Hill, J., Belfort, B.L., 2011. Global food demand and the sustainable intensification of agriculture. Proc. Natl. Acad. Sci. U.S.A. 108, 20260–20264.

Wirsenius, S., 2000. Human Use of Land and Organic Materials. Academic Thesis Chalmers university and Goteborg University.

Further reading

Deloitte, October 2013. The Shale Gas Revolution and its Impact on the Chemical Industry in the Netherlands, Addendum to VNCI Vision 2030–2050.

Lammens, T.M., Franssen, M.C.R., Scott, E.L., Sanders, J.P.M., 2012. Availability of Protein-Derived Amino Acids as Feedstock for the Production of Bio-Based Chemicals. Biomass and Bioenergy.

Chapter 3

Government regulation of bio-based fuels and chemicals

David J. Glass
D. Glass Associates, Inc., Needham, MA, United States

3.1 Introduction

As is described throughout this volume, biological systems are increasingly being developed and used for the production of renewable fuels or bio-based chemicals. Many of the strategies depend on the use of naturally occurring or genetically modified microorganisms, including bacteria, yeasts, microalgae, and cyanobacteria, as production organisms. In addition, production of bio-based products generally requires the use of biomass of some kind as the feedstock or carbon source for the process, and while this usually entails the use of naturally occurring plant biomass or some kind of organic waste, there has also been interest in the development of specially bred or genetically modified plants to serve as the feedstock for bio-product manufacture.

The use of biological methods of manufacturing commodity or specialty products that have historically been made from petrochemical feedstocks promises to make an important contribution to the reduction of global carbon emissions and the movement to more sustainable industrial activities. Microbiological methods have long been used for the production of ethanol and other industrial chemicals, but the increasingly common use of genetically modified microorganisms provides potentially significant advantages over traditional methods, such as improved productivity, decreased operational costs, the ability to use a more diverse range of feedstocks, and possibly more favorable carbon footprints. And the use of advanced biotechnology to develop "designer" feedstocks may also lead to increased productivity and other benefits.

Although bio-based products are generally considered to be more environmentally friendly and to have a more favorable carbon footprint than the petrochemical products they might replace, bio-based fuels and chemicals and their methods of manufacture will nevertheless be subject to regulation in most countries around the world (Philp et al., 2013). This article reviews the regulations that might apply to the biological manufacture of fuels and chemicals,

Biobased Products and Industries. https://doi.org/10.1016/B978-0-12-818493-6.00003-8

particularly chemical regulation in the United States and other countries, requirements for testing or certification of new automotive or aviation fuels, the possible requirements under programs to promote the use of biofuels and other renewably produced products, and regulations affecting the use of genetically modified microorganisms or plants in bio-based manufacturing. The major regulatory programs discussed in this chapter are shown in Table 3.1.

3.2 Regulation affecting bio-based chemicals

3.2.1 US chemical regulation under the Toxic Substances Control Act

Manufacture and sale of bio-based chemical products in the United States would fall under the jurisdiction of the Toxic Substances Control Act (TSCA), which is the major U.S. legislative and regulatory program that covers the introduction into commerce of new chemicals (15 US Code Sections 2601–2629). TSCA, first enacted by Congress in 1976 and amended in 2016, is a statute that requires notification to the Environmental Protection Agency prior to the importation or the use in commerce of new chemicals that are not subject to regulation under other federal laws. In spite of the name, it does not exclusively regulate toxic substances, nor does it carry the implication that any substance subject to the Act is in fact toxic: instead, the Act is designed to give EPA the ability to screen new chemicals before they are used commercially in the United States, to identify potentially hazardous compounds that might require regulation.

Detailed summaries of TSCA's history, scope, and its implementing regulations can be found elsewhere (Bergeson et al., 2000; Markell, 2010). The regulations implementing TSCA can be found in the Code of Federal Regulations at 40 CFR, Chapter I, Subpart R (Parts 700 ff.). Importantly, TSCA does not cover chemicals used as pesticides, drugs, foods, food additives, or cosmetics, all of which are regulated under different federal laws. However, certain chemical intermediates, notably in the manufacture of pesticides, may be subject to TSCA jurisdiction. It is clear that, in most cases, biologically produced fuels and those chemicals intended for industrial markets would potentially be subject to TSCA, as would certain novel enzymes intended for industrial processing.

The centerpiece of TSCA is the TSCA Inventory, which is a listing of all chemicals that were known to be used in commerce in the United States at the time of the passage of the legislation, along with all new chemicals that have been notified to EPA under the Act and which have begun to be used in commerce. Naturally occurring compounds are inherently considered to be on the Inventory. A company wishing to commercialize a new chemical compound that is not found on the TSCA Inventory is required to notify EPA through the filing of a Premanufacture Notice (PMN) at least 90 days before

TABLE 3.1 Regulatory programs discussed in this chapter.

Regulation	United States	European Union	Other Countries
Industrial Chemicals	• Premanufacture reporting to EPA under the Toxic Substances Control Act.	• Registration and reporting to ECHA under EU REACH.	• South Korea, China, Japan have laws resembling EU REACH. • Canada's law resembles U.S. TSCA.
Fuel Certification	• Civilian automotive fuels regulated by EPA. • Commercial aviation fuels certified by FAA through ASTM process. • Military branches have conducted their own evaluations and certifications.	• Renewable fuels must meet the requirements of the Fuel Quality Directive, which also mandates usage levels of renewable fuels in transportation.	
Promotion of Renewable, Bio-Based Products	• Mandated volumes and economic incentives for renewable fuels under U.S. Renewable Fuel Standard and state Low Carbon Fuel Standards. • Favorable treatment for bio-based products under USDA BioPreferred Program.	• Renewable fuels must meet the requirements of the Renewable Energy Directive, which also mandates usage levels of renewable fuels across all energy sectors.	• Many countries have mandates or targets for use of renewable fuels in their transportation sector.
Use of Genetically Modified Microorganisms	• Contained manufacture requires compliance with EPA biotechnology rules under the Toxic Substances Control Act.	• Contained manufacture requires compliance with national laws based on EU Contained Use Directive.	• Contained manufacture may require compliance with biosafety laws based on Cartagena Protocol on Biosafety.

Continued

TABLE 3.1 Regulatory programs discussed in this chapter.—cont'd

Regulation	United States	European Union	Other Countries
Use of Genetically Modified Plants as Feedstocks	• Growth and commercial use require compliance with USDA biotechnology regulations.	• Growth and commercial use require compliance with national laws based on EU Deliberate Release Directive.	• Growth and commercial use may require compliance with biosafety laws based on Cartagena Protocol on Biosafety.

importing the chemical into the country or beginning commercial use, subject to eligibility for certain exemptions described below (Bergeson et al., 2000). The TSCA Inventory is largely public, so that companies can themselves determine if the chemical is already on the Inventory. However, certain chemicals are listed on the Inventory but their identities have been claimed as confidential by the manufacturer. To determine if a new chemical is listed on this so-called Confidential Inventory, a manufacturer would need to formally ask EPA, stating a bona fide intent to manufacture the compound (this is accordingly known as a "bona fide" request) (Bergeson et al., 2000). In February 2019, the EPA released an updated version of the TSCA Inventory, which now identifies those chemicals that the agency believes are actively being manufactured, processed, and imported in the United States, which constitute only about half of the total number of chemicals listed on the Inventory (U.S. Environmental Protection Agency, 2019c).

Under the regulations, a PMN for a chemical not on the Inventory must include information on the identity and structure of the chemical, its intended use in commerce, the site(s) of manufacture and the anticipated production volumes, information on potential exposure to the chemical by workers and consumers, and what measures are being taken to minimize worker exposure and release of the compound to the environment. Although there is no mandatory toxicology test or other tests that are required under TSCA, applicants are required to submit to EPA any data or information in their possession that is relevant to the health or safety effects of the substance.

Upon submission of a PMN, EPA has 90 days to review the substance and the potential risks it may carry. In conducting its risk assessment, EPA relies upon the information provided by the submitter and also carries out a structure-activity-relation analysis of the chemical structure to determine the likelihood that the compound might have hazardous properties. EPA's review

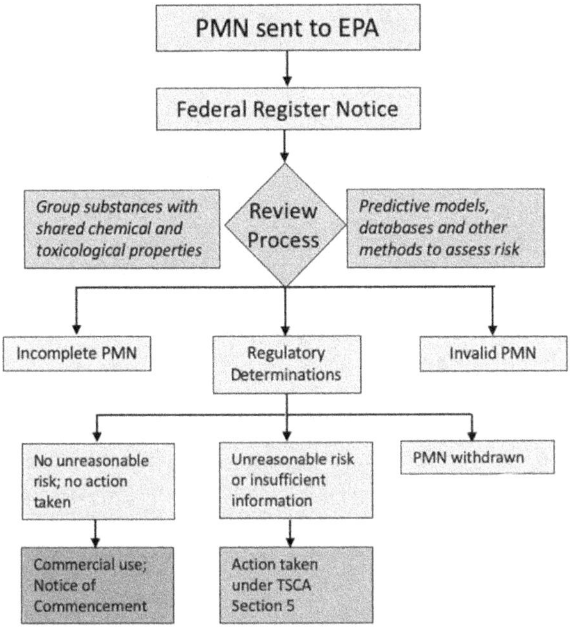

FIGURE 3.1 US EPA review process for Premanufacture Notices for new chemicals under the Toxic Substances Control Act. *Adapted from U.S. Environmental Protection Agency, 2018a. EPA's Review Process for New Chemicals. Retrieved from: https://www.epa.gov/reviewing-new-chemicals-under-toxic-substances-control-act-tsca/epas-review-process-new-chemicals.*

process is shown in Fig. 3.1. Under TSCA's original provisions, for those chemicals EPA determined to have low risk, the agency simply needed to allow the 90-day review period to lapse without taking any action, and it only took action within that period of time for chemicals it wished to subject to further restriction or where more data might be requested. However, in June 2016, after many years of hearings and negotiations, legislation was enacted to revise TSCA for the first time in 40 years, to make major changes in the regulation of chemicals in the United States (Rosen, 2016). This bill, H.R. 2576, was formally entitled "The Frank R. Lautenberg Chemical Safety for the 21st Century Act" (U.S. Congress, 2016). The amended legislation now requires EPA to reach one of the following three determinations for each Premanufacture Notice: either (a) a finding that the substance presents an unreasonable risk; or (b) a finding that either the information is insufficient to permit a reasoned evaluation of the substance; or in the absence of sufficient information and evaluation, the substance may present an unreasonable risk; or the substance may be produced in substantial quantities or may enter the environment in substantial quantities, which may lead to substantial human exposure; or (c) a finding that the substance is not likely to present an unreasonable risk. In spite of this added requirement, it would be expected that most bio-based products would fall into the third category and not be subject to any additional restrictions.

There are several exemptions which may be available for some bio-based manufacturers. First, TSCA is a commercial statute, and so most activities in which chemicals are manufactured solely for R&D or evaluation are exempt from reporting. But there are also exemptions for test marketing and low-volume production, which might require notice or application to the Agency; and in addition, manufacture of a new chemical solely for export would generally be exempt from the need to file a PMN (Bergeson, Campbell, PC 2019).

When the United States adopted its framework for biotechnology regulation in 1986 (see below), it extended the scope of TSCA to cover certain genetically modified microorganisms used for purposes within TSCA's statutory scope, or used to produce chemicals within this scope. This aspect of the TSCA regulations is discussed below, but in certain cases a manufacturer might need to file a PMN for a new chemical while also filing the required notice for the microorganism used in its manufacture (Microbial Commercial Activity Notice, or MCAN, described in detail below). This is especially true for the use of microorganisms to produce chemicals classified as Class 2 substances under TSCA.

Chemical compounds that can be represented by a definite chemical structural diagram are known under TSCA as Class 1 substances, and in most cases, if a Class 1 substance has been used in commerce and is on the TSCA Inventory, producing that substance by a novel production process, such as by a novel microorganism, would not require filing a new PMN. Examples of bio-based products that would be considered Class 1 substances under TSCA include fuels like ethanol or butanol, and chemicals such as succinic acid and other organic acids. PMNs would not be required for the manufacture of such compounds by any novel production process.

However, TSCA also defines another class of substances, Class 2 substances, as those that are of "unknown or variable composition, complex reaction products, [or] biological materials" (known as UVCB substances), which cannot be easily represented by a structural diagram. As described in Bergeson et al., 2012, there are several types of Class 2 substances, which are generally listed on the TSCA inventory by a definition specifying the source from which the substance has been derived. In many cases, production of a Class 2 substance by a new microbiological method will require filing a new PMN in addition to an MCAN, since the source of the substance will differ from the source defined on the Inventory listing. For example, a substance comprising a range of alkanes or alkyl molecules for use in diesel fuel would likely require a new PMN as a UVCB substance if produced by a novel microbial method. This is discussed in more detail in Bergeson et al., 2012, but it should be noted that many biofuels or bio-based chemicals that fall under the Class 2 category under TSCA may require the filing of both PMNs for the chemical and MCANs for the microorganism even if the ultimate chemical product is already used in commerce, because of the new manufacturing process.

Overall, the TSCA regulations would not be considered overly burdensome for the production of most bio-based products. The needed time, resources, and data to prepare and submit PMNs are generally fairly modest, and most bio-based products would not be likely to have hazardous properties or to exhibit structural features that may trigger potential safety concerns under EPA's models, so that EPA reviews should be straightforward.

3.2.2 European Union regulation of chemicals

There is no specific EU legislation or regulation for bio-based chemicals, but these products are potentially subject to the European Union chemical regulation program known as REACH, which is formally known as "Regulation (EC) No 1907/2006 of the European Parliament and of the Council of 18 December 2006 concerning the Registration, Evaluation, Authorisation and Restriction of Chemicals" (European Union, 2006; Philp et al., 2013). REACH is administered at the EU level by the European Chemicals Agency, known by the acronym ECHA[1]. The REACH regulations require manufacturers and importers of chemical substances to register, submit information, and provide evidence that adequate controls of risks are in place throughout the supply chain for each regulated chemical. The needed registrations cover all life cycle stages of the production and use of the substance, as well as waste disposal, emissions and other avenues of release and exposure. It is important to note that non-EU entities cannot themselves submit REACH registrations, but must instead work through an EU representative (Applegate, 2008; Luit et al., 2017).

The REACH regulations came into effect in 2007, and the reporting and registration requirements became mandatory in three phases, depending on the annual production volume of the substance. The third and final phase took effect on June 1, 2018, so that registration is now required for any chemical produced or imported into the EU above an annual volume of 1 tonne; volumes below this amount are exempt from reporting, unless the product is a part of a supply chain for a chemical for which reporting is required. REACH places additional requirements on certain chemical substances having "very high concern" due to potential negative impacts on human health or the environment, but these regulations are unlikely to apply to bio-based products.

Under the regulations, the importer or manufacturer of a substance is obligated to provide information to the ECHA on the substance identity, information on its intended use, and information on its chemical properties, such as physical

1. At this writing, the fate of the United Kingdom's planned exit from the European Union is uncertain, but most likely if the exit is accomplished, all laws that the U.K. adopted to comply with EU legislation would remain in force, although the country would no longer have the legal obligation to remain in compliance in the future. Because REACH is implemented at the EU level rather than nationally, the U.K. may need to put its own chemical regulations in place if it leaves the EU.

chemical specifications, *in vitro* toxicity, ecotoxicity, biodegradability, and in vivo toxicity. The specific information requirements depend on the anticipated production volume of the substance, as specified in Annexes VI to X of REACH. For chemicals with production or import volumes exceeding 10 tonnes, the applicant must also provide a chemical safety assessment of the substance. REACH also includes provisions requiring companies manufacturing the same substances to cooperate through data sharing to minimize the need for animal testing (vom Berg et al., 2018). Unlike the situation with TSCA in the United States, under REACH the burden of proof for the safety of compounds is placed on companies submitting registration dossiers, and also places the responsibility for understanding and managing risks on manufacturers and importers (Applegate, 2008; vom Berg et al., 2018).

Any bio-based chemical that is manufactured in the EU and/or intended to be used in the EU in quantities greater than 1 tonne per year that has not previously been registered under REACH must undergo registration as described above. A manufacturer or importer of a new chemical must first submit what is known as an inquiry dossier to the ECHA to determine if the substance has already been registered; if so, the manufacturer would have the opportunity to arrange to share safety data with the prior registrants, but if not, the new manufacturer would have to conduct safety testing and submit its own registration (Chemical Inspection and Regulation Service, 2012; European Chemicals Agency, 2018). This process might require 1–2 months for the inquiry and up to 12 months for the new registration, and cost between 6000 and 25,000 euros (Chemical Inspection and Regulation Service, 2012). Each submission is reviewed both by the ECHA and the regulatory authorities of the applicable member state. The review focuses on evaluation of the quality of the registration dossiers, the suitability of any testing proposed in the submission, and ultimately the determination of whether the substance might present a risk to human health or the environment. The review process under REACH is shown in Fig. 3.2.

Compliance with REACH can be time-consuming, difficult, and costly, especially for small companies. The possible burdens bio-based companies may face in complying with REACH are discussed in vom Berg et al., 2018.

3.2.3 Chemical regulation elsewhere in the world

Regulatory programs for new chemicals are also found elsewhere in the world, several of which are based on, or are similar to, EU REACH. South Korea maintains a program usually referred to as K-REACH (Ha et al., 2016). First enacted in 2013 and amended in 2018, K-REACH requires registration and annual reporting for chemicals that are manufactured, imported, or sold in South Korea at greater than or equal to 1 ton per year, subject to certain exemptions. Among the information required to be submitted are identification of the manufacturer or importer, identity of the chemical substance and its

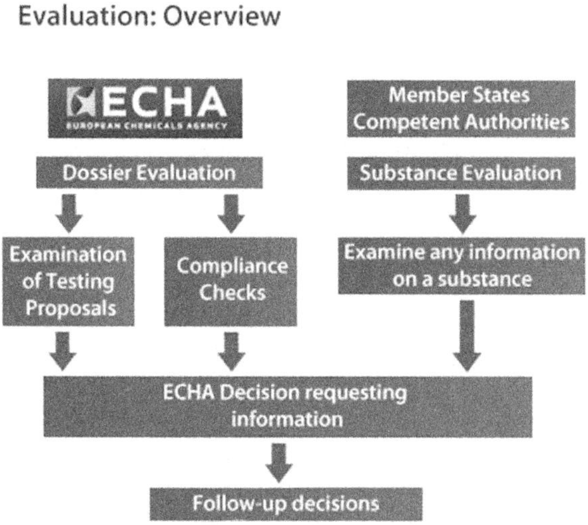

FIGURE 3.2 Evaluation process of chemicals under European REACH legislation. *Source: European Chemicals Agency, 2019. Evaluation Process. Retrieved from: https://echa.europa.eu/regulations/reach/evaluation/evaluation-procedure.*

intended uses, classification and labeling, physical and chemical properties of the substance, hazard and risk information, safety data, and use-related exposure information. The law requires certain testing for new chemicals based on annual production volume (Ha et al., 2016). Under the 2018 amendments, which came into effect on January 1, 2019 (Chemical Inspection and Regulation Service, 2019), registration may not be required for chemicals where manufacture or import is below 100 kg per year, although new chemicals below this limitation will require reporting to the National Institute of Environmental Research (ChemSafetyPro, 2019).

Japan instituted the "Act on the Evaluation of Chemical Substances and Regulation of Their Manufacture" (generally known as the "Chemical Substances Control Law") in 1973, with the most recent amendment coming in 2009 (ChemSafetyPro, 2015; Ministry of Economy Trade and Industry, 2016). Under this law, premanufacture evaluation and reporting are required for new chemical substances, while manufacturers or importers of existing substances must report their quantity and uses annually for volumes of manufacture or importation that exceed specified thresholds. The law also allows the government to designate certain chemicals as requiring priority risk assessment or even to prohibit the use or importation of certain chemicals. Note, however, that chemicals such as medicines and pesticides that are subject to other Japanese laws are outside the scope of the Chemical Substances Control Law. Under the Law, notifications for new chemicals must be made to the Ministry of Economy, Trade and Industry (METI), the Ministry of Labor and Welfare, and the Ministry of the Environment at least 3 months before the intended date of manufacture or importation. The annual reporting of existing substances is made to METI, with differing requirements for "General Chemical Substances" and "Priority Assessment Chemical Substances" (ChemSafetyPro, 2015; Ministry of Economy Trade and Industry, 2016).

China maintains a program formally called "The Measures for Environmental Administration of New Chemical Substances" that is often called "China REACH." This regulation, MEP Order 7, was issued by China's Ministry of Environmental Protection in January 2010 and became effective on October 15, 2010. Similarly to EU REACH, this regulation requires that manufacturers and importers of new substances file notifications and receive approval from the Ministry of Environmental Protection before production or importation, subject to certain exemptions (e.g., for naturally occurring substances). The notification requirement applies to new substances used as ingredients or intermediates for pharmaceuticals, pesticides, cosmetics, food additives, and feed additives, in addition to chemicals intended for direct use in commerce (ChemSafetyPro, 2018; Jarvis and Richmond, 2011).

Canada's program of chemical regulation, on the other hand, resembles that of the United States under TSCA. Under the Canadian Environmental Protection Act of 1999, any company or individual that intends to import or manufacture a "new" or "flagged" substance in Canada is required to notify the government under the New Substances Notification Regulations (NSNR) (Government of Canada, 2018). New substances are those that are not found on the Domestic Substances List (DSL) that lists chemicals and other substances known to be used in commerce in Canada; flagged substances are those that have been assigned Significant New Activity notice or a Reduced Regulatory Requirement. There is also a list known as the Non-domestic Substance List (NDSL), which lists substances that are not on the DSL (i.e., not in commerce in Canada) but are listed on the U.S. EPA TSCA Inventory. A company intending import or manufacture of a substance on the NDSL must file a notification, but with reduced data requirements (Government of Canada, 2018).

The NSNR regulations apply to chemicals, polymers, biochemicals, biopolymers, as well as living organisms: the regulations affecting the use of microorganisms to produce bio-based products are discussed below. There are differing data requirements for the different types of substance, but in general the type of data required for inclusion in the notification is similar to the information required in premanufacture notices under TSCA in the United States (Keener et al., 1999; Witt and Kirchof, 2008). The requirements for filing notifications differ among the different product categories, but most notifications must be filed at least 30 days before the intended date of manufacture or importation of the substance.

3.2.4 Government programs to promote use of bio-based products

Programs have been created in several countries either to provide some governmental support or incentive for the development and use of bio-based chemicals and other products, or to create some preferred labeling options for sustainably derived products. One example is the U.S. Department of Agriculture's "BioPreferred" program (U.S. Department of Agriculture, 2019). Under this program, federal agencies and their contractors are required to purchase bio-based products within certain categories that have been established by the USDA. Although the application process entails many steps for a company to become a qualified vendor for the federal government, the basis of the process is a calculation of the extent to which the product is bio-based: a calculation of the ratio of "new" organic carbon (derived from plants, trees, or agricultural sources) to the total organic carbon ("new" plus fossil-derived) in the product (U.S. Department of Agriculture, 2017). ASTM test method D6866 (ASTM International, 2019h) is to be used for this calculation, and this can be done by the company or by an independent third party. In order to qualify for one of the 109 different categories USDA has established and be accepted into the program, the product must meet the minimum bio-based content specified for that category. The agency also maintains a Voluntary Labeling Initiative, whereby qualified products can be labeled as meeting the bio-based standard (U.S. Department of Agriculture, 2019).

3.3 Regulations and government programs affecting bio-based fuels: certification and sustainability

3.3.1 Overview: fuel certification

In the United States and other countries around the world, new fuels or fuel additives may need to be certified or registered before commercial sale and use. In most countries, these regulations are distinct from laws or programs under which a fuel might be certified as "renewable" or to be sustainably produced (see below)—these regulations apply to all fuels and fuel additives, both traditional and renewable. The goal of these fuel certification regulations is to ensure that a new fuel has the appropriate chemical composition and that it is suitable for use in the range of engine types for which it is intended. Certification may also include emissions testing to assess the potential environmental impact of the fuel.

Generally, these regulatory regimes involve oversight over at least three distinct features of the fuel. These are fuel composition (chemical make-up and physical/chemical properties); engine suitability and performance testing; and health effects of the emissions from fuel combustion. A fourth criteria in the evaluation of novel aviation fuels is a consideration of the production pathway used to create the fuel, to ensure consistency and quality of the manufactured fuel product. Fuel composition and properties are often ascertained through the use of standard testing protocols to confirm that the fuel meets the accepted specification. Both the standards and the test protocols are often those certified by ASTM International or the equivalent standards issued by military branches or other international organizations. Engine testing may not be required in all cases, especially where the fuel has been shown to meet the applicable chemical specifications so as to be identical to a previously approved fuel, but where it is required the needed testing is often extensive. Similarly, assessment of health effects of emissions may be required under certain laws such as the U.S. Clean Air Act, but these requirements may be waived for fuels whose composition is identical to fuels currently on the market. In the United States, responsibility for certification of fuels has been shared to some extent by the EPA and the individual branches of the military. Table 3.2 shows how responsibility for fuel certification has historically been shared among different parts of the government in the United States.

3.3.2 US certification of civilian automotive fuels

Ethanol, butanol, and isobutanol. All fuels and fuel additives must be certified by the U.S. EPA Office of Transportation and Air Quality (OTAQ), under EPA regulations found in 40 CFR Part 79, before they can be sold in the United States. The most common use for alcohols like ethanol and butanol is blending with gasoline, and when used in this way they are regulated as fuel additives, subject to blending limits specifying the maximum concentrations permitted in gasoline. For ethanol, the longstanding maximum concentration has been 10% by volume; however in 2011, EPA revised its regulations after protracted public debate to allow ethanol to be blended up to 15% for autos and light trucks of model year 2001 or later (U.S. Environmental Protection Agency, 2015a). Because this approval was contingent upon the blended gasoline fuel having a Reid Vapor Pressure (RVP) of less than 9.0 pounds per

TABLE 3.2 Fuel certification programs in the United States.

Fuel Type	Application/Use	Agency/Military Branch with Responsibility
Ethanol, butanol, isobutanol	All (regulated as fuel additives)	EPA Office of Transportation and Air Quality
Diesel, biodiesel, renewable diesel, renewable gasoline and other "drop-in" fuels	Civilian (auto, truck, nautical, small engines)	EPA Office of Transportation and Air Quality
	Military (ground, nautical)[a]	Navy (including Coast Guard and Marines)
Jet fuel	Civilian (commercial aviation)	FAA, through the ASTM standards process
	Military[a]	Army, Navy, Air Force

[a]*These programs are currently unfunded or inactive.*

square inch, this limited the ability of E15 to be sold during the summer months; in March 2019, EPA published a proposed regulation that would amend the rule to allow sale of E15 year-round (U.S. Environmental Protection Agency, 2019b). Under the Clean Air Act, the maximum concentration for butanol was set at 12.5%, and in 2018 the Agency granted its approval for blending of isobutanol produced by Butamax Advanced Biofuels LLC at concentrations up to 16% (U.S. Environmental Protection Agency, 2018c).

The approval process for companies wishing to begin selling ethanol or butanol additives is fairly simple and primarily requires conducting analytical testing to show that the additive meets the applicable ASTM standard, which would be D-4806 for ethanol (ASTM International, 2019e) or D-7862 for butanol (ASTM International, 2019d). Applicants are also required to subscribe to publicly available applicable health effects studies for the relevant fuel. However, there are additional requirements for approval of E15 ethanol. A company wishing to certify an ethanol fuel for blending up to 15% must indicate that it is relying on a package of health effects data, developed by the Renewable Fuels Association (RFA) and Growth Energy and approved by EPA in February 2012; must adopt a Misfueling Mitigation Plan to ensure that consumers do not accidently use E15 in vehicles older than model year 2001, which can be met by the use of a model mitigation program developed by RFA, and to agree to participate in an EPA-approved survey of compliance with several E15-related requirements, including pump labeling and fuel specifications, which can be met by the use of an EPA-approved survey plan that is available for all E15 manufacturers or retailers (U.S. Environmental Protection Agency, 2015a).

Diesel and gasoline for civilian use. These fuels also must be certified by EPA OTAQ under the Part 79 regulations, but the process for fuels is potentially more complicated than it is for well-known additives like ethanol, and may require more extensive testing.

The Part 79 regulations establish two families of conventional fuel (diesel and gasoline), as well as four families of alternative fuels, with each family being defined by the specifications of a base fuel (e.g., traditional diesel for the diesel family). Each of the two conventional families are further subdivided into three categories, referred to as "baseline," "non-baseline," and "atypical". Baseline fuels are those that contain no elements other than those permitted in the category's base fuel, and which fully meet the accepted specifications for the base fuel. Nonbaseline fuels also contain no elements other than those permitted in the category's base fuel but may deviate from one or more of the limitations in the base fuel specification. Atypical fuels contain elements other than those permitted in the base fuel specification or otherwise do not meet the applicable standards.

The regulations specify different tiers of data that might be needed to support the registration of a new fuel. There are basic application requirements (found in 40 CFR Part 79.11), which include the name of the fuel, the results of analytical testing, and basic registration data (40 CFR Part 79.59(b)) which includes projected production volumes and other information. Beyond that, there are three potential tiers of testing: Tier 1 (40 CFR Part 79.52) includes basic emissions characterization data plus a literature search of potential health effects of fuel emissions; Tier 2 (40 CFR Part 79.53) includes toxicity, carcinogenicity, and other tests, and Tier 3 (40 CFR Part 79.54) includes other data EPA may require based on the results of data in the other tiers.

Although many manufacturers may need to submit the Tier 1 and Tier 2 data, there is an important exemption available under the regulations for small businesses (companies having less than $50 million in total annual sales for the 3 years prior to the application date). Small business companies proposing the registration of either a baseline or a nonbaseline fuel are exempt from the need for any Tier 1 or Tier 2 testing; and companies proposing registration of an atypical fuel are exempt from Tier 2 requirements (but not Tier 1) if their annual sales are less than $10 million. Generally, if the proposed new fuel meets the applicable ASTM specifications for that fuel family, which are D-975 for diesel (ASTM International, 2019f) and D-4814 for gasoline (ASTM International, 2019a), it would be considered either a baseline or a nonbaseline fuel. So, if a small business entity can show compliance with the applicable specification, it would be exempt from the Tier 1 and 2 testing requirements. If an applicant does not qualify for this exemption, it may be possible for that company to obtain access to the required test data by subscribing to a preexisting group that has generated the data to EPA's satisfaction (e.g., in recent years, health effects data has been available for biodiesel through such a group).

3.3.3 US certification of civilian/commercial aviation fuels

The U.S. regulatory agency for commercial (i.e., civilian) aviation is the Federal Aviation Administration (FAA), but its approvals are not for the fuels per se — FAA approves aircraft, which are given "type certificates" as part of the certification process. Type certificates are given when the agency determines that the aircraft meets applicable regulations, and they usually specify the type design of the aircraft, its operating limitations, applicable regulations and the other approved conditions for which the aircraft can be used, including which fuel types are compatible with the plane's engine. FAA regulations require that type certificate applicants identify the fuel grade or specifications that are to be used in their aircrafts. Upon approval, the specified fuels become part of the type certificate data sheet and the airplane flight manual (Rumizen and Edwards, 2017).

FAA's policy is that any fuel meeting the applicable standards of ASTM International would be acceptable for use in any engine certified for such fuels (U.S. Federal Aviation Administration, 2016). The specification for conventional petrochemical-derived Jet A fuel is D-1655 (ASTM International, 2019c), and the specification for Jet B is D-6615 (ASTM International, 2019g).

The specification for "aviation turbine fuel that consists of conventional and synthetic blending components" is D-7566 (ASTM International, 2019b). Recent standards-setting activity has led to the creation of three annexes under D-7566: synthesized paraffinic kerosene (SPK) blendstocks, which are covered in Annex A1 (FT SPK), "Hydroprocessed Esters and Fatty Acids" (HEFA) fuel derived from organic oils, covered in Annex A2 (HEFA SPK), synthetic fuels from fermentable sugars using synthesized isoparaffins in Annex A3 (SIP), synthesized paraffinic kerosene plus aromatics in Annex A4 (SPK/A), or alcohol-to-jet fuel in Annex A5 (ATJ) (Rumizen and Edwards, 2017; U.S. Federal Aviation Administration, 2016).

It is important to note that these ASTM standards are specific for feedstock and production process, so that a developer of a renewable or alternative jet fuel that is produced by a new or different process or feedstock will in all likelihood need to go through the full ASTM process to have such a fuel pathway certified for commercial aviation use.

ASTM International follows a standard-setting process which is uniform throughout the organization. This process is summarized in detail elsewhere (Mahwood et al., 2016; Rumizen and Edwards, 2017). In order to create a new standard, or an annex to an existing standard, it is necessary to develop a body of testing data which would be evaluated by a task force within a standing ASTM committee or subcommittee. The ASTM subcommittee responsible for the evaluation and approval of new aviation fuels is Committee D.02, Petroleum and Lubricants, Subcommittee J. Applicants proposing the inclusion of a new fuel would carry out the needed testing, and the results would be reviewed by the appropriate task force. The task force would then create a research report describing the results of the testing, which would be made available to the applicable subcommittee and committee members. After appropriate periods for review and comment, the standard must be adopted by balloting first at the subcommittee level and then at the full committee level. Ballot approvals must be unanimous, which requires that any objections raised at either the subcommittee or committee level be addressed, or the objections withdrawn, before the standard can be adopted. Once a new standard is adopted, the fuel is qualified for use with the specified engine types.

3.3.4 US certification of military fuels

The US military branches are prodigious consumers of fuel, particularly aviation fuel, and as such have had a long interest in adopting alternative fuels, particularly during the Obama Administration. The military branches have participated in the ASTM fuel certification process described above, although as pointed out by Rumizen and Edwards, 2017, each of the military branches had their own requirements for fuel that had to be addressed separately. The branches of the U.S. military, including the Army, Navy and the Air Force, at one time maintained programs for the development, testing and certification of alternative fuels, to address the costs and logistics of obtaining and transporting the very large amounts of fuels the military branches use, as well as an interest in reducing the carbon footprint, especially for aviation fuels (U.S. Government Accountability Office, 2015). To a large extent, these programs are no longer active, due to budget terminations and political factors.

At this writing, the U.S. Navy has no active program to test or qualify alternative fuels. Its program was previously managed by the Naval Air Systems Command at its fuel laboratory in Patuxent River, Maryland (U.S. Government Accountability Office, 2015). The Navy currently uses fuel meeting commercial specifications for its onshore transportation needs and uses JP-5 jet fuel for offshore uses. Adoption of specific fuels may be carried out on a case-by-case basis.

The U.S. Air Force and Army programs for alternate fuel testing are no longer funded. The Air Force has certified its entire fleet to utilize two different alternative aviation fuel blends. These are a synthetic fuel produced using the Fischer-Tropsch process and a biofuel produced by hydroprocessing esters and fatty acids, either of which would be blended with traditional JP-8 or Jet A at a ratio of up to 50/50 for use with Air Force Vehicles. The Air Force has stated that it is able to purchase and use alternative aviation fuels that are drop-in replacements and cost competitive with traditional petroleum-based jet fuels (U.S. Air Force, 2017).

3.3.5 US renewable fuel standard

The United States and other countries, as well as some US states and one Canadian province, maintain governmental programs that provide economic and other incentives to promote the development and use of biofuels and other renewable or low-carbon fuels. Although these programs do not require traditional premarket regulatory approval, these laws generally create certain obligations and legal requirements regarding R&D and manufacture for certain companies in the market.

The US Renewable Fuel Standard (RFS) is a nationwide program enacted under the Energy Policy Act of 2005 (Public Law 109-58), and later amended by the Energy Independence and Security Act (EISA) of 2007 (Public Law 110−140), with the goal of requiring increased use of renewable fuels through 2022 and beyond. Because the current law is based on this amended legislative authority, the law is sometimes referred to as "RFS2" (Congressional Research Service, 2019). EPA's regulations implementing the RFS can be found at 40 CFR Part 80. The goal of the RFS is to support and provide incentives for U.S. renewable fuel production by creating a mandatory market for qualifying fuels. The law requires fuel blenders to incorporate minimum volumes of renewable fuels in their annual transportation fuel sales. By guaranteeing a market for biofuels that is largely independent of pricing and fuel costs, RFS2 was intended to reduce the risk associated with biofuels production, and to provide an indirect subsidy for capital investment in the construction of biofuels plants.

To qualify for the RFS, a fuel must be produced from a renewable feedstock, which must fall within one of the seven categories defined as "renewable biomass" in 40 CFR Part 80.1401, and must also meet other requirements. RFS2 established four categories of renewable fuels and set yearly minimum volumes for each category. In its rulemaking, EPA calculated greenhouse gas (GHG) emissions savings for specific fuel types which were assigned into one of these four categories, while fuels produced through production pathways not specified in the original law or regulations could undergo a petition process to be qualified to be placed into one of the four categories. These four categories are defined in the law and regulations based on their feedstock and method of production, as well as the threshold levels of reduction of GHG emissions that a fuel must meet to qualify (GHG emission reductions are measured relative to the "baseline" emissions of the applicable conventional fuel—gasoline or diesel—that the renewable fuel would replace). These categories overlap somewhat, with some categories nested within others. The four categories are: "renewable fuels," "advanced biofuel," "cellulosic biofuel," and "biomass-based diesel."

"Renewable fuels," the broadest category, requires the fuel to be produced from a renewable feedstock and to reduce GHG emissions by at least 20%. (Corn-starch ethanol falls into this category.) "Advanced biofuels" is a subset of "renewable fuels," where the GHG emission reduction must be at least 50%, but where corn-starch ethanol is explicitly excluded. The most common fuel falling into this category is ethanol produced from sugar cane. "Cellulosic biofuel" must be produced from a cellulosic feedstock and reduce GHG emissions by at least 60%. The fourth category is "biomass-based biodiesel" which includes diesel fuel made from any renewable feedstock where the GHG reductions are at least 50%.

The original legislation required the use of 4.0 billion gallons of renewable fuel in 2006, with escalating obligations up to 36.0 billion gallons by 2022 (Congressional Research Service, 2019). The 2007 RFS2 legislation set mandated volumes through 2022 for each fuel category. EPA was given some authority to change the yearly mandates if required by prevailing circumstances, as well as the authority to set volume mandates beyond 2022 by rulemaking. Establishing the annual volume obligations has become contentious, because of opposition to the RFS from some sectors of the petroleum-based fuel industry and because the mandates for cellulosic fuels set in the legislation proved to be unattainable because the development of commercial production technology for cellulosic fuels has taken much longer than originally anticipated. As shown in Fig. 3.3, most recent rulemaking as of this writing set the volumes for 2019 at 19.92 billion gallons of total renewable fuels, including no more than 15 billion gallons of conventional biofuel and at least 4.92 billion gallons of total advanced biofuels. The latter must include at least 2.1 billion gallons of biomass-based diesel fuel and 0.418 billion gallons

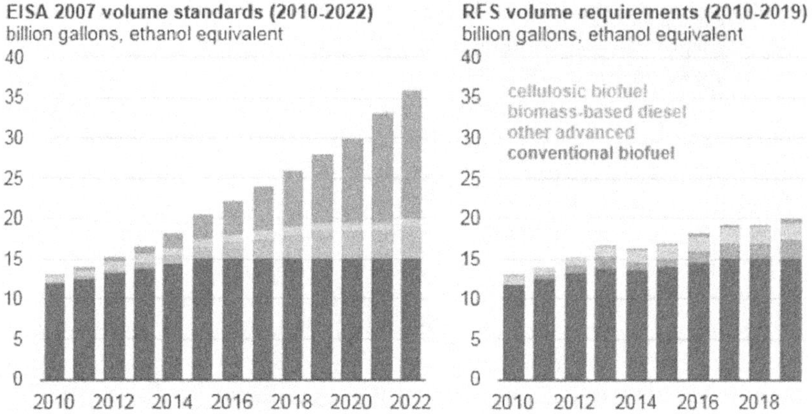

FIGURE 3.3 Volume requirements under original RFS legislation (left) and as amended by EPA rulemaking (right). *Source: U.S. Energy Information Administration, 2018. EPA Finalizes Renewable Fuel Standard for 2019, Reflecting Cellulosic Biofuel Shortfalls. Retrieved from: https://www.eia.gov/todayinenergy/detail.php?id=37712.*

of cellulosic biofuel (well below the target of 8.5 billion gallons set in the legislation) (Congressional Research Service, 2019).

Compliance with the RFS by fuel producers and sellers is tracked using a system wherein each gallon of fuel that qualifies as a renewable fuel is assigned an identification number called a Renewable Identification Number (RIN) (U.S. Environmental Protection Agency, 2017b). Each RIN is a unique 38-character number that is generated by the entity that manufactures or imports the fuel, in a format specified in the regulations. Each gallon of fuel within a batch is assigned its own RIN. Companies must report the RINs they create to EPA, generally on a quarterly basis. When the producer sells the fuel to another party such as a blender or distributor, the RINs are also transferred along with the fuel, although when the fuel is blended for retail sale or readied for export, the RINs are considered to be "separated" from the fuel, and at this point the RIN gains economic value on the open market (U.S. Environmental Protection Agency, 2017b).

The yearly minimum volume mandates for the individual fuel categories are passed along on a prorated basis to obligated parties, which are defined as the refiners that produce gasoline or diesel, along with companies that import gasoline or diesel, in the lower 48 states and Hawaii (i.e., excluding Alaska and the territories). To determine how much renewable fuel each obligated party is required to sell, EPA estimates the total volume of transportation fuels projected to be used in the coming year, and then calculates the percentage of that total that must be met by each of the four categories of renewable fuel in order for total fuel sales to achieve the volume mandates. EPA then applies these percentages to each obligated party, requiring each party to possess enough RINs to account for these percentages of each of the four fuel types in the fuels they sell in that year. This is called the "renewable volume obligation." Each provider can acquire these RINs either by producing biofuels itself or by purchasing RINs on the open market. Some of the required RINs would be acquired with the purchase of fuels (since RINs travel with fuels), but others would be acquired on the open market. Obligated parties that fail to meet their renewable volume obligation are subject to penalties under the law.

RINs are able to be traded on the open market and RINs for the different categories of renewable fuels (e.g., cellulosic biofuel, advanced biofuel, etc.) typically trade at different prices and may respond differently to prevailing market pressures. A discussion of RIN pricing dynamics and economics is outside the scope of this article, although a recent discussion can be found in a 2015 EPA document (Burkholder, 2015).

One aspect of the RFS program creates both an opportunity and a regulatory requirement for developers of new bio-based fuels: the need to petition to have new fuel pathways qualify as renewable under the RFS. EPA defines a fuel pathway as a specific combination of three components: (1) feedstock, (2) production process, and (3) fuel type (U.S. Environmental Protection Agency, 2017a). EPA created the pathway petition program in 2010, but after review of the program in 2014, the agency developed and published standardized guidance to assist fuel developers in preparing and submitting petitions, and this information is available on the EPA website (U.S. Environmental Protection Agency, 2018b). Petitioners must describe the fuel, its production method, and the feedstock used, as well as some other technical information. The petition must include a mass and energy balance and a calculation of the estimated GHG reduction that would be achieved by the production and use of the fuel, and although EPA will use the applicant's data to make its own life cycle analysis calculation, this analysis is critical to be sure that the fuel pathway can meet the statutory GHG emission levels in the legislation. EPA's website includes lists of approved and pending pathway petitions. At this writing, EPA has approved several dozen new fuel pathways under this petition process (U.S. Environmental Protection Agency, 2019a).

3.3.6 Low carbon fuel standards

Several U.S. states or regions, as well as one Canadian province, have adopted, or have considered adopting, their own regulations to promote the use of transportation fuels having more favorable carbon intensity (i.e., reduced greenhouse gas emissions) than traditional fuels. These are typically called "Low Carbon Fuel Standards" (LCFS), and they differ in several ways from the U.S. Renewable Fuel Standard. Most notably is that these laws typically apply to all transportation fuels, not only fuels produced from renewable feedstocks, but also in that they don't rely on tiered categories of fuels as is the case with the RFS, but instead utilize a system where each fuel is assigned a regulatory and economic value based on its carbon intensity, calculated through life cycle analysis.

The State of California maintains Low Carbon Fuel Standard (LCFS) regulations as a result of a sweeping climate change initiative known as Assembly Bill 32 adopted in 2006 that set an absolute statewide limit on greenhouse gas emissions. The LCFS regulations took effect in January 2010, and can be found in the California Code of Regulations at 17 CCR §§ 95,480–95,490. Under these regulations, fuel providers are required to reduce the carbon intensity of the fuels they sell, with the goal of achieving a 10% reduction in carbon intensity by 2020, through annually declining target levels relative to the "baseline" level of 2010 carbon intensity, with carbon intensity calculated as the fuel's life cycle GHG emissions per unit of energy. In September 2018, the state's Air Resources Board adopted amendments to the regulation that extended its requirements for an additional 10 years and created a more ambitious target of 20% reduction in carbon intensity from 2010 levels by 2030 (Witcover, 2018).

Fuel providers must determine and report to the Air Resources Board the carbon intensity of all the fuels they provide, with the carbon intensity of some fuels found in a "lookup" table based on life cycle analyses conducted by the staff, and the carbon intensity for other fuels established through a petition process. Petitions for new fuel pathways need to provide detailed information regarding the production pathway and its life cycle analysis, to allow a calculation of its carbon intensity. If a proposed pathway is approved, it is added to the lookup table, and becomes available to all regulated parties. More importantly, from the perspective of biofuel developers, the approval of a pathway means that the fuel can qualify for credits under the program. Fuels with low carbon intensities (i.e., lower than the applicable target under the LCFS) generate credits, which can be sold or transferred to fuel producers or providers with deficits: that is, having carbon intensities that are greater than the targets, thus creating an economic incentive for biofuel and renewable fuel developers.

The California LCFS regulations are generally considered as successful, and have reportedly resulted in a decline in the average carbon intensity of fuels sold in the state of almost 5% from 2010 to 2018, resulting in a reduction of over 38 million tons of carbon, with the share of alternative fuels used in California growing from 6.1% in 2011 to 8.5% percent in 2017 (Sperling and Murphy, 2018; Witcover, 2018). Because of the size of the market in California, these regulations have created a significant incentive for the use of low-carbon renewable fuels that have had an impact on national markets well beyond the state.

At this writing, Oregon is the only other US state with an operational Low Carbon Fuel Standard, although its path to this point was far from straightforward. Known as the Clean Fuels Program, it was first adopted by the Oregon legislature in 2009, although the process of developing regulations under the law did not begin until 2012. The original law had a "sunset" date of 2015, which led to legislative skirmishing that year which almost led to the law's repeal.

But the state legislature ultimately adopted a bill in 2015 which extended the law and allowed the Department of Environmental Quality (DEQ) to begin implementing it in 2016 (Oregon Department of Environmental Quality, 2017; Portland Business Journal, 2018). The law survived a lawsuit in a September 2018 ruling by a federal appeals court upholding the law and allowing it to be implemented (Portland Business Journal, 2018).

The regulations set a series of targets culminating in a 10% reduction of carbon intensity by 2025 relative to the baseline year of 2015. Regulated parties under the law (i.e., importers into the state of gasoline, diesel, ethanol, and biodiesel) must register with the DEQ and keep certain records of their fuel use relative to the target carbon intensity levels. A fuel can receive a carbon intensity value in one of three ways: it can use a carbon intensity value approved by the California Air Resources Board (adjusted for the different transportation distance between California and Oregon); the fuel producer can apply for a carbon intensity value from DEQ; or the fuel can be given a temporary carbon intensity value that can be used for up to two calendar quarters until DEQ approves an individual value for that fuel (Oregon Department of Environmental Quality, 2017).

British Columbia is the only Canadian province to adopt a low-carbon fuel regulation. The Renewable and Low Carbon Fuel Requirements Regulation (RLCFRR) took effect in January 2010 and is intended to reduce British Columbia's reliance on nonrenewable fuels, help reduce the environmental impact of transportation fuels, and contribute to a low-carbon economy. The RLCFRR provides a regulatory framework to enable the province to set benchmarks for the amount of renewable fuel in British Columbia's transportation fuel blends, reduce the carbon intensity of transportation fuels, and meet its commitment to adopt a low-carbon fuel standard. The province's overall goal is to lower provincial greenhouse gas emissions by 33% by 2020. The RLCFRR has two major requirements: the Renewable Fuel Requirement, requiring 5% renewable content in gasoline and diesel beginning in 2010; and the Low Carbon Fuel Requirement, which requires a 10% reduction in carbon intensity by 2020 (British Columbia Ministry of Energy Mines and Petroleum Resources, 2019). Under the regulations, fuel producers can apply for a unique carbon intensity based on a specific life cycle analysis. Once the carbon intensity is approved, the approved carbon intensity and corresponding British Columbia low carbon fuel code must be cited by any user of the fuel. The regulations include a list of approved carbon intensities and fuel codes (British Columbia Ministry of Energy Mines and Petroleum Resources, 2019).

At this writing, it is reported that programs similar to the California LCFS are being developed in Canada (a Clean Fuel Standard to cover transportation, industry, and building sectors) and Brazil (the RenovaBio program focused on renewable liquid fuels and biogas) (Witcover, 2018), and Washington State is again debating adoption of a Clean Fuel Standard, a policy first proposed in 2014, with one house of its legislature passing the legislation in March 2019 (Voegele, 2019).

3.3.7 European Union: fuel certification and sustainability

The European Union has maintained several programs that together provide ambitious goals and mandates meant to promote the use of renewable and low-carbon fuels as part of an overall strategy of reducing greenhouse gas emissions across the continent. The EU adopted two programs as part of its Energy and Climate Change Package, which was initially intended to run from 2010 through 2020. These are the Renewable Energy Directive (RED), aimed at promoting use of renewable fuels throughout all energy sectors, and the Fuel Quality Directive (FQD), focused more specifically on reducing GHG emissions associated with the transportation sector. The RED was put in place to ensure that all member states of the European Union achieved specified targets for use of renewable fuels and reduction of greenhouse gas emissions across all energy sectors, with specific requirements for the subset of fuels used for transportation (European Commission, 2019c). The FQD has additional, complementary requirements for GHG reductions within the transport sector. These directives place certain obligations on EU member states, but also create requirements with which developers or producers of renewable fuels must comply in order for their fuels to qualify as "renewable" under the regulations (European Commission, 2019a).

The Renewable Energy Directive, formally known as Directive 2009/28/EC of the European Parliament and of the Council, dated April 23, 2009, addresses the adoption of renewable energy within overall energy markets in EU member states (i.e., electricity, heating and cooling, transport). European Union legislation takes the form of "Directives," which are adopted centrally by the European Commission and which are binding upon all EU member states, which must then adopt national laws that conform with the provisions of the directive. As originally enacted, RED sets targets for the use of renewable fuels which member states were obligated to meet under national laws. These targets were (a) to derive 20% of overall energy consumption, across all sectors, from renewable sources by 2020; (b) to derive 10% of energy consumption within the transport sector from renewable sources by 2020; and (c) to achieve greenhouse gas emission reductions of at least 35%, relative to fossil fuels, by mid-2010, with this target rising to 50% in 2017 and 60% in 2018, for fuels produced in 2017 (European Commission, 2019c; USDA Global Agricultural Information Network, 2018b).

In its early years, the RED attracted controversy from groups that felt the program was promoting the increased use of biofuels derived from conventional biofuels (those produced using food crops as feedstock), which the groups maintained were less environmentally friendly and were a factor in increasing food prices. To address this concern, a 2015 amendment (which also applied to the FQD) was adopted that limited the share of conventional biofuels that could be applied to meet the 10% blending target to 7% while requiring a minimum of 0.5% of the transportation sector's fuel use attributable to advanced biofuels

(fuels produced from nonfood feedstocks) (USDA Global Agricultural Information Network, 2018b). In June 2018, the EU reached an agreement to extend RED's mandates through 2030. By that new end date, the overall target for contribution of renewable fuels to the overall energy sector would rise to 32%, the target for the transportation sector would increase to 14%, and the cap on conventional biofuels would be kept at 7% but with the minimum contribution of advanced biofuels rising to 3.5% (European Commission, 2019b; USDA Global Agricultural Information Network, 2018b). This proposal, which is being called "RED II," was reported to have been approved by the European Parliament in November 2018 (Glystra, 2018) and is expected to take effect on January 1, 2021 (USDA Global Agricultural Information Network, 2018b).

The Fuel Quality Directive, formally known as Directive 2009/30/EC of the European Parliament and of the Council, and which amended Directive 98/70/EC, was also originally adopted on April 23, 2009. It established the specifications (standards) for transportation fuels to be used across the EU. The Directive also required that all fuel suppliers (e.g., oil companies) were to meet a 6% reduction of GHG emissions by 2020, relative to 2010 baseline levels, across all fuel categories. This reduction in emissions could be achieved using any low-carbon fuel options, such as hydrogen or electricity, but it was generally expected that the use of biofuels will account for most of the targeted reductions. The target of 6% was designed to be consistent with the use of 10% biofuel with an average 60% carbon saving to comply with the Renewable Energy Directive, as described above. The FQD also established that ethanol may be blended into gasoline (petrol) up to a limit of 10% v/v, although this is subject to national laws in the member states (European Commission, 2019a). The amendments and extensions discussed above for the RED would apply to the FQD as well.

Although the obligations in the Directives to meet the specified targets are placed on national governments, these obligations are passed down to the entities that sell fuel to the public, as well as the companies that manufacture or import fuels for eventual sale in the EU. Specifically, there are defined requirements that a fuel and its production pathway must meet in order for the fuel to be considered a "renewable fuel" that qualifies to count toward fulfillment of the targets set in the Directives. Just as the U.S. Renewable Fuel Standard defines renewable fuels as those derived from "renewable biomass," the RED specifies that "renewable fuels" are not to be produced from raw materials obtained from land having a "high biodiversity status" or "high carbon content." In addition, if agricultural crops are used to produce the fuel, they must be grown in compliance with EU environmental regulations governing agriculture.

Furthermore, in order for a fuel to be considered as "renewable," it must show a reduction in GHG emissions over the life cycle of its production, such that the carbon intensity of the fuel can be determined and applied toward the national targets. The European Commission calculated default carbon intensities for a number of specific biofuel production pathways, which can be found in Annex V of the RED, and these can be used in reporting by regulated entities without providing any additional information to any national government. However, for fuel production pathways not included within this list, carbon intensities must be calculated using acceptable methods for developing life cycle analyses. In addition, renewable fuels must be produced sustainably; that is, in order for a fuel to be considered as "renewable" under the RED, it must be analyzed and certified to be in compliance with sustainability criteria established in the Directive. The EU Directives provide considerable detail about the factors that must be considered in the life cycle analysis and the sustainability assessment required under the rules. These analyses are often carried out by independent (e.g., nongovernmental) organizations, or in some cases by national voluntary systems (USDA Global Agricultural Information Network, 2018b). Because the EU sustainability criteria go beyond technical issues and encompass social issues, this imposes a requirement that is broader in scope than what is required in the United States under the Renewable Fuel Standard, and so this may be somewhat burdensome to biofuel producers, especially smaller, newer companies.

3.3.8 Biofuel mandates

In addition to programs to incentivize development of renewable fuels, many countries or regions also maintain laws or regulations mandating minimum levels or specifying target levels of use of biofuels in their transportation sector. These government mandates do not impose premarket regulatory burdens on fuel manufacturers, but they are governmental programs of relevance to the production of biofuels, and certain of them may require fuels to meet certain standards or specifications in order for their usage to count toward the mandated levels. Generally speaking, these programs either specify minimum percentages of ethanol or biodiesel that must be blended into the country's fuels or simply state aspirational targets for such biofuel use. These programs exist in too many countries of the world to summarize in this chapter, but a recent summary of biofuel mandates in the EU can be found in USDA Global Agricultural Information Network, 2018a, and another useful summary of mandates in many countries around the world is published every year by the online newsletter *The Digest* (The Digest, 2019).

3.4 Regulations affecting the use of genetically modified organisms in bio-based manufacturing

3.4.1 Overview: applications of genetic modification to bio-based manufacturing

As is amply described in the other papers in this volume, much of today's commercial activity using advanced biotechnology for biofuel or bio-based chemical production focuses on the creation, selection, or improvement of strains of desired microorganisms having enhanced properties for functions important for the production process. Microbiological methods for producing ethanol, fuels, or other chemicals generally rely on the use of one or more selected microbial strains to catalyze biosynthesis of the desired compound, generally through a traditional fermentation process. Historically, these methods have made use of naturally occurring or classically selected microorganisms, but in recent years the power of the new biotechnologies to develop enhanced strains is being investigated or used by numerous companies.

The wide variety of different strategies to improve microorganisms, cyanobacteria, and algae for bio-based product production are described in other papers in this volume, as well as numerous other reviews and original research articles in the literature. Notably, a 2018 special issue of the journal *Metabolic Engineering* (Alper, 2018) included several review papers describing progress in modifying common industrial microorganisms for fuel or chemical production. Other recent reviews of approaches for modifying industrially useful microorganisms for improved bio-based product production include Cho et al., 2015; de Farias Silva and Bertucco, 2016; Savakis and Hellingwerf, 2015 and Chen and Nielsen, 2016, among many others. See also Glass, 2015a for citations and discussion of a number of earlier review articles.

There is also interest in using genetic engineering or other advanced biological technologies to improve the plants that are used as feedstocks for production of ethanol or other fuels. Some efforts are being directed at food crops like corn: even though corn ethanol is generally considered to be a transitional fuel, the techniques for genetic engineering of corn are well known and easily practiced. The first transgenic plant product approved for commercial use in biofuel production is a variety of corn engineered to express alpha-amylase, sold under the tradename Enogen by Syngenta. Use of this corn as the feedstock in an ethanol plant eliminates the need to add a liquid form of the amylase enzyme, and enables a reduced viscosity of the corn mash (Syngenta, 2019). Efforts are also ongoing to engineer the plant species that might be alternative biofuel feedstocks, including trees, switchgrass, oil-rich crops like canola or jatropha and others (Salehi Jouzani et al., 2018). The prospects for modifying plant lignin content to facilitate production of cellulosic fuels have been recently reviewed by Aro, 2016; Badhan and McAllister, 2016 and Welker et al., 2015. Earlier reviews can also be found in Sticklen, 2006; Torney et al., 2007 and Weng et al., 2008.

As powerful and useful as the methods of recombinant DNA, synthetic biology, gene-editing, and others may be, their use may often subject the process to additional government oversight. Due to concerns over the safety and ethics of genetic manipulation in the early days of the biotechnology era, many countries around the world have adopted biotechnology or biosafety laws that may affect commercial exploitation of modern genetic technologies.

3.4.2 US regulation of modified microorganisms

Regulations that would affect uses of advanced biotechnology for production of biofuels and bio-based chemicals evolved in the United States and elsewhere following the initial debates beginning in the late 1970s over the safety of recombinant DNA technologies. This early history is well documented in many publications, including Glass, 1991, 2003; 2015a,b; and Krimsky, 1985. In the United States early recombinant DNA guidelines were created by the U.S. National Institutes of Health, but these guidelines were binding only on federally funded research. As the biotechnology industry developed in the 1980s, the focus of regulatory concern shifted not only to the larger-scale uses inherent in commercial application of this new technology, but also to deal with the intended use of engineered plants, animals, and microorganisms for use outside the lab, in the open environment (e.g., in agriculture). In the United States, the outcome of several years of public policy discussions was the adoption of a "Coordinated Framework" for biotechnology regulation in 1986 (Office of Science and Technology Policy, 1986). Under this framework, it was decided that the commercial products of biotechnology would be regulated under existing laws and regulations and that it was not necessary to enact a specific law broadly covering all biotechnology activities. This is in contrast to most other countries in the world, which have generally created a single national biotechnology ("biosafety") law, often in the wake of the adoption of, and in compliance with, the Cartagena Protocol on Biosafety (see below).

Thus, the use of biotechnology in the United States to produce drugs, vaccines, diagnostic products, foods, and food additives would be regulated by the Food and Drug Administration (FDA), using existing regulatory authority; biotechnology-derived pesticides would be governed by existing rules of the Environmental Protection Agency (EPA); and most other agricultural products would be regulated by the U.S. Department of Agriculture (USDA). New regulations were developed to allow USDA appropriate authority over transgenic plants, and to allow EPA the basis to regulate anticipated industrial uses of microorganisms that were not regulated elsewhere in the government. Both these regulations are described below. More detailed descriptions of the history and content of the Coordinated Framework can be found elsewhere, including Glass, 2003; Wozniak et al., 2012b; Wrubel et al., 1997 and others.

The use of certain genetically modified microorganisms, including algae or cyanobacteria in biofuel or bio-based chemical production may be subject to regulations promulgated by the US Environmental Protection Agency under TSCA (Bergeson et al., 2014; Glass, 2003). As mentioned above, TSCA is a law requiring companies or individuals to notify EPA before commencing manufacture or importation of any new chemical, not already in commerce in the United States, and which is intended to be used for a purpose not subject to federal regulation as a pesticide or under the food and drug laws. In the Coordinated Framework of June 1986 (Office of Science and Technology Policy, 1986),

EPA proposed to use TSCA in the same way as it is used for chemicals, to capture those genetically modified microorganisms to be used in commerce but that were not regulated by other federal agencies. The primary areas which were expected to become subject to the TSCA biotechnology regulations were (a) microorganisms used for production of nonfood-additive industrial enzymes, other specialty chemicals, and in other bioprocesses; (b) microorganisms used as, or considered to be, pesticide intermediates; (c) microorganisms used for nonpesticidal agricultural purposes (e.g., nitrogen fixation); and (d) microorganisms used for other purposes in the environment, such as bioremediation. As the field of industrial biotechnology has developed, production of biofuels and bio-based chemicals has become the most prominent "bioprocessing" application that might fall subject to TSCA (Bergeson et al., 2014; Glass, 2003, 2015a,b).

Although EPA established an interim policy of TSCA regulation under the 1986 Coordinated Framework, a final rule was not adopted until 1997 (Glass, 2015a,b). This rule created a new section of the Code of Federal Regulations (40 CFR Part 725) to specify the procedures for EPA oversight over commercial use and research activities involving microorganisms subject to TSCA. The rule instituted reporting requirements specific for microorganisms (but which paralleled the commercial notifications for new chemicals), while also creating new requirements to provide suitable oversight over outdoor uses of those genetically modified microorganisms subject to TSCA jurisdiction. The adoption of the 2016 Lautenberg Act had only a minimal impact on EPA's biotechnology programs, primarily through the requirement for EPA to make and publish definitive determinations of risk for each microbial notice it reviews (Bergeson and Auer, 2016; Glass, 2016a).

The biotechnology rule requires advance reporting for new organisms intended for commercial use. The rule potentially covers all genera of microorganisms, including microalgae and cyanobacteria. The final rule defines a "new organism" as "a microorganism that is formed by the deliberate combination of genetic material originally isolated from organisms of different taxonomic genera," commonly referred to as an "intergeneric organism." This is the same definition originally proposed in EPA's interim TSCA biotechnology policy. The history and rationale for this definition has been described elsewhere (Glass, 2015a,b). Under this definition, microorganisms that are not intergeneric are considered not to be new, and such organisms, including naturally occurring and classically mutated or selected microorganisms, as well as microorganisms modified only through gene deletions, directed evolution, or gene-editing techniques, are exempt from reporting requirements under TSCA.

Commercial uses of "new microorganisms" used for a "TSCA purpose" (that is, not regulated elsewhere in the federal government) may require notification to EPA at least 90 days in advance of commercial use or importation. This notification takes the form of a Microbial Commercial Activity Notice (MCAN). An individual MCAN is needed for each modified strain intended to be commercialized, although EPA maintains procedures to facilitate submission and review of "consolidated" MCANs covering related strains of similar genetic make-up.

The information and other data that applicants need to submit in the MCAN are listed in Section 725.155 of the regulations and are summarized in Glass, 2015a. Applicants must describe the taxonomy and biological traits of the modified microorganism along with a detailed description of how it was constructed. Information is also to be submitted on the proposed use of the microorganism, the proposed production process, the containment and control procedures to be used, the likelihood for worker exposure and the steps taken to control exposure, and an assessment of the potential environmental effects of the microorganism should it be released from the facility. It is important to note that, in MCANs or other submissions to EPA under the biotechnology rule, the applicant can claim much or all of the submitted information as "confidential business information," which the agency must keep confidential and which cannot be released to the public, but the applicant must provide EPA with the justification for the confidentiality claim (in fact, this justification must be included within the MCAN filing). EPA has provided guidance documents that can be used to help applicants prepare MCAN submissions.

EPA review of most MCANs can be expected to be fairly straightforward and will focus on the potential risks and benefits of the commercial use of the modified microorganism. Most of EPA's prior reviews of MCANs have taken place within the 90-day period specified in the regulations, although EPA has the power to unilaterally extend the review period by an additional 90 days or to ask the applicant to voluntarily suspend the review if the Agency decides it needs more time to complete its review. MCANs for the contained use of new microorganisms in bio-based manufacturing have generally not caused any concerns or significant issues in EPA's review, and most have been routinely cleared for commercial use without any delays or difficulties (Bergeson et al., 2014). MCANs (like chemical PMNs) are not "approved" per se but if no issues emerge they are cleared for commercialization if EPA takes no action within 90 days. As mentioned above, under the Lautenberg amendments, EPA must now publish an explanation of its risk assessment findings when making this final determination. Once an MCAN clears the review process, the applicant must file a Notice of Commencement within 30 days of beginning commercial use or importation of the microorganism, a notice that requires submission only of minimal information.

The biotechnology rule has been in effect since 1997. The EPA website lists 102 MCANs filed from that date through June 2016 (U.S. Environmental Protection Agency, 2018d). All MCANs filed since the passage of the Lautenberg Act in June 2016 are listed on a different page on the EPA website (U.S. Environmental Protection Agency, 2019d). As of this writing, there are 107 MCANs listed, not all of which have completed review, but it is notable that in the nearly 3 years since the passage of the Lautenberg Act, there have been as many MCANs filed as in the previous 20 years under the Final Biotechnology Rule. This increased level of activity can be seen in Fig. 3.4, which shows the numbers of biotechnology filings through Fiscal Year 2015. Since then, in each of Fiscal Years (2017), 2018 there were over 40 MCANs filed, although many were consolidated MCANs or submissions of multiple MCANs at one time to cover multiple related strains. The greatly enhanced frequency of MCAN filings is clearly due to increased interest in the use of modified microorganisms for production of fuels, chemicals, and enzymes. Most of these MCANs cover modified microorganisms intended for the production of industrial enzymes or MCANs for the use of *Saccharomyces cerevisiae* strains for ethanol production.

The biotechnology rule provides certain exemptions from MCAN reporting. These are the so-called "tiered exemptions" available for certain uses of modified strains of well-studied, common industrial microorganisms. To qualify for these exemptions the host, or recipient, organism must be one that is included on the list found in Section 725.420 of the regulations. This list includes many well-studied industrially useful species including *E. coli* K12, *S. cerevisiae, Bacillus subtilis,* and others. The organism must also meet the requirements of Section 725.421 regarding the introduced DNA and how well it is characterized. Finally, in accordance with Section 725.422, the organism must be used under conditions and controls designed to minimize the possibility that the engineered microorganism might inadvertently be released from the facility. If an organism meets the first two sets of criteria (the "biological" criteria) and the applicant can certify that it will use the microorganism in strict compliance with the provisions of Section 422, the process is eligible for a "Tier I" exemption and the microorganism can be used commercially merely upon 10 days advance notice to EPA. For microorganisms meeting the biological criteria but which are intended for use under conditions less strict than Section 422 procedures, the applicant can submit a petition for a "Tier II" exemption 45 days before intended manufacture. EPA would approve the Tier II request if it felt that the proposed containment and control procedures, although not identical to Section 422 procedures, were sufficient for the organism in question.

In addition to the tiered exemptions, the TSCA regulations also provide a procedure by which companies can apply for an exemption for test marketing purposes. This requires submitting certain information to EPA 45 days in advance of the proposed activity. According to (Bergeson et al., 2014), from 1997 through 2013, EPA had received and approved 118 Tier I and two Tier II exemption requests, as well as one request for a test marketing exemption for microorganisms under TSCA.

As noted above, the TSCA regulations include an exemption for "small quantities" of new chemicals used solely for R&D. This exemption was largely carried over into the biotechnology rule, except that EPA made the distinction that microorganisms, because they are self-replicating, could not be considered

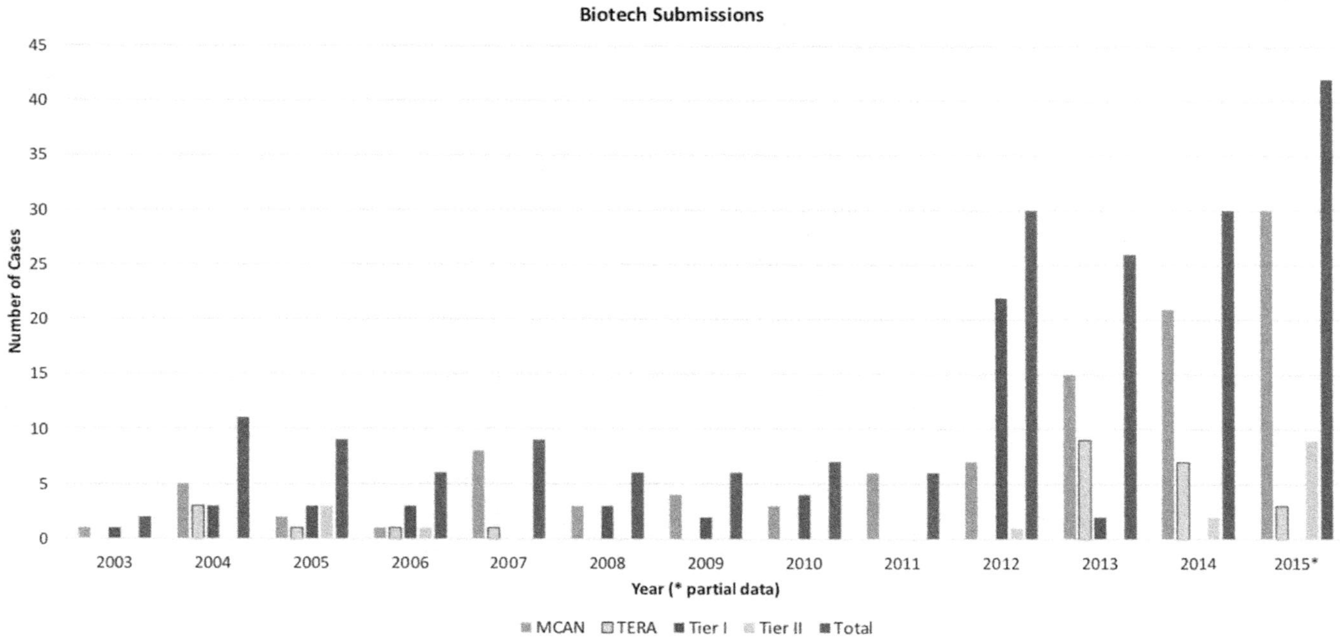

FIGURE 3.4 Numbers of biotechnology filings under EPA TSCA regulations, 2003–15. *Source: U.S. Environmental Protection Agency, 2015b. U.S. Environmental Protection Agency Biotechnology Algae Project. Retrieved from: https://www.epa.gov/sites/production/files/2015-09/documents/biotechnology_algae_project.pdf.*

to ever be used solely in "small quantities" unless certain restrictions were placed on how they were used. Thus, under 40 CFR Sections 725.234 and 725.235, new microorganisms used solely for R&D purposes could qualify for the exemption only if they are used under suitably contained conditions (i.e., in "contained structures"). Activities in contained structures such as most industrial fermenters and photobioreactors would qualify for the small quantities exemption if conducted "solely for research and development" and meeting other procedural requirements, such as carrying out the R&D under the supervision of a technically qualified individual, the adoption of specific containment procedures, record-keeping requirements, and worker notification. R&D meeting these requirements can be conducted with no EPA oversight or prior notice (in fact, entities determine for themselves if they are in compliance). Under this definition, it is likely that most laboratory research in biofuels or bio-based chemicals would be exempt from commercial reporting. In addition, many uses of engineered microorganisms in biofuel or bio-based chemical pilot plants or demonstration plants could also qualify for this exemption, as long as the microorganism were used solely for research and development and neither the organism or its product is used or sold commercially, and as long as the required criteria are meet.

However, the use of open-pond reactors for the cultivation of modified strains of algae would not qualify as contained structures: R&D use of intergeneric microorganisms in the open environment or in vessels or facilities judged not to be suitably contained would require notification to EPA at least 60 days before the proposed use, under an application known as a TSCA Experimental Release Application (TERA). The TERA process provides an expedited review procedure for small-scale field tests and other outdoor R&D uses of new microorganisms. Applicants proposing such uses must file a TERA with the EPA at least 60 days in advance of the proposed activity. The data requirements for TERAs are outlined in Sections 725.255 and 725.260 of the regulations (see Glass (2015a) for details), and these requirements address the key issues which should be considered in environmental risk assessments.

EPA is required to review the submitted information and decide whether or not to approve the proposed outdoor R&D activity within 60 days, although the agency could extend the review by an additional 60 days. If EPA determines that the proposed activity does not present an unreasonable risk of injury to health or the environment, it will notify the applicant in writing that the TERA has been approved. When a TERA is approved, the applicant must carry out the testing under the conditions and limitations described in the TERA application document, but also in accordance with any requirements or conditions included in EPA's written approval. In most cases, it is likely that EPA will require applicants to conduct some form of monitoring, to detect the possible spread or dispersal of the microorganism from the test site, or to detect any other potential adverse environmental effects. EPA's approval is legally binding on the applicant, and the Agency has the additional authority to modify or revoke the approval upon receipt of evidence that raises significant questions about the potential risk of the activity.

As of this writing, there has only been limited experience with TERAs, with only 33 TERAs filed since the biotechnology rule was put into place in 1997 (U.S. Environmental Protection Agency, 2018d). Beyond field testing, any proposed commercial use of an intergeneric microorganism in an open-pond reactor (i.e., for manufacturing a bio-based product) would require submission of an MCAN. EPA's review of such MCANs would focus on the potential environmental impacts of the proposed use, and would be expected to be a more thorough review than EPA has conducted for MCANs for contained uses.

3.4.3 US regulation of genetically modified plants

The US Department of Agriculture (USDA) maintains regulations at 7 CFR Part 340 that have been the major US government rules that have covered uses of transgenic plants in agriculture as well as the uses of plants for other industrial purposes, such as production of pharmaceuticals, industrial products, and phytoremediation (McHughen and Smyth, 2012; Wozniak et al., 2012a). A small number of modified agricultural microorganisms have also fallen under this regulation. Its applicability to bio-based products would primarily arise if genetically modified plants were to be grown as feedstocks for bio-based production.

This regulation was put into place in 1987 as an immediate outgrowth of the 1986 Coordinated Framework (Office of Science and Technology Policy, 1986). USDA proposed to use its existing statutory authority under a law then known as the Plant Pest Act to regulate certain genetically engineered plants intended for field testing and eventual commercial use in the open environment, to assess the potential environmental effects of such uses. The basis for this rule was the possibility (however remote) that such engineered plants might pose a plant pest risk, based on the presence of nucleic acid sequences arising from genera listed in the rule. These regulations were finalized in June 1987, and have been administered by a dedicated biotechnology office within USDA's Animal and Plant Health Inspection Service (APHIS) known as Biotechnology Regulatory Services.

Under this rule, "regulated articles" are defined to include "organisms that are or contain plant pests," which has been interpreted to cover only those plants (or microorganisms) engineered to contain nucleic acid sequences from plant and insect viruses and certain specific microbial, plant, and animal genera that contain species that were considered to be potential plant pests. The regulations include a fairly broad list of such genera, and this has the practical effect of causing most transgenic plants to be captured by the regulations; because the genus *Agrobacterium* was on the list, and in practice, DNA sequences from *Agrobacterium tumefaciens* were almost universally used in plant transformation and would be present in the resulting plant, causing it to be subject to regulation under this rule. The list of known or potential plant pest species is contained in 7 CFR Part 340.2.

Although the USDA rule initially required submission of permit applications for all proposed outdoor uses of organisms covered under the regulation, the rules were substantially relaxed, first in 1993, and then again in 1997, with the creation of a much simpler notification process for those plant species deemed to have low potential risks. Under the current version of the regulations, transgenic varieties of most common agricultural crops and other familiar plant species can be used in research field tests simply upon 30 days advance notice to APHIS, and the submission of only minimal information about the modified plants and the proposed field use. Uses of less-familiar transgenic plants are now required to undergo the longer permitting process. Commercial use of a transgenic plant variety of any species requires USDA review and approval of a petition for a determination that the variety qualifies for "nonregulated status." Such decisions by USDA clear the organism for commercial use, but in recent years have required the agency to first prepare Environmental Assessments justifying such actions, lengthening and complicating the approval process.

USDA has made several attempts to revise these regulations, in part because the Plant Pest Act, the law on which the Part 340 regulations was based, was combined in 2000 with other statutes to create a new law, the Agriculture Risk Protection Act, which includes language that could give USDA the ability to regulate modified organisms based on potential invasiveness or weediness. In 2008 and again in 2017, USDA published possible options and proposed new rules to accomplish this, but in both cases the agency withdrew those rules in response to public comments, and expressed the desire to reengage in a fresh dialogue with its stakeholders. In June 2018, USDA again opened a public comment period regarding its intent to revise the biotechnology rule, and in June 2019, the Department issued a proposed rule that, if adopted in final form, would substantially overhaul these regulations.

The USDA biotechnology regulations would affect developers or biofuels of bio-based chemicals only if the feedstock for production was a genetically modified plant (except in the rare case where a microbial production process made use of a genetically modified organism of a species found on the "plant pest" list). In the case of a transgenic plant, the burden of the regulation would fall on the developer of the plant variety, and in most cases the chemical manufacturer would purchase or otherwise obtain plants from the developer once the needed regulatory approvals were in hand.

3.4.4 International biotechnology regulations: Cartagena Protocol

Biotechnology regulations exist throughout the world, although they have developed differently than in the United States. Many other industrialized countries or regions, particularly the European Union, Canada, Australia, and

Japan, implemented biotechnology laws and regulations in the early days of the growth of the industry (i.e., the 1980s and 1990s), in ways that were consistent with the regulatory approaches of these jurisdictions. More recently, many other countries around the world have adopted biotechnology or "biosafety" laws and regulations based on the principles of an international convention adopted in 2000— the Cartagena Protocol on Biosafety. Countries taking this route generally have a single biotechnology law that, in principle, is applicable to all research and industrial uses of genetically modified organisms, although much of the focus of the Cartagena Protocol is on the cross-boundary movement or open-environment uses of modified organisms. More details on international regulations can be found in Glass, (2015a).

The Cartagena Protocol on Biosafety was adopted on January 29, 2000, as a supplementary agreement to the Convention on Biological Diversity, and took effect on September 11, 2003 (Eggers and Mackenzie, 2000). Those national biosafety laws that are modeled on the Cartagena Protocol will in general require government approvals for importation of living modified organisms (LMOs) into the country, and for many industrial activities including "contained uses" or "environmental uses". Such approvals may often require a risk assessment of the LMO and its proposed use, often guided by the principles of the Protocol.

Under the Cartagena Protocol, the term "LMO" is defined as any living organism that possesses a novel combination of genetic material obtained through the use of modern biotechnology, with "modern biotechnology" defined to include *in vitro* nucleic acid techniques as well as "fusion of cells beyond the taxonomic family." Although definitions vary, most countries use a definition such as this, with the United States being one exception; the definitions in the U.S. EPA TSCA regulations and the USDA regulations are narrower, with EPA's limited to "intergeneric" microorganisms and USDA's requiring the presence of nucleic acids from suspected plant pest species.

The Cartagena Protocol is primarily intended to ensure that national authorities are notified of any proposal to introduce LMOs into their countries, particularly for the purpose of deliberate release into the environment or for use in food or feed, and further to ensure that information about uses of LMOs is provided to the public and to other countries and interested parties. A key provision of the Protocol is to require there to be "Advance Informed Agreements" (AIAs) when LMOs are shipped across national boundaries, to ensure that the recipient nation is notified of the proposed shipment, and to allow the recipient nation to conduct needed risk assessments.

In most countries, uses of microorganisms within contained manufacturing will generally be subject to a lower level of oversight. "Contained Use" is defined in Article 3 of the Protocol as "any operation, undertaken within a facility, installation or other physical structure, which involves living modified organisms that are controlled by specific measures that effectively limit their contact with, and their impact on, the external environment." However, Article 6(2) of the Protocol provides an exemption from the AIA procedures for shipments of LMOs intended solely for contained use. Some national laws may institute permit requirements for commercial "contained uses" over and above the notification and labeling requirements under the Protocol. The Protocol can best be viewed as establishing minimum requirements for applicants proposing to use LMOs in contained commercial manufacturing, such as a notification requirement, with the understanding that national laws may impose additional requirements in certain countries.

Note, however, that the production of a bio-based product using an algae production organism in an open-pond reactor would likely be handled differently under laws based on the Protocol, because it would be judged to be an environmental release rather than a contained use. In many cases, the recipient national government would need to be notified and would need to conduct a risk assessment. An Advance Informed Agreement would almost certainly be required and the risk assessment would almost certainly be more rigorous. Annexes I and III of the Protocol provide specific guidance for how risk assessments are to be conducted. In many countries, a permit or some affirmative government permission would be needed before the LMO could be used in the open environment. Such proposals may also engender public or community interest and perhaps opposition.

In 2010, another supplementary agreement to the Convention on Biological Diversity, known as the Nagoya Protocol, was adopted. Its purpose was to help ensure the fair and equitable sharing of benefits arising from the utilization of genetic resources (Convention on Biological Diversity, 2019). The Nagoya Protocol took effect in 2014, and it does not appear that it will necessitate significant changes to the provisions of national biosafety laws that might affect bio-based manufacturing, although in some cases it may require companies to document the country of origin of its production organisms and the nucleic acid sequences introduced into those organisms.

3.4.5 International biotechnology regulations: European Union

The EU has adopted two directives to cover biotechnology—one covering contained uses of modified organisms, and the other covering uses of modified plants and other genetically modified organisms in the open environment (Enzing et al., 2012). Each EU member state is obligated to adopt national laws corresponding to EU directives, and so each EU member has its own biotechnology laws or regulations that mirror the provisions of the two EU directives.

Uses of modified microorganisms in contained manufacturing would require national government notification, with advance approval also required in some countries, in accordance with the EU "Contained Use" Directive 2009/41/EC (European Union, 2009). Article 2 of the directive defines "contained use" in a way that gives an applicant proposing to use a genetically modified microorganism in Europe a fair amount of leeway in determining that a system or process is "contained." Article 4 of the directive requires the user to carry out a risk assessment of the microorganism, using considerations set forth in Annex III of the directive, after which the user would determine the containment measures appropriate for the organism, in accordance with Annex IV of the directive. It is likely that most microorganisms used for fuel or chemical production would be classified within the lowest level of containment. Article 6 of the directive requires users to notify the designated government agency before a facility is to be used with modified microorganisms for the first time. Annex V specifies the information required to be submitted with such notifications, and for organisms in the lowest class of risk, the necessary information is fairly minimal. The laws of individual EU nations should conform to these provisions, although in some cases, national laws may require government review and approval of contained manufacturing activities, beyond the notifications required under the directive.

Uses of modified algae or other microorganisms for manufacturing in open ponds would be covered by national laws corresponding to EU Directive 2001/18/EC on "Environmental Release" (European Union, 2001). Generally speaking, any outdoor activities with LMOs in Europe, including small-scale field testing, would require approval from the country in which the activity is to take place. Applications for commercial use are more complicated, in that all EU member states have some say in commercial approvals granted by individual countries. This directive would also apply to research and commercial uses of transgenic plants, for example, for use as feedstocks in bio-based manufacturing. Although the field testing and commercial use of transgenic plants, particularly for food-producing species, have been extremely controversial or even prohibited in most EU nations, the same may not necessarily be true for uses of modified microorganisms in open-pond manufacturing.

3.4.6 International biotechnology regulations: Canada

Uses of modified microorganisms in biofuel or bio-based chemical manufacturing in Canada may require approval from Environment Canada under the New Substances Notification regulations under the Canadian Environmental Protection Act (Darch and Shahsavarani, 2012). These regulations are somewhat similar to the US TSCA biotechnology regulations and cover the use of any microorganism that is new to commercial use in Canada. Unlike the United States, however, the Canadian regulations cover unmodified microorganisms as well as modified microbial strains that have not previously been used in Canada. The regulations potentially cover both contained and open-environment use of microorganisms, with a greater level of scrutiny dedicated for the latter (Glass, 2015a).

3.4.7 Other international biotechnology regulations

Biotechnology or biosafety laws are maintained in many other countries. Space does not allow a detailed description of these other regulatory regimes, but an overview of the laws and regulations in Brazil, Japan, Australia, and China can be found in Glass (2015a).

3.5 Other regulations affecting bio-based manufacturing

3.5.1 Use of spent microbial biomass in animal feed

It is common for companies developing fermentation processes for bio-based manufacturing to seek to gain some economic value from the large volumes of inactivated microorganisms that remain at the end of each production run. One potential use for which there is long-standing precedent is to use the microbial biomass in animal feed, the way inactivated yeast from beverage or fuel ethanol production has been historically used in distillers' grains for animal feed (U.S. Grains Council, 2012). However, in the United States and in most other countries, the regulatory paths to obtain approval for new ingredients in animal food, particularly for use with the major food-producing animal species, can be time-consuming and difficult and expensive to navigate. Moreover, in the United States, an applicant would need to choose between several different pathways, pursuing Generally Recognized as Safe status, gaining approval as a feed additive, or obtaining a new ingredient definition from the Association of American Feed Control Officials. Beyond pointing out the challenges companies may face, a complete description of animal feed regulation is beyond the scope of this chapter. More details on US regulation of animal feed ingredients can be found in Glass, 2015a; Smedley, 2013 or Glass, 2016b.

3.5.2 US chemical facility anti-terrorism standards

Following the terrorist attacks of September 11, 2001, the US Congress passed a law entitled "Support Anti-terrorism by Fostering Effective Technologies Act of 2002" (Now Sections 441–444 of title 6, United States Code). The goal of this law was to protect chemical facilities from being terrorist targets, and the US Department of Homeland Security promulgated regulations (6 CFR Part 27) to implement this law. The regulations require chemical facilities possessing quantities of any of about 300 specified "Chemicals of Interest" above certain threshold levels ("Screening Threshold Quantities") to register with the Department and to report the chemicals and the amounts possessed. The law underlying these regulations has been extended several times, most recently in January 2019, and is now in effect until April 2020 (Gottron, 2019). The regulations define "chemical facility" quite broadly, and so many bio-based fuel or chemical production plants might be covered by these regulations. The procedural requirements can be quite onerous if any Chemicals of Interest are possessed at the plant above the threshold levels, but requirements would be minimal if all such chemicals are possessed in quantities below the thresholds.

3.5.3 US fuel ethanol plant permits

US bio-based production facilities manufacturing ethanol for fuel use would need a permit from the US Alcohol, Tobacco and Firearms: Tax and Trade Bureau (known as "TTB"). TTB is a branch of the US Department of the Treasury, the primary purpose of which is to regulate the production of alcohol in the United States, to ensure that appropriate taxes are paid on alcohol produced for food and beverage use. The legal basis for the TTB permit program is 26 U.S.C. 5181, "Distilled Spirits for Fuel Use," and the applicable regulations are found in 27 CFR Part 19, Subpart X. There are three different levels of permit, depending on the expected production output of the plant, all of which require application to the TTB. Requirements are more stringent for the permit level corresponding to the highest production volume, which requires the posting of a bond and other requirements. Once granted, the permits require yearly reporting to the TTB (Alcohol and Tobacco Tax and Trade Bureau, 2015). General background on the TTB regulatory program and instructions for applying for permits can be found at the agency's website (Alcohol and Tobacco Tax and Trade Bureau, 2015).

3.5.4 US facility registration requirements

Several of the US laws mentioned above require that production facilities be registered with the government, in addition to any requirements for approval or regulatory review of the fuel or chemical itself. For example, the EPA fuel certification regulations (40 CFR Parts 79 and 80) require facilities to be registered, and manufacturing facilities for fuels qualified as renewable under the RFS must also be registered. The latter process is more complicated because of the need to ensure that the manufacturing process is in compliance with the regulations and is expected to yield the required reductions of greenhouse gas emissions. Facility registration under the RFS is a multi-step process and requires that an independent third party conduct an engineering review of the company's facility(ies).

3.5.5 US construction and operating permits

Finally, construction and operation of biorefineries will likely require permits or government approvals of various kinds. In the United States, these typically relate to environmental concerns like air emissions, groundwater discharges, and waste disposal, and in many cases these permits are administered by state or local governments rather than any federal agency. In addition, the permits that may be needed during construction may in some cases differ from the permits needed for the ongoing operation of the plant. Similar types of permits may be required in other countries as well, and may also feature a mix of national, regional, and local authorizations.

A comprehensive summary of these types of permits is beyond the scope of this chapter. There are some guidance documents available on the Internet, including two documents published in the 2000s by one of EPA's regional offices, summarizing the permit requirements that might apply for ethanol plants (U.S. EPA Region 7, 2007) or biodiesel plants (U.S. EPA Region 7, 2008). In addition, companies can often rely on their architects and consulting engineers for guidance with local permits for any given construction project.

3.6 Conclusion

In today's highly regulated world, virtually all commercial activity will be subject to government regulation of some kind. Bio-based products are no exception, in spite of the environmental and climate benefits that they are expected to bring. Chemicals, fuels, and other bio-based products will be subject to regulation in most countries in the world, as will the use of genetically modified microorganisms or plants in manufacture. However, in most cases, the applicable regulatory requirements will not be too burdensome, and what burden there may be might be counterbalanced by those government programs meant to encourage or provide economic benefit to bio-based or renewable products. In any event, with proper advance planning, the needed government approvals of bio-based products can be obtained in acceptable timeframes.

References

Alcohol and Tobacco Tax and Trade Bureau, 2015. Requirements for Establishing an Alcohol Fuel Plant (AFP). Retrieved from: https://www.ttb.gov/industrial/alcohol-fuel-plants.shtml.

Metabolic engineering host organism special issue. In: Alper, H. (Ed.), Metab. Eng. 50, 1–208.

Applegate, J.S., 2008. Synthesizing TSCA and REACH: practical principles for chemical regulation reform. Ecol. Law Q. 35 (4). https://doi.org/10.2139/ssrn.1183942.

Aro, E.-M., 2016. From first generation biofuels to advanced solar biofuels. Ambio 45 (1), 24–31. https://doi.org/10.1007/s13280-015-0730-0.

ASTM International, 2019a. Standard Specification for Automotive Spark-Ignition Engine Fuel. Retrieved from: https://www.astm.org/Standards/D4814.htm.

ASTM International, 2019b. Standard Specification for Aviation Turbine Fuel Containing Synthesized Hydrocarbons. Retrieved from: https://www.astm.org/Standards/D7566.htm.

ASTM International, 2019c. Standard Specification for Aviation Turbine Fuels. Retrieved from: https://www.astm.org/Standards/D1655.htm.

ASTM International, 2019d. Standard Specification for Butanol for Blending With Gasoline for Use as Automotive Spark-Ignition Engine Fuel. Retrieved from: https://www.astm.org/Standards/D7862.htm.

ASTM International, 2019e. Standard Specification for Denatured Fuel Ethanol for Blending With Gasolines for Use as Automotive Spark-Ignition Engine Fuel. Retrieved from: https://www.astm.org/Standards/D4806.htm.

ASTM International, 2019f. Standard Specification for Diesel Fuel Oils. Retrieved from: https://www.astm.org/Standards/D975.htm.

ASTM International, 2019g. Standard Specification for Jet B Wide-Cut Aviation Turbine Fuel. Retrieved from: https://www.astm.org/Standards/D6615.htm.

ASTM International, 2019h. Standard Test Methods for Determining the Biobased Content of Solid, Liquid, and Gaseous Samples Using Radiocarbon Analysis. Retrieved from: https://www.astm.org/Standards/D6866.htm.

Badhan, A., McAllister, T., 2016. Designer plants for biofuels: a review. Curr. Metab. 4 (1), 49–55.

Bergeson, L.L., Auer, C.M., 2016. An analysis of TSCA reform provisions pertinent to industrial biotechnology stakeholders. Ind. Biotechnol. 12 (4), 197–203. https://doi.org/10.1089/ind.2016.29042.llb.

Bergeson, L.L., Auer, C.M., Hernandez, O., 2014. Creative adaptation: enhancing oversight of synthetic biology under the toxic substances control Act. Ind. Biotechnol. 10 (5), 1–10. https://doi.org/10.1089/ind.2014.1532.

Bergeson, L.L., Auer, C.M., Peveler, R.D., 2012. TSCA and the regulation of renewable chemicals. Ind. Biotechnol. 8 (5), 262–272. https://doi.org/10.1089/ind.2012.1539.

Bergeson, L.L., Campbell, L.M., Rothenberg, L., 2000. TSCA and the future of chemical regulation. EPA Adm. Law Rep. 15 (4), 15–17.

Bergeson, Campbell, P.C., 2019. The Toxic Substances Control Act Frequently Asked Questions. Retrieved from: http://www.lawbc.com/other_pdfs/tsca-faq.pdf.

British Columbia Ministry of Energy Mines and Petroleum Resources, 2019. Renewable and Low Carbon Fuel Requirements Regulation Summary: 2010–2017. Information Bulletin RLCF-007-2017. Retrieved from: https://www2.gov.bc.ca/assets/gov/farming-natural-resources-and-industry/electricity-alternative-energy/transportation/renewable-low-carbon-fuels/rlcf007_-_2017_summary_2010-17v2.pdf.

Burkholder, D., 2015. A Preliminary Assessment of RIN Market Dynamics, RIN Prices, and Their Effects. Retrieved from: https://www.regulations.gov/document?D=EPA-HQ-OAR-2015-0111-0062.

Chemical Inspection and Regulation Service, 2012. New Chemical Substance Notification in EU – REACH Inquiry. Retrieved from: http://www.cirs-reach.com/REACH/REACH_Registration_Inquiry.html.

Chemical Inspection and Regulation Service, 2019. Revised K-REACH – the Act on the Registration and Evaluation of Chemicals. Retrieved from: http://www.cirs-reach.com/news-and-articles/revised-korea-reach—the-act-on-the-registration-and-evaluation-of-chemicals.html.

ChemSafetyPro, 2015. Japan Chemical Substance Control Law (Japan CSCL). Retrieved from: https://www.chemsafetypro.com/Topics/Japan/Japan_CSCL_Chemical_Substance_Control_Law.html.

ChemSafetyPro, 2018. China REACH (MEP order 7). Retrieved from: https://www.chemsafetypro.com/Topics/China/China_REACH_MEP_Order_7_New_Substance_Notification.html.

ChemSafetyPro, 2019. How to Comply with Amended K-REACH Regulation 2019. Retrieved from: https://www.chemsafetypro.com/Topics/Korea/Korea_REACH.html.

Chen, Y., Nielsen, J., 2016. Biobased organic acids production by metabolically engineered microorganisms. Curr. Opin. Biotechnol. 37, 165–172. https://doi.org/10.1016/J.COPBIO.2015.11.004.

Cho, C., Choi, S.Y., Luo, Z.W., Lee, S.Y., 2015. Recent advances in microbial production of fuels and chemicals using tools and strategies of systems metabolic engineering. Biotechnol. Adv. 33 (7), 1455–1466. https://doi.org/10.1016/J.BIOTECHADV.2014.11.006.

Congressional Research Service, 2019. Renewable Fuel Standard (RFS): Overview and Issues. Report No. R43325, Version 38, Updated January 23, 2019. Washington, DC.

Convention on Biological Diversity, 2019. About the Nagoya Protocol. Retrieved from: https://www.cbd.int/abs/about/.

Darch, H., Shahsavarani, A., 2012. The Regulation of Organisms Used in Agriculture Under the Canadian Environmental Protection Act, 1999. In: Wozniak CA, McHughen A, eds. Regulation of Agricultural Biotechnology: The United States and Canada: Springer Netherlands; 137–145.

de Farias Silva, C.E., Bertucco, A., 2016. Bioethanol from microalgae and cyanobacteria: a review and technological outlook. Process Biochem. 51 (11), 1833–1842. In: https://doi.org/10.1016/J.PROCBIO.2016.02.016.

Eggers, B., Mackenzie, R., 2000. The Cartagena protocol on biosafety. J. Int. Econ. Law 3, 525–543.

Enzing, C., Nooijen, A., Eggink, G., Springer, J., Wijffels, R., 2012. Algae and Genetic Modification. Technopolis. Retrieved from: https://library.wur.nl/WebQuery/wurpubs/428629.

European Chemicals Agency, 2018. Inquiry – ECHA. Retrieved from: https://echa.europa.eu/regulations/reach/registration/data-sharing/inquiry.

European Chemicals Agency, 2019. Evaluation Process. Retrieved from: https://echa.europa.eu/regulations/reach/evaluation/evaluation-procedure.

European Commission, 2019a. Fuel Quality. Retrieved from: https://ec.europa.eu/clima/policies/transport/fuel_en.

European Commission, 2019b. Renewable Energy – Recast to 2030 (RED II). Retrieved from: https://ec.europa.eu/jrc/en/jec/renewable-energy-recast-2030-red-ii.

European Commission, 2019c. Renwable Energy Directive. Retrieved from: https://ec.europa.eu/energy/en/topics/renewable-energy/renewable-energy-directive.

European Union, 2001. Directive 2001/18/EC of the European Parliament and of the Council of 12 March 2001 on the Deliberate Release Into the Environment of Genetically Modified Organisms. Retrieved from: http://eur-lex.europa.eu/legal-content/EN/TXT/?uri=CELEX:32001L0018.

European Union, 2006. Regulation (EC) No 1907/2006 of the European Parliament and of the Council of 18 December 2006. Retrieved from: https://eur-lex.europa.eu/eli/reg/2006/1907/2018-05-09.

European Union, 2009. Directive 2009/41/EC of the European Parliament and of the Council of 6 May 2009 on the Contained Use of Genetically Modified Micro-Organisms. Retrieved from: http://eur-lex.europa.eu/LexUriServ/LexUriServ.do?uri=OJ:L:2009:125:0075:0097:EN:PDF.

Glass, D.J., 1991. Impact of government regulation on commercial biotechnology. In: Ono, D. (Ed.), Business of Biotechnology. Newnes, Boston, pp. 169–198.

Glass, D.J., 2003. Regulation of the commercial uses of microorganisms. In: Encyclopedia of Environmental Microbiology. John Wiley & Sons, Inc. https://doi.org/10.1002/0471263397.env018.

Glass, D.J., 2015a. Government regulation of the uses of genetically modified algae and other microorganisms in biofuel and bio-based chemical production. In: Prokop, A. (Ed.), Algal Biorefineries: Volume 2: Products and Refinery Design. Springer International Publishing, pp. 23–60. https://doi.org/10.1007/978-3-319-20200-6_2.

Glass, D.J., 2015b. Pathways to obtain regulatory approvals for the use of genetically modified algae in biofuel or biobased chemical production. Ind. Biotechnol. 11 (2). https://doi.org/10.1089/ind.2015.1503

Glass, D.J., 2016a. Impact of TSCA reform on EPA regulation of industrial biotechnology. Ind. Biotechnol. 12 (4). https://doi.org/10.1089/ind.2016.29041.djg.

Glass, D.J., 2016b. Streamlining US regulatory reviews of modified microbial ingredients for use in animal feed. Ind. Biotechnol. 12 (5), 272–279. https://doi.org/10.1089/ind.2016.29049.djg.

Glystra, C., 2018. EU Parliament Approves Revised Renewable Energy Directive. Retrieved from: https://www.spglobal.com/platts/en/market-insights/latest-news/agriculture/111318-eu-parliament-approves-revised-renewable-energy-directive.

Gottron, F., 2019. Chemical Facility Anti-Terrorism Standards: Congressional Research Service in Focus Report. Retrieved from: https://fas.org/sgp/crs/terror/IF10853.pdf.

Government of Canada, 2018. New Substances Program of the Chemicals Management Plan. Retrieved from: https://www.canada.ca/en/health-canada/services/chemical-subikstances/chemicals-management-plan/initiatives/new-substances.html.

Ha, S., Seidle, T., Lim, K.-M., 2016. Act on the registration and evaluation of chemicals (K-reach). Environ. Health Toxicol. 31. Article ID: e2016026. Retrieved from: https://doi.org/10.5620/eht.e2016026.

Jarvis, D.S.L., Richmond, N., 2011. Regulation and governance of nanotechnology in China: regulatory challenges and effectiveness. Eur. J. Law Technol. 2 (33), 1–11. https://doi.org/10.1590/S0103-50532010000100005.

Keener, R.L., Jourdan, L.A., Weiler, E.D., 1999. Chemical control laws on polymers: a case for harmonization. Regul. Toxicol. Pharmacol. 29 (3), 319–326. https://doi.org/10.1006/rtph.1999.1297.

Krimsky, S., 1985. Genetic Alchemy : the Social History of the Recombinant DNA Controversy. The MIT Press, Cambridge, Mass.

Luit, R.J., Waaijers-van der Loop, S.L., Heugens, E.H.W., 2017. REACHing Out to the Bio-Based Economy: Perspectives and Challenges of EU Chemicals Legislation. Retrieved from: https://www.rivm.nl/bibliotheek/rapporten/2016-0178.pdf.

Mahwood, R., Gazis, E., de Jong, S., Hoefnagels, R., Slade, R., 2016. Production pathways for renewable jet fuel: a review of commercialization status and future prospects. Biofuels Bioproducts Biorefining 10, 462–484.

Markell, D., January 2010. An overview of TSCA , its history and key underlying assumptions, and its place in environmental regulation. Wash. Univ. J. Law Pol. 32, 333. https://doi.org/10.2139/ssrn.1616674.

McHughen, A., Smyth, S.J., 2012. Regulation of genetically modified crops in USA and Canada: American overview. In: Wozniak, C., McHughen, A. (Eds.), Regulation of Agricultural Biotechnology: The United States and Canada. Springer Netherlands, pp. 35–56.

Ministry of Economy Trade and Industry, 2016. What is CSCL (Chemical Substances Control Law). Retrieved from: http://www.meti.go.jp/policy/chemical_management/english/cscl/about.html.

Office of Science and Technology Policy, 1986. Coordinated framework for regulation of biotechnology. Fed. Regist. 51, 23302–23393.

Oregon Department of Environmental Quality, 2017. Overview of the Clean Fuels Program. Retrieved from: https://www.oregon.gov/deq/FilterDocs/cfpoverview.pdf.

Philp, J.C., Ritchie, R.J., Allan, J.E.M., 2013. Biobased chemicals: the convergence of green chemistry with industrial biotechnology. Trends Biotechnol. 31 (4), 219–222. https://doi.org/10.1016/j.tibtech.2012.12.007.

Portland Business Journal, 2018. Appeals Court Backs Oregon's Clean Fuels Program. Retrieved from: https://www.bizjournals.com/portland/news/2018/09/10/appeals-court-backs-oregons-clean-fuels-program.html.

Rosen, A.S., 2016. The Lautenberg Act : Chemical Safety Overhaul of the Toxic Substances Control Act. Law Lines, (Aug. 5, 2016). Retrieved from: http://digitalcommons.pace.edu/lawfaculty/1040/.

Rumizen, M., Edwards, T., 2017. Certification of alternative fuels. In: Nelson, E.S., Reddy, D.R. (Eds.), Green Aviation: Reduction of Environmental Impact Through Aircraft Technology and Alternative Fuels. CRC Press, pp. 295−306. https://doi.org/10.1201/b20287-12.

Salehi Jouzani, G., Sharafi, R., Soheilivand, S., 2018. Fueling the future; plant genetic engineering for sustainable biodiesel production. Biofuel Res. J. 5 (3), 829−845. https://doi.org/10.18331/BRJ2018.5.3.3.

Savakis, P., Hellingwerf, K.J., 2015. Engineering cyanobacteria for direct biofuel production from CO_2. Curr. Opin. Biotechnol. 33, 8−14. https://doi.org/10.1016/J.COPBIO.2014.09.007.

Smedley, K.O., 2013. Comparison of Regulatory Management of Authorized Ingredients, Approval Processes, and Risk-Assessment Procedures for Feed Ingredients. Retrieved from: https://ifif.org/wp-content/uploads/2018/06/IFIF-Comparison-Project.pdf.

Sperling, D., Murphy, C.W., 2018. How (Almost) Everyone Came to Love Low Carbon Fuels in California. Retrieved from: https://www.forbes.com/sites/danielsperling/2018/10/17/how-almost-everyone-came-to-love-low-carbon-fuels-in-california/#5515333f5e84.

Sticklen, M., 2006. Plant genetic engineering to improve biomass characteristics for biofuels. Curr. Opin. Biotechnol. 17 (3), 315−319. https://doi.org/10.1016/j.copbio.2006.05.003.

Syngenta, 2019. Enogen Corn for Ethanol. Retrieved from: http://www.syngenta-us.com/corn/enogen.

The Digest, 2019. Biofuel Mandates Around the World 2019. Retrieved from: https://www.biofuelsdigest.com/bdigest/2019/01/01/biofuels-mandates-around-the-world-2019/.

Torney, F., Moeller, L., Scarpa, A., Wang, K., 2007. Genetic engineering approaches to improve bioethanol production from maize. Curr. Opin. Biotechnol. 18 (3), 193−199. https://doi.org/10.1016/j.copbio.2007.03.006.

U.S. Air Force, 2017. Energy Flight Plan 2017−2036.

U.S. Congress, 2016. H.R. 2576. Retrieved from: https://www.congress.gov/114/bills/hr2576/BILLS-114hr2576enr.pdf.

U.S. Department of Agriculture, 2017. Understanding Biobased Content. Retrieved from: https://www.biopreferred.gov/BPResources/files/UnderstandingBiobasedContent_2017.pdf.

U.S. Department of Agriculture, 2019. Updating Biotechnology Regulations. Retrieved from. https://www.aphis.usda.gov/aphis/ourfocus/biotechnology/biotech-rule-revision.

U.S. Department of Agriculture, 2019. What is BioPreferred? Retrieved from: https://www.biopreferred.gov/BioPreferred/faces/pages/AboutBioPreferred.xhtml.

U.S. Energy Information Administration, 2018. EPA Finalizes Renewable Fuel Standard for 2019, Reflecting Cellulosic Biofuel Shortfalls. Retrieved from: https://www.eia.gov/todayinenergy/detail.php?id=37712.

U.S. Environmental Protection Agency, 2015a. Regulation History of the E15 Partial Waivers Under the Clean Air Act; Report No. EPA-420-F-15-044. Washington, DC.

U.S. Environmental Protection Agency, 2015b. U.S. Environmental Protection Agency Biotechnology Algae Project. Retrieved from: https://www.epa.gov/sites/production/files/2015-09/documents/biotechnology_algae_project.pdf.

U.S. Environmental Protection Agency, 2017a. Fuel Pathways Under Renewable Fuel Standard. Retrieved from: https://www.epa.gov/renewable-fuel-standard-program/fuel-pathways-under-renewable-fuel-standard.

U.S. Environmental Protection Agency, 2017b. Renewable Identification Numbers (RINs) Under the Renewable Fuel Standard Program. Retrieved from: https://www.epa.gov/renewable-fuel-standard-program/renewable-identification-numbers-rins-under-renewable-fuel-standard.

U.S. Environmental Protection Agency, 2018a. EPA's Review Process for New Chemicals. Retrieved from: https://www.epa.gov/reviewing-new-chemicals-under-toxic-substances-control-act-tsca/epas-review-process-new-chemicals.

U.S. Environmental Protection Agency, 2018b. How to Submit a Complete Petition for an Approved Pathway for Renewable Fuel Standards. Retrieved from: https://www.epa.gov/renewable-fuel-standard-program/how-submit-complete-petition-approved-pathway-renewable-fuel.

U.S. Environmental Protection Agency, 2018c. Registration of isobutanol as a gasoline additive: opportunity for public comment. Fed. Regist. 83 (61), 13460–13463.

U.S. Environmental Protection Agency, 2018d. TSCA Biotechnology Notifications Status for Cases Reviewed Prior to June 22, 2016. Retrieved from: https://www.epa.gov/regulation-biotechnology-under-tsca-and-fifra/tsca-biotechnology-notifications-status-cases-reviewed.

U.S. Environmental Protection Agency, 2019a. Approved Pathways for Renewable Fuel. Retrieved from: https://www.epa.gov/renewable-fuel-standard-program/approved-pathways-renewable-fuel.

U.S. Environmental Protection Agency, 2019b. EPA Proposes Rule to Allow E15 Waiver and to Improve RIN Market Transparency. Retrieved from: https://www.epa.gov/newsreleases/epa-proposes-rule-allow-e15-waiver-and-improve-rin-market-transparency.

U.S. Environmental Protection Agency, 2019c. EPA Releases First Major Update to Chemicals List in 40 Years. Retrieved from: https://www.epa.gov/newsreleases/epa-releases-first-major-update-chemicals-list-40-years.

U.S. Environmental Protection Agency, 2019d. Microbial Commercial Activity Notices (MCANs) Table. Retrieved from: https://www.epa.gov/reviewing-new-chemicals-under-toxic-substances-control-act-tsca/microbial-commercial-activity.

U.S. EPA Region 7, 2007. Environmental Laws Applicable to Construction and Operation of Ethanol Plants. Retrieved from: https://archive.epa.gov/ncea/biofuels/web/pdf/ethanol_plants_manual-2.pdf.

U.S. EPA Region 7, 2008. Environmental Laws Applicable to Construction and Operation of Biodiesel Production Facilities. Retrieved from: https://archive.epa.gov/ncea/biofuels/web/pdf/biodiesel_manual.pdf.

U.S. Federal Aviation Administration, 2016. Special Airworthiness Information Bulletin: Engine Fuel and Control – Semi-Synthetic Jet Fuel Report NE-11-56R2.

U.S. Government Accountability Office, 2015. DEFENSE ENERGY Observations on DOD ' s Investments in Alternative Fuels, (July 2015), pp. 1–38.

U.S. Grains Council, 2012. A Guide to Distiller's Dried grains with Solubles (DDGS). Retrieved from: https://www.canr.msu.edu/uploads/236/58572/cfans_asset_417244.pdf.

USDA Global Agricultural Information Network, 2018a. Biofuel Mandates in the EU by Member State in 2018. Retrieved from: https://gain.fas.usda.gov/Recent GAIN Publications/Biofuel Mandates in the EU by Member State in 2018_Berlin_EU-28_6-19-2018.pdf.

USDA Global Agricultural Information Network, 2018b. EU Biofuels Annual 2018. GAIN Report NL8027. Retrieved from: https://gain.fas.usda.gov/Recent GAIN Publications/Biofuels Annual_The Hague_EU-28_7-3-2018.pdf.

Voegele, E., 2019. Washington House Passes Bill to Establish Clean Fuels Program. Retrieved from: http://ethanolproducer.com/articles/16052/washington-house-passes-bill-to-establish-clean-fuels-program?utm_source=Ethanol&utm_campaign=12e2030eb8-EMAIL_CAMPAIGN_2019_03_19_05_07&utm_medium=email&utm_term=0_1b2531b996-12e2030eb8-95162381.

vom Berg, C., Dammer, L., Vos, J., Pfau, S., 2018. Roadmap for the Chemical Industry in Europe towards a Bioeconomy: Deliverable D2.1: Report on Regulatory Barriers.

Welker, M.C., Balasubramanian, K.V., Petti, C., Rai, M.K., DeBolt, S., Mendu, V., 2015. Engineering Plant Biomass Lignin Content and Composition for Biofuels and Bioproducts. Energies. https://doi.org/10.3390/en8087654.

Weng, J.-K., Li, X., Bonawitz, N.D., Chapple, C., 2008. Emerging strategies of lignin engineering and degradation for cellulosic biofuel production. Curr. Opin. Biotechnol. 19 (2), 166–172. https://doi.org/10.1016/j.copbio.2008.02.014.

Witcover, J., 2018. Status Review of California's Low Carbon Fuel Standard, 2011–2018 Q1 September 2018 Issue. Research Report UCD-ITS-RR-18-25.

Witt, I.E., Kirchof, C.E., 2008. Preparing for implementation of new international chemical regulations: REACH in EU and NSNR 2005 in Canada. In: SPE International Conference on Health, Safety, and Environment in Oil and Gas Exploration and Production. Society of Petroleum Engineers, Nice, France. https://doi.org/10.2118/111812-MS.

Wozniak, C.A., Wagonner, A.F., Reilly, S., 2012a. An introduction to agricultural biotechnology regulation in the U.S. In: Wozniak, C., McHughen, A. (Eds.), Regulation of Agricultural Biotechnology: The United States and Canada. Springer Netherlands, pp. 1–14.

Wozniak, C., McClung, G., Gagliardi, J., Segal, M., Matthews, K., 2012b. Regulation of genetically engineered microorganisms under FIFRA, FFDCA and TSCA. In: Wozniak, C., McHughen, A. (Eds.), Regulation of Agricultural Biotechnology: The United States and Canada. Springer Netherlands, pp. 57–94.

Wrubel, R., Krimsky, S., Anderson, M., 1997. Regulatory oversight of genetically engineered microorganisms: has regulation inhibited innovation? Environ. Manag. 21 (4), 571–586.

Chapter 4

Biofuels

Leticia Casas-Godoy[1,a], Iliana Barrera-Martínez[1,a], Neydeli Ayala-Mendivil[2], Oscar Aguilar-Juárez[2], Luis Arellano-García[1], Ana Laura Reyes[2,b], Andrés Méndez-Zamora[2], Georgina Sandoval[2]
[1]*CONACYT-CIATEJ, Guadalajara, Jalisco, México;* [2]*CIATEJ, Guadalajara, Jalisco, Mexico*

Abbreviations

ABE acetone-butanol-ethanol
ASBR anaerobic sequential batch reactor
ATJ alcohol to jet
C/N carbon/nitrogen ratio
CSTR continuous stirred tank reactor
EGSB expanded granular sludge bed
FAO Food and Agricultural Agency
GHG Greenhouse gases
HEFA Hydroprocessed esters and fatty acids
HMF hydroxymethylfurfural
HRT hydraulic retention time
HVLV high-value-low-volume
HVO hydrotreated vegetable oil
IEA International Energy Agency
IRENA International Renewable Energy Agency
LCA life cycle assessment
LVHV low-value-high-volume
OTJ oil to jet
PHB poly-3-hydroxybutyric acid
SPK synthetic paraffinic kerosene
TBR trickling bed reactor
UASB upflow anaerobic sludge blanket reactor
VFA volatile fatty acids

4.1 Introduction

4.1.1 Definitions, feedstocks, and applications

Nowadays, oil is the main source of fuels. Nevertheless, fossil fuels are nonrenewable and they cause environmental problems like global warming. Because of that, the biofuels should be an environmentally friendly option to substitute fossil ones. Biofuels can be defined as liquid fuels produced from

a. The author contributed equally.
b. Current affiliation: INIFAP, Tuxtla Chico, Chis. Mexico.

Biobased Products and Industries. https://doi.org/10.1016/B978-0-12-818493-6.00004-X

biomass for either transport or burning purposes. They can be produced from agricultural and forest products and the biodegradable portion of industrial and municipal waste. The use of biomass to generate energy reduces problems like pollution and greenhouse gas emissions. Plant oils were the original choice of Otto Diesel in his early engine, and Henry Ford preferred grain ethanol to power his early vehicles. Many years later, from the total renewable energy consumed in 2017, 50% came from modern bioenergy (excluding the traditional use of biomass), which provided four times the contribution of solar, photovoltaic, and wind energies combined. And in the forecast period 2018–23, bioenergy leads the growth in renewables consumption. Indeed, around 30% of the growth in renewables is expected to come from modern bioenergy in the form of solid, liquid, and gaseous biofuel (IEA, 2018).

Liquid biofuels described in this chapter include alcohols (ethanol and butanol), biodiesel (esters of fatty acids), renewable diesel, and aviation biofuels. The production and applications of gaseous biofuels (biomethane and hydrogen) and solid biofuels are also presented.

Fig. 4.1 shows bioenergy feedstocks and pathways. Feedstocks and production processes can be combined to obtain products appropriate for the conversion routes that meet the final use requirements, such as transportation or combustion. Certain products and production processes are feedstock-specific and Fig. 4.1 is merely illustrative. It does not imply that all feedstocks are suitable for meeting all final use requirements in an efficient and cost-effective way. Biofuels can meet all the final energy uses: transport, electricity, and heat generation, but also the coproduction of valuable

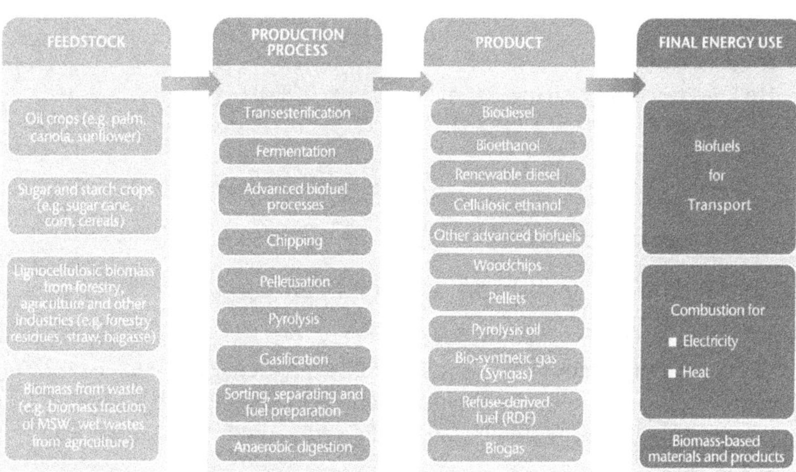

FIGURE 4.1 Bioenergy feedstocks and pathways. *Credit: From IEA-FAO, 2017. Bioenergy: roadmap develpment and implementation How2Guide for. Retrieved from: https://webstore.iea.org/how2guide-for-bioenergy.*

bio-based materials and chemicals could be achieved when biofuels are produced in biorefineries (see Section 3.5).

The International Energy Agency (IEA) considers advanced biofuels as "sustainable fuels produced from non-food crop feedstocks, which are capable of significantly reducing lifecycle GHG emissions compared with fossil fuel alternatives, and which do not directly compete with food and feed crops for agricultural land or cause adverse sustainability impacts" (IEA, 2018).

Some key advantages of advanced biofuels are the emission reductions of 80% compared to fossil-fuel-powered engines and reduction of emissions in heavy-freight, shipping, and aviation. But, to reverse pollution trends, the advanced biofuels production needs a strong and accelerated investment in advanced biofuel plants. Additionally, biofuels production must be substantially increased (IRENA, 2019a).

4.1.2 Targets and economical aspects

Biofuels implantation has been generally driven by targets and policies, frequently accompanied by incentives and subsidies. Table 4.1 presents

TABLE 4.1 Targets (T), mandates, and policies of selected countries[a].

Country	Ethanol and biodiesel	Policies
USA	136 billion L by 2022	Low Carbon Fuel Standard in California and Oregon
European Union	10% renewable energy in transport by 2020 (T) with 7% for conventional biofuels (% energy)	Provisional agreement for 14% renewable energy in transport in 2030
Brazil	27% ethanol and 10% biodiesel (%vol.)	RenovaBio signed into law, 10% GHG reduction by 2028 (T)
Finland	30% biofuel supply obligation by 2030 (% energy)	
Thailand	32% ethanol and 25% biodiesel (%vol.) by 2036 (T)	
China	10% ethanol (%vol.)	10% ethanol mandate to extend nationwide in 2020
India	5% ethanol	Biofuels policy expands approved feedstocks for ethanol production
Argentina	12% ethanol, 10% biodiesel (%vol.)	

[a]*Data from IEA, 2018. Renewables 2018: analysis and forecast to 2023 market report series. Retrieved from: https://webstore.iea.org/market-report-series-renewables-2018.*

targets, mandates, and policies of selected countries. Brazil maintains the highest mandate for ethanol and Thailand the highest target for both biodiesel and ethanol; whereas, Finland has a mandate of 30% biofuel by 2030. Europe also has ambitious targets. Growing economies such as China and India have mandates for 10% and 5% of ethanol, respectively. China currently produces about 3000 million liters of ethanol and about 1140 million liters of biodiesel per year (van Dyk et al., 2016). Although without mandates Mexico and Turkey do show signs of growing deployment (IEA, 2018).

Biofuels are an important part of the bioeconomy; they generate 3.045 billion jobs worldwide. From which 1931 thousand are from liquid biofuels, 779.5 thousand from biogas, and 334.3 thousand from solid biofuels (IRENA, 2019b). Regarding costs, although there is a high dependency on feedstocks prices, biofuels are becoming an attractive alternative to increasing fossil fuel prices. This feedstock price dependency applies to conventional biofuels. Advanced biofuels have the advantage of using residues and wastes as feedstocks. Prices vary between countries, but a good source to compare biofuels prices is the Alternative Fuels Data Center (AFDC, 2019). The global biofuel market was valued at USD 168 billion in 2016, and is expected to reach USD 218.7 billion in 2022 (ZMR, 2018).

4.2 Transportation biofuels

The liquid fuels more used are biodiesel and ethanol, but in the last years butanol and higher alcohols took relevance. Liquid biofuels, including both conventional and advanced forms of ethanol and biodiesel, could represent 10% of transport sector energy use by 2030, which is more than triple that in 2017 (IRENA, 2019a). Biofuels accounted for 92% of renewable energy in transport in 2017, due to their compatibility with existing internal combustion engine vehicles and fueling facilities (IEA, 2018).

4.2.1 Gasoline substitutes

Ethanol and butanol are considered in this section, as well as the feedstocks, technologies, and microorganisms to produce them. Bioethanol (C_2H_5OH) is the most studied and produced biofuel in the world (Trindade and Santos, 2017). The chemical characteristic (calorific value 26.83 MJ/kg and oxygen content of 33.63%) of bioethanol is ideal to use as a biofuel or a gasoline additive. The bioethanol production is environmentally friendly and easier than methanol production; the bioethanol fermentation yield is higher than butanol (10–30 times). The agroindustry and lignocellulosic residues, energy crops, and some edible crops are used to produce bioethanol. The technology to achieve higher yields of bioethanol is mature, but finding sustainable processes still requires a lot of research. The general process to get bioethanol is pretreatment (chemical, physical, biological, and combination of those) to remove lignin, hydrolysis (chemical or enzymatic) to produce fermentable sugars as hexoses and pentoses, and fermentation. As in all biotechnological process, there are separation and purification steps involved.

There are two general strategies to obtain bioethanol: separate hydrolysis and fermentation (SHF), simultaneous saccharification and fermentation (SSF) (Mejía-Barajas et al., 2018), and the combination of both called semi-simultaneous saccharification and fermentation (hydrolysis is carried out under optimized conditions followed by fermentation without removing the hydrolysate) (Weber et al., 2010; Zheng et al., 2009). One of the most used microorganisms to convert the sugars to bioethanol is *Saccharomyces cerevisiae* because it has an extraordinary bioethanol tolerance, nevertheless *S. cerevisiae* cannot ferment pentoses. To overcome this drawback, recombinant strains of *E. coli* (KO11) (Kim et al., 2008), *Saccharomyces cerevisiae* 1400 (pLNH33), and *Zymomonas mobilis* with the capacity to metabolize hexoses and pentoses have been produced. Moreover, Lee et al. (Lee et al., 2017a) reported *S. cerevisiae* with genes from *Chaetomium thermophilum, Chrysosporium lucknowense, Trichoderma reesei*, and *Saccharomycopsis fibuligera*, they produced 0.6–2.0 g/L of cellulases, then the cost of adding hydrolytic enzymes is saved.

Mono and cocultures of several bacteria such as *Thermoanaerobacterium saccharolyticum* (Raftery and Karim, 2017), *Thermoanaerobacter mathranii* (Paulova et al., 2015), *T. ethanolicus*, and *Geobacillus thermoglucosidasius* (Jiang et al., 2017) and *K. marxianus* (Sandoval-Nuñez et al., 2018) have been studied to produce advanced bioethanol.

The general composition of the biomass is lignin (15%–25%), cellulose (38%–50%), and hemicellulose (23%–32%). In advanced bioalcohol production process, the lignin must be removed to access the cellulose and hemicellulose, then a hydrolysis must be performed to break the polymers into cellulose and xylose (mainly). Then, the general process to generate fermentation sugars is (1) pretreatment (chemical, physical, biological, or a combination of them), and (2) hydrolysis (chemical or enzymatic).

The pretreatments (process to remove lignin) can be categorized in six processes: pyrolysis, hydrogenolysis, oxidation, gasification, combustion, and biological (Barrera et al., 2016). Nevertheless, there are some emerging technologies for the pretreatment, for example: ultrasound, microwave, gamma ray, electron beam, high hydrostatic pressure, high pressure homogenization, and pulse electric field (Hassan et al., 2018). The hydrolysis can be enzymatic (cellulases and hemicellulases) or chemical (using acids), but to have a cleaner process the enzymatic process is the most used, although the cost is higher. Another kind of advanced bioethanol uses algae as feedstock, but the yields are still very low in comparison with conventional and lignocellulosic bioethanol (El-Dalatony et al., 2017).

The feedstock was categorized to agroindustry residues, energy crops, and agriculture residues, the characteristics of which are shown in Fig. 4.1. The bioethanol produced from cellulosic feedstocks listed in Fig. 4.1 is considered second generation or advanced biofuel. Forecast for cellulosic ethanol is 0.84 billion liters by 2023 (IEA, 2018). Nevertheless, in the main ethanol producer countries (USA and Brazil) the ethanol is produced from edible crops (maize and sugar cane, respectively) and therefore the current bioethanol is a first generation or conventional biofuel the production of which has reached a plateau except some accelerate cases (Fig. 4.2).

Butanol is a linear alcohol with four carbons (C_4H_9OH) and needs less energy than ethanol to be produced. It can be obtained by petroleum or nonpetroleum sources. Butanol has an inferior latent heat of vaporization, superior cetane number, flash point, and higher caloric value compared with lower alcohols. The calorific value of butanol (29 MJ/L) is higher than calorific value of ethanol (19.6 MJ/L). Butanol can be additive to fuel and fuel source, causes minimal corrosive damage, has lower vapor pressure, and is more hydrophobic than ethanol (Bellido et al., 2014). Moreover, butanol could be blending with diesel due to its oxygen content (higher than biodiesel) (Trindade and Santos, 2017).

The acetone-butanol-ethanol (ABE) fermentation is the most popular process to produce butanol. The genus used is *Clostridium*, and now the researchers are looking for increasing the butanol yield. Because of scientific advances, butanol can be produced using genetically improved strains of clostridia species, *E. coli*, ionic liquids, gas stripping, super critical extraction, and pervaporation to increase the yield (Ezeji et al., 2007), although, the biological butanol production presents a low yield (10–30 times lower than ethanol fermentation process).

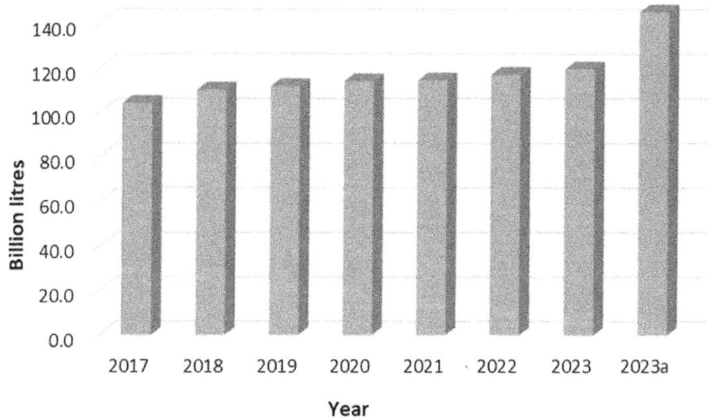

FIGURE 4.2 Conventional ethanol production forecast. 2023a-accelerated case to 2023 worldwide. An example of accelerated case is 10% nationwide ethanol blending fulfilled in China. *Credit: Based on data from IEA, 2018. Renewables 2018: analysis and forecast to 2023 market report series. Retrieved from: https://webstore.iea.org/market-report-series-renewables-2018.*

The ABE fermentation (García et al., 2011) was reported for the first time by Pasteur in 1861 and now there are several research teams working on that. The microorganisms more used to produce butanol are from the bacterial genus *Clostridium;* according to literature *C. acetobutylicum*, *C. beijerinckii*, and *C. saccharoperbutylacetonicum* (Berezina et al., 2009). The high toxicity of butanol against the microorganism is the main problem in the fermentation process, because it causes a self-inhibition. According with the metabolic pathway of clostridia species the main products of fermentation are: acetone, butanol, ethanol, lactic acid, acetic acid, butyric acid, carbon dioxide, and hydrogen (Xue, et al. 2013; Zheng et al., 2009). The reactions are as follows: glucose is transformed into organic acids (acetic and butyric acids); then as the pH goes down the acidic stress and the inhibition of the metabolic pathway occurs. Then the butyric acid is transformed into butanol (Kumar and Gayen, 2011). Generally, clostridia are very sensitive to the medium composition and fermentation conditions.

The main sources for conventional biobutanol production are hexose sugars from maize, wheat, rice, and cassava. Prior to using them, a hydrolysis must be performed to break starch or cellulose into simple sugars (as glucose). For example, Li et al. reported a butanol yield of 6.66 g/L using cassava starch and *C. beijerinckii* (Li et al., 2013) and Ezeji et al. achieved 26 g/L with maize meal using *C. beijerinckii* BA101(Ezeji et al., 2004).

The feedstock used to produce advanced butanol does not compete with edible crops and offers an intelligent way to use residues. Moreover, the low prices and abundance are some characteristics of the biomass that made their use convenient to produce biofuels.

Cellulose served as a carbon source using *C. acetobutylicum* genetically modified with genes from *C. cellulovorance* in order to degrade directly the cellulose (Jin et al., 2011). In some other studies barley silage liquor with gelatinized barley grain was used and butanol yield achieved was 0.20–0.17 g/g monosaccharide (Yang et al., 2014). *C. beijerinckii* TISTR was used to produce butanol from trunk fiber, the yield was 0.41 g/g (Komonkiat and Cheirsilp, 2013). Cai et al. (2013) used hydrolyzed sweet sorghum bagasse and the concentration of butanol was 12.3 g/L. Rice straw was the carbon source of *C. sporogenes*, and the concentration of butanol was 5.32 g/L (Gottumukkala et al., 2013). Using spoilt date palm fruit with a mixed culture of *C. acetobutylicum* ATCC 824 and *Bacillus subtilis* DMS 4451 the butanol yield was 0.42 g/g (Abd-Alla and Elsadek El-Enany, 2012). The butanol production with pineapple waste juice and *C. beijerinckii* was 3.14 g/L (Sanguanchaipaiwong and Leksawasdi, 2018).

Other feedstocks to produce advanced butanol are the algae residues remaining after oil and some other molecules extraction, which has been used as a carbon source. An acid hydrolysis of mixed microalgae was used as a carbon source by Castro et al. (Castro et al., 2015); the final butanol concentration was 3.74 g/L using *Clostridium saccharoperbutylacetonicum* N1-4.

After biodiesel production with microalgae, the residues were used with *C. acetobutylicum* and they produced 3.86 g/L of butanol (Cheng et al., 2015). These results show that algae can be used to produce butanol, but the yield has to be higher in order to make the process economically viable.

4.2.2 Diesel substitutes

Biodiesel is a type of alternative biofuel to blend with fossil diesel or to use instead of diesel (in some engines) that has a long future due to its physicochemical characteristics, which is considered to be less environmentally toxic, has high combustion efficiency, high cetane index, lower inflation point, better inflation point, among others (Knothe, 2009; Marchetti, 2012). From a technical point of view, to produce biodiesel, the method to choose depends on the composition, quality, availability of the raw material and inputs, as well as the economic ceiling that supports the production chain. In this sense, to produce 100% renewable biodiesel, strategies must consider new sources of biomass as raw material, biological resources such as biocatalysts and optimization of reaction conditions that allow high yields of conversion and production scaling. Therefore, in this chapter, conventional and emerging methods of biodiesel production will be addressed, as well as emerging diesel substitutes such as green diesel and hydrogenated vegetable oil (HVO) (Fig. 4.3).

Biodiesel is derived from fatty acids or from its acylglycerols through esterification or transesterification reactions with an alcohol (Fig. 4.4).

Vegetable oils are the main feedstock to produce biodiesel. Oleaginous plants are widely distributed throughout the world, and oils such as soybean (*Glycine max*), canola (*Brassica napus*), palm (*Elaeis guineensis*), and

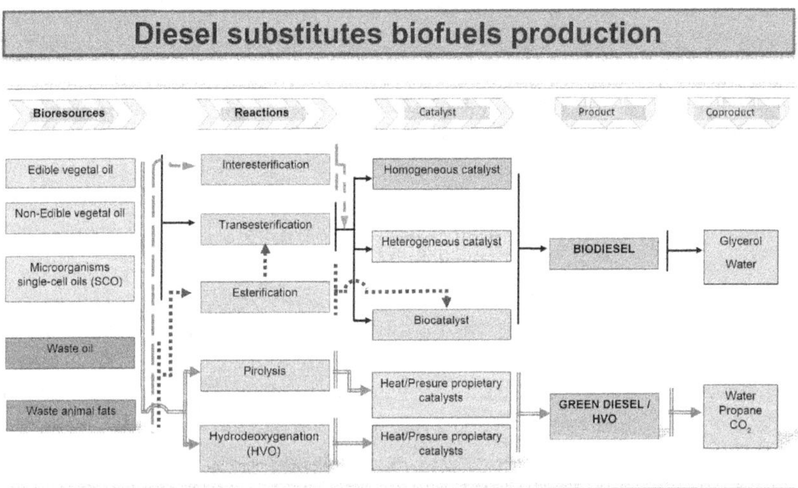

FIGURE 4.3 Diesel substitutes biofuels production. *Credit: This work.*

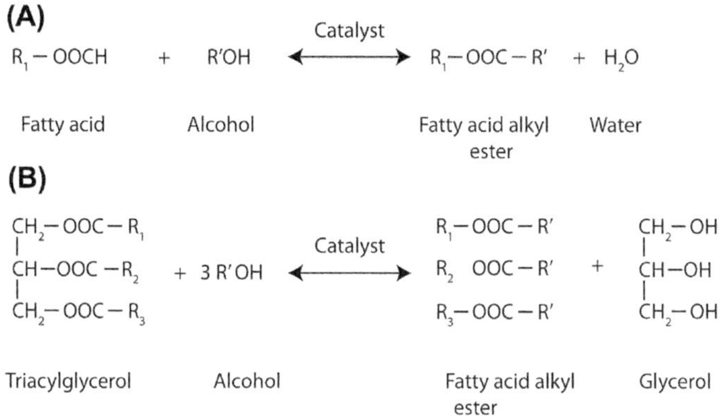

FIGURE 4.4 Esterification (A) and transesterification (B) reactions to produce biodiesel (fatty acid alkyl esters). *Credit: From Sandoval, G., Casas-Godoy, L., Bonet-Ragel, K., Rodrigues, J., Ferreira-Dias, S., Valero, F., 2017. Enzyme-catalyzed production of biodiesel as alternative to chemical-catalyzed processes: advantages and constraints. Curr. Biochem. Eng. 4, 109–141.*

sunflower (*Helianthus annuus*), among others, have been used to produce biodiesel (Singh and Singh, 2010). However, the use of such oleaginous plants is a major controversy as these crops are for human consumption. For this reason, some research focused on the use of nonedible vegetable oils derived from plants such as Mexican pinion (*Jatropha curcas*), moringa (*Moringa oleifera*), and beaver (*Ricinus communis*). These are crops widely adapted to different climatic regions of the world that present high contents of oils with desirable physicochemical characteristics (Rodrigues et al., 2016; Sudhanshu and Behera, 2019).

Waste fats and oils could also reduce biodiesel production costs and decrease GHG.

Emissions (Haas, 2005; IRENA, 2016). Besides vegetable and waste oils, microalgal oils were considered a possible feedstock for biodiesel. However, these algae productivity values were rather optimistic. Indeed, the high investment costs along with high energy demand for harvesting algae biomass have been some constraints on the industrial scale-up of a cost-effective biodiesel production from algal oils. Algae cultivation costs are also very high. For instance, only the costs of the salts used to obtain 1 kg of algal biodiesel are similar to the price of 1 kg of fossil diesel (Petkov et al., 2012).

It is in this way, to complement the other conventional sources, the study of oleaginous microorganisms represents a novel and sustainable option (Niehus et al., 2018a). Microbial lipids are a type of bioresource that take advantage of the timber agroresidues, using lignocellulosic biomass like carbon source during the culture of the microorganisms. In this way, unlike raw vegetable oils, they are not part of the human diet (Ayala-Mendivil Neydeli, 2018). Another advantage that represents the use of microbial lipids as raw material with respect to vegetable and residual oils is the purity and quality of the lipid.

Recently, with tools such as metabolic engineering and genetic modification, it was possible to overexpress native genes of the yeast *Yarrowia lipolytica* so that it could efficiently use xylose as a carbon source for the production of lipids. The results are promising, since lipid yields greater than 65% were achieved, using agave bagasse hydrolysates as substrate. With this work, it is demonstrated that it is possible to take advantage of technological tools and agroindustrial wastes (Niehus et al., 2018b).

Composition of oil as feedstock also impacts biodiesel quality. For instance, polyunsaturated oils are highly prone to oxidation which is reflected in a low stability of the obtained biodiesel. In biodiesel quality standards, the European (DIN EN 14214) and US legislations (ASTM D 6751), a minimum Oxidative Stability Index of 6 h/110°C or 3 h/110°C is required, respectively. The choice of raw material to produce biodiesel is also crucial for the choice of the type of reaction and catalyst to be used during the conversion of biofuel.

Regarding the alcohol used, methanol is the most commonly used alcohol. For this reason, biodiesel is currently referred as FAMEs (fatty acid methyl esters). However, ethanol has shown to be a real alternative to methanol, in some biodiesel standards (e.g., Brazilian biodiesel). When ethanol is used, biodiesel is referred as FAEEs (Fatty Acid Ethyl Esters). Concerning reactions to produce biodiesel, they proceed in the presence of a catalyst (see Fig. 4.3), basic catalysts being the most commonly used (generally, sodium or potassium hydroxide or methoxide), but when free fatty acids are present acid neutralization is needed, usually with hydrochloric acid or sulfuric acid, or more recently with lipases as biocatalysts. Reaction could also be uncatalyzed when supercritical alcohol is used.

Table 4.2 presents a comparison of reaction conditions to produce biodiesel using various catalysts. Only performances >96.5% of esters are present, as required by the EN 142414 biodiesel standard.

It can be observed that acid and heterogeneous catalysts and in particular biocatalysts require more reaction time, but they have other advantages such as noncorrosiveness, biodegradability, and recycling (for immobilized enzymes). Additional advantages are mild reaction conditions and that free fatty acids are also esterified and soap formation is avoided when lipases are used as biocatalysts (Sandoval et al., 2017; Vargas et al., 2018).

Reactor types for biodiesel production depends on the catalyst. Heterogeneous catalysts are usually used in agitated reactors, but ultrasonic, cavitation, fluidized-bed, packed-bed, and supercritical reactors have been explored, some of them also for biocatalysts (Sandoval et al., 2017). Table 4.3 shows some examples of the results of biocatalytic reactors in biodiesel production.

Supercritical methods to produce biodiesel have taken enough interest due to several benefits. One of the main benefits of this technology is its rapid reaction speed for the conversion of raw material to biodiesel, even in the absence of a catalyst. Likewise, the temperatures and operating pressures used for this technology allow to increase the solubility between the oil and alcohol

TABLE 4.2 Comparison between reaction conditions for various types of catalysts.

Source	Catalyst	Catalyst (wt%)	Alcohol	Temperature (°C)/Time (min)	Performance (%)	References
Homogeneous base catalyst						
Sunflower oil	NaOH	1	Methanol (6:1)	60/120	97.1	(Martınez et al., 2014)
Calophyllum oi	KOH	1	Methanol (9:1)	0.75/30	98.53	Silitonga et al. (2014)
Homogeneous acid catalyst						
Waste cooking oil	H_2SO_4	41.8	Methanol (245:1)	70/4140	99	(Zheng et al., 2006)
Sunflower oil	HCl	1.85	Methanol (85:1)	100/60	95.2	(Sagiroglu et al., 2011)
Heterogeneous base catalyst						
Jatropha oil	Li/CaO	5	Methanol (6:1)	65/120	99	Leung and Guo (2006)
Canola oil	MgO	3	Methanol (20:3)	190/120	98.2	Helwani et al. (2009)
Soybean oil	Na_2ZrO_3	3	Methanol (3:1)	65/180	98.3	Santiago-Torres et al. (2014)

Continued

TABLE 4.2 Comparison between reaction conditions for various types of catalysts.—cont'd

Source	Catalyst	Catalyst (wt%)	Alcohol	Temperature (°C)/Time (min)	Performance (%)	References
Heterogeneous acid catalyst						
Oleic acid	Chlorosulfonic zirconia	3	Methanol (8:1)[a]	100/720	100	Zhang et al. (2014)
Polanga oil	Sulfonated carbon	7.5	Methanol (30:1)	180/300	99	Abbaszaadeh et al. (2012)
Soybean oil	Ce/HUSY zeolite	0.001 mol	Ethanol (30:1)	200/1440	99.8	Silitonga et al. (2014)
Biocatalyst						
Jatropha oil	Lipase *Pseudomonas cepacia*	1	Ethanol (4:1)	50/480	98	Shweta Shah (2007)
Waste activated bleaching earth	Lipase *Candida cylindracea*	10	Methanol (4:1)	37/180	97	Kojima et al. (2004)
Safflower oil	Lipase *Pseudomona fluorescens* (immobilized)	0.1	1-Propanol (3:1)	50/300	100	Iso et al. (2001)

TABLE 4.3 Examples of processes for the lipase-catalyzed production of biodiesel in different bioreactors[a].

Reactor type	Alcohol/ Ester	Conversion/ Reutilization	Reference
Stirred	Methanol	98.0%/24 h	Park et al. (2008)
Stirred	Methyl acetate	99.8%/5 cycles	Ognjanovic et al. (2009)
Packed-bed reactor	Methanol	99.0%/20 cycles	Hama et al. (2011)
Two-stage packed-bed reactor	Ethanol	97.3%/$t_{1/2}$=1540 h	Costa e Silva et al. (2016)
Fluidized-bed reactor	Ethanol	98.1%/15 days	Fidalgo et al. (2016)

[a]*Adapted from Sandoval, G., Casas-Godoy, L., Bonet-Ragel, K., Rodrigues, J., Ferreira-Dias, S., Valero, F., 2017. Enzyme-catalyzed production of biodiesel as alternative to chemical-catalyzed processes: advantages and constraints. Curr. Biochem. Eng. 4, 109–141.*

phases; in this way, the concerns related to the limitation of mass transfer are discarded. In addition, the elimination of the catalytic system in the reaction process makes the separation and purification of the product relatively simple and cost-effective. However, the synthesis of biodiesel using current technologies requires severe reaction conditions to achieve a high yield of esters and a high-quality biodiesel. Therefore, high operative energy consumption conditions would not only increase the cost of the process, but also damage the performance of biodiesel due to the phenomenon of thermal decomposition. To minimize the reaction temperature necessary for the supercritical alcoholysis reactions, the addition of gaseous and liquid cosolvents is made spindle. Likewise, it is attributed that the addition of gaseous carbon dioxide as a cosolvent in a reaction mixture could improve the solubility of alcohol and oil. Likewise, it will facilitate the separation processes of the products (Avhad and Marchetti, 2015). Santana et al. (Santana et al., 2012) report on the production of biodiesel through the catalytic supercritical transesterification reaction; verifying that the use of a catalyst in this technology helps to reduce the high reaction conditions and, therefore, avoid the thermal degradation of the reactants and products.

Hydrotreated vegetable oils (HVO) are also known as "renewable diesel." Chemically HVO are mixtures of paraffinic hydrocarbons and are free of sulfur and aromatics. In the production process of HVO, crude vegetable oils, waste oils, and fats can be used as raw material. Cold properties of HVO can be adjusted to meet the local normative requirements by adjusting the severity of the process or by additional catalytic processing. Cetane number of HVO is very high, and other properties are very similar to the biomass-to-liquid or green diesel fuels (Aatola et al., 2008). It is forecasted that biodiesel and HVO output could increase 15.5 billion L, with Brazil India and Indonesia the key contributors (IEA, 2018).

The quality parameters of this type of biofuel are governed by the standard (EN 590, ASTM D 975). For the hydrotreatment of vegetable oil, different catalysts are used, such as: NiMo/γ-Al_2O_3 or CoMo/γ-Al_2O_3. Hydrodeoxygenation will generally produce saturated long chain alkanes from which the remainder of the ester group has been removed. Hydrodeoxygenation is a process by which a raw material containing double bonds and oxygen remains is converted into hydrocarbons by saturation of the double bonds and the elimination of oxygen (decarboxylation, decarbonylation, and dehydration). Likewise, the reaction requires presence of hydrogen (Maier et al., 1982).

The decarboxylation of inactivated carboxylic acids, although exothermic, requires high transition state energies (Maier et al., 1982), which are reflected in reaction conditions such as elevated temperature and pressure. Water formed during hydrotreating can decrease activity in the reaction (Huber and Corma, 2007; Senol et al., 2005). Reaction metallurgy hinders the processing of vegetable oils with high acidity in standard hydrotreating reactors (Huber and Corma, 2007). Therefore, it seems that some of the same quality restrictions that affect the production of biodiesel by transesterification also influence the production of renewable diesel by hydrodeoxygenation.

The catalysts may have different activities in the structure of the raw material (Senol et al., 2007). The acidity of the support influences the formation of active sites for the discharge and hydrogenation of carboxyl groups (Centeno et al., 1995). Regarding emissions produced by HVO, a recent report shows that engine injection control can influence the formation of NOx and particulate matter significantly. Only when optimizing the fuel injection, reductions in NOx emissions and particulate matter were achieved with HVO compared to fossil diesel (Bohl et al., 2018).

Pyrolysis is a process of splitting oil molecules and rapidly forming a mixture of hydrocarbons with properties similar to those of diesel fuel (Santos et al., 2010). The thermal cracking process (involving a decarboxylation reaction) is carried out at elevated temperatures, either in the presence or absence of a catalyst. The thermal cracking of vegetable oils or animal fats takes place in two successive and distinct stages: the first stage involves the formation of an acidic species through the breakdown of triglyceride molecules (TAGs) in which the breakdown of the C—O bonds take place within the glycerides, while the second stage is characterized by the degradation of free fatty acids (FFA) produced in the first stage, leading to the formation of hydrocarbons called "pyrolysis oils" or sometimes "green diesel." A schematic layout of the pyrolysis process is presented in Fig. 4.5. An advantage of pyrolysis oils is that they can produce less contaminants than fossil diesel, but a drawback is that the process requires high temperature (700—900°C) (Kalargaris et al., 2017).

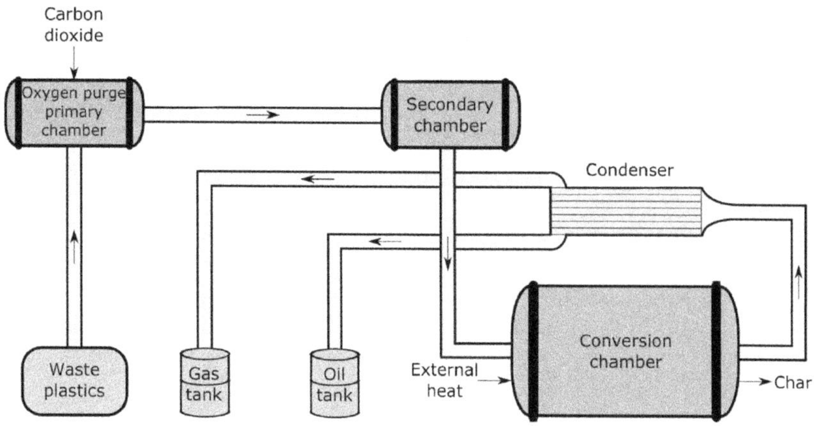

FIGURE 4.5 Schematic layout of the pyrolysis process. *Credit: From Kalargaris, I., Tian, G., Gu, S., 2017. Experimental evaluation of a diesel engine fuelled by pyrolysis oils produced from low-density polyethylene and ethylene–vinyl acetate plastics. Fuel Process. Technol. 161, 125–131. https://doi.org/10.1016/j.fuproc.2017.03.014 under the terms of the Creative Commons Attribution License (CC BY).*

4.2.3 Aviation biofuels

There is a need of sustainable aviation fuel, but production remains low, as accessing sufficient low-carbon feedstocks and pursuing technological development remain the main challenges to achieve the necessary production volumes. Regarding costs, the aviation biofuels are 1.5–3 times more expensive than fossil jet kerosene. Therefore, stronger policies and incentives are required to facilitate its implementation and development. With enhanced policies and mobilization of fuel supply, aviation biofuels could meet 2% of international aviation fuel demand in the medium term (IEA, 2018). In the right way, the International Civil Aviation Organization has committed to carbon-neutral growth from 2020 and to reducing CO_2 emissions by 50% from the 2005 level by 2050. But it is a big challenge, considering that the aviation activity is increasing every year (5 billion passengers by 2023). This will increase the demand for aviation fuels to 403 billion L by 2023 (IEA, 2018), with alternative jet fuels market valued at 246.52 USD billion by 2024 (RSB, 2019). The long-term decarbonization goals of the International Civil Aviation Organization could be met by blending sustainable aviation fuel with fossil jet kerosene, which implies that sustainable aviation fuel must be "drop-in" liquid hydro carbons with the following characteristics: (1) suitable for aircraft use without technical modifications, (2) comply fully with the same jet fuel standards, (3) demonstrate lifecycle GHG emission reductions compared with fossil jet fuel, (4) meet certification sustainability criteria such as RSB (RSB, 2019).

TABLE 4.4 Aviation biofuel production pathways.

Pathway	Feedstocks	Technology[a]
Oil to jet (OTJ)	Fatty acid and esters	HEFA
	Triglyceride-rich oils	Catalytic hydrothermolysis
	Wastes and lignocellulloses	Hydrogenated pyrolysis
Alcohol to jet (ATJ)	Alcohols derived from fermentation	ATJ
Gas to jet (GTJ)	Agricultural wastes	Fisher-Tropsch synthesis
Sugar to jet (STJ)	Sugars	Direct sugar to hydrocarbons
	Sugars or lignocellulosic hydrolysates	Catalytic upgrading

[a]*Technologies described in (Wang and Tao, 2016).*

Aviation biofuel production pathways are presented in Table 4.4. However, only biofuel derived from hydroprocessed esters and fatty acids to produce synthetic paraffinic kerosene (HEFA-SPK) is currently technically mature and commercialized, and therefore the forecast as the principal biofuel used by 2013 (IEA, 2018).

Hydrotreatment of oils produce high cetane number and straight chain alkanes ranging from C9 to C18 that can be used in the aviation industry (Liu et al., 2013b). The hydrotreating process uses high temperatures and pressures (350–450°C, 40–150 atm), and sulfided $NiMo/Al_2O_3$ catalysts. To isomerize the alkanes, other catalysts such as molecular sieves or zeolites must be used (Liu et al., 2013b).

Private companies have developed proprietary technologies. Honeywell UOP developed the technology HEFA-SPK, Bio-based Synthetic Paraffinic Kerosene (Bio-SPK), or Green Jet Fuel, from hydroprocessed esters and fatty acids. UOP's process obtained aromatics from pyrolysis oil by Rapid Thermal Processing and the resulting fuels meet the key freeze point, flash point, and density specifications of ASTM D7566. The pyrolysis oil was mainly generated from two types of cellulosic biomass widely available in USA (corn stover and woody materials). In Rapid Thermal Processing, biomass is rapidly heated in the absence of oxygen, vaporized, and then rapidly cooled to generate high yields of pyrolysis oil. Dry woody materials yield 65–75 wt% pyrolysis oils. As UOP has experience in petroleum industry, a similar fluid catalytic cracking circulating transported fluidized bed reactor system is used. Heterogeneous catalysts such as sulfided CoMo and NiMo-based catalysts are used at moderate temperature with high pressure H_2. Valuable coproducts as green diesel can be obtained also in this process.

Another technology that uses fatty acids or fatty acid methyl esters was developed by the Tianjin University (registered as C–L process). It leads to C5–C15 jet biofuel. In this technology the main steps are: (1) the feedstock (fatty acids or fatty acid methyl esters) is hydrotreated to eliminate double bonds. (2) Through Kolbe reaction, the products from the hydrotreating process are converted into long chain hydrocarbons. (3) The hydrocracking process is used to adjust chain length of hydrocarbons into desired jet fuels. The C-L process could lower H_2 pressure if required and less hydrogen is consumed because hydrogen is a byproduct of the Kolbe electrosynthesis (Liu et al., 2013b).

A third technology of OTJ is biosynfining SPK. In this process, fatty acid chains are converted into n-paraffins via exothermic hydrogenation and deoxygenation reactions in a hydrotreater. As for previously described OTJ technologies, in the last step of the process, chain-length of paraffins is adjusted by are hydrocracked into shorter branched paraffins. The hydrocracked products have C15–C18n paraffin compositions.

Regarding emission of sustainable aviation fuel, they can also be beneficial by reducing air pollutant emissions, such as sulfur dioxide and particulate matter, but a recent study concluded that more studies are needed to evaluate the performances of new bio-based jet fuels such as synthesized iso-paraffins and ATJ (Yang et al., 2019).

4.3 Gaseous biofuels

4.3.1 Biomethane

Biomethane is defined as methane produced from biomass (ISO DIS 15669, in preparation), with properties close to natural gas. It can be produced by thermochemical conversion (see bio-SNG) or biochemical conversion (see below biomethane from biogas upgrading) (Thrän et al., 2014). Biomethane as a biofuel has wide application potential; like any source of bioenergy, it has advantages and disadvantages. Biomethane has a wide range of applications since it is a versatile reservoir of energy that can be used for thermal, transport, and electricity production purposes. For this biogas (biologically produced gas equivalent to natural gas), there are many industrial applications such as domestic boilers and services in general. Nowadays, the use of biogas as fuel in the transport of cargo, public and private, is becoming popular. Although biomethane is more commonly used for heat and power generation (see Section 3.4), but consumption in transport can be expected to increase in many countries due to subsidies and targets (IEA, 2018).

As in most bioenergy sources, the conversion of solar radiation and several inert elements that synthesize organic matter result in a concentration of energy, which mainly through photosynthesis concentrates this energy in carbohydrates, lipids, and proteins. This accumulated energy is integrated into economic and productive systems through the production of food for humans and animals; as a raw material in production processes, and as a source of thermal energy. In this way, the main raw materials that will be used as a source of biomethane are mainly the waste from productive activities, such as agroindustrial waste related to the processing of both human and animal food; the metabolic waste of animals that produce meat, milk, and eggs (cattle, pigs, horses, sheep, birds, etc.), the organic fraction of urban solid waste, but also organic waste from fruit trees, forestry, etc.

At this time, the thermodynamic alternatives for use are based on the energy content of these substrates restricted by the moisture content. Dry organic waste has greater feasibility for use in combustion processes and waste with high moisture content in fermentation processes and biological digestion. The process of biodegradation of organic matter to produce biomethane is called anaerobic digestion, which involves a series of syntrophic microbiological processes (Sieber et al., 2010).

The flow of energy that occurs in any catabolic process involves the use of complex molecules that through hydrolytic processes, usually performed by exoenzymes and fermentative bacteria, transform complex molecules (polymers) into simple molecules (monomers). In fact, the hydrolytic processes are fundamental to expand and improve the use of organic matter as bioenergy to produce biodiesel, bioethanol, biohydrogen, and biomethane (Fig. 4.6).

After hydrolysis, different microorganisms will transform monomers into volatile fatty acids, in the process known as *acidogenesis*. Subsequently, the different volatile fatty acids as well as the interaction with the medium and low

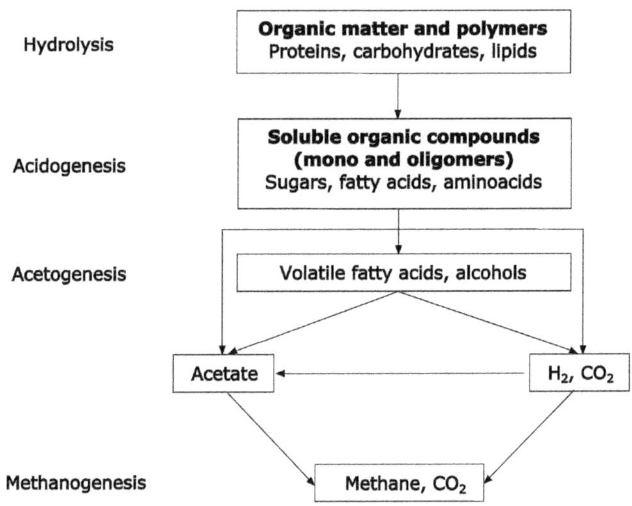

FIGURE 4.6 Simplified biochemical reaction diagram for anaerobic degradation of organic matter. *Credit: This work.*

factors resulting from this biological activity will constitute a critical stage in the production of biomethane, since there are reactions that require energy to be carried out, in addition to depending on the a balance between the concentration of easily hydrolysable substrates, the buffer capacity of the medium, and the establishment of viable bacterial consortiums, among other factors such as the partial pressure of the hydrogen present in the medium, which is a byproduct of the fermentation process; there are ample risks of accumulation of volatile fatty acids and the consequent decrease of pH to values close to 4, there is a risk that significantly affects methane production. If the environmental and ecological conditions are favorable, the metabolic routes will produce volatile fatty acids of low molecular weight that converge in acetate, in a process known as *acetogenesis*.

At this point, and depending on the heterogeneity of the medium, there are different alternatives for the conversion of products, such as the production of acetic acid from carbon dioxide and hydrogen and vice versa, depending on the conditions of the medium, made by microorganisms homoacetogens (Diekert and Wohlfarth, 1994). The following kinetics can be carried out with different types of archaebacteria, where hydrogenotrophic archaebacteria will thermodynamically predominate briefly if there is hydrogen in the medium, since these microorganisms generally produce 30%–50% of the methane, while other archaebacteria called acetoclastic archaebacteria with lower growth capacity but more stable will generate from acetate methane and carbon dioxide (Kotsyurbenko et al., 2004). It is important to note that sulfur or nitrogen contents will favor the development of microorganisms that will compete for substrates and potentially affect the conversion performance to methane, or undesirable compounds will be generated for the use of biogas such as hydrogen sulfide, processes that take place due to sulfate-reducing bacteria (Harada et al., 1994) or nitrate-reducing bacteria (Scholten and Stams, 1995).

In the anaerobic digestion of organic waste, there are several critical points that must be considered to maximize the methane production. In macroscopic terms, some basic considerations in the use of organic waste to produce biomethane are its availability, accessibility, biodegradability, moisture content, among other factors.

In intrinsic terms of the process and the residue to be treated, there are several factors to consider such as the quality of the substrates (due to the ease of hydrolysis thereof), the C/N ratio, the pH, the acidity of the medium, the redox potential, the temperature, the presence of inhibitors or inhibition by substrates/products, as well as bacterial competition, among other factors (Khalid et al., 2011).

In terms of waste availability, the seasonality waste production should be considered; with the understanding that a continuous production of biogas is desirable, it will not necessarily be guaranteed if the substrate to be used come from agricultural activities and therefore depends on production stations, which will force an intermittent bioenergy production that could restrict its exploitation viability; such is the case of seasonal fruits and vegetables agroindustrial waste. In the livestock waste case, availability will not necessarily be a restriction, but accessibility, will. An important aspect to consider in the energy balance is the maximum substrate concentration at a place with minimum transport costs. If this is possible the process will be profitable, being this is mainly a scale factor. In the case of landfills, the size of the site to be managed is determined by the size of the population and the degree of development of the community, since these factors will influence the final methane production at the site rather than the availability and accessibility conditions. Different governments have implemented various laws and regulations to enhance waste management. A waste hierarchy involving recycling and reuse has been proposed, but improvements on treatment technologies are still indispensable. At present, only a limited number of technologies are widely applied. Landfill is most commonly used and accounts for approximately 95% of the total collected MSW worldwide (Dong et al., 2014; Johari et al., 2014; Lee et al., 2017b; Malinauskaite et al., 2017).

The biodegradability of the waste as well as the moisture content in macroscopic terms are aspects that directly influence the profitability and valorization of these residues, since the biodegradable fractions determine the yield potential of biomethane and are basically associated with the content of inorganic material and hardly biodegradable fractions associated with lignin. The higher the lignin content, the lower the potential biogas yield and the higher the content of easily biodegradable fractions, the greater the control requirements will be necessary. The moisture content is also an important factor in macroscopic terms since the volumetric biogas yield may be affected to the downside by the simple fact of the availability of biodegradable biomass per unit volume of the reactor or the site will be diminished without that. However, it will be a parameter to be considered both in sanitary landfills and in agroindustrial, urban, and livestock wastewater.

In addition to the biodegradability of the above mentioned substrate, it is also necessary to contemplate the availability of nutrients in the medium that can be completed by codigestion of different substrates in order to ensure the necessary trace elements as well as the proper carbon/nitrogen ratio (C/N).

Different operating conditions influence the yield of biogas to be produced either in a sanitary landfill (Zamorano et al., 2007), in a biodigester (Solano et al., 2016), or in a bioreactor (Lay et al., 1997). One of the key parameters is the pH, which as noted, a significant variation of which can inhibit the production of biogas, where the greatest risk occurs before the acidification of the environment because of the accumulation of volatile fatty acids. Some measures that have been implemented, for example, in sanitary landfills, are the control of the compatibility of the residues deposited on the site; or in the case of water treatment plants, by means of the operation of bioreactors in two stages, one of hydrolysis—acetogenesis and the second of methanogenesis. In more specific terms related to a certain extent with pH are the acidity of the medium and the redox potential; parameters that will indicate the robustness of the system and the effectiveness of the process, which as noted, archaebacteria are sensitive to the ionization state of acetate to produce methane. Temperature is also an important factor that influences the speed of production and potential yield of biogas; however, it will be for some processes a performance factor as in the case of landfills and biodigesters, but also a control parameter in bioreactors.

Finally, some of the factors that influence the performance of biomethane are the inhibitors that can be toxic substances present in the environment, since when dealing with waste, the disposal or dumping of toxic or reactive substances can modify the performance of the system; the substrates or intermediate products themselves can affect the production yield of biogas. Therefore, it is necessary to identify operational factors that will maximize the biogas yield such as particle size, density, cell filling speed, moisture content, among other factors in landfills and parameters such as the time of hydraulic and cellular residence, substrate concentration, codigestion of products, type of reactors, etc., in the treatment of domestic and agroindustrial wastewater.

4.3.2 Biohydrogen

Although to date most of molecular hydrogen is obtained chemically through methane reformation with water steam, hydrogen can also be produced biologically based on the microbial transformation of carbohydrate rich substances, including waste and waste treatment effluents.

Hydrogen is an interesting energy carrier due to its energetic density in mass (122 kJ/g), which is higher than gasoline (45 kJ/g) or ethanol (37 kJ/g). Nonetheless, other factors such as net efficiency transferred to, for example, kinetic energy on the wheels in cars plays a major role when calculating delivered energy yield. Around 80%–90% of produced hydrogen worldwide is used in oil refineries, the rest is consumed by other processes such ammonia, methanol, and other chemicals manufacture. Also, metal, food, electronics, and glass making industries use hydrogen as feedstock. Future demand is expected to increase exponentially since heavier oil and tar sands would need more hydrogen to attain more stringent regulations on sulfur concentration in gasoline and diesel (Levin and Chahine, 2010).

Biohydrogen might be produced through direct and indirect biophotolysis of water by cyanobacteria and algae. Hydrogen is generated also through biological degradation of organic compounds such as volatile fatty acids (VFA) in presence of light by photosynthetic and nonphotosynthetic organisms in photofermentation. Also, hydrogen can be produced through light-independent pathways such as dark (or heterotrophic) fermentation and the microbial-mediated electrolysis of VFA. Dark fermentation is especially interesting among other biological methods due to hydrogen yields, besides not requiring the presence of light and the wide variety of carbohydrate-containing substrates including waste and waste treatment effluents that can be used.

In dark fermentation of glucose H_2-producing bacteria transform glucose to pyruvate through glycolytic pathways as initial stage. Pyruvate is then oxidized to acetyl coenzyme A (acetyl-CoA), carbon dioxide (CO_2), and H_2 by pyruvate ferredoxin oxidoreductase and hydrogenase. An alternate route for pyruvate is the conversion to acetyl-CoA and formate, which then convert into H_2 and CO_2. On the contrary, dark fermentation of a mixture of carbohydrates with a mixed microbial community might lead to a wide variety of metabolites and intermediates depending on conditions such as pH, temperature, and other operation parameters (Ghimire et al., 2015; Jung et al., 2011a).

In general, from pure glucose hydrogen yields oscillate between 2 and 4 mol H_2/mol hexose as shown in Eqs. (4.1) and (4.2).

$$C_6H_{12}O_6 + 2H_2O \rightarrow 2\ CH_3COOH + 2CO_2 + 4H_2 \tag{4.1}$$

$$C_6H_{12}O_6 + 2H_2O \rightarrow CH_3CH_2CH_2COOH + 2CO_2 + 4H_2 \tag{4.2}$$

Carbohydrates such as glucose, sucrose, or fructose have been shown to be the best substrates for hydrogen production by dark fermentation; however, its cost is higher compared to carbohydrates from waste. Besides they bear the ethical discussion about producing energy from food. To date hydrogen yields from pure carbohydrates in dark fermentation range from 0.75 to 2.80 mol H_2/mol-hexose when utilizing glucose, fructose, or sucrose (Jung et al., 2011a). In comparison, highest yields on mixtures on waste-originated carbohydrates range from 0.7 to 2.4 mol H_2/mol-hexose (Table 4.5).

Hydrogen production through dark fermentation of organic wastes, including residual lignocellulosic biomass, is recognized as an environmentally friendly, cost-effective sustainable process for energy production along with waste and biomass valorization. Hydrogen has been produced in continuous reactors from hydrolysates obtained from lignocelullosic materials after hydrothermal, acid, enzymatic, bacterial, and other pretreatments (Table 4.5). Hydrogen yields from these lignocellulosic-based hydrolysates up to 2.38 mol H_2/mol hexose have been obtained from materials such as wheat, rice and oat straw, wheat starch and agave bagasse, and grass with hydraulic retention times (HRT) between 4 and 72 h (Nissilä et al., 2014).

Direct utilization of waste, with no previous pretreatment ,as substrates for dark fermentation, such as molasses and brewery wastewater, and starch manufacturing have been utilized to produce hydrogen (Table 4.6).

Gaseous hydrogen as it is produced in biological processes might represent up to 70% of total composition, balanced with CO_2, water vapor, and hydrogen sulfide traces. Despite operation costs, hydrogen production by dark fermentation might be competitive with chemically produced hydrogen when process is scaled up, besides utilizing biomass that fixated atmospheric CO_2 as feedstock. Additional biological processes such as algal reactors are being applied to

TABLE 4.5 Hydrogen yields from hydrolysates in continuous mode bioreactors.[a]

Pretreatment method	Substrate	Reactor type	HRT (h)	Yield (mol H_2/mol hexose)	Reference
Hydrothermal	Wheat straw	CSTR	72	1.43	Kongjan et al. (2010)
	Wheat straw	UASB	24	1.59	Kongjan et al. (2010)
Diluted acid	Ground wheat starch	nr	24	0.97	Sagnak et al. (2010)
	Oat straw	TBR	12	2.00	Arriaga et al. (2011)
Concentrated acid	Rice straw	CSTR	4	0.69	(Liu et al., 2013a)
Bacterial	Starch	CSTR	12	2.38	(Chen et al., 2009)
Enzymatic	Agave bagasse	ASBR	18	1.90	Toledo-Cervantes et al. (2018)
	Oat straw	TBR	8	2.30	Arreola-Vargas et al. (2015)
	Agave bagasse	TBR	4	1.53	Contreras-Dávila et al. (2017)

ASBR, anaerobic sequential batch reactor; *CSTR*, continuous stirred tank reactor; *nr*, not reported; *TBR*, Trickling bed reactor; *UASB*, upflow anaerobic sludge blanket reactor.
[a]*Adapted from Nissilä, M.E., Lay, C.-H., Puhakka, J.A., 2014. Dark fermentative hydrogen production from lignocellulosic hydrolyzates – a review. Biomass Bioenergy 67(0), 145–159. https://doi.org/10.1016/j.biombioe.2014.04.035.*

reduce the costs of purification of hydrogen from dark fermentation (Choix et al., 2017; Franco-Morgado et al., 2017).

So far biohydrogen production fits the biorefinery concept (see Section 3.5) by means of covering the first steps of anaerobic degradation of organic waste, that is, hydrolysis, acidogenesis, and acetogenesis (Fig. 4.6). As main byproducts residual carbohydrates and a mixture of VFAs are obtained, which can be fed to a variety of biological processes to obtain a myriad of endpoints such as methane in a two-step system (Montiel Corona and Razo-Flores, 2018), lipids (Vajpeyi and Chandran, 2015), biofuels, and bioplastics (Strazzera et al., 2018).

TABLE 4.6 Hydrogen yields from organic waste in continuous mode bioreactors.[a]

Substrate	Temperature (°C)	Reactor type	HRT (h)	Yield (mol H_2/mol hexose)	Reference
Sugary wastewater	60	CSTR	0.5–3	2.52	(Ueno et al., 1996)
Coffee drink manufacture wastewater	35	CSTR	6–12	0.32	Jung et al. (2010)
Cheese whey wastewater	35	CSTR	24	0.83	Venetsaneas et al. (2009)
Rice winery wastewater	20–55	UASB	2–24	2.14	Yu et al. (2002)
Coffee drink manufacture wastewater	35	UASB	6–12	1.78	(Jung, 2011b; Jung, 2011c)
Molasses wastewater	-	EGSB	1–6	1.95	Guo et al. (2008)
Tequila vinasse	35	TBR	12	1.30	Buitrón et al. (2014)
Organic fraction of municipal solid waste with shredded paper	60	CSTR	29	2.40	Ueno et al. (2007)

ASBR, anaerobic sequential batch reactor; *CSTR*, continuous stirred tank reactor; *EGSB*, Expanded granular sludge bed; *nr*, not reported; *TBR*, Trickling bed reactor; *UASB*, upflow anaerobic sludge blanket reactor.
[a]*Adapted from Jung, K.-W., Kim, D.-H., Kim, S.-H., Shin, H.-S., 2011a. Bioreactor design for continuous dark fermentative hydrogen production. Bioresour. Technol. 102(18), 8612–8620. https://doi.org/10.1016/j.biortech.2011.03.056.*

4.4 Biofuels for renewable electricity and heat

4.4.1 Heat

Biomass has been used for heating and cooking since man first discovered fire and continues to do so today. And solid biofuels have been traditionally used to produce heat in rural zones or in developing countries, but in this section, we refer to "modern" solid biofuels as described by ISO 16559 for terminology, definitions, and descriptions for solid biofuels; ISO 17225 for specifications and classifications of solid biofuels. And the standard EN 15234 for quality assurance of solid biofuels is also important in this scope.

There are numerous benefits to adhering to ISO standards, market adoption of these will: (1) facilitate domestic and international trade, (2) enhance uptake of new technologies, (3) promote public safety and contribute to a more sustainable industry, (4) minimize emissions of pollutants, and (5) facilitate quality assessment of solid biomass resources.

According to ISO 16559 standard solid biofuels can be raw and processed material originating from forestry and arboriculture, agriculture and horticulture, and aquaculture. Raw and processed material includes woody, herbaceous, fruit, and aquatic biomass from the sectors mentioned. Moreover, the classification of solid biofuels is described on the ISO 17225 standard and is based on origin and source of biomass: (1) woody biomass, (2) herbaceous biomass, (3) aquatic biomass, and (4) blends and mixtures. These groups are further classified into several other groups and subgroups enabling a complete classification of a feedstock which may be derived from specific production chain or may be collected as residual biomass from distinct processes (Table 4.7).

In order to obtain high energy and high-quality solid biofuels, the raw biomass pretreatment is necessary. Pretreatment of solid biomass is used to improve the physical and chemical properties such as particle size, moisture content, density and energy content. In addition, biomass pretreatment eases the handling, storage, and transport of solid biofuels in the whole bioenergy chain. Pelleting, briquetting, torrefaction, and hydrothermal carbonization are the most important biomass pretreatments and pathways for the production of improved solid biofuel (Christoforou and Fokaides, 2019; Tumuluru et al., 2011). Table 4.8 presents the biomass sources and pretreatments used for solid biofuels in recent works.

TABLE 4.7 Classification of biomass feedstock resources for solid biofuels production (ISO 17225: 2014).

Classification/ origin	Group	Subgroup
1. Woody biomass	1.1 Forest plantation 1.2 Byproducts and residues 1.3 Used wood 1.4 Blends and mixtures	1.1.1 Whole trees w/o roots 1.1.2 Whole trees w/roots 1.1.3 Stemwood 1.1.4 Logging residues 1.1.5 Stumps and roots 1.1.6 Bark 1.1.7 Segregated wood from gardens, parks, roadside maintenance, etc. 1.1.8 Blends and mixtures

Continued

TABLE 4.7 Classification of biomass feedstock resources for solid biofuels production (ISO 17225: 2014).—cont'd

Classification/ origin	Group	Subgroup
		1.2.1 Chemically untreated wood byproducts and residues 1.2.2 Chemically treated wood byproducts, residues, fibers, and wood constituents 1.2.3 Blends and mixtures 1.3.1 Chemically untreated used wood 1.3.2 Chemically treated used wood 1.3.3 Blends and mixtures
2. Herbaceous biomass	2.1 Herbaceous biomass from forestry and agriculture 2.2 Byproducts and residues from food and herbaceous processing industry 2.3 Blends and mixtures	2.1.1 Cereal crops 2.1.2 Grasses 2.1.3 Oil seed crops 2.1.4 Root crops 2.1.5 Legume crops 2.1.6 Flowers 2.1.7 Segregated herbaceous biomass from gardens, parks, roadside maintenance, etc. 2.1.8 Blends and mixtures 2.2.1 Chemically untreated herbaceous residues 2.2.2 Chemically treated herbaceous residues 2.2.3 Blends and mixtures
3. Fruit biomass	3.1 Orchard and horticulture fruit 3.2 Byproduct residues from food and fruit processing industry 3.3 Blends and mixtures	3.1.1 Berries 3.1.2 Stone/kernel fruits 3.1.3 Nuts and acorns 3.1.4 Blends and mixtures 3.2.1 Chemically untreated fruit residues 3.2.2 Chemically treated fruit residues 3.2.3 Blends and mixtures
4. Aquatic biomass	4.1 Algae 4.2 Water hyacinth 4.3 Lake and sea weed 4.4 Reeds 4.5 Blends and mixtures	4.1.1 Microalgae 4.1.2 Macroalgae 4.2.3 Blends and mixtures 4.3.1 Lake weed 4.3.2 Sea weed 4.3.3 Blends and mixtures 4.4.1 Common reed 4.4.2 Other reed 4.4.3 Blends and mixtures

TABLE 4.8 Biomass feedstock, pretreatment, and properties of solid biofuels.

Biomass	Traded form/ processing	Relevant properties (density and high heating value)	References
Rice husk and bran	Briquetting	460 kg/m^3 16.08 MJ/kg	Yank, 2016
Furniture wood waste	Codensification/ Briquetting	400–900 m^3 16.7–17.5 MJ/kg	Moreno et al. (2016)
Cotton stalk and wood sawdust	Hydrothermal and briquetting	838.5–1170.5 kg/m^3 17.3–27.9 MJ/kg	Wu et al. (2018)
Wheat straw, rape straw, and maize straw	Pellets	386–572 kg/m^3 15.3–16.2 MJ/kg	Niedziolka et al. (2015)
Birch, spruce, and reed canary grass chips	Pellets	1240 kg/m^3	(Huang et al., 2017)
Biomass—sludge mixed	Pellets	863–1217 kg/m^3 15.04–18.96 MJ/kg	(Jiang et al., 2016)
Pine sawdust and glycerol	Torrefaction and pelletization	244–499 kg/m^3 17.2–20.4 MJ/kg	(García et al., 2018)
Grape marc	Hydrothermal carbonization	19.8–24.1 MJ/kg	Basso, 2016

Pellets are small densified cylindrical granules produced by compression of solid biomass with standardized properties and characteristics. A pelletizer is similar to extrusion process and commercial ones are available with capacities in the range of 0.2–8 ton/hr. The conventional pelletization process includes: grinding, drying, pelleting, and cooling. The pellet obtained must have a minimum density of 600 kg/m^3, moisture content less than 10%, and high calorific value greater or equal to 16.5 MJ/kg (ISO 17225:2014). In order to reduce the energy consumption during pelletizing process, Tumuluru et al. (Tumuluru et al., 2014) preheated and pelletized high-moisture biomass instead of dried biomass. Following that drying step at the end reduced the energy consumption and cost of the process.

As with pelletizing process, biomass briquetting uses pressure agglomeration methods to enable the formation of solid biofuels which present lower moisture content, higher energy density, and improved mechanical and physical properties compared with raw material. Biomass briquetting process is usually performed using hydraulic, mechanical, or roller presses and includes grinding, drying, and briquetting steps.

Although all kinds of woody biomass can be exploited as possible raw material for pelletizing and briquetting, the different characteristics of materials play an important role to their selection. Industrial chips, wood shavings, sawdust, and other byproducts from wood and forestry industry are the most used biomass for pellets and briquettes production. Moreover, the use agricultural biomass and alternative blends and mixtures for solid biofuels production have been extensively researched in the last years. Those materials include fibers, leaves, husks, grasses, and mixtures with municipal and industry solid wastes. Woody is the most used raw material for pellet production today, while agricultural residues and grasses are more commonly used for briquetting processes (Stelte et al., 2012).

Other methods for biomass pretreatment are torrefaction and hydrothermal carbonization. Torrefaction is a mild pyrolysis process where biomass is subjected to temperatures around 220–330°C in the absence of oxygen and atmospheric pressure conditions using low heating rates (Chen et al., 2015). The torrefaction process includes initial heating, predrying, postdrying, torrefaction, and cooling (Stelt et al., 2011). During the residence of biomass in the reactor, the material is partly decomposed, resulting in a solid fuel called biochar, bio-oil, and a mixture of combustible gases. Biochar is reported to contain 41%–90% of the initial biomass, whereas an 80%–95% of the initial energy is retained (Yang et al., 2017).

Hydrothermal carbonization, also called *wet torrefaction*, is a thermochemical conversion technique which is attractive due to its ability to transform wet biomass into energy and chemicals without predrying. The solid product obtained is called hydrochar with high mass and energy density and improved performance combustion. Hydrochar can be produced from several types of feedstocks ranging from municipal solid wastes, wood and forestry residues, and agro-industrial wastes (Benavente et al., 2015). In others pretreatment methods, literature reported an increase in energy recovery with an increase of the lignocellulosic content of the biomass, while the hydrochar was improved with mixtures richer in low molecular weight carbohydrates (Christoforou and Fokaides, 2019).

Combustion is the most used process for energy utilization of solid biofuels because of its low costs and high reliability. Different combustion technologies are used both for domestic use and industrial applications. During combustion, solid biofuel is burned in excess air and produces hot gases which may be used for direct heating purposes or by using a secondary conversion process into power. The combustion of solid biofuels generally consists of four stages: (1) drying, (2) pyrolysis and reduction or volatilization, (3) combustion of the volatile gases, and (4) char combustion. Although biomass combustion presents lower greenhouse gas emissions compared to solid fossil fuels, these emissions during biomass combustion stages can be minimized with deoptimization of some parameters such as heat transfer, combustion unit design, fuel properties, combustion temperature, and air-fuel ratio.

The majority of works about biofuels Life Cycle Assessment (LCA) focus on the environmental impact of biodiesel and bioethanol production systems. Hence, the lack of solid biofuels LCA studies is highlighted in literature. Christoforou and Fokaides (Christoforou and Fokaides, 2019) gathered information about specific studies of LCA of biofuels pretreatment and combustion, and they concluded that significant carbon reduction can be achieved by replacing conventional fuels with biomass in heat and power systems. In addition, biomass cofiring can positively contribute for the improvement of the efficiency of existing systems. On the other hand, with regard to process pretreatment the transportation of feedstock and product, as well as the production process, are two critical factors which significantly impact the LCA; consequently, more detailed studies are required. Throughout the past decades, there has been an increase in trading of processed solid biofuels for both the production of process heat and electrical power (Dafnomilis et al., 2017).

Regarding biogas, one of its advantages is versatility, where the application in heating is also quite broad. For example, in the cities there is a great potential for the use of biogas in heating and in the production of electricity, as is the case in Sweden and France. In the same way, biogas is used to heat greenhouses for the production of vegetables for domestic consumption in Argentina; another application is the heating of livestock farms in Peru. Domestic use to replace other fuels such as wood in developing countries is also becoming important to improve women's health by avoiding exposure to volatile organic compounds. That is, both in developed and developing countries in agricultural, livestock, domestic, and industrial applications the use of biogas is widespread.

4.4.2 Electricity

Despite the possibility of directly burning the biohydrogen generated from dark fermentation as fuel, to obtain mainly heat, biohydrogen use might be further valorized by upgrading through the use of membranes or other refining methods, to obtain hydrogen with a higher purity. This high-grade hydrogen may be used in the industrial processes or in fuel cells to obtain electricity with energy yield between 40% and 50% compared to 15%–20% energy yield of combustion.

Concerning biogas, one of its future applications is precisely the use as a reservoir of energy, as is currently done under the concept "power to gas" (Collet et al., 2017) where the surplus of energy captured by renewable energies such as wind turbines or photovoltaic panels through the storage of renewable energy and CO_2. However, the highest renewable power from biomass comes from solid biofuels, with a small contribution from liquid biofuels (Fig. 4.7). Due to that solid biofuels present lower heating value compared to liquid biofuels, cofiring or cocombustion of biomass with fossil fuel is considered as attractive process in the power generation for industry.

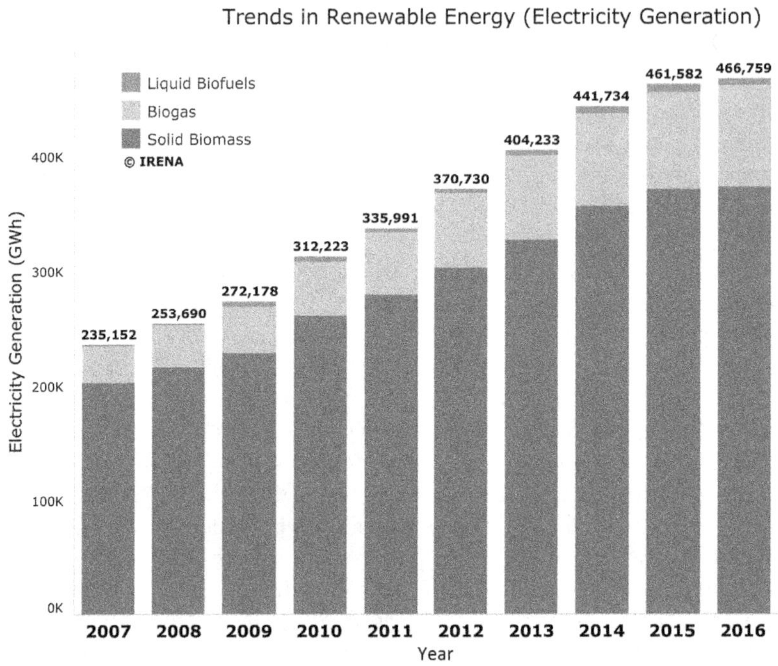

FIGURE 4.7 Trends in renewable energy (electricity generation). *Credit: From IRENA, 2019c. Trends in renewable energy (electricity generation). Retrieved from: https://www.irena.org/bioenergy).*

This arrangement leads to the reduction of solid-based CO_2 emissions and the process has an advantage in the disposal of waste products and fuel cost reduction in power plants. Renewable electricity is encouraged in many countries and biofuels have nowadays a great contribution to this sector, providing in 2017 as much as solar, hydropower, and wind combined (IEA, 2018).

4.5 Biorefinery and bio-based chemicals

4.5.1 Biorefinery

As mentioned before, biofuels are an important alternative to fossil fuels. Alternative energy sources should be readily available, cheap, greenhouse gas emission–neutral, and should not endanger water, land, and food availability (Michailos and Webb, 2019; Saini et al., 2015). Therefore, biofuels obtained through an integrated production platform known as biorefinery could have a better economical and environmental outcome. A biorefinery can be defined as the complex processing methods required to transform renewable resources to multiple products (Fernando et al., 2006). The International Energy Agency Bioenergy Task 42 defined biorefinery as "the sustainable processing of biomass into a spectrum of bio-based products (food, feed, chemicals, materials) and energy (fuels, power, heat)" (IEA, 2012).

In biorefineries, the production of high-value-low-volume (HVLV) and low-value-high-volume (LVHV) products can be achieved. HVLV products increase the profitability of the process, while LVHV are converted into energy (Fernando et al., 2006; Naik et al., 2010). LVHV products include biofuels, while products for pharmaceutical, cosmetics, or fine chemicals have high market prices (Apprich et al., 2014; Moncada et al., 2016). In addition, biorefineries increase their sustainability by maximizing the use of raw material and reducing the amount of residues to a minimum (Moncada et al., 2016).

Biorefineries are the first step in a multistep technology in feedstock selection, followed by pretreatment of the biomass. After this pretreatment, a combination of biological and/or chemical processes is applied to obtain multiple building blocks for further processing (Moncada et al., 2016). Generally three types of biorefineries are identified, which are classified based on the type of feedstock, conversion technologies, and products (Maity, 2015; Naik et al., 2010).

4.5.2 Biorefinery feedstocks

Feedstocks are the raw materials that can be used in biorefinery; biomass is the feedstock of biorefineries. Plants are responsible of biomass production via photosynthesis for the conversion of carbon dioxide and water into sugars that are then converted into complex materials (Cherubini, 2010). An efficient biorefinery system requires a renewable, sustainable, consistent, and regular supply feedstock (Maity, 2015). In biorefineries raw materials are mainly obtained from agriculture (dedicated crops or residues), forestry, industrial (process residues), households (municipal solid waste and wastewaters), and aquaculture (algae and seaweeds) (Cherubini, 2010). Biomass is also divided into three main groups: (i) carbohydrates and lignin, (ii) fats and oils, and (iii) mixed organic residues (see also Fig. 4.1).

Carbohydrates are the most common biomass component of plants feedstocks. They are composed of six carbon sugars like glucose, galactose, and mannose, and five carbon sugars like xylose and arabinose. Sugar cane and sugar beets are the most important sugar crops and in biorefineries are used for ethanol production (Cherubini, 2010). Starch is a glucose polymer, with 20–25 %wt. amylose and 75–80%wt. amylopectin (Maity, 2015). In bio-based processes starch is hydrolyzed before fermentation. The most important starch crops are wheat and corn that can be used for ethanol and chemical production. Lignocellulosic biomass is composed of cellulose, hemicellulose, and lignin. Cellulose represents between 30% and 50% of lignocellulosic dry matter and has long chains of glucose that differs from starch in the types of bonds that link the glucose molecules. Hemicellulose is the second most important component of lignocellulosic biomass (20%–40% dry matter), is composed by sugars with five and six carbons, and is easier to break than cellulose. Lignin is the component responsible for giving structure to plants and trees and is made of phenolic polymers; therefore, it is not used in fermentation processes. Lignocellulosic biomass includes agricultural residues.

Fats and oils are the second group of biomass feedstock (Cherubini, 2010; Maity, 2015). They are triglycerides formed by a glycerol molecule joined to three fatty acids of different carbon length and saturation degree. This feedstock is obtained from vegetables and animals. The vegetable oils with major worldwide production include palm, soybean, rapeseed, and sunflower seed. The sustainability of triglyceride-based products such as biodiesel requires a constant production of nonedible oils and the valorization of waste oils.

Mixed organic residues include municipal waste, sewage sludge, and a variety of industrial residues (Maity, 2015). These types of residues have different characteristics and go thorough different conversion processes that include anaerobic digestion and energy recovery (Cherubini, 2010).

Biorefinery raw materials can also be classified according to its generation. First-generation feedstocks include food crops like soybean, sunflower, rapeseed, sugarcane, corn stover, sorghum, and wheat; consequently, they have economic, social, and environmental challenges (Moncada et al., 2016). Second-generation feedstock is composed of lignocellulosic biomass, solid waste, wood, animal fat, forest and crop residues, fibers, and bagasse. Third-generation feedstocks are microorganisms capable of accumulating oils like algae and yeast. But generation classification could cause confusion and the "advanced" biofuel or biorefinery has become widely accepted (see Section 3.1.1).

Another classification used for biorefinery feedstock includes energy crops, agricultural residues and waste, forestry waste and residues, and industrial and municipal wastes (Maity, 2015). Energy crops are high-yielding with low costs and low maintenance, including switchgrass, alfalfa, bamboo, sweet sorghum, and reed canary grass (Cherubini, 2010; Maity, 2015). In addition, energy crops include woody energy crops (silver maple hybrid poplar, eastern cottonwood, green ash, sycamore), agricultural crops (palm soybean, jatropha, sunflower, maize, rye, sugarcane), and aquatic crops (algae, seaweed) (Maity, 2015). Agricultural residues include stalks, leaves, husks, cobs, hulls, and bagasses that represent leftovers of the process and have no commercial application.

4.5.3 Types of biorefinery

Table 4.9 shows the types of biorefineries. The first type of biorefinery (phase I) has fixed processing capabilities and it uses dried grain as feedstock to

TABLE 4.9 Types of biorefineries.

Type of biorefinery	Processing capabilities	Feedstock	Products
Phase I	Fixed	Dried grain	Ethanol, feed products, dried distilled grain, and carbon dioxide.
Phase I	Flexible	Wet biomass (grains, cereals, or green grass)	Ethanol, starch, high fructose corn syrup, corn oil, dried distilled grains, and carbon dioxide.
Phase III			
Whole Crop	Flexible	Crops (rye, maize, wheat, triticale, rice, soya, rapeseed, barley)	Meal, oil, starch, PHB, HMF, dimethyl furan, sorbitol, glycerin, ethylene glycol, propylene glycol, carboxymethyl starch, fatty acid esters, bioplastics, biopolymers, binder, adhesive, cement, syngas.
Green	Flexible	Green grass, closure fields, lucerne, clover, switchgrass, immature cereals	Cellulose, starch, dyes, pigment pellets for fodder, solid fuel, syngas, hydrocarbons, biogas, fibers, chemicals, proteins, organic acids, dyes, carbohydrates, enzymes and minerals, organic acids, amino acids, proteins, enzymes, and ethanol.
Lignocellulose Feedstock (LCF)	Flexible	Straw, grass, forest biomass, wood, paper-waste, cereals, lignocellulosic biomass, and cellulosic solid waste	Hemicellulose, xylite, Nylon 6, thickeners, adhesives, emulsifiers, stabilizers, glucose, softeners, lubricants, chemicals, binders, adhesives, coal, and solid fuel.

produce ethanol, feed products, dried distilled grain, and carbon dioxide. The second type (phase II) has more flexibility than phase I biorefineries; it can produce ethanol, starch, high fructose corn syrup, corn oil, dried distilled grains, and carbon dioxide from wet biomasses, such as grains, cereals, or green grass (Maity, 2015; Naik et al., 2010).

The third type of biorefinery (phase III), has more advanced technology that allows it to use a mix of biomass as feedstock (Kamm et al., 2016; Maity, 2015). Phase III biorefineries use crops, crop and forest residues, green grasses, lignocellulosic biomass, and industrial wastes (Elmekawy et al., 2013; Fernando et al., 2006; Kamm and Kamm, 2004; Naik et al., 2010; Nonato et al., 2001; Scholey et al., 2016). This last type of biorefineries can produce gaseous biofuels (syngas, biogas, hydrogen, biomethane), liquid biofuels (bioethanol, biodiesel, biokerosene, biooil), organic acids (lactic, succinic, fumaric, malic, pyruvic), polymers (biodegradable plastics, phenol resins), fertilizers, food, and feed (Cherubini, 2010; Elmekawy et al., 2013; Kamm and Kamm, 2004; Menon and Rao, 2012; Nonato et al., 2001). Phase III biorefineries can be subdivided in three categories: (i) Whole Crop, (ii) Green and (iii) Lignocellulose Feedstock (LCF) Biorefinery.

Whole crop biorefineries use feedstock such as rye, maize, wheat, triticale, rice, soya, rapeseed, and barley (Abdulkhani et al., 2017; Cherubini, 2010; Elmekawy et al., 2013; Kamm et al., 2016; Kamm and Kamm, 2004; Luo et al., 2011). The first step is the mechanical separation of the grains into seed and straw. The seed can be converted into starch, ground to produce meal or separated to produce oil (Abdulkhani et al., 2017; Kamm et al., 2016; Maity, 2015). Starch can then be processed through biotechnological conversion, chemical conversion, or plasticization. Biotechnological conversion via glucose can generate products such as ethanol, syngas, methanol, poly-3-hydroxybutyric acid (PHB), hydroxymethylfurfural (HMF), and dimethyl furan (Abdulkhani et al., 2017; Bhatia et al., 2015; González-García et al., 2011; Jang et al., 2012; Kawaguchi et al., 2017; Nonato et al., 2001; Scholey et al., 2016). Chemical conversion includes hydrogenation, etherification, esterification, and amination to produce sorbitol, glycerin, ethylene glycol, propylene glycol, carboxymethyl starch, and fatty acid esters, among others (Jang et al., 2012; Kamm et al., 2016). Plasticization generates bioplastics and biopolymers (Kamm et al., 2016; Kamm and Kamm, 2004; Mohammadi Nafchi et al., 2013). Meal can be transformed into binder, adhesive, and cement (Amini et al., 2013; Imam et al., 2013; Isik and Ozkul, 2014; Kamm et al., 2016). The straw can be transformed in a lignocellulose feedstock biorefinery, through decomposition into lignin, hemicellulose, and cellulose. In addition, straw can be used for the production of syngas via pyrolysis (Beneroso et al., 2014; Huang et al., 2016; Huang et al., 2013; Neumann et al., 2015).

Green biorefineries use green biomass as feedstock. Green grass, closure fields, lucerne, clover, switchgrass, and immature cereals, are processed in this multiproduct biorefinery (Cherubini, 2010; Kamm et al., 2016). Green crops are generally used as forage and as source of leafy vegetables, but through a process called wet fractionation of green biomass, they can be used to simultaneously produce food and nonfood items (Carlsson, 1994; Kamm et al., 2016). These systems deal with their refinery cuts, fractions, and products depending on the physiology of each plant material (Kamm et al., 2016). In this type of biorefineries the wet raw materials arrive from the field and are treated through wet biorefinery to separate the contents into fiber-rich press cake and nutrient rich green juice. The press cake is rich in cellulose, starch, dyes, and pigment, and is used to produce pellets for fodder and solid fuel, and raw material for syngas, hydrocarbons, biogas, fibers, and chemicals (Fernando et al., 2006; Kamm et al., 2016; Kamm and Kamm, 2004). The green juice is rich in proteins, organic acids, dyes, carbohydrates, enzymes, and minerals, among others. The green juice can be used as carbon source for the production of organic acids, amino acids, proteins, enzymes, and ethanol (Fernando et al., 2006; Kamm et al., 2016; Kamm and Kamm, 2004). Green biorefineries yield high biomass per hectare and cheap feedstock (Kamm et al., 2016).

LCF is the most promising type of biorefinery due to its versatility. They can use straw, grass, forest biomass, wood, paper-waste, cereals, lignocellulosic biomass, and cellulosic solid waste as feedstock and the cost of the raw materials is low. Lignocellulose raw materials are cleaned and broken into hemicellulose/polyoses, cellulose, and lignin, using chemical or enzymatic digestion (Fernando et al., 2006; Kamm et al., 2016). Hemicellulose contains different sugar units such as glucose, mannose, galactose, xylose, and arabinose. Hemicellulose is processed to obtain bio-based products like sugar substitutes (xylite), furan residues, Nylon 6, thickeners, adhesives, emulsifiers, and stabilizers (Kamm et al., 2016). Cellulose is a glucose polymer that can be chemically or enzymatically hydrolyzed to produce glucose that can be used in fermentation or for the production of HMF and levulinic acid that can be transformed to softeners, lubricants, and other chemicals. Lignin is a phenol polymer that can be used for the production of natural binders, adhesives, coal, and solid fuel.

4.6 Concluding remarks

Sustainable biofuels are a requisite for decarbonization of the energy system and have a key role in climate-change mitigation. Bioenergy represents the largest portion of renewable energy with an increase of 30% to 2023. Other benefits of biofuels include job creation, waste elimination and rural development. The liquids biofuels are a very good environmentally friendly and a sustainable option of renewable fuels and in some cases, the only option for fossil decarbonization (i.e., in aviation fuels). The use of residues and energetic crops represents a very intelligent way to solve two problems, their correct disposal and the generation of products with high added value and friendly to the environment. The use of residues and energy crops is an intelligent way of not competing with food crops and to solve an environmental problem of waste disposal. In addition, the social impact caused using agroindustry waste, especially in developing countries, is important in offering an alternative for the use of waste generated in their communities.

The use of biogas from organic waste contributes to several modern concepts, such as the circular economy, the use of renewable energies, the storage of energy, etc., where the scale of the facilities and the application of biofuels are very broad. From greenhouse gas to alternative energy, the use of biogas is a great contribution to the sustainability of our society. Promoting generation of biohydrogen at industrial scale from waste might facilitate a massive use of hydrogen as fuel at competitive prices, possible so far after government subsidies strongly supported on environment preservation. Bio-based production in biorefineries is a sustainable alternative to fossil fuels. However, integrated processes are required to maximize valorization of all feedstock components, simultaneous generation of products, and low environmental impact. In general, it is necessary to continue researching to achieve more efficient processes and with less environmental impact, and it is important that all the fractions of the biomass are used, to achieve the concept of biorefinery and circular bioeconomy.

References

Aatola, H., Larmi, M., Sarjovaara, T., Mikkonen, S., 2008. Hydrotreated vegetable oil (HVO) as a renewable diesel fuel: trade-off between NOx, particulate emission, and fuel consumption of a heavy duty engine. ETIP Bioenergy. Retrieved from: http://www.etipbioenergy.eu/images/SAE_Study_Hydrotreated_Vegetable_Oil_HVO_as_a_Renewable_Diesel_Fuel.pdf.

Abbaszaadeh, A., Ghobadian, B., Omidkhah, M.R., Najafi, G., 2012. Current biodiesel production technologies: a comparative review. Energy Convers. Manag. 63, 138–148. https://doi.org/10.1016/j.enconman.2012.02.027.

Abd-Alla, M.H., Elsadek El-Enany, A.W., 2012. Production of acetone-butanol-ethanol from spoilage date palm (Phoenix dactylifera L.) fruits by mixed culture of *Clostridium acetobutylicum* and *Bacillus subtilis*. Biomass Bioenergy. https://doi.org/10.1016/j.biombioe.2012.03.006.

Abdulkhani, A., Alizadeh, P., Hedjazi, S., Hamzeh, Y., 2017. Potential of Soya as a raw material for a whole crop biorefinery. Renew. Sustain. Energy Rev. 75, 1269–1280. https://doi.org/10.1016/j.rser.2016.10.082.

AFDC, 2019. Fuel Prices. Retrieved from. https://afdc.energy.gov/fuels/prices.html.

Amini, M.H.M., Hashim, R., Hiziroglu, S., Sulaiman, N.S., Sulaiman, O., 2013. Properties of particleboard made from rubberwood using modified starch as binder. Compos. B Eng. 50, 259–264. https://doi.org/10.1016/j.compositesb.2013.02.020.

Apprich, S., Tirpanalan, Ö., Hell, J., Reisinger, M., Böhmdorfer, S., Siebenhandl-Ehn, S., et al., 2014. Wheat bran-based biorefinery 2: valorization of products. LWT – Food Sci. Technol. 56 (2), 222–231. https://doi.org/10.1016/j.lwt.2013.12.003.

Arreola-Vargas, J., Alatriste-Mondragón, F., Celis, L.B., Razo-Flores, E., López-López, A., Méndez-Acosta, H.O., 2015. Continuous hydrogen production in a trickling bed reactor by using triticale silage as inoculum: effect of simple and complex substrates. J. Chem. Technol. Biotechnol. 90 (6), 1062–1069. https://doi.org/10.1002/jctb.4410.

Arriaga, S., Rosas, I., Alatriste-Mondragón, F., Razo-Flores, E., 2011. Continuous production of hydrogen from oat straw hydrolysate in a biotrickling filter. Int. J. Hydrogen Energy 36 (5), 3442–3449. https://doi.org/10.1016/j.ijhydene.2010.12.019.

Avhad, M.R., Marchetti, J.M., 2015. A review on recent advancement in catalytic materials for biodiesel production. Renew. Sustain. Energy Rev. 50, 696–718. https://doi.org/10.1016/j.rser.2015.05.038.

Ayala-Mendivil Neydeli, S.G., 2018. Bioenergía a partir de residuos forestales y de madera. Madera y Bosques, vol. 24.

Barrera, I., Amezcua-Allieri, M.A., Estupiñan, L., Martínez, T., Aburto, J., 2016. Technical and economical evaluation of bioethanol production from lignocellulosic residues in Mexico: case of sugarcane and blue agave bagasses. Chem. Eng. Res. Des. https://doi.org/10.1016/j.cherd.2015.10.015.

Basso, D., Patuzzi, F., Castello, D., Baratieri, M., Rada, E.C., Weiss-Hortala, E., Fiori, L., 2016. Agro-industrial waste to solid biofuel through hydrothermal carbonization. Waste Manag. 47, 114–121.

Bellido, C., Loureiro Pinto, M., Coca, M., González-Benito, G., García-Cubero, M.T., 2014. Acetone-butanol-ethanol (ABE) production by *Clostridium beijerinckii* from wheat straw hydrolysates: efficient use of penta and hexa carbohydrates. Bioresour. Technol. https://doi.org/10.1016/j.biortech.2014.06.020.

Benavente, V., Calabuig, E., Fullana, A., 2015. Upgrading of moist agro-industrial wastes by hydrothermal carbonization. J. Anal. Appl. Pyrolysis 113, 89–98.

Beneroso, D., Bermúdez, J.M., Arenillas, A., Menéndez, J.A., 2014. Integrated microwave drying, pyrolysis and gasification for valorisation of organic wastes to syngas. Fuel 132, 20–26. https://doi.org/10.1016/j.fuel.2014.04.064.

Berezina, O.V., Brandt, A., Yarotsky, S., Schwarz, W.H., Zverlov, V.V., 2009. Isolation of a new butanol-producing *Clostridium* strain: high level of hemicellulosic activity and structure of solventogenesis genes of a new *Clostridium saccharobutylicum* isolate. Syst. Appl. Microbiol. https://doi.org/10.1016/j.syapm.2009.07.005.

Bhatia, S.K., Shim, Y.-H., Jeon, J.-M., Brigham, C.J., Kim, Y.-H., Kim, H.-J., et al., 2015. Starch based polyhydroxybutyrate production in engineered *Escherichia coli*. Bioproc. Biosyst. Eng. 38 (8), 1479–1484. https://doi.org/10.1007/s00449-015-1390-y.

Bohl, T., Smallbone, A., Tian, G., Roskilly, A.P., 2018. Particulate number and NOx trade-off comparisons between HVO and mineral diesel in HD applications. Fuel 215, 90–101. https://doi.org/10.1016/j.fuel.2017.11.023.

Buitrón, G., Prato-Garcia, D., Zhang, A., 2014. Biohydrogen production from tequila vinasses using a fixed bed reactor. Water Sci. Technol. 70 (12), 1919–1925.

Cai, D., Zhang, T., Zheng, J., Chang, Z., Wang, Z., Qin, P.y., Tan, T.w., 2013. Biobutanol from sweet sorghum bagasse hydrolysate by a hybrid pervaporation process. Bioresour. Technol. https://doi.org/10.1016/j.biortech.2013.02.094.

Carlsson, R., 1994. Sustainable primary production–Green crop fractionation: effects of species, growth conditions, and physiological development. In: Pessarakli, M. (Ed.), Handbook of Plant and Crop Physiology. Marcel Dekker, New York, pp. 941–963.

Castro, Y.A., Ellis, J.T., Miller, C.D., Sims, R.C., 2015. Optimization of wastewater microalgae saccharification using dilute acid hydrolysis for acetone, butanol, and ethanol fermentation. Appl. Energy. https://doi.org/10.1016/j.apenergy.2014.11.045.

Centeno, A., Laurent, E., Delmon, B., 1995. Influence of the support of CoMo sulfide catalysts and of the addition of potassium and platinum on the catalytic performances for the hydrodeoxygenation of carbonyl, carboxyl, and guaiacol-type molecules. J. Catal. 154 (2), 288–298. https://doi.org/10.1006/jcat.1995.1170.

Chen, S.-D., Lo, Y.-C., Lee, K.-S., Huang, T.-I., Chang, J.-S., 2009. Sequencing batch reactor enhances bacterial hydrolysis of starch promoting continuous bio-hydrogen production from starch feedstock. Int. J. Hydrogen Energy 34 (20), 8549–8557.

Chen, W.-H., Peng, J., Bi, X.T., 2015. A state-of-the-art review of biomass torrefaction, densification and applications. Renew. Sustain. Energy Rev. 44, 847–866. https://doi.org/10.1016/j.rser.2014.12.039.

Cheng, H.-H., et al., 2015. Biological butanol production from microalgae-based biodiesel residues by *Clostridium acetobutylicum*. Bioresour. Technol. 184, 379–385.

Cherubini, F., 2010. The biorefinery concept: using biomass instead of oil for producing energy and chemicals. Energy Convers. Manag. 51 (7), 1412–1421. https://doi.org/10.1016/j.enconman.2010.01.015.

Choix, F.J., Snell-Castro, R., Arreola-Vargas, J., Carbajal-López, A., Méndez-Acosta, H.O., 2017. CO_2 removal from biogas by *Cyanobacterium leptolyngbya* sp. CChF1 isolated from the lake Chapala, Mexico: optimization of the temperature and light intensity. Appl. Biochem. Biotechnol. 183 (4), 1304–1322.

Christoforou, E., Fokaides, P.A., 2019. Solid Biomass Pretreatment Processes. Springer International Publishing, Cham.

Collet, P., Flottes, E., Favre, A., Raynal, L., Pierre, H., Capela, S., Peregrina, C., 2017. Techno-economic and life cycle assessment of methane production via biogas upgrading and power to gas technology. Appl. Energy 192, 282–295.

Contreras-Dávila, C.A., Méndez-Acosta, H.O., Arellano-García, L., Alatriste-Mondragón, F., Razo-Flores, E., 2017. Continuous hydrogen production from enzymatic hydrolysate of *Agave tequilana* bagasse: effect of the organic loading rate and reactor configuration. Chem. Eng. J. 313, 671–679. https://doi.org/10.1016/j.cej.2016.12.084.

Costa e Silva, W., Freitas, L., Oliveira, P.C., de Castro, H.F., 2016. Continuous enzymatic biodiesel production from coconut oil in two-stage packed-bed reactor incorporating an extracting column to remove glycerol formed as by-product. Bioproc. Biosyst. Eng. 39 (10), 1611–1617. https://doi.org/10.1007/s00449-016-1636-3.

Dafnomilis, I., Hoefnagels, R., Pratama, Y.W., Schott, D.L., Lodewijks, G., Junginger, M., 2017. Review of solid and liquid biofuel demand and supply in Northwest Europe towards 2030—A comparison of national and regional projections. Renew. Sustain. Energy Rev. 78, 31–45.

Diekert, G., Wohlfarth, G., 1994. Metabolism of homoacetogens. Antonie van Leeuwenhoek 66 (1), 209–221. https://doi.org/10.1007/bf00871640.

Dong, J., Chi, Y., Zou, D., Fu, C., Huang, Q., Ni, M., 2014. Energy–environment–economy assessment of waste management systems from a life cycle perspective: model development and case study. Appl. Energy 114, 400–408. %@ 0306-2619.

El-Dalatony, M., Salama, E.S., Kurade, M., Hassan, S., Oh, S.E., Kim, S., Jeon, B.H., 2017. Utilization of microalgal biofractions for bioethanol, higher alcohols, and biodiesel production: a review. Energies 10, 2110.

Elmekawy, A., Diels, L., De Wever, H., Pant, D., 2013. Valorization of cereal based biorefinery byproducts: reality and expectations. Environ. Sci. Technol. 47 (16), 9014–9027. https://doi.org/10.1021/es402395g.

Ezeji, T.C., Qureshi, N., Blaschek, H.P., 2004. Acetone butanol ethanol (ABE) production from concentrated substrate: reduction in substrate inhibition by fed-batch technique and product inhibition by gas stripping. Appl. Microbiol. Biotechnol. https://doi.org/10.1007/s00253-003-1400-x.

Ezeji, T.C., Qureshi, N., Blaschek, H.P., 2007. Production of acetone butanol (AB) from liquefied corn starch, a commercial substrate, using *Clostridium beijerinckii* coupled with product recovery by gas stripping. J. Ind. Microbiol. Biotechnol. https://doi.org/10.1007/s10295-007-0253-1.

Fernando, S., Adhikari, S., Chandrapal, C., Murali, N., 2006. Biorefineries: current status, challenges, and future direction. Energy Fuels 20 (4), 1727–1737. https://doi.org/10.1021/ef060097w.

Fidalgo, W.R.R., Ceron, A., Freitas, L., Santos, J.C., de Castro, H.F., 2016. A fluidized bed reactor as an approach to enzymatic biodiesel production in a process with simultaneous glycerol removal. J. Ind. Eng. Chem. 38, 217–223. https://doi.org/10.1016/j.jiec.2016.05.005.

Franco-Morgado, M., Alcántara, C., Noyola, A., Muñoz, R., González-Sánchez, A., 2017. A study of photosynthetic biogas upgrading based on a high rate algal pond under alkaline conditions: influence of the illumination regime. Sci. Total Environ. 592, 419–425.

García, R., González-Vázquez, M.P., Pevida, C., Rubiera, F., 2018. Pelletization properties of raw and torrefied pine sawdust: effect of co-pelletization, temperature, moisture content and glycerol addition. Fuel 215, 290–297. https://doi.org/10.1016/j.fuel.2017.11.027.

García, V., Päkkilä, J., Ojamo, H., Muurinen, E., Keiski, R.L., 2011. Challenges in biobutanol production: how to improve the efficiency? Renew. Sustain. Energy Rev. https://doi.org/10.1016/j.rser.2010.11.008.

Ghimire, A., Frunzo, L., Pirozzi, F., Trably, E., Escudie, R., Lens, P.N., Esposito, G., 2015. A review on dark fermentative biohydrogen production from organic biomass: process parameters and use of by-products. Appl. Energy 144, 73–95.

González-García, Y., Rosales, M.A., González-Reynoso, O., Sanjuán-Dueñas, R., Córdova, J., 2011. Polyhydroxybutyrate production by *Saccharophagus degradans* using raw starch as carbon source. Eng. Life Sci. 11 (1), 59–64. https://doi.org/10.1002/elsc.201000118.

Gottumukkala, L.D., Parameswaran, B., Valappil, S.K., Mathiyazhakan, K., Pandey, A., Sukumaran, R.K., 2013. Biobutanol production from rice straw by a non acetone producing *Clostridium sporogenes* BE01. Bioresour. Technol. https://doi.org/10.1016/j.biortech.2013.01.046.

Guo, W.-Q., Ren, N.-Q., Chen, Z.-B., Liu, B.-F., Wang, X.-J., Xiang, W.-S., Ding, J., 2008. Simultaneous biohydrogen production and starch wastewater treatment in an acidogenic expanded granular sludge bed reactor by mixed culture for long-term operation. Int. J. Hydrogen Energy 33 (24), 7397–7404.

Haas, M.J., 2005. Improving the economics of biodiesel production through the use of low value lipids as feedstocks: vegetable oil soapstock. Fuel Process. Technol. 86 (10), 1087–1096. In: https://doi.org/10.1016/j.fuproc.2004.11.004.

Hama, S., Tamalampudi, S., Yoshida, A., Tamadani, N., Kuratani, N., Noda, H., et al., 2011. Process engineering and optimization of glycerol separation in a packed-bed reactor for enzymatic biodiesel production. Bioresour. Technol. 102 (22), 10419–10424. https://doi.org/10.1016/j.biortech.2011.08.073.

Harada, H., Uemura, S., Momonoi, K., 1994. Interaction between sulfate-reducing bacteria and methane-producing bacteria in UASB reactors fed with low strength wastes containing different levels of sulfate. Water Res. 28 (2), 355–367.

Hassan, S., Gwilym, A., Williams, A., Jaiswal, K., 2018. Emerging technologies for the pretreatment of lignocellulosic biomass. Bioresour. Technol. 216, 310–318.

Helwani, Z., Othman, M.R., Aziz, N., Fernando, W.J.N., Kim, J., 2009. Technologies for production of biodiesel focusing on green catalytic techniques: a review. Fuel Process. Technol. 90 (12), 1502–1514. https://doi.org/10.1016/j.fuproc.2009.07.016.

Huang, Y.F., Chiueh, P.T., Kuan, W.H., Lo, S.L., 2013. Microwave pyrolysis of rice straw: products, mechanism, and kinetics. Bioresour. Technol. 142, 620–624. https://doi.org/10.1016/j.biortech.2013.05.093.

Huang, Y.-F., Chiueh, P.-T., Lo, S.-L., 2016. A review on microwave pyrolysis of lignocellulosic biomass. Sustain. Environ. Res. 26 (3), 103–109. https://doi.org/10.1016/j.serj.2016.04.012.

Huang, Y., Finell, M., Larsson, S., Wang, X., Zhang, J., 2017. Biofuel pellets made at low moisture content – influence of water in the binding mechanism of densified biomass. Biomass Bioenergy 98, 8–14. https://doi.org/10.1016/j.biombioe.2017.01.002.

Huber, G.W., Corma, A., 2007. Synergies between bio- and oil refineries for the production of fuels from biomass. Angew. Chem. Int. Ed. 46 (38), 7184–7201. https://doi.org/10.1002/anie.200604504.

IEA, 2012. Bio-Based Chemicals: Value Added Products From Biorefineries | Bioenergy. Retrieved from. https://www.ieabioenergy.com/publications/bio-based-chemicals-value-added-products-from-biorefineries/.

IEA, 2018. Renewables 2018: Analysis and Forecast to 2023 Market report Series. Retrieved from: https://webstore.iea.org/market-report-series-renewables-2018.

IEA-FAO, 2017. Bioenergy: Roadmap Develpment and Implementation How2Guide for. Retrieved from: https://webstore.iea.org/how2guide-for-bioenergy.

Imam, S.H., Bilbao-Sainz, C., Chiou, B.-S., Glenn, G.M., Orts, W.J., 2013. Biobased adhesives, gums, emulsions, and binders: current trends and future prospects. J. Adhes. Sci. Technol. 27 (18–19), 1972–1997. https://doi.org/10.1080/01694243.2012.696892.

IRENA, 2016. Roadmap For A Renewable Energy Future. In: Agency, I.R.E. (Ed.). Retrieved from: http://www.irena.org/remap/.

IRENA, 2019a. Liquid Biofuels. Retrieved from: https://www.irena.org/transport/Liquid-Biofuels.

IRENA, 2019b. Renewable Energy Employment by Technology. Retrieved from: http://resourceirena.irena.org/gateway/dashboard/?topic=7&subTopic=10.

IRENA, 2019c. Trends in Renewable Energy (Electricity Generation). Retrieved from: https://www.irena.org/bioenergy.

Isik, I.E., Ozkul, M.H., 2014. Utilization of polysaccharides as viscosity modifying agent in self-compacting concrete. Constr. Build. Mater. 72, 239–247. https://doi.org/10.1016/j.conbuildmat.2014.09.017.

Iso, M., Chen, B., Eguchi, M., Kudo, T., Shrestha, S., 2001. Production of biodiesel fuel from triglycerides and alcohol using immobilized lipase. J. Mol. Catal. B Enzym. 16 (1), 53–58. https://doi.org/10.1016/S1381-1177(01)00045-5.

Jang, Y.S., Kim, B., Shin, J.H., Choi, Y.J., Choi, S., Song, C.W., et al., 2012. Bio-based production of C2-C6 platform chemicals. Biotechnol. Bioeng. 109 (10), 2437–2459. https://doi.org/10.1002/bit.24599.

Jiang, Y., Xin, F., Lu, J., Dong, W., Zhang, W., Zhang, M., Wu, H, Ma, J, Jiang, M., 2017. State of the art review of biofuels production from lignocellulose by thermophilic bacteria. Bioresource Technology. 245, 1498–1506.

Jiang, L., Yuan, X., Xiao, Z., Liang, J., Li, H., Cao, L., Wang, H., 2016. A comparative study of biomass pellet and biomass-sludge mixed pellet: energy input and pellet properties. Energy Convers. Manag. 126, 509–515. https://doi.org/10.1016/j.enconman.2016.08.035.

Jin, C., Yao, M., Liu, H., Lee, C.F.F., Ji, J., 2011. Progress in the production and application of n-butanol as a biofuel. Renew. Sustain. Energy Rev. https://doi.org/10.1016/j.rser.2011.06.001.

Johari, A., Alkali, H., Hashim, H., Ahmed, S.I., Mat, R., 2014. Municipal solid waste management and potential revenue from recycling in Malaysia. Mod. Appl. Sci. 8 (4), 37.

Jung, K.-W., Kim, D.-H., Kim, S.-H., Shin, H.-S., 2011a. Bioreactor design for continuous dark fermentative hydrogen production. Bioresour. Technol. 102 (18), 8612–8620. https://doi.org/10.1016/j.biortech.2011.03.056.

Jung, K.-W., Kim, D.-H., Shin, H.-S., 2010. Continuous fermentative hydrogen production from coffee drink manufacturing wastewater by applying UASB reactor. Int. J. Hydrogen Energy 35 (24), 13370–13378.

Jung, K.-W., Kim, D.-H., Shin, H.-S., 2011b. Fermentative hydrogen production from *Laminaria japonica* and optimization of thermal pretreatment conditions. Bioresour. Technol. 102 (3), 2745–2750.

Jung, K.-W., Kim, D.-H., Shin, H.-S., 2011c. A simple method to reduce the start-up period in a H_2-producing UASB reactor. Int. J. Hydrogen Energy. 36 (2), 1466–1473.

Kalargaris, I., Tian, G., Gu, S., 2017. Experimental evaluation of a diesel engine fuelled by pyrolysis oils produced from low-density polyethylene and ethylene–vinyl acetate plastics. Fuel Process. Technol. 161, 125–131. In: https://doi.org/10.1016/j.fuproc.2017.03.014.

Kamm, B., Gruber, P.R., Kamm, M., 2016. Biorefineries-Industrial Processes and Products Ullmann's Encyclopedia of Industrial Chemistry, pp. 1–38.

Kamm, B., Kamm, M., 2004. Biorefinary – systems. Chem. Biochem. Eng. Q. 18 (1), 1–6.

Kawaguchi, H., Ogino, C., Kondo, A., 2017. Microbial conversion of biomass into bio-based polymers. Bioresour. Technol. 245 (Pt B), 1664–1673. https://doi.org/10.1016/j.biortech.2017.06.135.

Khalid, A., Arshad, M., Anjum, M., Mahmood, T., Dawson, L., 2011. The anaerobic digestion of solid organic waste. Waste Manag. 31 (8), 1737–1744.

Kim, T.H., Taylor, F., Hicks, K.B., 2008. Bioethanol production from barley hull using SAA (soaking in aqueous ammonia) pretreatment. Bioresour. Technol. https://doi.org/10.1016/j.biortech.2007.10.055.

Knothe, G., 2009. Improving biodiesel fuel properties by modifying fatty ester composition. Energy Environ. Sci. 2 (7), 759–766. https://doi.org/10.1039/b903941d.

Kojima, S., Du, D., Sato, M., Park, E.Y., 2004. Efficient production of fatty acid methyl ester from waste activated bleaching earth using diesel oil as organic solvent. J. Biosci. Bioeng. 98 (6), 420–424. https://doi.org/10.1016/S1389-1723(05)00306-3.

Komonkiat, I., Cheirsilp, B., 2013. Felled oil palm trunk as a renewable source for biobutanol production by *Clostridium* spp. Bioresour. Technol. https://doi.org/10.1016/j.biortech.2013.07.067.

Kongjan, P., Angelidaki, I., 2010. Extreme thermophilic biohydrogen production from wheat straw hydrolysate using mixed culture fermentation: effect of reactor configuration. Bioresour. Technol. 101 (20), 7789–7796.

Kongjan, P., O-Thong, S., Kotay, M., Min, B., Angelidaki, I., 2010. Biohydrogen production from wheat straw hydrolysate by dark fermentation using extreme thermophilic mixed culture. Biotechnol. Bioeng. 105 (5), 899–908.

Kotsyurbenko, O.R., Chin, K.J., Glagolev, M.V., Stubner, S., Simankova, M.V., Nozhevnikova, A.N., Conrad, R., 2004. Acetoclastic and hydrogenotrophic methane production and methanogenic populations in an acidic West-Siberian peat bog. Environ. Microbiol. 6 (11), 1159–1173.

Kumar, M., Gayen, K., 2011. Developments in biobutanol production: New insights. Applied Energy 88 (6), 1999–2012.

Lay, J., Li, Y., Noike, T., Endo, J., Ishimoto, S., 1997. Analysis of environmental factors affecting methane production from high-solids organic waste. Water Sci. Technol. 36 (6–7), 493–500.

Lee, U., Han, J., Wang, M., 2017b. Evaluation of landfill gas emissions from municipal solid waste landfills for the life-cycle analysis of waste-to-energy pathways. J. Clean. Prod. 166, 335–342.

Lee, C.R., Sung, B.H., Lim, K.M., Kim, M.J., Sohn, M.J., Bae, J.H., Sohn, J.H., 2017a. Co-fermentation Using Recombinant *Saccharomyces cerevisiae* Yeast Strains Hyper-Secreting Different Cellulases for the Production of Cellulosic Bioethanol/631/61/252/631/326/2522/82/80 Article. Scientific Reports. https://doi.org/10.1038/s41598-017-04815-1.

Leung, D.Y.C., Guo, Y., 2006. Transesterification of neat and used frying oil: optimization for biodiesel production. Fuel Process. Technol. 87 (10), 883–890. https://doi.org/10.1016/j.fuproc.2006.06.003.

Levin, D.B., Chahine, R., 2010. Challenges for renewable hydrogen production from biomass. Int. J. Hydrogen Energy 35 (10), 4962–4969. https://doi.org/10.1016/j.ijhydene.2009.08.067.

Li, L., Ai, H., Zhang, S., Li, S., Liang, Z., Wu, Z.Q., et al., 2013. Enhanced butanol production by coculture of *Clostridium beijerinckii* and *Clostridium tyrobutyricum*. Bioresour. Technol. https://doi.org/10.1016/j.biortech.2013.06.023.

Liu, C.-M., Chu, C.-Y., Lee, W.-Y., Li, Y.-C., Wu, S.-Y., Chou, Y.-P., 2013a. Biohydrogen production evaluation from rice straw hydrolysate by concentrated acid pre-treatment in both batch and continuous systems. Int. J. Hydrogen Energy 38 (35), 15823–15829.

Liu, G., Yan, B., Chen, G., 2013b. Technical review on jet fuel production. Renew. Sustain. Energy Rev. 25, 59–70. https://doi.org/10.1016/j.rser.2013.03.025.

Luo, G., Talebnia, F., Karakashev, D., Xie, L., Zhou, Q., Angelidaki, I., 2011. Enhanced bioenergy recovery from rapeseed plant in a biorefinery concept. Bioresour. Technol. 102 (2), 1433–1439. https://doi.org/10.1016/j.biortech.2010.09.071.

Maier, W.F., Roth, W., Thies, I., Schleyer, P.V.R., 1982. Hydrogenolysis, IV. Gas phase decarboxylation of carboxylic acids. Chem. Ber. 115 (2), 808–812. https://doi.org/10.1002/cber.19821150245.

Maity, S.K., 2015. Opportunities, recent trends and challenges of integrated biorefinery: part I. Renew. Sustain. Energy Rev. 43, 1427–1445. https://doi.org/10.1016/j.rser.2014.11.092.

Malinauskaite, J., Jouhara, H., Czajczyńska, D., Stanchev, P., Katsou, E., Rostkowski, P., et al., 2017. Municipal solid waste management and waste-to-energy in the context of a circular economy and energy recycling in Europe. Energy 141, 2013–2044.

Marchetti, J.M., 2012. A summary of the available technologies for biodiesel production based on a comparison of different feedstock's properties. Process Saf. Environ. Protect. 90 (3), 157–163. https://doi.org/10.1016/j.psep.2011.06.010.

Martínez, G., Sánchez, N., Encinar, J.M., González, J.F., 2014. Fuel properties of biodiesel from vegetable oils and oil mixtures. Influence of methyl esters distribution. Biomass and Bioenergy 63, 22–32. https://doi.org/10.1016/j.biombioe.2014.01.034.

Mejía-Barajas, J.A., Alvarez-Navarrete, M., Saavedra-Molina, A., Campos-García, J., Valenzuela-Vázquez, U., Amaya-Delgado, L., Arellano-Plaza, M., 2018. Second-generation bioethanol production through a simultaneous saccharification-fermentation process using *Kluyveromyces marxianus* thermotolerant yeast. In: Yüksel, E. (Ed.), Special Topics in Renewable Energy Systems: IntechOpen. https://doi.org/10.5772/intechopen.78052.

Menon, V., Rao, M., 2012. Trends in bioconversion of lignocellulose: biofuels, platform chemicals & biorefinery concept. Prog. Energy Combust. Sci. 38 (4), 522–550. https://doi.org/10.1016/j.pecs.2012.02.002.

Michailos, S.E., Webb, C., 2019. Biorefinery Approach for Ethanol Production From Bagasse Bioethanol Production from Food Crops, pp. 319–342.

Mohammadi Nafchi, A., Moradpour, M., Saeidi, M., Alias, A.K., 2013. Thermoplastic starches: properties, challenges, and prospects. Starch – Stärke 65 (1–2), 61–72. https://doi.org/10.1002/star.201200201.

Moncada, B.,J., Aristizábal, M.,V., Cardona, A.,C.A., 2016. Design strategies for sustainable biorefineries. Biochem. Eng. J. 116, 122–134. https://doi.org/10.1016/j.bej.2016.06.009.

Montiel Corona, V., Razo-Flores, E., 2018. Continuous hydrogen and methane production from *Agave tequilana* bagasse hydrolysate by sequential process to maximize energy recovery efficiency. Bioresour. Technol. 249, 334–341. https://doi.org/10.1016/j.biortech.2017.10.032.

Moreno, A.I., Font, R., Conesa, J.A., 2016. Physical and chemical evaluation of furniture waste briquettes. Waste Manag. 49, 245–252. https://doi.org/10.1016/j.wasman.2016.01.048.

Naik, S.N., Goud, V.V., Rout, P.K., Dalai, A.K., 2010. Production of first and second generation biofuels: a comprehensive review. Renew. Sustain. Energy Rev. 14 (2), 578–597. https://doi.org/10.1016/j.rser.2009.10.003.

Neumann, J., Binder, S., Apfelbacher, A., Gasson, J.R., Ramírez García, P., Hornung, A., 2015. Production and characterization of a new quality pyrolysis oil, char and syngas from digestate – introducing the thermo-catalytic reforming process. J. Anal. Appl. Pyrolysis 113, 137–142. https://doi.org/10.1016/j.jaap.2014.11.022.

Niedziolka, I., Szpryngel, M., Kachel-Jakubowska, M., Kraszkiewicz, A., Zawislak, K., Sobczak, P., Nadulski, R., 2015. Assessment of the energetic and mechanical properties of pellets produced from agricultural biomass *. Renew. Energy 76, 312–317. https://doi.org/10.1016/j.renene.2014.11.040.

Niehus, X., Casas-Godoy, L., Rodríguez-Valadez, F.J., Sandoval, G., 2018a. Evaluation of *Yarrowia lipolytica* oil for biodiesel production: land use oil yield, carbon, and energy balance. J. Lipids 2018. https://doi.org/10.1155/2018/6393749, 6393749-6393749.

Niehus, X., Crutz-Le Coq, A.-M., Sandoval, G., Nicaud, J.-M., Ledesma-Amaro, R., 2018b. Engineering *Yarrowia lipolytica* to enhance lipid production from lignocellulosic materials. Biotechnol. Biofuels 11. https://doi.org/10.1186/s13068-018-1010-6, 11-11.

Nissilä, M.E., Lay, C.-H., Puhakka, J.A., 2014. Dark fermentative hydrogen production from lignocellulosic hydrolyzates – a review. Biomass Bioenergy 67 (0), 145–159. https://doi.org/10.1016/j.biombioe.2014.04.035.

Nonato, R.V., Mantelatto, P.E., Rossell, C.E., 2001. Integrated production of biodegradable plastic, sugar and ethanol. Appl. Microbiol. Biotechnol. 57 (1–2), 1–5. https://doi.org/10.1007/s002530100732.

Ognjanovic, N., Bezbradica, D., Knezevic-Jugovic, Z., 2009. Enzymatic conversion of sunflower oil to biodiesel in a solvent-free system: process optimization and the immobilized system stability. Bioresour. Technol. 100 (21), 5146–5154. https://doi.org/10.1016/j.biortech.2009.05.068.

Park, E.Y., Sato, M., Kojima, S., 2008. Lipase-catalyzed biodiesel production from waste activated bleaching earth as raw material in a pilot plant. Bioresour. Technol. 99 (8), 3130–3135. https://doi.org/10.1016/j.biortech.2007.05.059.

Paulova, L., Patakova, P., Branska, B., Rychtera, M., Melzoch, K., 2015. Lignocellulosic ethanol: technology design and its impact on process efficiency. Biotechnol. Adv. 33 (6, Part 2), 1091–1107. https://doi.org/10.1016/j.biotechadv.2014.12.002.

Petkov, G., Ivanova, A., Iliev, I., Vaseva, I., 2012. A critical look at the microalgae biodiesel. Eur. J. Lipid Sci. Technol. 114 (2), 103–111. https://doi.org/10.1002/ejlt.201100234.

Raftery, J.P., Karim, M.N., 2017. Economic viability of consolidated bioprocessing utilizing multiple biomass substrates for commercial-scale cellulosic bioethanol production. Biomass Bioenergy. https://doi.org/10.1016/j.biombioe.2017.05.012.

Rodrigues, J., Canet, A., Rivera, I., Osório, N.M., Sandoval, G., Valero, F., Ferreira-Dias, S., 2016. Biodiesel production from crude Jatropha oil catalyzed by non-commercial immobilized heterologous *Rhizopus oryzae* and *Carica papaya* lipases. Bioresour. Technol.

RSB, 2019. Alternative Aviation Fuels -A Sustainables Future is Taking Off. Retrieved from. https://rsb.org/wp-content/uploads/2018/09/RSB-Alternative-Aviation-Fuels-A-Sustainable-Future-is-Taking-Off.pdf.

Sagiroglu, A., Selen, I., Ozcan, M., Paluzar, H., Toprakkiran, N., 2011. Comparison of biodiesel productivities of different vegetable oils by acidic catalysis. Chem. Ind. Chem. Eng. Q. 17 (1), 53–58. https://doi.org/10.2298/ciceq100114054s.

Sagnak, R., Kapdan, I.K., Kargi, F., 2010. Dark fermentation of acid hydrolyzed ground wheat starch for bio-hydrogen production by periodic feeding and effluent removal. Int. J. Hydrogen Energy 35 (18), 9630–9636.

Saini, J.K., Saini, R., Tewari, L., 2015. Lignocellulosic agriculture wastes as biomass feedstocks for second-generation bioethanol production: concepts and recent developments. 3 Biotech 5 (4), 337–353. https://doi.org/10.1007/s13205-014-0246-5.

Sandoval, G., Casas-Godoy, L., Bonet-Ragel, K., Rodrigues, J., Ferreira-Dias, S., Valero, F., 2017. Enzyme-catalyzed production of biodiesel as alternative to chemical-catalyzed processes: advantages and constraints. Curr. Biochem. Eng. 4, 109–141.

Sandoval-Nuñez, D., Arellano-Plaza, M., Gschaedler, A., Arrizon, J., Amaya-Delgado, L., 2018. A comparative study of lignocellulosic ethanol productivities by *Kluyveromyces marxianus* and *Saccharomyces cerevisiae*. Clean Technol. Environ. Policy 20 (7), 1491–1499. https://doi.org/10.1007/s10098-017-1470-6.

Sanguanchaipaiwong, V., Leksawasdi, N., 2018. Butanol production by *Clostridium beijerinckii* from pinneapple waste juice. Energy Procedia 153, 231–236.

Santana, A., Maçaira, J., Larrayoz, M.A., 2012. Continuous production of biodiesel from vegetable oil using supercritical ethanol/carbon dioxide mixtures. Fuel Process. Technol. 96, 214–219. https://doi.org/10.1016/j.fuproc.2011.12.021.

Santiago-Torres, N., Romero-Ibarra, I.C., Pfeiffer, H., 2014. Sodium zirconate (Na_2ZrO_3) as a catalyst in a soybean oil transesterification reaction for biodiesel production. Fuel Process. Technol. 120, 34–39. https://doi.org/10.1016/j.fuproc.2013.11.018.

Santos, A.L.F., Martins, D.U., Iha, O.K., Ribeiro, R.A.M., Quirino, R.L., Suarez, P.A.Z., 2010. Agro-industrial residues as low-price feedstock for diesel-like fuel production by thermal cracking. Bioresour. Technol. 101 (15), 6157–6162. https://doi.org/10.1016/j.biortech.2010.02.100.

Scholey, D.V., Burton, E.J., Williams, P.E., 2016. The bio refinery; producing feed and fuel from grain. Food Chem. 197 (Pt A), 937–942. https://doi.org/10.1016/j.foodchem.2015.11.063.

Scholten, J.C.M., Stams, A.J.M., 1995. The effect of sulfate and nitrate on methane formation in a freshwater sediment. Antonie Van Leeuwenhoek 68 (4), 309–315. https://doi.org/10.1007/bf00874141.

Senol, O.İ., Ryymin, E.M., Viljava, T.R., Krause, A.O.I., 2007. Effect of hydrogen sulphide on the hydrodeoxygenation of aromatic and aliphatic oxygenates on sulphided catalysts. J. Mol. Catal. A Chem. 277 (1–2), 107–112. https://doi.org/10.1016/j.molcata.2007.07.033.

Senol, O.İ., Viljava, T.R., Krause, A.O.I., 2005. Hydrodeoxygenation of aliphatic esters on sulphided NiMo/γ-Al_2O_3 and CoMo/γ-Al_2O_3 catalyst: the effect of water. Catal. Today 106 (1–4), 186–189. https://doi.org/10.1016/j.cattod.2005.07.129.

Shweta Shah, M.N.G., 2007. Lipase catalyzed preparation of biodiesel from *Jatropha* oil in a solvent free system. Process Biochem. 42 (3), 6. In: https://doi.org/10.1016/j.procbio.2006.09.024.

Sieber, J.R., McInerney, M.J., Plugge, C.M., Schink, B., Gunsalus, R.P., 2010. Methanogenesis: syntrophic metabolism. In: Timmis, K.N. (Ed.), Handbook of Hydrocarbon and Lipid Microbiology. Springer Berlin Heidelberg, Berlin, Heidelberg, pp. 337–355.

Silitonga, A.S., Ong, H.C., Mahlia, T.M.I., Masjuki, H.H., Chong, W.T., 2014. Biodiesel conversion from high FFA crude *Jatropha curcas*, *Calophyllum inophyllum* and *Ceiba pentandra* oil. Energy Procedia 61, 480–483. https://doi.org/10.1016/j.egypro.2014.11.1153.

Singh, S.P., Singh, D., 2010. Biodiesel production through the use of different sources and characterization of oils and their esters as the substitute of diesel: a review. Renew. Sustain. Energy Rev. 14 (1), 200–216. https://doi.org/10.1016/j.rser.2009.07.017.

Solano, O.R., Vargas, M.F., Watson, R.G., 2016. Biodigesters: chemical, physical and biological factors related to their productivity. Tecnología en Marcha 30 (1), 47–53.

Stelt, M.J.C.V.D., Gerhauser, H., Kiel, J.H.A., Ptasinski, K.J., 2011. Biomass upgrading by torrefaction for the production of biofuels: a review. Biomass Bioenergy 35, 3748–3762. https://doi.org/10.1016/j.biombioe.2011.06.023.

Stelte, W., Sanadi, A.R., Shang, L., Holm, J.K., Ahrenfeldt, J., Henriksen, U.B., 2012. Recent developments in biomass pelletization – a review. BioResources 7, 4451–4490.

Strazzera, G., Battista, F., Garcia, N.H., Frison, N., Bolzonella, D., 2018. Volatile fatty acids production from food wastes for biorefinery platforms: a review. J. Environ. Manag. 226, 278–288. https://doi.org/10.1016/j.jenvman.2018.08.039.

Sudhanshu, S., Behera, R.C.R., 2019. Chapter 12 – forest bioresources for bioethanol and biodiesel production with emphasis on mohua (*Madhuca latifolia* L.) flowers and seeds. Sustainable Sources, Interventions, and Challenges. https://doi.org/10.1016/B978-0-12-813766-6.00012-6.

Thrän, D., Billig, E., Persson, T., Svensson, M., Daniel-Gromke, J., Ponitka, J., Seiffert, M., 2014. Biomethane: Status and Factors Affecting Market Development and Trade: A Joint Study. IEA Bioenergy.

Toledo-Cervantes, A., Arreola-Vargas, J., Elias-Palacios, S.V., Marino-Marmolejo, E.N., Davila-Vazquez, G., González-Álvarez, V., Méndez-Acosta, H.O., 2018. Evaluation of semi-continuous hydrogen production from enzymatic hydrolysates of *Agave tequilana* bagasse: insight into the enzymatic cocktail effect over the co-production of methane. Int. J. Hydrogen Energy 43 (31), 14193–14201.

Trindade, W.R.D.S., Santos, R.G.D., 2017. Review on the characteristics of butanol, its production and use as fuel in internal combustion engines. Renew. Sustain. Energy Rev. 69, 642–651. https://doi.org/10.1016/j.rser.2016.11.213.

Tumuluru, J.S., Cafferty, K.G., Kenney, K.L., 2014. Techno-economic analysis of conventional, high moisture pelletization and briquetting process. In: CSBE/SCGAB Annual International Meeting.

Tumuluru, J.S., Wright, C.T., Hess, J.R., Kenney, K.L., 2011. A review of biomass densification systems to develop uniform feedstock commodities for bioenergy application. Biofuels Bioprod. Biorefining 5, 683–707. https://doi.org/10.1002/bbb.

Ueno, Y., Fukui, H., Goto, M., 2007. Operation of a two-stage fermentation process producing hydrogen and methane from organic waste. Environ. Sci. Technol. 41 (4), 1413–1419.

Ueno, Y., Otsuka, S., Morimoto, M., 1996. Hydrogen production from industrialwastewater by anaerobic microflora in chemostat culture. J. Ferment. Bioeng. 82 (2), 194–197.

Vajpeyi, S., Chandran, K., 2015. Microbial conversion of synthetic and food waste-derived volatile fatty acids to lipids. Bioresour. Technol. 188, 49–55. https://doi.org/10.1016/j.biortech.2015.01.099.

van Dyk, J.S., Li, L., Barros-Leal, D., Hu, J., Zhang, X., Tan, T., Saddler, J., 2016. The Potential of Biofuels in China. Retrieved from: http://task39.sites.olt.ubc.ca/files/2013/05/The-Potential-of-biofuels-in-China-IEA-Bioenergy-Task-39-September-2016.pdf.

Vargas, M., Niehus, X., Casas-Godoy, L., Sandoval, G., 2018. Lipases as biocatalyst for biodiesel production. In: Sandoval, G. (Ed.), Lipases and Phospholipases: Methods and Protocols. Springer New York, New York, NY, pp. 377–390.

Venetsaneas, N., Antonopoulou, G., Stamatelatou, K., Kornaros, M., Lyberatos, G., 2009. Using cheese whey for hydrogen and methane generation in a two-stage continuous process with alternative pH controlling approaches. Bioresour. Technol. 100 (15), 3713–3717.

Wang, W.-C., Tao, L., 2016. Bio-jet fuel conversion technologies. Renew. Sustain. Energy Rev. 53, 801–822. https://doi.org/10.1016/j.rser.2015.09.016.

Weber, C., Farwick, A., Benisch, F., Brat, D., Dietz, H., Subtil, T., Boles, E., 2010. Trends and challenges in the microbial production of lignocellulosic bioalcohol fuels. Appl. Microbiol. Biotechnol. 87 (4), 1303–1315.

Wu, S., Zhang, S., Wang, C., Mu, C., Huang, X., 2018. High-strength charcoal briquette preparation from hydrothermal pretreated biomass wastes. Fuel Process. Technol. 171, 293–300. https://doi.org/10.1016/j.fuproc.2017.11.025.

Xue, C., Zhao, X.Q., Liu, C.G., Chen, L.J., Bai, F.W., 2013. Prospective and development of butanol as an advanced biofuel. Biotechnol. Adv. 31 (8), 1575–1584.

Yang, M., Kuittinen, S., Keinänen, M., Vepsäläinen, J., Romar, H., Tynjälä, P., et al., 2014. The use of (green field) biomass pretreatment liquor for fermentative butanol production and the catalytic oxidation of biobutanol. Chem. Eng. Res. Des. https://doi.org/10.1016/j.cherd.2014.01.015.

Yang, X., Wang, H., Strong, P., Xu, S., Liu, S., Lu, K., He, L., 2017. Thermal properties of biochars derived from waste biomass generated by agricultural and forestry sectors. Energies 10, 469.

Yang, J., Xin, Z., He, Q., Corscadden, K., Niu, H., 2019. An overview on performance characteristics of bio-jet fuels. Fuel 237, 916–936. https://doi.org/10.1016/j.fuel.2018.10.079.

Yank, A., Ngadi, M., Kok, R., 2016. Biomass and Bioenergy Physical properties of rice husk and bran briquettes under low pressure densification for rural applications. Biomass and Bioenergy, 84, 22–30. https://doi.org/10.1016/j.biombioe.2015.09.015.

Yu, H., Zhu, Z., Hu, W., Zhang, H., 2002. Hydrogen production from rice winery wastewater in an upflow anaerobic reactor by using mixed anaerobic cultures. Int. J. Hydrogen Energy. 27 (11–12), 1359–1365.

Zamorano, M., Perez, J.I.P., Paves, I.A., Ridao, A.R., 2007. Study of the energy potential of the biogas produced by an urban waste landfill in Southern Spain. Renew. Sustain. Energy Rev. 11 (5), 909–922. https://doi.org/10.1016/j.rser.2005.05.007.

Zhang, Y., Wong, W.-T., Yung, K.-F., 2014. Biodiesel production via esterification of oleic acid catalyzed by chlorosulfonic acid modified zirconia. Appl. Energy 116, 191–198. https://doi.org/10.1016/j.apenergy.2013.11.044.

Zheng, S., Kates, M., Dube, M.A., McLean, D.D., 2006. Acid-catalyzed production of biodiesel from waste frying oil. Biomass Bioenergy 30 (3), 267–272. https://doi.org/10.1016/j.biombioe.2005.10.004.

Zheng, Y.N., Li, L.Z., Xian, M.O., Ma, Y.J., Yang, J.M., Xu, X., He, D.Z., 2009. Problems with the microbial production of butanol. Journal of Industrial Microbiology and Biotechnology 36 (9), 1127–1138.

ZMR, 2018. Biofuels market analysis by type (bioethanol, biodiesel), and by form (solid, liquid, and gaseous) – global industry perspective, comprehensive analysis, and forecast, pp. 2016–2022. Retrieved from. https://www.zionmarketresearch.com/news/biofuels-market.

Chapter 5

Bio-based production of chemicals through metabolic engineering

Louis M.M. Mouterde, Florent Allais
URD Agro-Biotechnologies Industrielles, CEBB, AgroParisTech, Pomacle, France

5.1 Introduction

In a context of increasing scarcity and price of fossil resources, petro-based chemical sourcing is a central concern among both the scientific and industrial communities. Nowadays, 70% of crude oil is used as fuel and 30% by chemical industries (Nations, 2016). The demand keeps on growing from year to year, while the oil reserve is irremediably diminishing. This can result in excessively expensive crude oil and can provoke both an energetic and economic crisis, bringing unprecedented consequences to our planet and to our industrialized society.

In order to avoid these outcomes, the use of available biomass as renewable carbon sources has been considered and widely studied. Although it is very attractive, such an approach raises multiple questions. Is there enough feedstock? Is it sustainable? Are biomass prices going to increase because of the increasing consumption of this resource? The fact is: biomass is limited and most of the industrialized countries, such as USA and China, or regions such as Europe, North Africa, and Middle East, have already been considered as "biocapacity debtors" in 2014. This means that their footprints are larger than their biocapacities. Moreover, these countries represent less than 50% of Earth's land surface. In other words, the world's ecological footprint is higher than its biocapacity. Under such considerations, the conclusion that can be drawn is that the surface of a 1.7 Earth would be necessary to meet the needs of humanity. It is also clear that all energy usages cannot be compensated with the available biomass (Fig. 5.1) (Lin et al., 2018).

The fuel transition from fossil resources to biomass has already started with first-generation bioethanol that consists in fermenting sugars contained in starch, found in feedstocks such as cereals (e.g., wheat, barley, corn, and

Biobased Products and Industries. https://doi.org/10.1016/B978-0-12-818493-6.00005-1

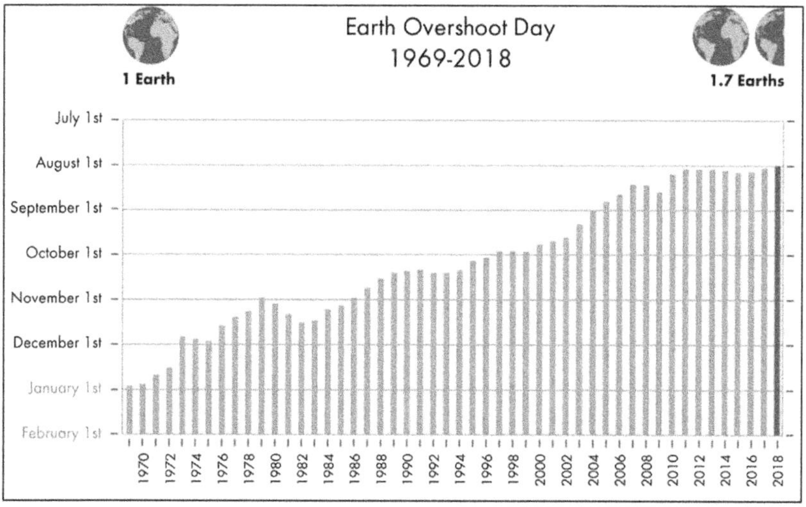

FIGURE 5.1 An overview of human activities global footprint (Lin et al., 2018).

sorghum), and saccharose (e.g., sugarcane, beetroot). A massive drawback of this approach is that it uses the eatable part of the feedstock, resulting in either the overexploitation of the land to produce them or in the potential destabilization of the existing food/feed sectors. Therefore, second-generation bioethanol has been developed where the sugar sources are cellulose and hemicelluloses, two nonfood resources. These polysaccharides are found in feedstocks such as wheat straw, corn stover, wood or agricultural residues, byproducts that otherwise would be of low value. As a continuation of those efforts, the possibility of replacing petro-based chemicals (e.g., alcohols, dicarboxylic acids, diols, alkanes, alkenes, aromatics, and thermoplastics) was explored. This transition must, however, meet quite drastic industrial requirements: (1) feedstocks must be readily available in large enough quantities and at a low price, (2) it is imperative that their developed processes are at high yields and purities, (3) their systems are robust and they are inserted into an existing industrial ecosystem.

Along with the need to substitute petro-based chemicals on account of their decreasing availability, industries are also highly motivated to realize this transition, forced by increasingly stringent regulations (e.g., REACH), and influenced by the growing demand for natural or bio-based products, especially for the pharmaceutical, cosmetic, and food industries. As a consequence, biorefineries that convert biomass into fuels, power, heat, and value-added chemicals have emerged. In synergy with existing industries, these biorefineries participate to a circular economy, also called "blue economy" in which the byproduct of one process becomes the starting material of another one.

While this biorefinery approach is very promising, there is still plenty of room for improvement. For instance, the scale-up of biotechnological processes is a real issue. As for chemical transformations, it is a difficult and long process to go from the laboratory to the industrial reactor. However, when it comes to fermentation, new constraints must be taken into consideration during the development of the process. Microorganisms must be robust, with a low contamination risk, in order to resist to harsh reactor conditions (e.g., temperature, pH, pressure, organic solvent). Fermentation conditions also need to be anticipated (aerobic, anaerobic) depending on the specifications of the industry involved. To address all these potential constraints, metabolic engineering has been developed. Such genetic and regulatory processes optimizations within microorganisms to increase the production of a specific chemical are already an industrial reality.

5.2 Metabolic engineering

The capacity of a microorganism to produce a specific molecule is defined by its genomic DNA. However, the corresponding metabolic pathways do not lead to all known petro-based chemicals and the natural productivity does not necessarily meet the industrial needs. *Metabolic engineering* is a scientific field that aims to overcome these limitations. It was defined by Bailey in 1981 as "the improvement of cellular activities by manipulating enzymatic, regulatory, and transport functions of the cell" (Bailey, 1991). Until nowadays, it relies on the principle of fermentation. A source of carbon, typically glucose, or another sugar, is made available to a particular microorganism to be catabolized into pyruvate. The latter will be converted into acetyl-CoA which "feeds" the Krebs cycle and the respiratory chain, allowing the production of energy in the form of adenosine triphosphate (ATP). The metabolites produced by these pathways are used for the biosynthesis of the necessary buildings blocks (e.g., amino acids, ribonucleotides, deoxyribonucleotides, carbohydrates, fatty acids, vitamins) needed to synthesize macromolecules or biopolymers (e.g., proteins, lipids, nucleic acids, polysaccharides).

Metabolic engineering consists in making cells overproduce a metabolite or a biosynthetic product of interest. The first technique used to improve strains consisted in inducing uncontrolled mutations in genomic DNA using carcinogen chemicals. However, the randomness of this method makes it time-consuming and the target phenotype is not necessarily reached. The progress that was made in recombinant DNA and gene knockout technology provided a powerful toolbox. It thus became possible to improve directly specific properties through the modification of particular biochemical reactions.

Metabolic engineering was brought to a new level with the advent of synthetic biology. This strategy consists in using computer-assisted biological engineering to (re)design and build biological components and systems that do not exist in nature. It differs from classical techniques by bringing an "engineering design and development" component (primarily computation modeling) to the thinking process. Using this strategy, a significant number of molecules have been produced by academic laboratories for a wide variety of industrial sectors, essentially agricultural, pharmaceutical, cosmetics, food, or energy industries.

5.3 Different carbon sources for the microorganisms

When producing high added value chemicals/polymers via metabolic engineering, a carbon source is required to feed the microorganisms of interest. The first generation of bioprocesses used sugars, mainly glucose, or molasses that had been fermented to give the target molecule. This is the case for first-generation bioethanol produced through natural strains (*Saccharomyces cerevisiae* or *Zymomonas mobilis*). While this approach is perfectly valid for small and medium-scale production, it is not applicable for a global industrial scale. Indeed, several times the production capacity of land-based agriculture would be necessary to completely substitute the quantity of fossil hydrocarbons that is currently extracted. Knowing that 80% of crude oil is used for fuels, first-generation bioethanol undoubtedly cannot meet this demand. Moreover, the overexploitation of the land with soybean, cereal, cane, and sugar beet crops leads to depletion of soils and water resources. At the same time, the use of herbicides and pesticides leads to the disappearance of insects and increases pollution. For that reason, significant efforts needed to be done in order to find an environmentally friendly alternative solution.

A second generation of carbon was found in biomasses. These plant and agricultural residues and other organic waste are highly abundant on the planet and very little valued. They are composed for the most part of cellulose and hemi-cellulose (~75%); polymers structured around six different buildings blocks: glucose, xylose, galactose, mannose, arabinose, and rhamnose. It is a massive reserve of renewable carbon that is only waiting to be used. However, most microorganisms cannot use biomass directly because of the complexity of the structures. Biomass must therefore be fractionated and decomposed beforehand to release the sugars that compose it.

There are two major approaches with the intent on deconstructing biomass. The first one is thermochemical degradation, such as gasification or liquefaction processes. These methods consist in roasting and gasification. The first one stabilizes the biomass, and the second gasifies the substrate under a controlled atmosphere, as a way to obtain a gas composed of simple organic molecules (carbon monoxide and hydrogen). The latter can be then treated by conventional chemical reactions or digested by microorganisms. Yet, this technology is energy-intensive and produces substantial carbon dioxide. It can thus only be considered in the case of a superabundant and low-cost biomass.

The second approach consists in decomposing the biomass without affecting the integrity of the building blocks. This approach is particularly relevant for the production of chemicals through metabolic engineering in which microorganisms use glucose, arabinose, or xylose as carbon source. Such an approach is generally based on a steam explosion pretreatment followed by an enzymatic hydrolysis. The sugars that are obtained through this process can then be directly used for fermentation. The next step in this strategy is to artificially manufacture a strain or a consortium of microbial strains capable of digesting complex carbon matrices and produce a finished compound in a single step.

Also, encouraging works are currently being performed toward a third generation of carbon, the carbon dioxide present in our atmosphere. There are two different strategies developed around this research topic: (1) microalgae that utilize light, atmospheric carbon dioxide, and photosynthesis as energy source, and (2) natural or genetically modified microorganisms that use hydrogen and atmospheric carbon dioxide as energy source.

5.4 Success in producing chemicals through metabolic engineering

Nowadays, microorganisms can produce a wide range of bio-based chemicals and polymers that are useful for various industries. These chemicals can have very different applications going from fuel and commodity plastics to pharmaceutical, cosmetics, and agricultural products.

5.4.1 Biofuels

Historically, lower alcohols, such as ethanol and butanol were the first chemicals produced by fermentation. There are two principal reasons for this particular interest. Firstly, studies on several microorganisms (*Escherichia coli, S. cerevisiae, Clostridium acetobutylicum,* or *Z. mobilis*) showed that natural metabolic pathways were leading to these chemicals. Secondly, and most importantly, these lower alcohols can be readily used as fuels.

Considerable amount of work was done on the production of bioethanol. Many microorganisms were studied, going from Gram-negative bacterium to yeast. It is probably, along with *n*-butanol, the molecule that gathers the most data in the literature when it comes to fermentation. The original metabolic engineering work that led to production of ethanol was realized in 1987 where Ingram et al. inserted the genes encoding essential enzymes of the fermentative pathway for ethanol production in *Z. mobilis* into *E. coli* (Ingram ct al., 1987). This modified strain, as well as the fermentation conditions, were deeply studied and optimized to obtain an efficient microorganism capable of producing ethanol from glucose and xylose at a titer of 48 and 40 g/L, respectively (Neale et al., 1988; Alterthum and Ingram, 1989; Ohta et al., 1991). Further work was realized to enhance ethanol tolerance in *E. coli* (Gonzalez et al., 2003) and improve the productivity (Trinh et al., 2008) or substitute the carbon source toward glycerol (Trinh and Srienc, 2009). Other bacteria strains (*C. phytofermentans, C. thermocellum, C. glutamicum, C. cellulovorans* DSM 743B, and *C. beijerinckii* NCIMB 8052 consortium, and *Z. mobilis*) (Tolonen et al., 2011; Balusu et al., 2004; Inui et al., 2005; Wen et al., 2017; Lee et al., 2010; Rogers et al., 2007) were also extensively studied. While none of them reach the productivity level of *E. coli*, some proved to be quite interesting on account of their abilities to digest directly cellulosic biomass into ethanol. Finally, some yeast strains (*Pichia stipitis* and *S. cerevisiae*) (Jeffries et al., 2007; Bai et al., 2008; Bro et al., 2006; Hjersted et al., 2007; Mo et al., 2009; Pereira et al., 2010) also demonstrated very good productivity (~125 g/L). The utilization of yeast for industrial purposes is particularly interesting as these robust microorganisms are capable of working in acidic conditions (pH 4), where risk of contamination is low.

Although metabolic engineering succeeded in building microorganisms capable of producing bioethanol at an industrial level, using it directly as fuel poses some critical issues. First, its efficiency is one-third lower than traditional fuel. Secondly, its high hygroscopic properties can damage the engine. Therefore, it is highly recommended to use it in a mix with traditional fuel.

In order to overcome the issues cited above, the scientific community put a great deal of effort toward the production of *n*-butanol, isobutanol, propanol, and isopropanol due to their better efficiency and lower hygroscopic properties. The first reported work was that of Charles Weizmann who discovered a microorganism capable of producing acetone and *n*-butanol from maize or other grain starch (Weizmann, 1919). Unfortunately, the name of the microorganism and the productivity data do not appear in his patent document. As for ethanol, *E. coli* was largely studied and it resulted in engineered strains that produce *n*-butanol with relatively high productivity (~30 g/L) (Shen and Liao, 2008; Atsumi et al., 2008a; Inui et al., 2008; Gulevich et al., 2012; Saini et al., 2015; Shen et al., 2011). Other studied microorganisms provided lower accumulation levels but used different carbon source such as mannitol or xylose (Wen et al., 2017; Yu et al., 2011; Xiao et al., 2012). Although the accumulation levels cannot be compared, interesting work was done on cyanobacteria that were capable of producing *n*-butanol using CO_2 as carbon source (Lan and Liao, 2011). However, it has to be noted that, in the case of *n*-butanol, metabolic engineering did not succeed in producing the best strain in term of productivity yet. Still nowadays, the natural strain *C. acetobutylicum* is capable of reaching accumulation levels higher than 100 g/L (Lee et al., 2012). Unfortunately, metabolic engineered strains are far from this level of productivity (Table 5.1).

Other alcohols, such as 1-propanol, *iso*-butanol, and *iso*-propanol, have been also considered as biofuels and strains were engineered to give relatively good productivity levels. The metabolic engineered strains were especially efficient in the case of *iso*-butanol and *iso*-propanol, reaching productivity levels of 50 and 140 g/L, respectively (Atsumi et al., 2008b, Baez et al., 2011;

TABLE 5.1 List of biofuels produced by metabolic engineered strains.

Product	Organism	Carbon source	Maximum productivity (mg/L)
Ethanol 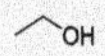	*E. coli*	Glucose	48,000
		Glycerol	810
	C. phytofermentans	Cellulose	1,150
	C. thermocellum	Cellulose biomass	3,210
	C. glutamicum	Glucose	9,200
	C. cellulovorans DSM 743B and C. beijerinckii NCIMB 8052	Deshelled corncobs	6,370
	Z. mobilis	Glucose	25,550
	P. stipitis	Xylose	36,000
	S. cerevisiae	Glucose	125,000
n-Butanol 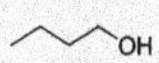	*E. coli*	Glucose	30,000
	C. tyrobutyricum	Mannitol	16,000
	C. beijerinckii	Xylose	16,910
	C. cellulovorans DSM 743B and C. beijerinckii NCIMB 8052	Deshelled corncobs	11,500
	Cyanobacteria	Carbon dioxide	14
1-Propanol 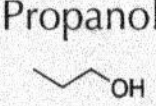	*E. coli*	Glucose	7,000
iso-Butanol 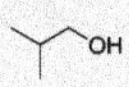	*E. coli*	Glucose	50,000
iso-Propanol 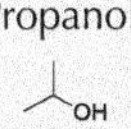	*E. coli*	Glucose	140,000

Continued

TABLE 5.1 List of biofuels produced by metabolic engineered strains.—cont'd

Product	Organism	Carbon source	Maximum productivity (mg/L)
Biodiesel	*Y. lipolytica*	Glycerol	3,000
	P. tricornutum	Carbon dioxide	400
	S. cerevisiae	Glucose	150

Inokuma et al., 2010). Furthermore, other defined biofuels were studied (e.g., hydrogen, *iso*-butyraldehyde, and triacylglycerol) but showed relatively low productivity levels (Yoshida et al., 2005; Atsumi et al., 2009; Santala et al., 2011).

Another approach for the production of biofuel is to focus on a mixture of combustible compounds rather than on a particular molecule. Biodiesel is such a mixture, and it can be obtained through the transesterification of vegetable oils (e.g., soybean, rapeseed, and palm) with methanol or ethanol. However, the production of biodiesel from these oils raises the same issues than the ones raised by first-generation bioethanol. Therefore, many researches have been carried out to produce biodiesel of second generation. Different microorganisms were identified (bacteria, yeasts, and microalgae). However, unlike second-generation bioalcohol production, the accumulation levels reached with engineered microorganisms were not high enough to consider economically relevant industrialization. This is mostly due to the fact that the lipases involved in the formation of the biodiesels require chaperones and/or post-translational modifications for proper refolding (Alnoch et al., 2018). The lack of knowledge on these mechanisms greatly hampered the development of efficient engineered strains. Nonetheless, the capacity of modified yeasts (*Yarrowia lipolytica* and *S. cerevisiae*) and microalgae (*Phrynosoma tricornutum*) to produce biodiesels at relatively interesting concentration levels has been shown with accumulation levels of 3000, 400, and 150 mg/L, respectively (Papanikolaou and Aggelis, 2002; Radakovits et al., 2011; Tang et al., 2013).

5.4.2 Polymers

The second largest industry depending on fossil resources, after the fuel industry, is probably the plastic industry that manufactures polymer materials. This industry is very much criticized because of the ecological concerns it raises (e.g., pollution, toxicity). It is subjected to fossil resources that are not renewable and their disposals and recyclability are expensive and quite complicated. And most of the time polymers end up in the ocean, causing great threat for marine wildlife. Furthermore, some additives that confer specific properties to polymers are known to be toxic. For instance, polyvinyl chloride (PVC), the third-most widely produced polymer after polyethylene (PE) and polypropylene (PP) (Allsopp and Vianello, 2012), contains phthalates that have neurotoxic and carcinogenic potentials. Other plastics contain bisphenol-A (BPA), a fossil-based chemical that has been identified as endocrine disruptor. Since plastics are known to release their monomers and additives over time because of different factors (e.g., temperature, light), the presence of these molecules in the polymer matrices falls under a public health and environment issue.

Therefore, there is an urgent need to substitute these petrosourced polymers with nontoxic alternatives that are recyclable and/or biodegradable. Nowadays, there are strains capable of producing monomers identical to those of fossil origin (e.g., 5-aminovaleric acid, ethylene, styrene, isobutene) (Park et al., 2013; Wang et al., 2010; Pirkov et al., 2008; McKenna and Nielsen, 2011; Leeuwen et al., 2012). Although the bioproduction of these chemicals through microorganisms is encouraging, it does not solve the problems of biodegradability of their respective polymers. Fortunately, there are strains capable of producing new materials with interesting properties. Some of them are still under development, such as 3-hydroxybutyric acid (Liu et al., 2011a; Tseng et al., 2009; Slater et al., 1999), but others are already an industrial reality. The most studied monomers are probably the lactic and succinic acids that can be accumulated by modified microorganisms at levels of 122 and 146 g/L, respectively (Table 5.2). (Fong et al., 2005; Sangproo et al., 2012; Ishida et al., 2006a,b; Chae et al., 2013; Lin et al., 2005; Xia et al., 2015; Okino et al., 2008; Raab et al., 2010).

Other monomers of interest can be produced by microorganisms, and more specifically by recombinant *E. coli*. 3-hydroxypropionic acid, listed in the top 10 most wanted building blocks by the DOE, can be obtained with high concentration level (185 g/L) (Yu et al., 2016). Production of muconic acid can reach accumulation of 37 g/L (Niu et al., 2002). Glucaric acid can be produced at concentration up to 4.85 g/L (Shiue and Prather, 2014). Chitinbiose, a precursor for chitinoligosaccharides, can be accumulated at 4 g/L (Cottaz and Samain, 2005). These new materials are usually copolymerized with diols that are also produced via fermentation. Among them, we can mention the four most used in the industry: 1,2-propanediol, 1,3-propanediol, 1,4-butanediol, and 2,3-butanediol (Jain et al., 2015; Lee et al., 2018; Hwang et al., 2014; Erian et al., 2018).

Considerable work has been done to obtain the monomers cited above. However, another strategy can be used in order to produce directly the

TABLE 5.2 Bioproduction of lactic and succinic acids by metabolic engineered strains.

Product	Organism	Carbon source	Maximum productivity (mg/L)
Lactic acid 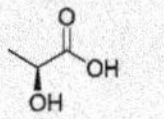	*E. coli*	Glucose	2,000
	Klebsiella oxytoca	Glucose	24,000
		Maltodextrin	34,000
	Leuconostoc citreum	Glucose	61,000
	S. cerevisiae	Cane juice-based medium	122,000
Succinic acid 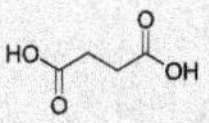	*E. coli*	Glucose	7,000
	E. coli consortium	Glucose and xylose	40,000
	S. cerevisiae	Glucose	3,620
	C. glutamicum	Glucose	146,000

corresponding polymers. Indeed, the progress of metabolic engineering gave access to strains capable of handling these relatively complex materials. Nowadays, modified microorganisms can efficiently produce unnatural polymers such as poly(lactic acid) (PLA) and poly(3-hydroxypropionic acid) (PHP) using *E. coli* or poly(hydroxyalkanoic acid) (PHA) using *Pseudomonas putida* (Jung et al., 2010; Zhou et al., 2011; Poblete-Castro et al., 2013).

5.4.3 Therapeutic molecules

Most of therapeutic molecules that are currently used in the pharmaceutical industry are either natural products or one of their analogs. It is very likely that microorganisms, having metabolic pathways to produce them, exist in nature. However, these microorganisms can be difficult to handle or the metabolic pathways associated to the molecule simply do not exist; this is mostly true for natural compound analogs. In any case, synthetic biology is particularly relevant for this industry. Firstly, the need in term of quantity is way smaller compared to other industries (e.g., biofuels or plastics). Secondly, the added value to the final target allows more flexibility in terms of cost of the production process. Therefore, although a high productivity process will be preferred, relatively average accumulation levels do not hinder a production of such molecules.

There are many therapeutic molecules produced through synthetic biology (Pauthenier and Faulon, 2013). Although not all these molecules are at the same level of development, some of them can be mentioned to illustrate the importance of metabolic engineering in this field of research. Penicillin (Fig. 5.2), the best known antibiotic, can be efficiently excreted by modified yeast, *Hansenula polymorpha*, enabling exploration of yeast-based cell factories for antibiotic production (Gidijala et al., 2009). Artemisinin (Fig. 5.2), an antimalarial drug, can be obtained by a tedious and costly total chemical synthesis (Schmid and Hofheinz, 1983). However, its production is highly simplified when artemisinic acid (Fig. 5.2) is used as precursor (Roth and Acton, 1989; Haynes and Vonwiller, 1994). The latter can be produced using a modified strain of *S. cerevisiae* with accumulation levels of 100 mg/L, thus contributing to a reliable and cost-effective source of artemisinin (Liu et al., 2011b; Yoon et al., 2004; Ro et al., 2006). Hydrocortisone (Fig. 5.2) can be produced by recombinant *S. cerevisiae* strains with accumulation levels ca. 10 mg/L (Szczebara et al., 2003). These illustrative examples show perfectly how a strain that would not be selected in fuel industry because of its relatively low productivity would be considered a hit in the pharmaceutical industry.

5.4.4 Food additives and supplements

Food additives and supplements are ubiquitous molecules in the food industry. Therefore, their production in large quantity and at reasonable price represents a major challenge. Chemical processes have been developed in order to meet the demand. However, the current trend among consumers is a return to "healthy eating" and there is also a growing concern for the environment, which translates into a demand for natural products, locally produced or in short circuits and in the logic of sustainable development. The biosourced character of food additives and supplements has, therefore, become an essential marketing argument for the food industry.

The most representative example of this category is vanillin, one of the most widely used flavoring agents. Because of its success in food industry, the demand became rapidly greater than the offer. Finding an alternative to the expensive and less available natural vanillin (Fig. 5.3) being a major stake,

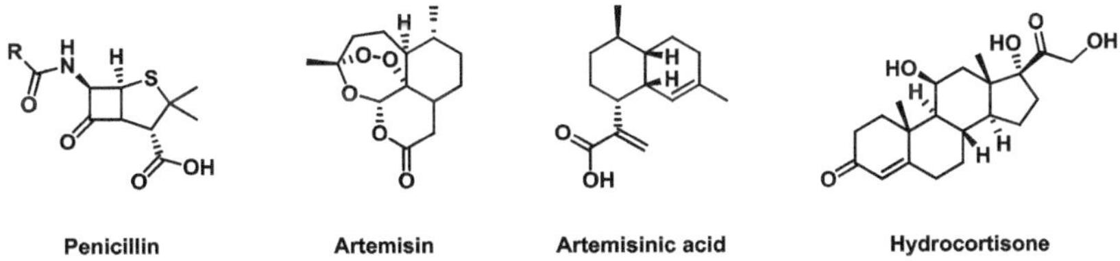

FIGURE 5.2 Therapeutic molecules produced by engineered microorganisms.

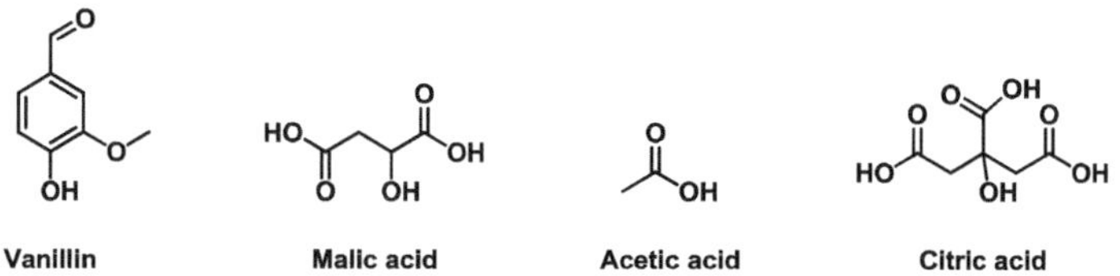

FIGURE 5.3 Food additives and supplements produced by engineered microorganisms.

processes to obtain cheaper synthetic vanillin in larger quantity were developed. However, the growing will for natural products forced the food industry to imagine new sustainable production processes. Metabolic engineering seemed to be the best solution to overcome the latter issues. Several modified microorganisms were thus designed to meet both industry and consumers' expectations. Among them, three microorganisms stood out. Hansen et al. modified *S. cerevisiae* and *S. pombe* strains and were able to report the first relatively high vanillin productivity with maximum concentration of 45 and 65 mg/L, respectively (Hansen et al., 2009). A year later, Brochado et al. improved the productivity of *S. cerevisiae* with a different approach. They targeted vanillin-β-D-glucoside and used a gene deletion strategy. Doing so, concentration up to 500 mg/L was reached (Brochado et al., 2010). Interesting works have been also reported by Ruzzi et al. that engineered a *P. fluorescens* strain capable of producing vanillin from ferulic acid at titer of 1.3 g/L (Di Gioia et al., 2011). However, modified *E. coli* outdid all other microorganisms and allowed the production of vanillin from isoeugenol with concentration up to 28.3 g/L (Yamada et al., 2008). Other major food additives and supplements, such as malic acid (34 g/L), acetic acid (78 g/L), or citric acid (87 g/L) can be produced by metabolic engineering (Fig. 5.3). (Zhang et al., 2011; Huang et al., 1998; Foerster et al., 2007).

5.5 Bio-based production of alkanes and alkenes through metabolic engineering

While significant progress has been done for alcoholic biofuels, as we previously have seen in this chapter, advances made in renewable hydrocarbon biofuels (aka "green" hydrocarbons, biohydrocarbons, "drop-in" biofuels, and sustainable or advanced hydrocarbon biofuels) are still limited. This is all the more unfortunate because these "drop-in" biofuels are similar to petroleum gasoline, diesel, or jet fuel in chemical makeup and are therefore considered infrastructure-compatible fuels (e.g., compatible with current engines). Faced with the challenges of their bioproduction, the metabolic pathways involved in their biosynthesis were studied in order to design an efficient microorganism applying metabolic engineering strategies.

5.5.1 Biosynthesis of hydrocarbons

The presence of alkanes and alkenes (e.g., hydrocarbons) was observed in nature in different types of organisms. For instance, these chemicals are used by plants as cuticular waxes, by insects as pheromones, and were found in a series of microorganisms such as *E. coli*, *R. rubrum*, or *Saccharomyces* sp (Samuels et al., 2008; Tillman et al., 1999; Ladygina et al., 2006). Most of the pathways described to date for the biosynthesis of hydrocarbons in these organisms use fatty acids as intermediary metabolites. These molecules are obtained from acetyl and malonyl CoA being the sources of acyl groups for the fatty acid synthase complex. Using acetyl and malonyl CoA-ACP transacylase, AT and MT, respectively, the acetyl group is added to the cysteine −SH group of the β-ketoacyl-ACP synthase (KS), and the malonyl group is added to the −SH group of 4′-phosphopantatheine of the ACP. The orientation of the fatty acid synthase places the malonyl and acetyl groups in close proximity. The following four steps take place iteratively to yield the saturated carbon chain of the fatty acid molecule: condensation by β-ketoacyl-ACP synthase (KS), reduction of the ketone by β-ketoacyl-ACP reductase (KR), dehydration by β-hydroxyacyl-ACP dehydratase (HD), and reduction of the double bond by enoyl-ACP reductase (ER) (Fig. 5.4). (Hanada, 2003) These steps are repeated until the carbon chain reaches the wanted size (usually 16 carbons).

Four metabolic pathways that convert fatty acids, or their derivatives, into straight-chain hydrocarbons have been identified and previously described by Wang et al. (Wang and Lu, 2013) The "elongation-decarboxylation" pathway (blue [dark gray in print version] pathway in Fig. 5.5) is probably the most widely discussed. It consists of a classical de novo synthesis of fatty acids followed by a decarboxylation step (Fig. 5.4). This route was first discovered

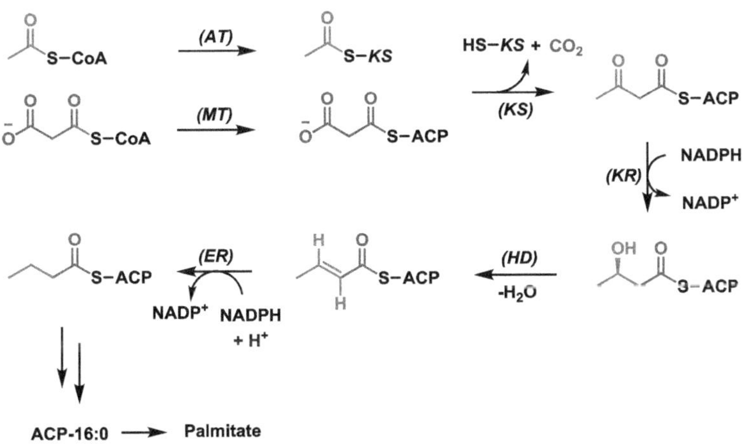

FIGURE 5.4 Biosynthesis of fatty acids.

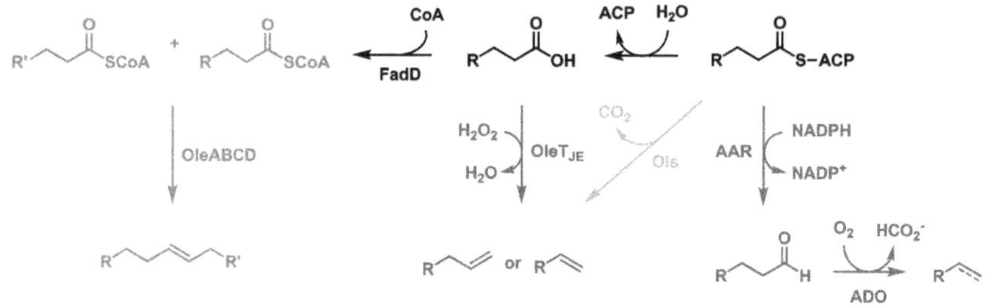

FIGURE 5.5 Straight-chain hydrocarbons biosynthesis pathways.

in the 60s while studying the metabolism of plants (Kolattukudy, 1967). It was later on found that *E. coli* would use this pathway as well (Naccarato et al., 1974). A big concern raised by this pathway was the high energy-demanding decarboxylation step that needs to be activated by a β-substituent. This question was solved in 2011 by Rude et al. who highlighted that the decarboxylation of fatty acids was catalyzed by $OleT_{JE}$, a P450 peroxygenase from the cyp152 family (Rude et al., 2011).

The "head-to-head condensation" pathway (green [light gray in print version] pathway in Fig. 5.5) is the second most widely discussed. It was first reported by Albro and Dittmer (1969). It is only 30 years later than the genes involved in this transformation were characterized. Bioinformatics and biochemical analysis allowed the identification of a three-gene cluster from *M. luteus* (Beller et al., 2010). The head-to-head condensation reaction requires four protein families (OleABCD). The first step, namely the OleA-catalyzed Claisen condensation, generates a β-ketoacid capable to spontaneously decarboxylate to yield the corresponding ketone (Frias et al., 2011). OleC and OleD then convert this product to olefins (Sukovich et al., 2010). It is still not clear what the role of OleB is, but its presence in the cluster suggests that it is necessary for olefins biosynthesis.

The third route, also called PKS pathway (orange [lighter gray in print version] pathway in Fig. 5.5), highly similar to the "elongation-decarboxylation" pathway, was elucidated thanks to Gu et al. who were interested in the enzyme *curM* responsible for the formation of the terminal double bond of Curacin A (Gu et al., 2009). In parallel, it has been shown that a marine cyanobacterium, *Synechococcus* sp. PCC 7002, was capable of producing 1-nonadecene and 1,14-nonadecadiene. Therefore, BLAST of *curM* sequence against the genome of *Synechococcus* sp. PCC 7002 was performed which resulted in the discovery of an open reading frame (ORF) encoding a protein sharing 45% homology with *curM*. This ORF was named olefin synthase (Ols). Ols is a part of a multidomain type I polyketide synthases (PKS). The fatty acid moiety of the acyl-ACP is transferred to the ACP1 of the PKS domain via the loading domain (LD). A condensation of the acyl-ACP1 with malonyl-CoA catalyzed by the central extension module (ketosynthase (KS), acyltransferase (AT), ketoreductase (KR)) form a β-hydroxy group that is activated via sulfation using the sulfotransferase (ST) domain. It has been hypothesized that subsequent dehydration and decarboxylation reactions could be catalyzed by the C-terminal thioesterase (TE) domain (Mendez-Perez et al., 2011).

A fourth route (red [gray in print version] pathway in Fig. 5.5), called AAR/ADO pathway, was identified when studying cyanobacteria. It has been established that Acyl-acyl carrier protein (ACP) can be reduced to the corresponding aldehyde by an acyl-ACP reductase (AAR) (Schirmer et al., 2010). The latter can be converted to hydrocarbons by an aldehyde deformylating oxygenase (ADO) (Li et al., 2012). The formation of formate alongside hydrocarbons is done via a redox oxygenation process. ADO was characterized as being a non-heme di-iron oxygenase, which requires a reducing system (NADPH) and O_2 to catalyze the reaction. It is noteworthy to mention that ADO activity is reversibly inhibited by H_2O_2. This inhibition can be solved by the use of a fusion catalase (Andre et al., 2013).

5.5.2 Production of hydrocarbons by engineered microorganisms

Attempts to improve the productivity of the natural pathways have been investigated. Most of the work has been done on long and very-long chain hydrocarbons, C9 to C31, and on two different host microorganisms, *E. coli* and *S. cerevisiae*. Two other microorganisms, *Synechocystis* sp. PCC6803 and *Streptomyces globisporus*, presented encouraging data as well.

5.5.2.1 "Elongation-decarboxylation" pathway

Few studies have been realized on this pathway in order to obtain good production of hydrocarbons. Among them, one engineered microorganism reached relatively high concentrations. In 2014, Li et al. reported the in vitro decarboxylation activity of $OleT_{JE}$ from *Jeotgalicoccus* sp. ATCC 8456 on fatty acids in presence of H_2O_2. Furthermore, RhFRED reductase domain from *Rhodococcus* sp. was fused to the C-terminus of $OleT_{JE}$ to see if it could act as a monooxygenase. Authors reported that the $OleT_{JE}$-RhFRED converted lauric acid to 1-undecene in presence of NADPH with half the efficiency of $OleT_{JE}$ in presence of O_2. Their insertion in different *E. coli* strains overproducing free fatty acids, and more specifically in the XL-100 strain (an *E. coli* BL21 fadD deletion mutant strain) resulted in overproduction of hydrocarbons (C11 to C17) with a maximum concentration at 97.6 mg/L (Liu et al., 2014) A year later, despite the significant lower productivity, interesting work has been realized on *S. cerevisiae* by Wook Chang et al. By inserting $OleT_{JE}$ gene from *Jeotgalicoccus* sp. ATCC 8456 into the yeast, deleting genes using H_2O_2 as cofactor (catalase A (CTA1), cytochrome C peroxidase (CCP1), and catalase T (CTT1)) to redirect the flux toward $OleT_{JE}$ and deleting the genes coding for enzymes converting free fatty acids to fatty acyl-CoAs (fatty acyl-CoA synthetases (FAA1 and FAA4)), the authors were able to reach accumulation of terminal alkenes up to 3.7 mg/L (Chen et al., 2015).

5.5.2.2 "Head-to-head condensation" pathway

Due to the relative complexity of this pathway involving four different enzymes, and because their involvement in hydrocarbons biosynthesis is not completely understood yet, the "head-to-head condensation" pathway has been very little studied. Nevertheless, Beller et al. attempted to engineer an *E. coli* strain overproducing long chain hydrocarbons via this route. The heterologous expression of a three-gene cluster from *Micrococcus luteus* (Mlut 13230, 13240, and 13250), coding for OleA, a fusion of OleB and OleC, and OleD, in a fatty acid-overproducing *E. coli* strain allowed the formation of long-chain alkenes (C27 and C29). Unfortunately, authors reported very low productivity with concentrations up to 40 μg/L (Beller et al., 2010).

5.5.2.3 PKS pathway

Very encouraging work has been done on the PKS pathway. Contrarily to the "head-to-head condensation" pathway, competitive productivity has been reported. In 2011, Shen et al. engineered a heptaene-overproducing *S. globisporus* strain while attempting to ameliorate the production of C-1027, a relatively complicated chromophore. In a previous study, authors identified three regulatory enzymes (SgcR1, SgcR2, and SgcR3), part of the Ols family, as being involved in C-1027 biosynthesis. The overexpression of SgcR1 in the *S. globisporus* wild-type strain resulted in overexpression of C-1027 as well as 1,3,5,7,9,11,13-pentadecaheptaene (reaching concentration of 66 mg/L for the latter) (Chen et al., 2010). Nonetheless, Shen et al. took a step further by studying the influence of two additional C-1027 regulatory enzymes, sgcE1 and sgcR. It resulted in a heptaene-overproducing *S. globisporus* strain which was obtained by deletion of sgsE1 and sgcR genes, and overexpression of sgcR1. By doing so, authors increased twofold the production of heptaene reaching concentration up to 129.3 mg/L (Chen et al., 2011) Few years later, Liu et al. implanted this pathway into *E. coli* in order to access pentadecaheptaene (PDH). Their initial in vitro results suggested that alkene production was linked to the SgcE10:SgcE1 ratio. Therefore, authors developed an *E. coli* strain with fine-tuning of sgsE10 expression by using synthetic promoters in order to control the quantity of targeted enzymes in the microorganism. Through fed-batch fermentation, accumulation of PDH reached concentration up to 140 mg/L. A flow chemical hydrogenation was used to get to the corresponding alkane (Liu et al., 2015).

5.5.2.4 AAR/ADO pathway

Researches on the AAR/ADO pathway seemed to have given the best results. Firstly, proofs of concept have been validated using yeast and cyanobacteria: *S. cerevisiae* and *Synechocystis* sp. PCC6803. Nielsen et al. focused on the modifications of *S. cerevisiae*'s metabolism to access long chain hydrocarbons (C10 to C17). Fatty acyl-ACP/CoA reductase (AAR) and a fatty aldehyde deformylating oxygenase (*Se*ADO), encoded by *Synechococcus elongatus* orf1594 and *S. elongatus* orf1593, respectively, were introduced into *S. cerevisiae* using plasmids. However, no production of hydrocarbons was observed despite the inserted heterologous biosynthetic pathway. It has been hypothesized that this lack of activity was due to the presence of an aldehyde dehydrogenase naturally present in yeast's metabolism. Deletion of the gene encoding for hexadecenal dehydrogenase (Hfd1) resulted in the formation of alkane at concentration up to 22 μg/L (Buijs et al., 2015).

Access to very-long chain alkanes (C27 to C31) was made possible by Joubes et al. that imported the alkane biosynthetic enzymes of *Arabidopsis thaliana* into *S. cerevisiae*. It has been shown that *A. thaliana* Eceriferum 1 (CER1), from the family of fatty aldehyde deformylating oxygenase (ADO), was an essential protein for wax alkane biosynthesis. However, its expression in heterologous systems did not result in the formation of alkanes (Bourdenx et al., 2011). CER1 was found to interact with *A. thaliana* Eceriferum 3 (CER3) and with the cytochrome b5 isoforms (CYTB5s). The coexpression of CER 1 and 3 allowed the conversion of fatty aldehydes into the corresponding alkanes, amplified in presence of CYTB5s. However, accumulation remained very low with a maximum concentration of 86 μg/g of dry weight (Bernard et al., 2012).

Interesting work has also been performed on a cyanobacteria strain, *Synechocystis* sp. PCC6803. It is, for modified AAR/ADO pathway, the second best microorganism in term of hydrocarbons productivity. It has been shown that the natural strain of *Synechocystis* sp. PCC6803 possesses the AAR/ADO pathway in their metabolism. The access to acyl-ACPs has proved to be difficult because of their incorporation to membrane lipids. However, the latter can release free fatty acids during its degradation that will be converted back to acyl-ACPs via an acyl-ACP synthetase (AAS). Recombinant plasmids were inserted into *Synechocystis* sp. PCC6803 in order to overexpress these key proteins (AAS, AAR, and ADO). This strategy redirected the carbon flux to acyl-ACPs and favored the production of hydrocarbons, with reported accumulations up to 26 mg/L (Wang et al., 2013).

Researches done on cyanobacteria allowed 1000 fold higher concentration of hydrocarbons, del Cardayre et al. greatly improved the productivity by using *E. coli* as host microorganism. Although the wild strain produced no detectable hydrocarbons, insertion of *S. elongatus* PCC7942_orf1593 and orf1594, coding for AAR and ADO, respectively, allowed a titers of 300 mg/L (~ 80% of excreted hydrocarbons) with the use of a modified mineral medium (Schirmer et al., 2010).

Other engineered *E. coli* have been reported in the literature, unfortunately with lower accumulation. Zhao et al. attempted to increase the productivity with the coexpression of a catalase aimed at withdrawing toxic byproducts of the culture media. They also introduced a type-I fatty acid synthase (FAS) from *C. ammoniagenes* to bypass regulatory mechanism of fatty acids production in *E. coli*. By doing so, authors were capable of producing hydrocarbons with maximum concentrations up to 57 mg/L (Coursolle et al., 2015) Zhu et al. proposed a similar modified microorganism than the one constructed by del Cardayre et al. Overexpression of AAR and ADO from *S. elongatus* PCC7942 along with ACP allowed an accumulation of hydrocarbons from 3.1 to 24 mg/L, depending on the expression strategy that was used. Authors improved greatly their strain by the identification of two genes: yqhD, an inherent aldehyde reductase in *E. coli*, and fadR, an activator for fatty acid biosynthesis. The deletion of the first and the overexpression of the second allowed a tenfold increase in hydrocarbons titer reaching concentration up to 255.6 mg/L (Song et al., 2016).

5.5.2.5 New pathways

Although very encouraging improvements have been obtained by using natural pathways, stimulating works have been developed around alternative pathways. Most of them have been conducted by substituting AAR, from AAR/ADO pathway, by another enzyme. Foo et al. designed a modified *S. cerevisiae* strain that utilized a fatty acid-α-dioxygenase (αDOX) from *Oryza sativa* coupled with a gene deletion strategy of FAA1 and FAA4 to avoid side consumption of FFAs. What differentiates this route from the AAR/AADO pathway is the precursor (FAA) and the cofactors involved; AAR and αDOX go through an $NADPH/NAPD^+$ system and an O_2/CO_2 system, respectively. Though αDOX is less energy-intensive due to its catalytic system, the accumulation levels of hydrocarbons were very low with maximum concentration of 73.5 μg/L (Foo et al., 2017).

While other studies focused primarily on hydrocarbons productivity, Akhtar et al. got interested in the substrate selectivity of these enzymes and identified a carboxylic acid reductase (CAR) from *Mycobacterium marinum* that converts fatty acids (C6−C18) into corresponding aldehydes. In vitro trials showed the catalytic conversion of fatty acids to fatty alkanes (C7−C15) in presence of CAR and ADO. Applying this strategy in vivo by inserting both genes encoding for CAR and a chain-length-specific thioesterase into *E. coli* BL21(DE3) revealed the production of alkanes (C7−C15). Unfortunately, low alkane productivity was reported with accumulation of 2 mg/L (Akhtar et al., 2013). Howard et al. followed the same strategy using a fatty acid reductase (FAR) from *P. luminescens* and an aldehyde decarbonylase (ADO) from *Nostoc punctiforme* inserted into *E. coli* and reported production of a wide range of alkanes, and more surprisingly alkenes, to concentration up to 5 mg/L (Howard et al., 2013).

Nielsen et al., inspired by the work of Akhtar et al., aimed to improve the hydrocarbons productivity by using a modified *S. cerevisiae* strain through

CAR/ADO pathway. It has been established that free fatty acids (FFAs) can be accumulated to much higher levels than fatty acyl-CoA (>200 fold higher) (Schjerling et al., 1996). Therefore, they first designed a platform strain of *S. cerevisiae* that redirects citrate to acetyl-CoA that is then converted in FFAs by endogenous or heterogeneous enzymes. Furthermore, reverse pathways were knocked down through the deletion of the corresponding genes (hFD1, FAA 1,4, and POX 1). Doing so, FFAs accumulated in the modified strain to a maximum concentration of 10.4 g/L. Authors used their previous AAR/FADO system to produce alkanes with this new strain but observed low productivity of 100 μg/L. To overcome this issue, an alternative has been found by using carboxylic acid reductase (MmCAR) expressed in *M. marinum* activated by 4′-phosphopantetheinyl transferase (NpgA) from *Aspergillus nidulans* (Akhtar et al., 2013). This pathway enabled a higher alkane production up to 480 μg/L. The identification and deletion of genes converting fatty aldehydes to fatty alcohols, aldehyde reductases (ALRs), and/or alcohol dehydrogenases (ADHs), coupled with the expression of *Se*ADO, permitted the intensification of the process. Finally, the additional expression of *N. punctiforme* (*Np*ADO) increased alkane production to 820 μg/L (Zhou et al., 2016).

In 2013, Lee and Choi used knock out gene strategy in order to minimize the potential intermediates loss. β-oxidation pathway was disabled by deleting the fadE gene thus avoiding the degradation of fatty acyl-CoAs. Then, the authors deleted the fadR gene in order to prevent upregulation of the fabA and fabB genes responsible for unsaturated fatty acids biosynthesis (Nunn et al., 1983). The absence of unsaturated fatty acids lifted the inhibition on 3-oxoacyl-ACP synthase (FabH) that favored the accumulation of medium-chain acyl-ACPs. The latter were converted into FFAs using a modified thioesterase ('tesA) that were then converted into the corresponding alkanes through a sequence composed of an *E. coli* fatty acyl-CoA synthetase (fadD), a *C. acetobutylicum* fatty acyl-CoA reductase (ACR), and an *A. thaliana* fatty aldehyde deformylating oxygenase (CER1). By using this *E. coli* strain platform, authors reported a maximum accumulation of alkanes of 580.8 mg/L, consisting of nonane (327.8 mg/L), dodecane (136.5 mg/L), tridecane (64.8 mg/L), 2-methyl-dodecane (42.8 mg/L), and tetradecane (8.9 mg/L). (Choi and Lee, 2013).

These last decades, many strategies have been developed to improve hydrocarbons productivity through microorganisms. Although modification or substitution of the enzymes directly involved in alkanes/alkenes biosynthesis (Fig. 5.5) allowed significant amelioration, it is clear that controlling the side metabolic pathways is essential in order to offer a competitive engineered microorganism. Major progresses have been made with the deletion of specific metabolic pathways, such as β-oxidation or fatty alcohols biosynthesis, and redirection of the metabolic flux toward fatty acyl-ACPs through acetyl-CoA (Fig. 5.6).

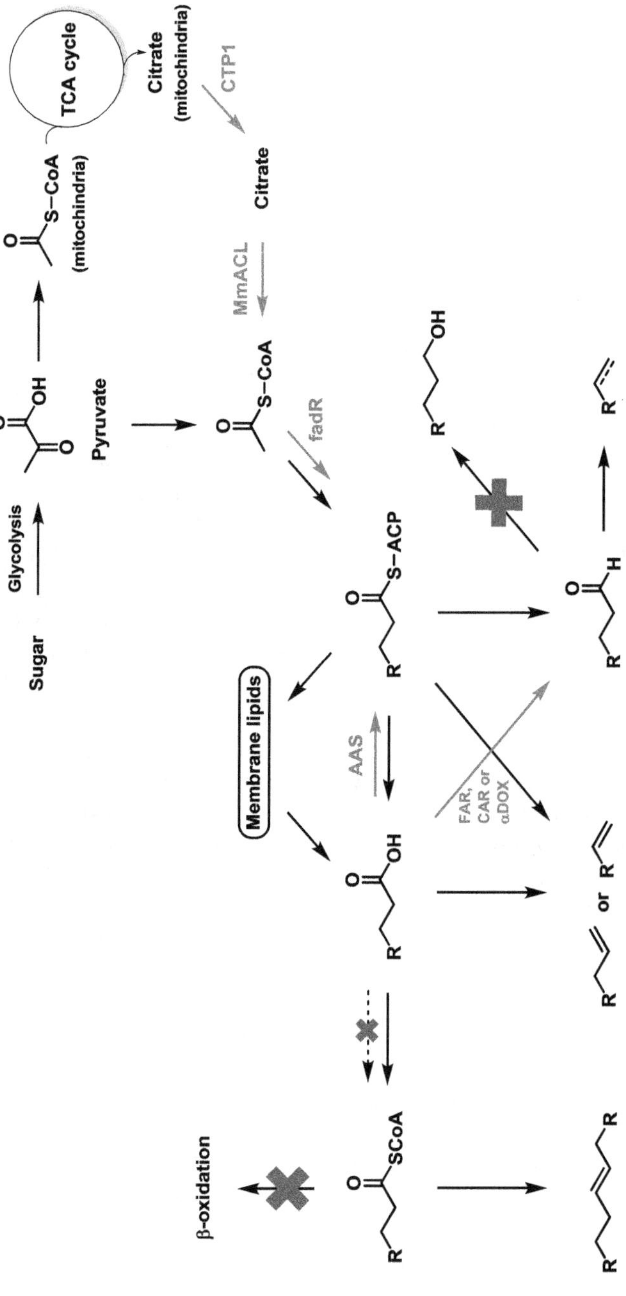

FIGURE 5.6 Metabolic flux optimization for hydrocarbons production.

5.5.3 Bottlenecks and solutions

Despite the promising results obtained with engineered microorganisms, the best strains do not present a sufficient hydrocarbons productivity for potential industrialization. Different factors could be studied to achieve it. Firstly, elements directly linked to enzymes (mostly their activities) can have a negative effect on the alkanes/alkenes productivity. The most efficient pathway (AAR/ADO) is hindered by the low activity of ADO that results in the accumulation of fatty alcohols from fatty aldehydes. This issue can partially be overcome via the overexpression of ADO gene to boost the productivity. Another strategy consists in using directed or random mutagenesis in order to ameliorate the specific activity of the enzyme. This approach allowed Khara et al. (2013) to enhance the activity of ADO from *P. marinus* toward short chain aldehydes by twofold. Metabolism of host microorganisms can also affect the productivity of the target molecules. Different studies, discussed previously, strongly suggest that a number of metabolic pathways lower the biosynthesis of hydrocarbons. Knockout gene strategy allowed researchers to redirect the metabolic flux toward the targeted route; however, this can be at the expense of the growth of the modified microorganisms and so of the productivity itself. Nevertheless, strains that were designed so far permitted to improve the accumulation of alkanes/alkenes.

Precursor sourcing can also be a problematic factor. Hydrocarbons biosynthesis being dependent on fatty acyl-ACPs and their derivatives (fatty acids and acyl-CoAs), any side reactions that consume them (fatty alcohols biosynthesis or β-oxidation) will directly affect the productivity. Deletion of the genes involved in these pathways seems to be mandatory to reach relatively good concentrations. In addition, it is also possible to intensify the production upstream. As seen previously in this chapter, acetyl- and malonyl-CoA are central intermediates in fatty acids biosynthesis; however, they are also used in many other metabolic pathways. Therefore, intensifying their production would allow both supporting metabolism and boosting fatty acids production. Koffas et al. followed this strategy and designed an *E. coli* strain that over-expresses each enzyme involved in fatty acid biosynthesis. Combining it with deletion strategy on the precursor enzyme of β-oxidation (fadD), authors were able to reach high productivity with concentration up to 8.6 g/L (Xu et al., 2013). It is easily conceivable to add the metabolic pathways yielding hydrocarbons to this microorganism and therefore improving the productivity.

While these approaches give positive preliminary results, it seems that environmental conditions greatly influence the hydrocarbons productivity as well. Temperature can play a crucial role in the process. Zhu et al. showed that growth of their modified strain was closely linked to this parameter. At 18°C, cell mass reached an optical density at 600 nm (OD600) of 8.5 with accumulation of hydrocarbons up to 5.3 mg/L. At 30°C, these values were quite similar with OD600 of 9.3 and an accumulation of 5.0 mg/L; however, they dropped at 37°C with an OD600 of 7.3 and accumulation of 4.2 mg/L. Surprisingly, the maximum titer was observed at 24°C with an OD600 of 19 and accumulation of 26 mg/L. This observation shows the importance of bringing to attention the details with a 3.5 fold improvement in productivity when incubated the culture at an "unconventional" temperature. The composition of the culture medium can also improve the accumulation of hydrocarbons. By using a modified mineral medium, del Cardayre et al. developed a strain that is capable of excreting hydrocarbons up to 240 mg/L. This discovery opened the door to continuous extraction which makes industrialization more plausible.

Other major concerns are linked to the concept of fermentation in general. When it comes to incorporating a promising engineered strain in an industrialized process, several questions can be raised. The cost of the culture medium, the accumulation time of a wanted product and the robustness of the process are the main obstacles to scale up. The question of the carbon source, discussed in the introduction, is also a major issue. Use of first-generation sugars has to be ruled out as it can destabilize an existing industry, especially the agricultural one. The valorization of biomasses allowed to take a step toward sustainability but is not sufficient when it comes to high productivity. Thus, the solution that offers metabolic engineering feeding from first- and second-generation sugars is only sustainable for low or medium production of molecules of specialty and fine chemicals. Using microorganisms that consume CO_2 as carbon source would allow considering large-scale fermentation, as for the production of biofuels. Moreover, industrial fermentation handles large amounts of water that must be recycled and reprocessed before being released. Therefore, the nature of the wanted product and the purification method to access it dictates the viability of a process. While approaches involving extraction with solvent have been developed, purification by distillation will be largely preferred when possible. Finally, parameters intrinsic to fermentation can hinder industrial viability such as contamination of cultures, formation of biofilms, or oxygenation of the tank during use of aerobic processes (Pauthenier and Faulon, 2018).

Since production of "drop-in" biofuels revealed to be quite difficult, approaches have been developed in order to access small building blocks that can be later transformed into hydrocarbons. These building blocks had to have reactive chemical functions, which precluded methane and small alkanes. However, unsaturated molecules such as ethylene have shown great interest. Three biosynthesis pathways have been reported in the literature. The first one, called the "ACC pathway" is mostly found in plants and uses methionine as precursor. The latter is transformed into *S*-adenosyl-methionine via SAM synthase. The intermediate is then converted into ACC and 5′-methylthioadenosine (MTA) via ACC synthase. Finally, ACC is converted into ethylene by an ACC oxidase. It has to be noted that MTA is regenerated into methionine by a sequence of four enzymes allowing the level of methionine to remain constant while ethylene accumulates in the medium (Cristescu et al., 2002; Stearns and Glick, 2003; Chae and Kieber, 2005). The second pathway, so called the "KMBA pathway" can be found in microorganisms and uses methionine as precursor as well. The first step consists of a transamination involving a α-keto acid via a transaminase. The intermediate thus formed (KMBA) is oxidized via a peroxidase or an oxidoreductase resulting in the formation of ethylene (Graham and Linderman, 1980; Fukuda et al., 1989; Ince and Knowles, 1986). This pathway has been found in various microorganisms such as *E. coli, S. albidus, or B. cinerea* (Fukuda et al., 1989; Ince and Knowles, 1986; Chague et al., 2002). The third pathway uses 2-oxoglutaric acid as precursor which is converted into ethylene via an "ethylene-forming enzyme" (EFE). This pathway has been found mostly in fungi such as *P. cyclopium, P. digitatum, or F. oxisporum* (Pazout and Pazoutova, 1989; Sprayberry et al., 1965; Hottiger and Boller, 1991). The comprehension and relative simplicity of these pathways allowed the development of modified microorganisms capable of good accumulation of ethylene. Therefore, Gill et al. performed metabolic engineering on an *E. coli* strain and reported accumulation of ethylene up to 5 g/L (Lynch et al., 2016).

Another interesting building block for the production of "drop-in" biofuel is isobutene. Its biosynthesis has been studied on a yeast strain called *Rhombophryne minuta* (Fujii et al., 1987). Whereas the enzymes involved in this pathway are not identified, the different intermediates have been identified as being α-ketoisocaproic acid, *iso*-valeryl-CoA, and 3-methylcrotonyl-CoA. Few published works are available on metabolic engineered strain capable of producing isobutene; however, Marliere et al. filed a patent that describes the production of isobutene from 3-methylcrotonyl-CoA (Allard and Marliere, 2017). No data concerning concentration or productivity is revealed in this patent; however, it is the founding stone of the company called Global Bioenergies. They recently published a press release announcing the "First production of isobutene from wheat straw at the scale of a demonstrator" which leaves no doubt on the efficiency of their process (https://www.global-bioene).

5.5.4 Conclusion

For many years, fossil resources have been overexploited to manufacture a wide variety of chemicals that are essential to everyday life. Ranging from polymers to therapeutic molecules or food additives, the applications are diverse. Nevertheless, the renewal of fossil resources is not fast enough for a global demand that keeps on growing exponentially. In order to overcome this major concern, bioproduction of such chemicals has been developed through metabolic engineering. This approach led to industrial success thanks to modified microorganisms capable of producing valuable chemical building blocks such as succinic acid, penicillin, or vanillin at the industrial scale and at relatively low cost. With regards to renewable energy, it is also possible to produce biofuels, such as bioethanol or biodiesel, using this approach. However, as bioethanol cannot be used directly in classic engine, many efforts are being directed toward the production of "drop-in" biofuels.

Different host microorganisms have been studied such as yeast or cyanobacteria; however, it seems that *E. coli* is the one which is the most suitable for hydrocarbon production. Although alkane/alkene accumulations up to 580.8 mg/L have been observed, these concentrations are still way below the required rate/titer. To compare, very efficient process for bioethanol production uses strains capable of accumulating ca. 125 g/L; however, a process can be profitable at lower concentration. In order to determine where the limit lies costs of goods sold (COGS) values must be taken into consideration. COGS of 0.6 USD/L have been estimated for a microbial biofuel production to be viable. Afterward, theoretical yields must be calculated to determine if the process could be profitable. These calculations have been realized on a particular biodiesel (oleyl ethyl ester) and can be compared to hydrocarbons, since their energetic yields are quite similar (ca. 0.82) (Caspeta and Nielsen, 2013). Foremost, even if some metabolic pathways involving cofactors such as NADPH or ATP will reduce the yield on glucose, it has been calculated that the productivity would only be reduced by ca. 10% compared to ethanol. Then, it has been calculated that a productivity of 7.5 g/L per h and a production scale of 50,000–75,000 tons/year with the sugar price not exceeding 0.07 USD/kg would allow staying below COGS of 0.6 USD/L. Consequently, hydrocarbons productivity using modified *E. coli* must be improved by at least tenfold.

References

Akhtar, M.K., Turner, N.J., Jones, P.R., 2013. Carboxylic acid reductase is a versatile enzyme for the conversion of fatty acids into fuels and chemical commodities. Proc. Natl. Acad. Sci. U.S.A. 110, 87–92. S87/1-S87/38.

Albro, P.W., Dittmer, J.C., 1969. The biochemistry of long-chain, nonisoprenoid hydrocarbons. 3. The metabolic relationship of long-chain fatty acids and hydrocarbons and other aspects of hydrocarbon metabolism in *Sarcina lutea*. Biochemistry 8, 1913–1918.

Allard, M., Marliere, P., 2017. Metabolic Pathways for Isobutene Production from 3-Methylcrotonyl-CoA. WO2017162738A1.

Allsopp, M.W., Vianello, G., 2012. Poly(Vinyl chloride). In: Ullmann's Encyclopedia of Industrial Chemistry, vol. 28. Wiley-VCH, W., pp. 441–468

Alnoch, R.C., Stefanello, A.A., Paula Martini, V., Richter, J.L., Mateo, C., Souza, E.M.D., Mitchell, D.A., Muller-Santos, M., Krieger, N., 2018. Co-expression, purification and characterization of the lipase and foldase of Burkholderia contaminans LTEB11. Int. J. Biol. Macromol. 116, 1222–1231.

Alterthum, F., Ingram, L.O., 1989. Efficient ethanol production from glucose, lactose, and xylose by recombinant *Escherichia coli*. Appl. Environ. Microbiol. 55, 1943–1948.

Andre, C., Kim, S.W., Yu, X.-H., Shanklin, J., 2013. Fusing catalase to an alkane-producing enzyme maintains enzymatic activity by converting the inhibitory byproduct H_2O_2 to the cosubstrate O_2. Proc. Natl. Acad. Sci. U.S.A. 110, 3191–3196.

Atsumi, S., Cann, A.F., Connor, M.R., Shen, C.R., Smith, K.M., Brynildsen, M.P., Chou, K.J.Y., Hanai, T., Liao, J.C., 2008. Metabolic engineering of *Escherichia coli* for 1-butanol production. Metab. Eng. 10, 305–311.

Atsumi, S., Hanai, T., Liao, J.C., 2008. Non-fermentative pathways for synthesis of branched-chain higher alcohols as biofuels. Nature 451, 86–89.

Atsumi, S., Higashide, W., Liao, J.C., 2009. Direct photosynthetic recycling of carbon dioxide to isobutyraldehyde. Nat. Biotechnol. 27, 1177–1180.

Baez, A., Cho, K.-M., Liao, J.C., 2011. High-flux isobutanol production using engineered *Escherichia coli*: a bioreactor study with in situ product removal. Appl. Microbiol. Biotechnol. 90, 1681–1690.

Bai, F.W., Anderson, W.A., Moo-Young, M., 2008. Ethanol fermentation technologies from sugar and starch feedstocks. Biotechnol. Adv. 26, 89–105.

Bailey, J.E., 1991. Toward a science of metabolic engineering. Science 252, 1668–1675.

Balusu, R., Paduru, R.M.R., Seenayya, G., Reddy, G., 2004. Production of ethanol from cellulosic biomass by *Clostridium thermocellum* SS19 in submerged fermentation: screening of nutrients using Plackett-Burman design. Appl. Biochem. Biotechnol. 117, 133–141.

Beller, H.R., Goh, E.-B., Keasling, J.D., 2010. Genes involved in long-chain alkene biosynthesis in *Micrococcus luteus*. Appl. Environ. Microbiol. 76, 1212–1223.

Bernard, A., Domergue, F., Pascal, S., Jetter, R., Renne, C., Faure, J.-D., Haslam, R.P., Napier, J.A., Lessire, R., Joubes, J., 2012. Reconstitution of plant alkane biosynthesis in yeast demonstrates that *Arabidopsis* ECERIFERUM1 and ECERIFERUM3 are core components of a very-long-chain alkane synthesis complex. Plant Cell 24, 3106–3118.

Bourdenx, B., Bernard, A., Domergue, F., Pascal, S., Leger, A., Roby, D., Pervent, M., Vile, D., Haslam, R.P., Napier, J.A., Lessire, R., Joubes, J., 2011. Overexpression of *Arabidopsis* ECERIFERUM1 promotes wax very-long-chain alkane biosynthesis and influences plant response to biotic and abiotic stresses. Plant Physiol 156, 29–45.

Bro, C., Regenberg, B., Foerster, J., Nielsen, J., 2006. In silico aided metabolic engineering of *Saccharomyces cerevisiae* for improved bioethanol production. Metab. Eng. 8, 102–111.

Brochado, A.R., Matos, C., Moeller, B.L., Hansen, J., Mortensen, U.H., Patil, K.R., 2010. Improved vanillin production in baker's yeast through in silico design. Microb. Cell Fact. 9. No pp. given.

Buijs, N.A., Zhou, Y.J., Siewers, V., Nielsen, J., 2015. Long-chain alkane production by the yeast *Saccharomyces cerevisiae*. Biotechnol. Bioeng. 112, 1275–1279.

Caspeta, L., Nielsen, J., 2013. Economic and environmental impacts of microbial biodiesel. Nat. Biotechnol. 31, 789–793.

Chae, H.S., Kieber, J.J., 2005. Eto Brute? Role of ACS turnover in regulating ethylene biosynthesis. Trends Plant Sci. 10, 291–296.

Chae, H.S., Lee, S.H., Lee, J.-H., Park, S.J., Lee, P.C., 2013. Use of a novel *Escherichia coli Leuconostoc* shuttle vector for metabolic engineering of *Leuconostoc citreum* to overproduce D-lactate. Appl. Environ. Microbiol. 79, 1428–1435.

Chague, V., Elad, Y., Barakat, R., Tudzynski, P., Sharon, A., 2002. Ethylene biosynthesis in Botrytis cinerea. FEMS Microbiol. Ecol. 40, 143–149.

Chen, Y., Yin, M., Horsman, G.P., Huang, S., Shen, B., 2010. Manipulation of pathway regulation in *Streptomyces globisporus* for overproduction of the enediyne antitumor antibiotic C-1027. J. Antibiot. 63, 482–485.

Chen, Y., Yin, M., Horsman, G.P., Shen, B., 2011. Improvement of the enediyne antitumor antibiotic C-1027 production by manipulating its biosynthetic pathway regulation in *Streptomyces globisporus*. J. Nat. Prod. 74, 420–424.

Chen, B., Lee, D.-Y., Chang, M.W., 2015. Combinatorial metabolic engineering of *Saccharomyces cerevisiae* for terminal alkene production. BioRxiv Microbiol. 22350/1–22350/34.

Choi, Y.J., Lee, S.Y., 2013. Microbial production of short-chain alkanes. Nature 502, 571–574.

Cottaz, S., Samain, E., 2005. Genetic engineering of *Escherichia coli* for the production of NI,NII-diacetylchitobiose (chitinbiose) and its utilization as a primer for the synthesis of complex carbohydrates. Metab. Eng. 7, 311–317.

Coursolle, D., Lian, J., Shanklin, J., Zhao, H., 2015. Production of long chain alcohols and alkanes upon coexpression of an acyl-ACP reductase and aldehyde-deformylating oxygenase with a bacterial type-I fatty acid synthase in *E. coli*. Mol. Biosyst. 11, 2464–2472.

Cristescu, S.M., De Martinis, D., te Lintel Hekkert, S., Parker, D.H., Harren, F.J.M., 2002. Ethylene production by *Botrytis cinerea* in vitro and in tomatoes. Appl. Environ. Microbiol. 68, 5342–5350.

Di Gioia, D., Luziatelli, F., Negroni, A., Ficca, A.G., Fava, F., Ruzzi, M., 2011. Metabolic engineering of *Pseudomonas fluorescens* for the production of vanillin from ferulic acid. J. Biotechnol. 156, 309–316.

Erian, A.M., Gibisch, M., Pflugl, S., 2018. Engineered *E. coli* W enables efficient 2,3-butanediol production from glucose and sugar beet molasses using defined minimal medium as economic basis. Microb. Cell Fact 17, 190.

Foerster, A., Aurich, A., Mauersberger, S., Barth, G., 2007. Citric acid production from sucrose using a recombinant strain of the yeast *Yarrowia lipolytica*. Appl. Microbiol. Biotechnol. 75, 1409–1417.

Fong, S.S., Burgard, A.P., Herring, C.D., Knight, E.M., Blattner, F.R., Maranas, C.D., Palsson, B.O., 2005. In silico design and adaptive evolution of *Escherichia coli* for production of lactic acid. Biotechnol. Bioeng. 91, 643–648.

Foo, J.L., Susanto, A.V., Keasling, J.D., Leong, S.S.J., Chang, M.W., 2017. Whole-cell biocatalytic and de novo production of alkanes from free fatty acids in *Saccharomyces cerevisiae*. Biotechnol. Bioeng. 114, 232–237.

Frias, J.A., Richman, J.E., Erickson, J.S., Wackett, L.P., 2011. Purification and characterization of OleA from *Xanthomonas campestris* and demonstration of a non-decarboxylative claisen condensation reaction. J. Biol. Chem. 286, 10930–10938.

Fujii, T., Ogawa, T., Fukuda, H., 1987. Isobutene production by *Rhodotorula minuta*. Appl. Microbiol. Biotechnol. 25, 430–433.

Fukuda, H., Takahashi, M., Fujii, T., Tazaki, M., Ogawa, T., 1989. An NADH: Fe(III) EDTA oxidoreductase from *Cryptococcus albidus*: an enzyme involved in ethylene production in vivo? FEMS Microbiol. Lett. 60, 107–111.

Gidijala, L., Kiel, J.A.K.W., Douma, R.D., Seifar, R.M., van Gulik, W.M., Bovenberg, R.A.L., Veenhuis, M., van der Klei, I.J., 2009. An engineered yeast efficiently secreting penicillin. PLoS One 4. No pp. given.

https://www.global-bioenergies.com/premiere-mise-en-production-disobutene-a-partir-de-paille-de-ble-a-lechelle-dun-demonstrateur/.

Gonzalez, R., Tao, H., Purvis, J.E., York, S.W., Shanmugam, K.T., Ingram, L.O., 2003. Gene array-based identification of changes that contribute to ethanol tolerance in ethanologenic *Escherichia coli*: comparison of KO11 (parent) to LY01 (resistant mutant). Biotechnol. Prog. 19, 612–623.

Graham, J.H., Linderman, R.G., 1980. Ethylene production by ectomycorrhizal fungi, *Fusarium oxysporum* f. sp. pini, and by aseptically synthesized ectomycorrhizae and *Fusarium*-infected Douglas-fir roots. Can. J. Microbiol. 26, 1340–1347.

Gu, L., Wang, B., Kulkarni, A., Gehret, J.J., Lloyd, K.R., Gerwick, L., Gerwick, W.H., Wipf, P., Hakansson, K., Smith, J.L., Sherman, D.H., 2009. Polyketide decarboxylative chain termination preceded by O-sulfonation in Curacin A biosynthesis. J. Am. Chem. Soc. 131, 16033–16035.

Gulevich, A.Y., Skorokhodova, A.Y., Sukhozhenko, A.V., Shakulov, R.S., Debabov, V.G., 2012. Metabolic engineering of *Escherichia coli* for 1-butanol biosynthesis through the inverted aerobic fatty acid β-oxidation pathway. Biotechnol. Lett. 34, 463–469.

Hanada, K., 2003. Serine palmitoyltransferase, a key enzyme of sphingolipid metabolism. Biochim. Biophys. Acta 1632, 16–30.

Hansen, E.H., Moeller, B.L., Kock, G.R., Bunner, C.M., Kristensen, C., Jensen, O.R., Okkels, F.T., Olsen, C.E., Motawia, M.S., Hansen, J., 2009. De novo biosynthesis of vanillin in fission yeast (*Schizosaccharomyces pombe*) and Baker's yeast (*Saccharomyces cerevisiae*). Appl. Environ. Microbiol. 75, 2765–2774.

Haynes, R.K., Vonwiller, S.C., 1994. Cyclic Peroxyacetal Lactone, Lactol and Ether Compounds. US 5310946.

Hjersted, J.L., Henson, M.A., Mahadevan, R., 2007. Genome-scale analysis of *Saccharomyces cerevisiae* metabolism and ethanol production in fed-batch culture. Biotechnol. Bioeng. 97, 1190–1204.

Hottiger, T., Boller, T., 1991. Ethylene biosynthesis in *Fusarium oxysporum* f. sp. *tulipae* proceeds from glutamate/2-oxoglutarate and requires oxygen and ferrous ions in vivo. Arch. Microbiol. 157, 18–22.

Howard, T.P., Middelhaufe, S., Moore, K., Edner, C., Kolak, D.M., Taylor, G.N., Parker, D.A., Lee, R., Smirnoff, N., Aves, S.J., Love, J., 2013. Synthesis of customized petroleum-replica fuel molecules by targeted modification of free fatty acid pools in *Escherichia coli*. Proc. Natl. Acad. Sci. U.S.A. 110, 7636–7641. S7636/1-S7636/6.

Huang, Y.L., Mann, K., Novak, J.M., Yang, S.-T., 1998. Acetic acid production from fructose by *Clostridium formicoaceticum* immobilized in a fibrous-bed bioreactor. Biotechnol. Prog. 14, 800–806.

Hwang, H.J., Park, J.H., Kim, J.H., Kong, M.K., Kim, J.W., Park, J.W., Cho, K.M., Lee, P.C., 2014. Engineering of a butyraldehyde dehydrogenase of *Clostridium saccharoperbutylacetonicum* to fit an engineered 1,4-butanediol pathway in *Escherichia coli*. Biotechnol. Bioeng. 111, 1374–1384.

Ince, J.E., Knowles, C.J., 1986. Ethylene formation by cell-free extracts of *Escherichia coli*. Arch. Microbiol. 146, 151–158.

Ingram, L.O., Conway, T., Clark, D.P., Sewell, G.W., Preston, J.F., 1987. Genetic engineering of ethanol production in *Escherichia coli*. Appl. Environ. Microbiol. 53, 2420–2425.

Inokuma, K., Liao, J.C., Okamoto, M., Hanai, T., 2010. Improvement of isopropanol production by metabolically engineered *Escherichia coli* using gas stripping. J. Biosci. Bioeng. 110, 696–701.

Inui, M., Kawaguchi, H., Murakami, S., Vertes, A.A., Yukawa, H., 2005. Metabolic engineering of Corynebacterium glutamicum for fuel ethanol production under oxygen-deprivation conditions. J. Mol. Microbiol. Biotechnol. 8, 243–254.

Inui, M., Suda, M., Kimura, S., Yasuda, K., Suzuki, H., Toda, H., Yamamoto, S., Okino, S., Suzuki, N., Yukawa, H., 2008. Expression of *Clostridium acetobutylicum* butanol synthetic genes in *Escherichia coli*. Appl. Microbiol. Biotechnol. 77, 1305–1316.

Ishida, N., Suzuki, T., Tokuhiro, K., Nagamori, E., Onishi, T., Saitoh, S., Kitamoto, K., Takahashi, H., 2006. D-lactic acid production by metabolically engineered *Saccharomyces cerevisiae*. J. Biosci. Bioeng. 101, 172–177.

Ishida, N., Saitoh, S., Ohnishi, T., Tokuhiro, K., Nagamori, E., Kitamoto, K., Takahashi, H., 2006. Metabolic engineering of *Saccharomyces cerevisiae* for efficient production of pure L-(+)-lactic acid. Appl. Biochem. Biotechnol. 129–132, 795–807.

Jain, R., Huang, J., Yuan, Q., Yan, Y., 2015. Engineering microaerobic metabolism of *E. coli* for 1,2-propanediol production. J. Ind. Microbiol. Biotechnol. 42, 1049–1055.

Jeffries, T.W., Grigoriev, I.V., Grimwood, J., Laplaza, J.M., Aerts, A., Salamov, A., Schmutz, J., Lindquist, E., Dehal, P., Shapiro, H., Jin, Y.-S., Passoth, V., Richardson, P.M., 2007. Genome sequence of the lignocellulose-bioconverting and xylose-fermenting yeast *Pichia stipitis*. Nat. Biotechnol. 25, 319–326.

Jung, Y.K., Kim, T.Y., Park, S.J., Lee, S.Y., 2010. Metabolic engineering of *Escherichia coli* for the production of polylactic acid and its copolymers. Biotechnol. Bioeng. 105, 161–171.

Khara, B., Menon, N., Levy, C., Mansell, D., Das, D., Marsh, E.N.G., Leys, D., Scrutton, N.S., 2013. Production of propane and other short-chain alkanes by structure-based engineering of ligand specificity in aldehyde-deformylating oxygenase. Chembiochem 14, 1204–1208.

Kolattukudy, P.E., 1967. Biosynthesis of paraffins in *Brassica oleracea*: fatty acid elongation-decarboxylation as a plausible pathway. Phytochemistry 6, 963–975.

Ladygina, N., Dedyukhina, E.G., Vainshtein, M.B., 2006. A review on microbial synthesis of hydrocarbons. Process Biochem. 41, 1001–1014.

Lan, E.I., Liao, J.C., 2011. Metabolic engineering of cyanobacteria for 1-butanol production from carbon dioxide. Metab. Eng. 13, 353–363.

Lee, K.Y., Park, J.M., Kim, T.Y., Yun, H., Lee, S.Y., 2010. The genome-scale metabolic network analysis of *Zymomonas mobilis* ZM4 explains physiological features and suggests ethanol and succinic acid production strategies. Microb. Cell Fact. 9, 94.

Lee, J., Jang, Y.-S., Choi, S.J., Im, J.A., Song, H., Cho, J.H., Seung, D.Y., Papoutsakis, E.T., Bennett, G.N., Lee, S.Y., 2012. Metabolic engineering of *Clostridium acetobutylicum* ATCC 824 for isopropanol-butanol-ethanol fermentation. Appl. Environ. Microbiol. 78, 1416–1423.

Lee, J.H., Lama, S., Kim, J.R., Park, S.H., 2018. Production of 1,3-propanediol from glucose by recombinant *Escherichia coli* BL21(DE3). Biotechnol. Bioproc. Eng. 23, 250–258.

Leeuwen, B.N.M., Wulp, A.M., Duijnstee, I., Maris, A.J.A., Straathof, A.J.J., 2012. Fermentative production of isobutene. Appl. Microbiol. Biotechnol. 93, 1377–1387.

Li, N., Chang, W.-c., Warui, D.M., Booker, S.J., Krebs, C., Bollinger, J.M., 2012. Evidence for only oxygenative cleavage of aldehydes to Alk(a/e)nes and formate by cyanobacterial aldehyde decarbonylases. Biochemistry 51, 7908–7916.

Lin, H., Bennett, G.N., San, K.-Y., 2005. Metabolic engineering of aerobic succinate production systems in *Escherichia coli* to improve process productivity and achieve the maximum theoretical succinate yield. Metab. Eng. 7, 116–127.

Lin, D., Hanscom, L., Murthy, A., Galli, A., Evans, M., Neill, E., Mancini, M.S., Martindill, J., Medouar, F.-Z., Huang, S., Wackernagel, M., 2018. Ecological footprint accounting for countries: updates and results of the national footprint accounts, 2012–2018. Resources 7, 58–80.

Liu, F., Jian, J., Shen, X., Chung, A., Chen, J., Chen, G.-Q., 2011. Metabolic engineering of *Aeromonas hydrophila* 4AK4 for production of copolymers of 3-hydroxybutyrate and medium-chain-length 3-hydroxyalkanoate. Bioresour. Technol. 102, 8123–8129.

Liu, B., Wang, H., Du, Z., Li, G., Ye, H., 2011. Metabolic engineering of artemisinin biosynthesis in *Artemisia annua* L. Plant Cell Rep. 30, 689–694.

Liu, Y., Wang, C., Yan, J., Zhang, W., Guan, W., Lu, X., Li, S., 2014. Hydrogen peroxide-independent production of α-alkenes by OleTJE P450 fatty acid decarboxylase. Biotechnol. Biofuels 7, 28/1-28/12, 12.

Liu, Q., Wu, K., Cheng, Y., Lu, L., Xiao, E., Zhang, Y., Deng, Z., Liu, T., 2015. Engineering an iterative polyketide pathway in *Escherichia coli* results in single-form alkene and alkane overproduction. Metab. Eng. 28, 82–90.

Lynch, S., Eckert, C., Yu, J., Maness, P.-C., Gill, R., 2016. Overcoming substrate limitations for improved production of ethylene in *E. coli*. Biotechnol. Biofuels 9, 3.

McKenna, R., Nielsen, D.R., 2011. Styrene biosynthesis from glucose by engineered *E. coli*. Metab. Eng. 13, 544–554.

Mendez-Perez, D., Begemann, M.B., Pfleger, B.F., 2011. Modular synthase-encoding gene involved in α-olefin biosynthesis in *Synechococcus* sp. strain PCC 7002. Appl. Environ. Microbiol. 77, 4264–4267.

Mo, M.L., Palsson, B.O., Herrgard, M.J., 2009. Connecting extracellular metabolomic measurements to intracellular flux states in yeast. BMC Syst. Biol. 3, 37.

Naccarato, W.F., Gilbertson, J.R., Gelman, R.A., 1974. Effects of different culture media and oxygen on lipids of *Escherichia coli* K-12. Lipids 9, 322–327.

Nations, U., Energy Statistics Yearbook 2016. United Nations Publication 2018.

Neale, A.D., Scopes, R.K., Kelly, J.M., 1988. Alcohol production from glucose and xylose using *Escherichia coli* containing *Zymomonas mobilis* genes. Appl. Microbiol. Biotechnol. 29, 162–167.

Niu, W., Draths, K.M., Frost, J.W., 2002. Benzene-free synthesis of adipic acid. Biotechnol. Prog. 18, 201–211.

Nunn, W.D., Giffin, K., Clark, D., Cronan Jr., J.E., 1983. Role for fadR in unsaturated fatty acid biosynthesis in *Escherichia coli*. J. Bacteriol. 154, 554–560.

Ohta, K., Beall, D.S., Mejia, J.P., Shanmugam, K.T., Ingram, L.O., 1991. Genetic improvement of *Escherichia coli* for ethanol production: chromosomal integration of *Zymomonas mobilis* genes encoding pyruvate decarboxylase and alcohol dehydrogenase II. Appl. Environ. Microbiol. 57, 893–900.

Okino, S., Noburyu, R., Suda, M., Jojima, T., Inui, M., Yukawa, H., 2008. An efficient succinic acid production process in a metabolically engineered *Corynebacterium glutamicum* strain. Appl. Microbiol. Biotechnol. 81, 459–464.

Papanikolaou, S., Aggelis, G., 2002. Lipid production by *Yarrowia lipolytica* growing on industrial glycerol in a single stage continuous culture. Bioresour. Technol. 82, 43–49.

Park, S.J., Kim, E.Y., Noh, W., Park, H.M., Oh, Y.H., Lee, S.H., Song, B.K., Jegal, J., Lee, S.Y., 2013. Metabolic engineering of *Escherichia coli* for the production of 5-aminovalerate and glutarate as C5 platform chemicals. Metab. Eng. 16, 42–47.

Pauthenier, C., Faulon, J.-P., 2013. Composé Produits Par Ingénierie Métabolique. Techniques de l'ingénieur. BIO 801.

Pauthenier, C., Faulon, J.L., 2018. Ingénierie métabolique et biologie de synthèse pour la chimie verte. Techniques de l'ingénieur. BIO 800.

Pazout, J., Pazoutova, S., 1989. Ethylene is synthesized by vegetative mycelium in surface cultures of *Penicillium cyclopium* westling. Can. J. Microbiol. 35, 384–387.

Pereira, F.B., Guimaraes, P.M.R., Teixeira, J.A., Domingues, L., 2010. Optimization of low-cost medium for very high gravity ethanol fermentations by *Saccharomyces cerevisiae* using statistical experimental designs. Bioresour. Technol. 101, 7856–7863.

Pirkov, I., Albers, E., Norbeck, J., Larsson, C., 2008. Ethylene production by metabolic engineering of the yeast *Saccharomyces cerevisiae*. Metab. Eng. 10, 276–280.

Poblete-Castro, I., Binger, D., Rodrigues, A., Becker, J., Martins dos Santos, V.A.P., Wittmann, C., 2013. In-silico-driven metabolic engineering of *Pseudomonas putida* for enhanced production of poly-hydroxyalkanoates. Metab. Eng. 15, 113–123.

Raab, A.M., Gebhardt, G., Bolotina, N., Weuster-Botz, D., Lang, C., 2010. Metabolic engineering of *Saccharomyces cerevisiae* for the biotechnological production of succinic acid. Metab. Eng. 12, 518–525.

Radakovits, R., Eduafo, P.M., Posewitz, M.C., 2011. Genetic engineering of fatty acid chain length in *Phaeodactylum tricornutum*. Metab. Eng. 13, 89–95.

Ro, D.-K., Paradise, E.M., Ouellet, M., Fisher, K.J., Newman, K.L., Ndungu, J.M., Ho, K.A., Eachus, R.A., Ham, T.S., Kirby, J., Chang, M.C.Y., Withers, S.T., Shiba, Y., Sarpong, R., Keasling, J.D., 2006. Production of the antimalarial drug precursor artemisinic acid in engineered yeast. Nature 440, 940–943.

Rogers, P.L., Jeon, Y.J., Lee, K.J., Lawford, H.G., 2007. *Zymomonas mobilis* for fuel ethanol and higher value products. Adv. Biochem. Eng. Biotechnol. 108, 263–288.

Roth, R.J., Acton, N., 1989. A simple conversion of artemisinic acid into artemisinin. J. Nat. Prod. 52, 1183–1185.

Rude, M.A., Baron, T.S., Brubaker, S., Alibhai, M., Del Cardayre, S.B., Schirmer, A., 2011. Terminal olefin (1-alkene) biosynthesis by a novel P450 fatty acid decarboxylase from *Jeotgalicoccus species*. Appl. Environ. Microbiol. 77, 1718–1727.

Saini, M., Hong Chen, M., Chiang, C.-J., Chao, Y.-P., 2015. Potential production platform of n-butanol in *Escherichia coli*. Metab. Eng. 27, 76–82.

Samuels, L., Kunst, L., Jetter, R., 2008. Sealing plant surfaces: cuticular wax formation by epidermal cells. Annu. Rev. Plant Biol. 59, 683–707.

Sangproo, M., Polyiam, P., Jantama, S.S., Kanchanatawee, S., Jantama, K., 2012. Metabolic engineering of *Klebsiella oxytoca* M5a1 to produce optically pure D-lactate in mineral salts medium. Bioresour. Technol. 119, 191–198.

Santala, S., Efimova, E., Kivinen, V., Larjo, A., Aho, T., Karp, M., Santala, V., 2011. Improved triacylglycerol production in *Acinetobacter baylyi* ADP1 by metabolic engineering. Microb. Cell Fact. 10, 36.

Schirmer, A., Rude, M.A., Li, X., Popova, E., del Cardayre, S.B., 2010. Microbial biosynthesis of alkanes. Science 329, 559–562.

Schjerling, C.K., Hummel, R., Hansen, J.K., Boersting, C., Mikkelsen, J.M., Kristiansen, K., Knudsen, J., 1996. Disruption of the gene encoding the acyl-CoA-binding protein (ACB1) perturbs acyl-CoA metabolism in *Saccharomyces cerevisiae*. J. Biol. Chem. 271, 22514–22521.

Schmid, G., Hofheinz, W., 1983. Total synthesis of qinghaosu. J. Am. Chem. Soc. 105, 624–625.

Shen, C.R., Liao, J.C., 2008. Metabolic engineering of *Escherichia coli* for 1-butanol and 1-propanol production via the keto-acid pathways. Metab. Eng. 10, 312–320.

Shen, C.R., Lan, E.I., Dekishima, Y., Baez, A., Cho, K.M., Liao, J.C., 2011. Driving forces enable high-titer anaerobic 1-butanol synthesis in *Escherichia coli*. Appl. Environ. Microbiol. 77, 2905–2915.

Shiue, E., Prather, K.L.J., 2014. Improving D-glucaric acid production from myo-inositol in *E. coli* by increasing MIOX stability and myo-inositol transport. Metab. Eng. 22, 22–31.

Slater, S., Mitsky, t.A., Houmiel, K.L., Hao, M., Reiser, S., Taylor, N.B., Tran, M., Valentin, H.E., Rodriguez, D.J., Stone, D.A., Padgette, S.R., Kishore, G., Gruys, K.J., 1999. Metabolic engineering of *Arabidopsis* and *Brassica* for poly(3-hydroxybutyrate-co-3-hydroxyvalerate) copolymer production. Nat. Biotechnol. 17, 1011–1016.

Song, X., Yu, H., Zhu, K., 2016. Improving alkane synthesis in *Escherichia coli* via metabolic engineering. Appl. Microbiol. Biotechnol. 100, 757–767.

Sprayberry, B.A., Hall, W.C., Miller, C.S., 1965. Biogenesis of ethylene in *Penicillium digitatum.* Nature 208, 1322–1323.

Stearns, J.C., Glick, B.R., 2003. Transgenic plants with altered ethylene biosynthesis or perception. Biotechnol. Adv. 21, 193–210.

Sukovich, D.J., Seffernick, J.L., Richman, J.E., Hunt, K.A., Gralnick, J.A., Wackett, L.P., 2010. Structure, function, and insights into the biosynthesis of a head-to-head hydrocarbon in *Shewanella oneidensis* strain MR-1. Appl. Environ. Microbiol. 76, 3842–3849.

Szczebara, F.M., Chandelier, C., Villeret, C., Masurel, A., Bourot, S., Duport, C., Blanchard, S., Groisillier, A., Testet, E., Costaglioli, P., Cauet, G., Degryse, E., Balbuena, D., Winter, J., Achstetter, T., Spagnoli, R., Pompon, D., Dumas, B., 2003. Total biosynthesis of hydrocortisone from a simple carbon source in yeast. Nat. Biotechnol. 21, 143–149.

Tang, X., Feng, H., Chen, W.N., 2013. Metabolic engineering for enhanced fatty acids synthesis in *Saccharomyces cerevisiae*. Metab. Eng. 16, 95–102.

Tillman, J.A., Seybold, S.J., Jurenka, R.A., Blomquist, G.J., 1999. Insect pheromones-an overview of biosynthesis and endocrine regulation. Insect Biochem. Mol. Biol. 29, 481–514.

Tolonen, A.C., Haas, W., Chilaka, A.C., Aach, J., Gygi, S.P., Church, G.M., 2011. Proteome-wide systems analysis of a cellulosic biofuel-producing microbe. Mol. Syst. Biol. 7, 461.

Trinh, C.T., Srienc, F., 2009. Metabolic engineering of *Escherichia coli* for efficient conversion of glycerol to ethanol. Appl. Environ. Microbiol. 75, 6696–6705.

Trinh, C.T., Unrean, P., Srienc, F., 2008. Minimal *Escherichia coli* cell for the most efficient production of ethanol from hexoses and pentoses. Appl. Environ. Microbiol. 74, 3634–3643.

Tseng, H.-C., Martin, C.H., Nielsen, D.R., Jones-Prather, K.L., 2009. Metabolic engineering of *Escherichia coli* for enhanced production of (R)- and (S)-3-hydroxybutyrate. Appl. Environ. Microbiol. 75, 3137–3145.

Wang, W., Lu, X., 2013. Microbial Synthesis of alka(e)nes. Front. Bioeng. Biotechnol 1, 10.

Wang, J.-P., Wu, L.-X., Xu, F., Lv, J., Jin, H.-J., Chen, S.-F., 2010. Metabolic engineering for ethylene production by inserting the ethylene-forming enzyme gene (efe) at the 16S rDNA sites of *Pseudomonas putida* KT2440. Bioresour. Technol. 101, 6404–6409.

Wang, W., Liu, X., Lu, X., 2013. Engineering cyanobacteria to improve photosynthetic production of alka(e)nes. Biotechnol. Biofuels 6, 69.

Weizmann, C., 1919. Production of Acetone and Alcohol by Bactebiological Processes. US1315585A.

Wen, Z., Minton, N.P., Zhang, Y., Li, Q., Liu, J., Jiang, Y., Yang, S., 2017. Enhanced solvent production by metabolic engineering of a twin-clostridial consortium. Metab. Eng. 39, 38–48.

Xia, T., Altman, E., Eiteman, M.A., 2015. Succinate production from xylose-glucose mixtures using a consortium of engineered *Escherichia coli*. Eng. Life Sci. 15, 65–72.

Xiao, H., Li, Z., Jiang, Y., Yang, Y., Jiang, W., Gu, Y., Yang, S., 2012. Metabolic engineering of D-xylose pathway in *Clostridium beijerinckii* to optimize solvent production from xylose mother liquid. Metab. Eng. 14, 569–578.

Xu, P., Gu, Q., Wang, W., Wong, L., Bower, A.G.W., Collins, C.H., Koffas, M.A.G., 2013. Modular optimization of multi-gene pathways for fatty acids production in *E. coli*. Nat. Commun. 4 ncomms2425, 8.

Yamada, M., Okada, Y., Yoshida, T., Nagasawa, T., 2008. Vanillin production using *Escherichia coli* cells over-expressing isoeugenol monooxygenase of *Pseudomonas putida*. Biotechnol. Lett. 30, 665–670.

Yoon, Y.J., Kim, E.S., Hwang, Y.S., Choi, C.Y., 2004. Avermectin: biochemical and molecular basis of its biosynthesis and regulation. Appl. Microbiol. Biotechnol. 63, 626–634.

Yoshida, A., Nishimura, T., Kawaguchi, H., Inui, M., Yukawa, H., 2005. Enhanced hydrogen production from formic acid by formate hydrogen lyase-overexpressing *Escherichia coli* strains. Appl. Environ. Microbiol. 71, 6762–6768.

Yu, M.-R., Zhang, Y.-L., Tang, I.C., Yang, S.-T., 2011. Metabolic engineering of *Clostridium tyrobutyricum* for n-butanol production. Metab. Eng. 13, 373–382.

Yu, S., Yao, P., Li, J., Ren, J., Yuan, J., Feng, J., Wang, M., Wu, Q., Zhu, D., 2016. Enzymatic synthesis of 3-hydroxypropionic acid at high productivity by using free or immobilized cells of recombinant *Escherichia coli*. J. Mol. Catal. B Enzym. 129, 37–42.

Zhang, X., Wang, X., Shanmugam, K.T., Ingram, L.O., 2011. L-Malate production by metabolically engineered *Escherichia coli*. Appl. Environ. Microbiol. 77, 427–434.

Zhou, Q., Shi, Z.-Y., Meng, D.-C., Wu, Q., Chen, J.-C., Chen, G.-Q., 2011. Production of 3-hydroxypropionate homopolymer and poly(3-hydroxypropionate-co-4-hydroxybutyrate) copolymer by recombinant *Escherichia coli*. Metab. Eng. 13, 777–785.

Zhou, Y.J., Buijs, N.A., Zhu, Z., Qin, J., Siewers, V., Nielsen, J., 2016. Production of fatty acid-derived oleochemicals and biofuels by synthetic yeast cell factories. Nat. Commun. 7, 11709.

Chapter 6

Algae for the production of bio-based products

Michail Syrpas, Petras Rimantas Venskutonis

Kaunas University of Technology, Department of Food Science & Technology, Radvilėnu pl. 19, Kaunas, Lithuania

6.1 Microalgae and macroalgae

The informal term *algae* includes a diverse category of photosynthetic (as well as secondarily nonphotosynthetic evolutionary descendants), oxygen-evolving entities from various phylogenetic groups (Metting, 1996). Based on their pigmentation algae can be broadly classified as Rhodophyta (red algae), Phaeophyta (brown algae), and Chlorophyta (green algae) (Cook, 1945). Whereas, depending on their size, which can vary from unicellular of 3–10 μm forms to giant kelps up to 70 m long, they can be divided into micro- and macroalgae, respectively. Although classification of algae into different taxonomical groups remains under discussion, it could be mentioned that there are at least seven distinct evolutionary lineages, which arose independently during geological time and that evolved at different rates that can be assembled by eukaryotic algal groups, some of which include protists, historically identified as fungi and protozoa (Andersen, 1992). Besides these eukaryotic organisms, phycology includes the prokaryotic cyanobacteria also known as *blue-green algae* and prochlorophytes (Metting, 1996). These biochemically diverse organisms are present in both aquatic (marine and freshwater) and terrestrial habitats. The exact number of different algal species remains unknown; it is characteristic that for the most abundant lineages most of them remain as undescribed species. For example, the Rhodophyta lineage has approximately 5000 recognized extant species and an estimated 500 to 15,000 undescribed species (Andersen, 1992). Similarly, the green algae lineage has almost 100,000 undescribed species and approximately 16,000 recognized species whereas the chomophyte lineage includes 15,000 recognized species and from one to 10 million uncharacterized species (Andersen, 1992).

Microalgae are unicellular species found in both freshwater and marine systems. It is estimated that approximately 20–800 thousands of species exist

Biobased Products and Industries. https://doi.org/10.1016/B978-0-12-818493-6.00006-3

with only a small number (40–50,000) of the species described (Oncel, 2013). Microalgae are primarily responsible for 40%–50% of the total photosynthetic primary production contributed by all algal species. Their chemical composition varies depending on the season, species culture conditions, etc. However, several microalgae species are characterized by their ability to accumulate high levels of lipids and proteins (Table 6.1).

The phylogenetic diversity of these microorganisms is depicted in an equally broad biochemical spectrum of pigments, photosynthetic storage products, fatty acids and lipids, oils, sterols and hydrocarbons, different cell wall materials and mucilage, as well as a plethora of biologically active secondary metabolites (Metting, 1996). Their size ranges from a few micrometers (0.2–2 μm and 2–20 μm for picoplankton and nanoplankton, respectively) to a few hundreds of micrometers (20–200 μm for microplankton) and unlike higher plants, they do not have roots, leaves, or stems (Sieburth et al., 1978).

Macroalgae, also known as seaweeds, are multicellular, macroscopic algae that can be found mainly in the marine environment. This polyphyletic group of organisms includes green macroalgae (i.e., *Enteromorpha* spp.), red macroalgae (i.e., *Porphyra* spp.), brown macroalgae (i.e., *Laminaria* spp.); and could include tuft-forming cyanobacteria (Suganya et al., 2016). They are generally observed in the seabed, where they form dense stands on the well-illuminated rocky margin of all continents (Lüning et al., 1990). Approximately, 10,000 named species were recorded in coastal marine habitats, out of which 200 species are used around the globe, and only ten are intensively cultivated (Lüning and Pang, 2003). Although the chemical composition of macroalgae depends on the species and growing conditions, on average macroalgae contain only 10%–15% of dry matter. The dehydrated macroalgae are characterized by large amounts of carbohydrates, like mannan, ulvan, carrageenan, agar, laminarin, mannitol, alginate, fucoidin, fucose, and uronic acid (Chen et al., 2015). Table 6.1 presents the elemental chemical composition of various microalgae and macroalgae species.

6.2 Types of cultivation

Depending on the primary energy source(s) microalgae are known to assume different types of metabolisms. Moreover, as a response to the changes of environmental conditions, several species can shift their metabolism (Mata et al., 2010). Photoautotrophic cultivation is the most commonly applied type of microalgae cultivation. In this form algae, utilize light and carbon dioxide (CO_2) as the primary energy and carbon source, respectively, which are subsequently converted into chemical energy via the photosynthetic reactions. The consumption of CO_2 can be considered as the significant advantage of autotrophic cultivation; thus, when CO_2 is utilized as a carbon source, the production site should be in the vicinity or adjacent to facilities

TABLE 6.1 The chemical composition of various microalgae and macroalgae species.

Species	Lipids %	Carbohydrates %	Proteins %
Microalgae			
Acutodesmus dimorphus (UTEX-1237)	18.8	38.6	28.1
Anabaena cylindrica	4–7	25–30	43–56
Aphanizomenon flos-aquae	3	23	62
Arthrospira maxima	6–7	13–16	60–71
Botryococcus braunii (UTEX-572)	34.4	18.5	39.9
Chlamydomonas rheinhardii	21	17	48
Chlorella vulgaris	14–22	12–17	51–58
Chlorella pyrenoidosa	2	26	57
Chlorella vulgaris	14–22	12–17	51–58
Dunaliella bioculata	8	4	49
Dunaliella salina	6	32	57
Dunaliella tertiolecta	17	31	43
Euglena gracilis	14–20	14–18	39–61
Heterochlorella luteoviridis	11	41	29
Nannochloropsis granulata (CCMP-535)	47.8	27.4	17.9
Neochloris oleoabundans (UTEX-1185)	15.4	37.8	30.1
Phaeodactylum tricornutum (CCMP-1327)	18.2	25.2	39.6
Porphyridium aerugineum (UTEX-755)	13.7	45.8	31.6
Porphyridium cruentum	9–14	40–57	28–39
Porphyridium cruentum	9–14	40–57	8–39
Prymnesium parvum	22–38	25–33	28–45
Scenedesmus dimorphus	16–40	21–52	8–18
Scenedesmus obliquus	12–14	10–17	50–56
Scenedesmus quadricauda	1.9	–	47
Spirogyra sp.	11–21	33–64	6–20

Continued

TABLE 6.1 The chemical composition of various microalgae and macroalgae species.—cont'd

Species	Lipids %	Carbohydrates %	Proteins %
Spirulina maxima	6–7	13–16	60–71
Spirulina platensis	4–9	8–14	46–63
Synechococcus sp.	11	15	63
Tetraselmis chuii (PLY-429)	12.3	25	46.5
Tetraselmis maculate	3	15	52
Macroalgae			
Acanthophora spicifera	10.0–12.0	11.6–13.2	12.0–13.2
Anadyomene brownii	6.2	25.8	9
Boergesenia forbesii	11.42	21.38	7.43
Caulerpa cupressoides	10.97	51.75	7.43
Caulerpa fergusonii	7.15	23.63	7.76
Caulerpa laetevirens	8.8	56.25	8.78
Caulerpa peltata	11.42	45	6.41
Caulerpa racemosa	2.3–10.5	16–33.75	6.8–12.5
Caulerpa sertularioides	6.99	49.5	9.11
Chaetomorpha aerea	8.5	31.5	10.13
Chaetomorpha antennina	11.45	27	10.13
Chaetomorpha linoides	12	27	9.45
Cladophora fascicularis	15.7	49.5	15.53
Codium adhaerens	7.4	40.5	7.26
Codium decorticatum	9	50.63	6.08
Codium tomentosum	3.6–7.15	29.25–32.8	5.06–18.8
Dictyosphaeria cavernosa	10.51	42.75	6
Enteromorpha compressa	11.45	24.75	7.26
Gracilaria gracilis	0.60	46.6	20.2
Grateloupia turuturu	2.2	43.2	22.5

TABLE 6.1 The chemical composition of various microalgae and macroalgae species.—cont'd

Species	Lipids %	Carbohydrates %	Proteins %
Halimeda opuntia	2.3–2.9	2.5–2.7	3.2
Halimeda macroloba	2.3–9.9	2.7–32.6	4.6–6.6
Hypnea valentiae	9.6–11.6	11.8–13.0	11.8–12.6
Laurencia papillosa	8.9–10.8	12.0–13.3	11.8–12.9
Microdictyon agardhianum	9.4	27	20.93
Neomeris van-bosseae	2.6–2.7	8.3–15.2	1.4–1.5
Osmundea pinnatifida	0.9	32.4	23.8
Saccorhiza polyschides	1.1	45.6	14.44
Sargassum muticum	1.45	49.3	16.9
Ulva lactuca	9.6–11.4	11.6–13.2	11.4–12.6
Ulva reticulate	8.5	16.88	12.83
Valoniopsis pachynema	9.09	31.5	8.78

With information from Becker (2007); Sahu et al. (2013); Tibbetts et al. (2015); Diprat et al. (2017); Rodrigues et al. (2015); Kaliaperumal et al. (2002).

capable of supplying the required large amounts for algal growth (Chen et al., 2011). This type of cultivation is typically applied in most outdoor scale-up microalgae cultivation systems such as open and raceway ponds (Mata et al., 2010).

Like bacteria, certain microalgae could also be grown in the dark, utilizing only a carbon source, in a process known as heterotrophic cultivation. Several *Chlorella* species, as well as other microalgae genus like *Tetraselmis* and *Neochloris*, are known to be able to assimilate both heterotrophic and autotrophic metabolism (Hu et al., 2018). The overall better control of cultivation process, the higher biomass production and productivity, the lower costs of biomass harvesting and more importantly the absence of light prerequisite are the significant advantages of heterotrophic over phototrophic cultivation (Medipally et al., 2015). Despite these advantages, there are several limitations associated with heterotrophic cultivation. First, only a small number of industrially relevant species can be grown under these conditions, while the number of species grown for biodiesel production is even more restricted.

Glucose, which is typically used as the organic substrate for heterotrophic production, is also used for human consumption; thus, alternative substrates based on lignocellulose and glycerol should be considered. From an environmental perspective, life cycle assessment is in favor of phototrophic cultivation where CO_2 is fixed during growth, whereas in heterotrophic cultivation CO_2 is generated, thus contributing negatively to CO_2 emissions (Hu et al., 2018).

Although these two cultivation types are the most common growth modes of microalgae, some species can also grow better under mixotrophic or photoheterotrophic conditions, which couple the advantages of autotrophic and heterotrophic systems and overcome their disadvantages (Zhan et al., 2017; Selvakumar and Umadevi, 2014). Although these two types of metabolism are not well distinguished, they could also be defined based on the energy source required (Chen et al., 2011). The main difference between these metabolisms is that photoheterotrophic involves light as the energy source, whereas mixotrophic cultivation can utilize various organic substrates for this purpose (Bhatnagar et al., 2011). Hence, photoheterotrophic cultivation requires both light and sugars simultaneously.

6.3 Microalgal cultivation systems

Culture systems vary significantly between macroalgae (seaweed) and microalgae. Due to their small size, microalgae require specially designed systems, while macroalgae can be grown directly in the open sea (van Iersel et al., 2009). Historically macroalgae have been cultivated for centuries, mainly in Asian countries, whereas dedicated microalgae culture systems only started to develop in the 1950s (van Iersel et al., 2009). Over the last 4 decades, commercial cultivation of algae has mainly been associated with certain microalgal species grown for different applications. Examples of commercially matured products in the market include growing of *Chlorella* and *Arthrospira* species as biomass for nutraceutical or food use, production of astaxanthin from *Haematococcus pluvialis*, β-carotene from *Dunaliella salina,* and other species grown as a feed for aquaculture.

Although each algal strain has specific demands for optimal growth, generally algae require relatively simple conditions to be grown. These conditions include a light source, a carbon source, water, nutrients, and controlled temperature. Even though these conditions are easily achieved, not each algal strain is appropriate for mass cultivation. Factors like the life cycle and growth rate of the strain, nutritional requirements, shear stress tolerance, genetic stability, and main yields of the desired product should also be considered (Suh and Lee, 2003).

Various open and closed cultivation systems that can meet these criteria, each with its own advantages and disadvantages, have been developed over the years. However, successful large-scale production of algal biomass remains rather complicated and at large unsophisticated (Borowitzka, 1999). Especially for microalgae, the main issue toward commercialization of novel algal products is the need for closed culture systems and the high capital investment required. Nevertheless, research focused on a deeper understanding of algal metabolism and biology, as well as progress in genetic engineering of algal strains will lead toward the design of sophisticated cultivation systems for specific applications.

6.3.1 Open systems

Open ponds are one of the most widely used systems for microalgae biomass production. These systems can be divided into two categories, natural water systems like lakes, ponds, etc. and artificial ponds like circular and raceway ponds. For many years naturally grown algae were harvested from lakes or lagoons and were used for food and feed purposes. Although this approach is still practiced by some food companies and aquaculture hatcheries, biomass productivity and final product quality cannot be assured (Lee, 2001).

Raceway ponds are shallow artificial ponds widely applied in microalgae cultivation. These systems are built as shallow closed loop channels that allow water circulation. Typically fertilizers are used to increase biomass productivity, and bypaddle wheels are used to agitate the culture (Lee, 2001). Cell concentrations of 0.5−1 kg/m^3 and average areal biomass productivity of approximately 25 g/m^2/day are obtained in these systems (Kumar et al., 2015). Two of the most common species grown in raceway ponds are *Chlorella* and *Arthrospira*. Productivities of 25 and 27 g/m^2/day were reported for *Chlorella* sp. and *Spirulina platensis* grown in facilities in the Czech Republic and Israel, respectively (Lee, 2001). Circular center-pivot ponds can be widely found in countries like Japan, Indonesia, and Taiwan. These typically shallow (<5 cm) systems require a rotating scraper to mix the culture. However, their size is restricted, for example, up to 50 m diameter of the most extensive known system, by the strain of water resistance on the rotating motor (Lee, 2001).

The major advantage of open systems is that they are easier to construct, operate, and maintain than most closed systems. They are relatively economic, thus making them a good option for mass cultivation of the selected algal strains. However, these systems also suffer from various limitations. As they are open systems, they are easily susceptible to contamination by other fast growing organisms, making it difficult to maintain a monoculture. Thus, these systems have been restricted for algal strains that can grow in highly selective environments (i.e., salinity, pH), which prevent the growth of contaminating organisms (Suh and Lee, 2003). Moreover, they require vast land mass and therefore the control of culture conditions is somewhat limited. Consequently, biomass production could be finite. In open-pond systems, the geographic location of the cultivation site has the highest impact on biomass productivity (Chisti, 2016). The chosen site should have such climatic conditions, mainly average temperature and annual irradiance level that allow consistent high-biomass yields throughout the production period (Chisti, 2016). Besides site location, both engineering and biological parameters can influence biomass productivity. Engineering factors include the pond depth, the delivery system of CO_2, the applied mixing method, and overall power consumption, whereas salinity, pH, and oxygen accumulation are among the biochemical factors (Kumar et al., 2015).

6.3.2 Photobioreactors

A photobioreactor (PBR) could be considered any kind of culture vessel, which has no direct gas or contaminant exchange with the environment and which is a naturally or artificially illuminated system designed and used for controlled biomass production. These systems minimize contamination risks and enable more effective axenic or monoculture algae cultivation. They provide higher control over critical culture conditions such as pH and temperature with minimal CO_2 and water losses, allow higher cell concentrations and the production of higher added value products such as complex biopharmaceuticals, etc (Singh and Sharma, 2012). There is a number of factors that should be taken into consideration for the efficient operation and design of a PBR for algal cultivation. Tsoglin et al. suggested the following factors in designing a PBR:

(i) possibility to cultivate various microalgal species;
(ii) uniform illumination of the reactor's surface and fast mass transfer of CO_2 and O_2;
(iii) design and material preventing or minimizing biofouling;
(iv) attaining high rates of mass transfer without damaging cells or suppressing their growth;
(v) ensuring the capacity to perform under conditions of intense foaming;
(vi) avoiding or minimizing nonilluminated parts (Tsoglin et al., 1996).

Furthermore, Vasumathi et al. reviewing the performance of different types of PBRs as well as the growth mechanism of microalgae reported that for efficient operation and design of a PBR system the following principles should further be considered:

(i) as high as possible spread area to culture volume ratio;
(ii) optimum cell concentration matching the highest rate;
(iii) harvesting the excess microalgae formed above this point for maintaining the optimal culture concentration;
(iv) maintaining a constant concentration of nutrients (Vasumathi et al., 2012).

Different types of PBRs have been designed and developed for the production of algae. For this reason, a variety of closed photobioreactors has been proposed to suit the particular characteristics of different microalgal strains. The designs range from horizontal tubular systems (Lüning and Pang, 2003; Chen et al., 2015), to helical tubular reactors (Becker, 2007), cascade reactors (Sahu et al., 2013), alveolar flat panels (Tibbetts et al., 2015), vertical flat panels (Diprat et al., 2017), or bubble columns (Bruton et al., 2009). The most scalable designs correspond to horizontal or helical tubular systems, as well as combinations of vertical flat panels and bubble columns, and therefore these types of PBRs have attracted the most interest.

Flat plate PBRs have been suggested for mass production of photoautotrophic organisms, and due to their large illumination surface area, it is believed they could become a standard reactor type for microalgae (Sierra et al., 2008). Construction of these systems dates back to the 1950s, which was followed-up with extensive works on various designs of vertical alveolar panels and flat plate reactors. Large illumination surface area with small light path, suitability for outdoor cultivation or immobilization of algal strains, low cost of construction and maintenance, high biomass yields, and low oxygen accumulation could be mentioned as the advantages of flat panel PBRs (Ugwu et al., 2008). However, the complexity for scaling-up due to the additional compartment and support materials required, some degree of wall growth, the limitation in temperature control, and the risks of high hydrodynamic stress to certain algal strains are their disadvantages (Ugwu et al., 2008). Smaller light paths of flat plate PBRs allow sufficient exposure time resulting in higher photosynthetic efficiency and higher density cultures.

Horizontal tubular PBRs are another prevalent form of the outdoor closed systems. They consist of an array of transparent tubes (that can be made either from glass or plastic), aligned horizontally, inclined, vertical, or as a helix. They also contain a gas exchange chamber with circulating cooling water (enables temperature control for introducing fresh culture, removing oxygen, and thereby avoiding potential buildup and photobleaching) (SundarRajan et al., 2019). In tubular PBRs microalgae culture is circulated through the tubes by a mechanical pump or an airlift pump (Hamed, 2016). The airlift pump can further allow gas (CO_2 and O_2) exchange between aeration gas and liquid medium (Medipally et al., 2015). Like flat plate PBRs, these systems have a high surface-to-volume ratio, which enables maximized exposure to light. Critical parameters of these systems are also flowing velocity and mixing efficiency in a radial direction. Main problems that are associated with tubular PBRs are an accumulation of dissolved O_2 and excessive power consumption, which is necessary to maintain the turbulent flow of high velocity (typically 20–50 m/s) to prevent biomass sedimentation (SundarRajan et al., 2019; Hamed, 2016). Scaling-up can be achieved by increasing the diameter and/or length of the tube. However, scaling-up of these systems is restricted by the length of the tubes, which depends on pH variation in the system, potential O_2 accumulation, and CO_2 depletion (Eriksen, 2008a). Currently, the largest existing closed PBRs are tubular, and examples include a 700 m^3 plant in

Klötze in Germany and a 25 m^3 plant at Mera Pharmaceuticals in Hawaii (Brennan and Owende, 2010).

Vertical tubular PBRs are systems that allow sufficient light penetration. They consist of vertical transparent tubing with air sparger at the base. The sparger's function is to convert sparged air into numerous tiny bubbles to prevent algal settling and thus improve mass transfer between gas and liquid medium (SundarRajan et al., 2019). These reactors allow the best control over culture conditions, the most efficient mixing system, and the highest volumetric mass transfer rates among the various PBR types (Medipally et al., 2015; Brennan and Owende, 2010). Although they have a low surface-to-volume ratio and relatively small illumination area, they can still attain comparable biomass concentration and growth rate due to efficient mixing. However, intense bubbling can induce shear stress on algal cells, and scaling-up attempts typically result in a further decrease in illumination surface (Ugwu et al., 2008). Based on their mode of liquid flow vertical tubular PBRs have two types of configuration namely, bubble column and airlift (Singh and Sharma, 2012). In bubble column the culture is mixed through all volume, while in airlift the volume is divided into two interconnected riser and downcomer zones and medium circulates between them (Hamed, 2016). The diameter of the bubble columns is usually twofold larger than their height. Bigger scale versions have perforated plates installed to disrupt and redistribute coalesced bubbles. Bubble column has no internal structure, but because light does not penetrate to the very center of culture, there is an internal dark zone and an outer illuminated zone. This means that photosynthesis efficiency depends on the gas flow rate as it directly influences light and dark cycle length. Also, if the gas flow rate is too low, the circulation pattern does not form at all (SundarRajan et al., 2019). In airlift reactors, the column is divided into two zones (riser and downcomer) by a baffle or a draft tube. In the riser zone, gas bubbles the liquid and drives it upward, whereas, in the downcomer zone, gas leaves the liquid, thus becomes denser again and moves down and back into the riser zone (SundarRajan et al., 2019).

Based on various PBR configurations, different technologies have been developed over the last years. The examples include more modern approaches like spectral shifting or internal illumination reactors, which aim to increase light capturing and distribution, plastic bag reactors, which provide a significant cost reduction, and membrane PBRs, which enable better mass transfers (Wang et al., 2012).

Hybrid cultivation systems consist of two distinct growth stages, combining both open and closed systems. In the first stage, microalgae culture is grown in closed PBR in a controlled environment. In the second stage, culture is transferred to an open system and exposed to nutritional stress to increase production of specific metabolites (SundarRajan et al., 2019; Hamed, 2016).

6.3.3 Macroalgae cultivation

Large-scale production of macroalgae started several decades ago, primarily in Asian countries, and after significant technology transfer it is now expanding in Europe and America (Hughes et al., 2012). Large-scale macroalgae cultivation is performed in coastal areas; thus, suitable site location requires a number of social, environmental, legislative, and financial criteria (i.e., depth, shipping traffic, and distance to ports) to be met for sustainable macroalgal biomass production (Thomas et al., 2019). Macroalgae cultivation offers specific advantages in comparison to biomass produced terrestrially. The main advantages is that it does not compete with other products for arable land nor does it require fresh water, fertilizers, or pest control products; however, it does compete for coastal and off-shore space with touristic and other marine activities (Thomas et al., 2019).

There are mainly two methods of macroalgal cultivation, one that is also known as indirect and is based on natural stocks and the direct approach where targeted genera are cultivated. The cheapest cultivation method is the indirect method, which requires very little human interference and input (Mooney-McAuley et al., 2016). This method involves wild, fertile material, which can naturally settle on ropes and which is harvested when it reaches a certain size. However, under this method the obtained biomass is usually a mixture of species, which will most likely be of a little commercial value (Mooney-McAuley et al., 2016). It is thus reasonable that this technique contributes a small (around 5%) of global seaweed production (Bruton et al., 2009). The alternative approach utilizes targeted macroalgal strains where firstly juveniles are grown in a hatchery and subsequently deployed at sea (Mooney-McAuley et al., 2016). From a financial perspective, this approach is undoubtedly more expensive; however, it allows more control over the timings of the cultivation season and development of targeted species tailored according to the end-product required (Mooney-McAuley et al., 2016). Macroalgal cultivation with the direct approach could be described with nine steps including:

(i) initial collection of fertile seaweed, which is followed by five steps performed at the hatchery, namely
(ii) the release of spores from fertile material,
(iii) culture development under controlled environment,
(iv) induction of reproduction,
(v) subsequent spraying of fertile cultures on the culture ropes, and
(vi) development of culture on the ropes under the controlled environment. The final three steps include
(vii) the grow out phase when the culture is deployed on long line (or other) systems at sea;
(viii) the seaweed development (typically 6–7 months), and
(ix) harvesting of the deployed culture (Ghadiryanfar et al., 2016).

Global macroalgal aquaculture production, which is heavily concentrated in Asia, shows an increasing trend and is estimated to approximately 24 million tons per year (Thomas et al., 2019). On the other hand, European seaweed production, which depends almost entirely on harvesting of natural stocks, has decreased by approximately a third from 2000 to 2012, to approximately 230,000 tons per year, primarily due to environmental concerns but is gaining momentum as research initiatives along with business develop the capacity for large-scale cultivation and processing of this feedstock (Thomas et al., 2019).

6.4 Biofuels from algae

The world faces the progressive depletion of its energetic resources which are currently primarily based on non-renewable fuels. At the same time, population growth and increasing affluence worldwide have resulted in higher energy demands thus higher waste generation. Nowadays, petroleum remains the primary source of energy and materials, which has led to severe implications for both energy security and the environment. Conversion of abundant renewable feedstocks into bioenergy and bio-based products has potential to contribute to meeting the ever-increasing demand for energy and products.

The prospect of biofuel production from algae appears to be promising. However, technoeconomic impediments limit the development of commercial processes utilizing algal biomass. Currently, a series of technologies for cost-effective biomass conversion into fuels and chemicals are under development. A critical evaluation of the available literature and generated knowledge indicates that the critical factor for successful commercialization of algae-based fuels is the economic viability of such process with regard to operational and maintenance cost minimization along with maximization of oil-rich algae production (Singh and Gu, 2010). The main challenges that need to be addressed, besides the fluctuating price of fossil-based fuels, include the cost reduction of algal cultivation systems, harvesting techniques (Uduman et al., 2010), and downstream processes (Leite et al., 2013).

6.4.1 Biodiesel

Biodiesel is an efficient, sustainable, environmentally friendly, and readily biodegradable natural energy alternative to fossil fuels. It is a mixture of methyl esters obtained by transesterification of oils or fats from natural origin (Bozbas, 2008). Over the years and depending on the feedstock used, biofuels and thus biodiesel were divided into three generations. First-generation biodiesel is produced from food crops, with the most common being rapeseed, soybeans, sunflower, or palm oil (Ahmad et al., 2011). Second-generation biofuels were developed to overcome certain limitations, mainly to reduce dependency on edible oils, of first-generation biofuels. They typically derive from nonfood crops such as wood, organic waste, or energy crops such as jatropha, mahua, jojoba oil, tobacco seed, etc. (Ahmad et al., 2011). The high cost, inefficiency, and unsustainability of first- and second-generation biodiesel feedstocks could be overcome by microalgae, which could potentially produce large volumes of biomass, and thus biofuels (Maity et al., 2014).

Presently, microalgae are receiving great interest because of their potential use as a feedstock for biodiesel production (Wijffels et al., 2010). According to some authors, microalgal biodiesel appears as a renewable biofuel with the potential to displace petroleum-derived fuels without having an impact on food supply or other crop products (Chisti, 2008). A vast amount of investments in research and development of this biofuel is held by both private and public organizations focusing on mass production of algal biodiesel, based upon effective and sustainable technologies (Mata et al., 2010). Besides, extended research on selection of appropriate strains (Rawat et al., 2013) and technologies for production, processing, and extraction of biofuels (Brennan and Owende, 2010), specific focus has been given on applying metabolic engineering of algal fatty acid biosynthesis for development of strains capable of high productivity and energy value with the lowest possible cost (Harun et al., 2010a; Blatti et al., 2013).

Oil content in microalgae varies and depends on a number of factors including the strain, culture type, and environmental factors such as light, temperature, or minerals, which can alter algal lipid metabolism (Guschina and Harwood, 2006). Commonly studied genera for biodiesel production include green microalgae like *Chlorella*, *Tetraselmis*, *Botryococcus*, *Dunaliella* diatoms such as *Phaeodactylum*, *Nitzschia,* and *Cylindrotheca,* as well as cyanobacteria (Mata et al., 2010; Parmar et al., 2011). Biodiesel production from macroalgae, although feasible, seems a less attractive choice as compared to high lipid content microalgae biomass (Chen et al., 2015).

Besides the importance of oil content and productivities, the fatty acid profile of the different microalgae species is also of major importance, as it influences the applicability and characteristics of biodiesel produced. In a study based on published data of microalgal fatty acid compositions and predictive fuel models, the fuel properties of microalgal biodiesel were predicted (Stansell et al., 2012). In most cases, microalgal biodiesels were predicted to have extremely poor oxidative stabilities and insufficient cold-flow properties, with the cetane number in most cases being out of specification (Stansell et al., 2012). Conclusively, these findings supported the belief that the composition of the saturated fatty acids is of great importance while feedstocks rich in monounsaturated fatty acids are desirable for biodiesel production (Stansell et al., 2012). Nevertheless, the produced algal biodiesel should comply with existing regional or national standards. For instance, EU Standards EN 14214 and EN 14213 for vehicle fuel and heating oil require a maximum iodine value of 120 and 130 g I_2/100 g biodiesel, respectively, with a maximum of 1% of FAME with four and more double bonds (Knothe, 2006).

Over the past years, there has been a lot of debate and contradictory research on whether microalgal biodiesel or biofuels, in general, are environmentally or financially sustainable. The fact remains that algae are not yet produced at a large scale for bulk applications. From a technological perspective, in order to reach this level, a number of bottlenecks need to be addressed by multidisciplinary approaches, which would include among others systems biology; metabolic modeling, design, and operation of cultivation systems; integrated production chain; and strain development (Wijffels et al., 2010). Attempts to perform life cycle assessment were mainly based on theoretical models with diverging impact assessment methods and energy balance computations (Collet et al., 2014). In 2012, scientists from MIT, ExxonMobil, and SGI published an assessment of algal biofuels which concluded that if key research hurdles are overcome, algal biofuels will have about 50% lower life cycle greenhouse gas emissions than petroleum-derived fuel (Vasudevan et al., 2012).

6.4.2 Bioethanol

Bioethanol is considered as a viable alternative to fossil fuels due to its low toxicity, biodegradability, and the ability to be used directly as a blend with gasoline without further engine modifications (Sánchez and Cardona, 2008). Several crops, like corn and sugarcane, have been successfully utilized for bioethanol production. However, as their primary utility remains food and feed, these crops cannot meet the global demand (Banerjee et al., 2010). Significant amounts of research efforts have been made over the last years for utilization of algal biomass, as an alternative to crop feedstocks, in bioethanol production (Jambo et al., 2016; Harun et al., 2014; Sirajunnisa and Surendhiran, 2016). Production of algal bioethanol can be accomplished through three potential routes. In the first one, the natural carbohydrates of algal biomass (starch/cellulose) can be converted into fermentable sugars (John et al., 2011). Although the overall process varies, depending on the type/strain of algae utilized, biomass conversion to bioethanol typically follows three main steps: (i) feedstock pretreatment; (ii) hydrolysis; and (iii) fermentation to bioethanol (Harun et al., 2014). The second path, involves the natural accumulation of ethanol from algal cells during dark fermentation, whereas the third part includes the genetic engineering of algal strains for direct ethanol production (John et al., 2011).

Many microalgae are considered as useful raw materials for ethanol production. For example, *Chlorella*, *Dunaliella*, *Chlamydomonas*, *Scenedesmus*, *Spirulina* are reported to contain a large amount (>50% of the dry weight) of starch and glycogen (John et al., 2011). Hirano et al. isolated more than 250 strains of marine microalgae and selected strains were further screened for their ethanol productivity (Hirano et al., 1997). The authors reported that several strains showed a high starch content ($\geq$20% of the dry weight) with *Chlorella vulgaris* IAM C-534 (37% starch) achieving a 65% ethanol conversion rate after saccharification and fermentation of starch (Hirano et al., 1997).

Macroalgae are promising feedstocks for bioethanol production with inherent advantages, including the absence of structural lignin, high growth and biomass rate, and no competition for land and fresh water. However, macroalgal utilization presents specific technical difficulties as existing ethanologenic microbes can neither degrade alginate nor assimilate alginate degradation products (Takeda et al., 2011; Ji et al., 2016). To tackle this issue, Takeda et al. developed a metabolically engineered bacterium, namely *Sphingomonas* sp. A1. After 3 days of incubation using alginate as the sole carbon source, it was shown that this strain could accumulate up to 13.0 g/L ethanol (Takeda et al., 2011). *Defluviitalea phaphyphila* Alg1 was the first characterized thermophilic bacterium capable of simultaneously utilizing mannitol, glucose, and alginate from brown algae to produce ethanol (0.47 g/g mannitol, 0.44 g/g glucose, and 0.3 g/g alginate) (Ji et al., 2016).

6.4.3 Biogas

Biogas, a mixture of primarily methane (55%−75%) and CO_2 (25%−45%), is a biofuel obtained through the anaerobic digestion of organic matter. The potential of micro- and macroalgae as substrates in processes of biogas production has recently been addressed only sporadically (Debowski et al., 2013). The main reason for a reduced interest in algae biomass as a biodegradable organic matter feedstock for biogas production systems is the difficulties with its use as a substrate (Debowski et al., 2013). Several authors have highlighted the key factors that could limit the effectiveness of methane fermentation processes and thus hinder or fully inhibit the mechanisms of biogas production from algal biomass. These limitations occur mainly by the complex cell walls of many algae species, which are resistant to degradation owing to the presence of cellulose or hemicellulose (Passos et al., 2014, 2016). For this reason, several thermal, mechanical, chemical, and biochemical pretreatment methods have been suggested to improve microalgal anaerobic biodegradability (Passos et al., 2014). Moreover, pretreatment byproducts of algal biomass such as furanic and phenolic compounds has been reported to act as inhibitors of both anaerobic digestion processes and especially dark fermentation (Monlau et al., 2014). In addition, the carbon to nitrogen ratio of biomass subjected to fermentation processes has been characterized as detrimental (Debowski et al., 2013; Mata-Alvarez et al., 2000). Thus, it is critical to determine the C:N ratio as low values can adversely influence the process through ammonia accumulation and an increase in pH, which can lead to inhibition of methanogenic archaea growth (Koutra et al., 2018). Despite the above-mentioned limitations, anaerobic digestion of microalgae is a prospective environmentally feasible option for creating a sustainable energy source for domestic and industrial needs (Ward et al., 2014). Several strains of microalgae have been reported in the literature for anaerobic digestion including *Scenedesmus* sp. (Yang et al., 2011; González-Fernández et al., 2012; Mussgnug et al., 2010), *Chlorella* sp (Lü et al., 2013; Mendez et al., 2013; Wang et al., 2013; Lakaniemi et al., 2011), *Dunaliella* sp. (Lakaniemi et al., 2011), *Arthospira* sp. (Rodríguez et al., 2018; Inglesby and Fisher, 2012; Ding et al., 2017), *Chlamydomonas* sp., *Durvillea* sp., and *Tetraselmis* sp. (Debowski et al., 2013). Production of biogas from anaerobic digestion is primarily affected by its organic loadings, temperature, pH, and retention time in the reactors. Essentially, high methane yields can be obtained by long solid retention time and high organic loading rate (Harun et al., 2010b). Moreover, anaerobic digestion can operate in both mesophilic (35°C) and thermophilic (55°C) conditions. Dual purpose integrated approaches that combine both algal cultivation and wastewater treatment for biogas production have been suggested as a more suitable approach for cost-effective and competitive production of biogas along with other high-added value nonfuel products (Olguin, 2012; Wang and Yin, 2018).

Macroalgae, which are also often rich in carbohydrates suitable for biogas production, could contain high metal concentrations, which can have an impact on conversion processes either by inhibiting or catalyzing the processes. Moreover, high nitrogen and sulfur contents are also typical to macroalgae and may be problematic in the production of biogas (Suutari et al., 2015).

6.4.4 Biohydrogen and syngas

Hydrogen is considered as the fuel of the future, mainly due to its high conversion efficiency and sustainable and environmentally friendly nature (Das and Veziroğlu, 2001). Several methods including pyrolysis, electrolysis gasification, and steam reforming have been suggested for hydrogen production (Rashid et al., 2013). However, hydrogen production via biological processes is recognized as a key factor for the development of sustainable energy supply and as a promising alternative to fossil fuels. Since biohydrogen production is achieved at ambient temperature and pressure, it is more energy efficient than thermochemical or electrochemical processes. Processes for the production of biological hydrogen can be classified as follows:

- **(i)** biophotolysis of water using algae and cyanobacteria;
- **(ii)** photodecomposition of organic compounds by photosynthetic bacteria;
- **(iii)** fermentative hydrogen production from organic compounds;
- **(iv)** hybrid systems using photosynthetic and fermentative bacteria (Das and Veziroğlu, 2001).

Due to the high protein content and low C/N ratio of microalgae, cofermentation with macroalgae, which are rich in carbohydrates and with a high C/N ratio has been suggested for the improved hydrogen production (Xia et al., 2016). Xia et al. demonstrated that cofermentation of *Arthrospira platensis* with *Laminaria digitata* could improve the performance of hydrogen production, with an optimal specific hydrogen yield of 85.0 mL/g VS obtained at an algal C/N ratio of 26.2 (Xia et al., 2016). Besides microalgae, red, green, and brown macroalgae are propitious candidates toward biohydrogen production (Shobana et al., 2017).

Synthesis gas (syngas) is a fuel gas mixture, typically produced by gasification, comprising of carbon monoxide, carbon dioxide, and primarily hydrogen. This mixture is considered as an ecofriendly fuel with high potential in the near future for commercial applications including transportation, heat, and mainly electricity generation via fuel cells (Raheem et al., 2017).

Despite extensive studies into various aspects of algal biofuels over the past few years, the conversion of microalgae feedstocks to these products has not received a lot of attention so far. However, due to a relatively high technology readiness level, conversion of algae to syngas and hydrogen via gasification seems to be a very promising strategy for the realization of algal energy soon, which would, in turn, allow for the development of algae cultivation and processing infrastructure. Nevertheless, biohydrogen or syngas production from algae seems a promising route. However, several technological issues, as well as cost minimization, should be addressed to achieve success on a commercial scale.

6.5 Pigments from algae

Pigments are substances, which reflect or absorb only specific wavelengths of light. Chlorophylls, carotenoids (carotenes and xanthophylls), and phycobilins are the primary pigments of algae. Algal pigments are responsible for light harvesting, CO_2 fixation, protection of algal cells against damage by excessive illumination, and, macroscopically, the coloration of the algal culture. In general, chlorophylls and carotenes are nonpolar molecules present in thylakoid, whereas phycobilins are water-soluble. Many industries rely on the use of synthetic colorants for their applications. However, artificial colorants are often linked to harmful health effects. Societal inclination toward usage of products with natural origin as well as increasing awareness of the harmful effects of synthetic dyes has created an opportunity for algae utilization as a source of natural colors. Besides their use as natural colorants algal pigments have recently found applications in biomedical research; as diagnostic tools; in pharmaceutical, textile, cosmetics, and mainly food industries.

6.5.1 β-Carotene

β-Carotene is the most common form of carotene found in plants and microalgae. It serves as a precursor for vitamin A biosynthesis in the human body, and it is a known antioxidant (Fig. 6.1). β-Carotene supplements are used for the prevention of sunburn, and they are known to be effective either alone or in combination with other antioxidant vitamins or carotenoids (Stahl and Sies, 2005). Consumption of β-carotene was also linked with anticancer (by inhibition and prevention of various types of tumors) and cardioprotective properties (by controlling cholesterol levels). However, this concept has been challenged as data from intervention studies are conflicting (Stahl and Sies, 2005). Several microalgal species like *Chlorella zofingiensis, Spirulina platensis,* and *Caulerpa taxifolia* are known to synthesize β-carotene at an average yield of 0.1%–2% of their dry biomass weight (Rammuni et al., 2019). However, the halophilic green biflagellate *Dunaliella salina*, which accumulates up to 13% of β-carotene on its dry biomass is the predominant source for commercial production of natural β-carotene (Rammuni et al., 2019). In fact, the first high-value product commercially produced from microalgae was β-carotene from *Dunaliella salina*. Large-scale production is typically performed in shallow open-air ponds with several stress factors such as high salinity and light intensity, temperature, and availability of nutrients known to influence the final yield (Borowitzka, 1999). In contrast to synthetic

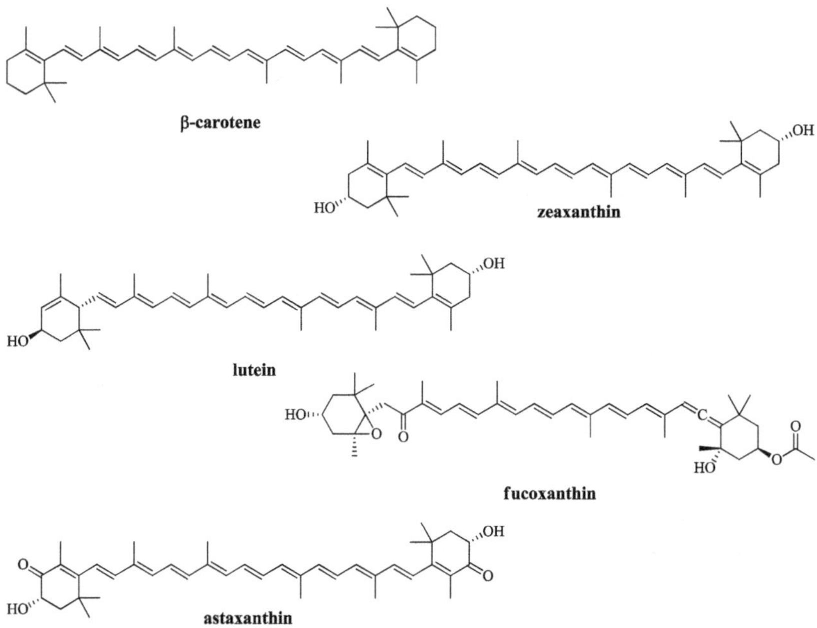

FIGURE 6.1 Chemical structures of commercially important algal carotenoids.

β-carotene, which is limited to its all-*trans* isomer, natural consists of a mixture of *cis-trans* isomers (9-*cis*-β-carotene isomer) which shows higher bioavailability, thus considered as a superior product (Raja et al., 2007). Natural β-carotene finds application as a food colorant to enhance appearance and consumer acceptability of products like margarine, cheese, fruit juices, baked goods, dairy products, canned foods, etc. (Begum et al., 2016).

6.5.2 Lutein

Lutein (β,ε-carotene-3,3′-diol) is a naturally occurring pigment belonging to the xanthophyll division of carotenoids (Fig. 6.1). As mammals cannot synthesize lutein, they rely on dietary intake. The role of this compound in human health and in particular visual function (lutein is accumulated in the macula) is well established from epidemiological, clinical, and interventional studies (Abdel-Aal et al., 2013). Moreover, there is an evidence that lutein may have biological effects that include anti-inflammatory and antioxidant properties and play a role in cognitive function (Johnson, 2014). Currently, marigold oleoresins are the major source of lutein, but several microalgae species have been suggested as potentially alternative sources (Lin, 2015). *Dunaliella*, *Muriellopsis*, *Scenedesmus*, and *Chlorella* accumulate high lutein content, which varies between 3.4 and 7.6 mg/g dry weight of algal biomass (Fernandez-Sevilla et al., 2010). Factors that are known to affect lutein content in microalgae include the pH, temperature, salinity, nitrogen availability, and mainly the specific growth rate of the cultured strain (Guedes et al., 2011). In the EU, plant origin lutein is allowed as a food and feed additive and finds applications as a color enhancer of poultry products. In 2015, the global market of lutein was estimated at 135 million US$, with a predicted annual growth rate of 5.3% until 2024 (Hu et al., 2018).

6.5.3 Astaxanthin

Astaxanthin (3,3′-dihydroxy-β,β-carotene-4,4′-dione) is a naturally occurring xanthophyll, which is responsible for the bright pink color of several marine organisms (shrimps, krill, salmon, etc.) and the feather color of birds (Higuera-Ciapara et al., 2006). Astaxanthin is generally recognized as safe (GRAS) and has been approved as a food colorant by The U.S. Food and Drug Administration (FDA) and by the European Food Safety Authority (E161), while its synthetic form is approved for only specific uses in animal feeds. Currently, commercial astaxanthin market is dominated by synthetic astaxanthin, which is significantly cheaper and a more stable source as compared to naturally derived astaxanthin (Guerin et al., 2003). From a structural perspective, microalgal astaxanthin is a mixture of (3S, 3′S) or (3R, 3′R) only isomers mostly esterified by fatty acids whereas synthetic astaxanthin is a mixture of three isomers, namely (3S, 3′S), (3R, 3′S), and (3R, 3′R), and is not esterified on its hydroxyl groups. Although synthetic astaxanthin is a steady source of large quantities of astaxanthin, there are still concerns about its biological functions and food safety whereas interest in natural sources of this pigment is increasing substantially. Thus, the market for naturally derived astaxanthin may increase in the near future (Li et al., 2011). Many studies have reported potential health-promoting effects of astaxanthin in the prevention and treatment of a wide range of diseases, including cancer, chronic inflammatory diseases, diabetes, diabetic nephropathy, cardiovascular diseases, gastrointestinal diseases, etc. (Yuan et al., 2011). However, it should be noted that research on the potential health benefits of astaxanthin is rather recent and has mostly been shown in vitro or at the preclinical level with humans (Higuera-Ciapara et al., 2006).

Several microalgae have been suggested for astaxanthin production including *Chlorella* sp., *Chlorococcum* sp., and *Scenedesmus* sp. Nevertheless, *Haematococcus pluvialis* is known to accumulate high amounts of astaxanthin ($\geq$50 mg/g). Thus, it is highly preferred for large-scale cultivation (Guedes et al., 2011). Due to its financial significance, metabolic engineering of wild type *Haematococcus pluvialis* strains for enhanced astaxanthin production has also received considerable attention, with reports of up to 67% higher astaxanthin accumulation in transformed strains than in wild-type (Galarza et al., 2018). Commercial production of astaxanthin by *Haematococcus* sp. has been implemented by several companies all over the world, that is, Cyanotech (USA), Alga-technologies (Israel), and Astareal (Japan). Typical production involves a two-stage system, which consists of a so-called green stage, where green biomass is grown under optimal growth conditions, followed by the red stage where the microalgae is exposed to stress conditions (nutrient limitation and high illumination) inducing astaxanthin accumulation (Guerin et al., 2003). High astaxanthin productivity can be achieved at a bench scale (11.5 mg/L/day), but average productivity is approximately 2.2 mg/L/day for large-scale facilities (Guedes et al., 2011). The major bottleneck for microalgal-derived astaxanthin is the high cost of production, with technoeconomic feasibility studies presenting contradicting results (Li et al., 2011; Panis and Carreon, 2016).

6.5.4 Fucoxanthin

Fucoxanthin (Fig. 6.1) is an abundant allenic carotenoid, which contributes more than 10% of the estimated total production of carotenoids in nature (Peng et al., 2011). Fucoxanthin has shown a wide range of biological properties, including antioxidant, anticancer, antiangiogenic, anti-inflammatory, and antimalarial activities as well as protective effects on liver, skin, bones, and eyes (Peng et al., 2011). However, the increasing popularity of this carotenoid is primarily due to its antiobesity and antidiabetic effects, demonstrated mainly by murine studies (Miyashita et al., 2011). Fucoxanthin can be found in edible brown seaweeds such as *Sargassum fulvellum, Undaria pinnatifida, Laminaria japonica,* and *Hijikia fusiformis* (D'Orazio et al., 2012). Moreover, several diatom species such as *Phaeodactylum tricornutum, Chaetoseros* sp., *Cylindrotheca closterium,* and *Odontella aurita* are known to accumulate high amounts of this carotenoid (Peng et al., 2011).

6.5.5 Zeaxanthin

Zeaxanthin is another natural xanthophyll, which is found in plants, algae, and other microorganisms (Zhang et al., 2018). This carotenoid is known to hold a vital role in the prevention of age-related eye diseases such as macular degeneration and cataracts (Zhang et al., 2018). In addition, due to its reported antioxidant and anticancer properties, zeaxanthin has found applications in the food, pharmaceutical, and nutraceutical industries (Zhang et al., 2018). Currently, zeaxanthin extraction relies on plant materials; however, several microalgae such as *Scenedesmus almeriensis* (Granadolorencio et al., 2009), *Microcystis aeruginosa* (Chen et al., 2005), or *Chlorella saccharophila* (Singh et al., 2013) could provide alternative routes for zeaxanthin production.

6.6 Phycobiliproteins and mycosporine-like amino acids from algae

6.6.1 Phycobiliproteins

Phycobiliproteins are water-soluble, highly fluorescent macromolecules of the photosynthetic light-harvesting antenna complexes of cyanobacteria and certain eukaryotic algae, which carry covalently attached linear tetrapyrrole pigments (Glazer, 1994). Phycobiliproteins can be divided depending on their protein structure, bilin content, natural color, and absorbance properties, into three main groups: phycocyanin (PC) (with λ_{max} 610–620 nm), phycoerythrin (with λ_{max} 540–570 nm), and allophycocyanin (with λ_{max} 650–655 nm) (Soni et al., 2006). To distinguish between biliproteins from different origins the prefixes R, B, and C are used for Rhodophyta, Bangiales, and cyanobacteria, respectively. Phycobiliproteins do not fluoresce in their native state with excitation energy efficiently transferred to chlorophyll molecules for further utilization in the photosynthetic processes (Hermanson and Hermanson, 2013). However, upon purification, excitation energy is released as strong luminosity from phycobiliproteins, with very high (0.98) fluorescent quantum efficiencies (Hermanson and Hermanson, 2013). Besides these high fluorescent quantum efficiencies, purified phycobiliproteins have several distinct features such as a large Stokes shift, broad absorption in the visible light spectrum, high extinction coefficient, and very little fluorescence quenching (Yen et al., 2013). As a result, these macromolecules are employed in fluorescent immunoassays, flow cytometry, fluorescent microscopy, immunohistochemistry, and other biomedical research purposes as fluorescent labeling reagents (Yen et al., 2013). Interest in these molecules could also be depicted by the number of patents that have been granted by 2009 for phycobiliproteins, with 236 patents on applications utilizing the fluorescence properties of phycobiliproteins, 55 patents on their production, and 30 patents on other industrial applications (Sekar and Chandramohan, 2008). Complementary to their use as natural pigments in various food products, cosmetics, and dyes, phycobiliproteins have shown potential health benefits including antioxidative, antiviral, antitumor, immunity enhancing, and anti-inflammatory effects (Li et al., 2019).

Algal phycobiliprotein content can reach as high as 20% of dry algal material. For example, in red algae its content reaches approximately 50% of total water-soluble proteins (Dumay and Morançais, 2016). Phycobiliproteins can be found in Cyanobacteria, Rhodophyceae, and Cryptophyceae. Currently, the production of this pigments relies heavily on open-pond cultures of various *Arthrospira* species (i.e., *platensis, maxima*). Besides *Arthrospira* species, several other cyanobacterial and microalgae species such as *Phormidium* sp., *Synechococcus* sp., *Nostoc* sp., and *Aphanizomenon* sp. have been evaluated for their potential for phycobiliprotein production (Kuddus et al., 2013). Phycobiliprotein content can be influenced by many factors. Eriksen in his review on phycobiliprotein production reported that depending on the cyanobacterial strain used, growth conditions (i.e., photoautotrophic, heterotrophic, or mixotrophic) and the reactor type the phycobiliprotein yield can range from 7 to 131 mg/g of dry biomass (Eriksen, 2008b). Extraction of PC usually relies on aqueous buffer systems with a combination of various techniques such as ion exchange chromatography, size exclusion chromatography, and/or precipitation applied for further purification of the crude extract. The ratio of Abs_{620}/Abs_{280} is used as an indicator of PC purity; at ≤ 0.7 it is considered as food grade, between 0.7 and 3.9 as reagent grade, and ≥ 4.0 as analytical grade (Fernández-Rojas et al., 2014). In 1997 the global market was evaluated to be around US$50 million, with market prices ranging from US$3/mg to US$25/mg depending on their purity (Yen et al., 2013). In 2018, SPIRALG a BBI-JU project coordinated by Greensea in France was granted 4 million € to demonstrate the sustainable feasibility of biorefining EU produced *Spirulina* biomass in the agrofood and health sectors and complete the whole value chain, from sustainably produced biomass to high-value marketable products (Spiralg, 2019). Currently, Far East Bio-Tec Co. Ltd. (Taiwan), with the product brand name of Flogen and ProZyme, Inc. (Canada), are the leading manufacturers and suppliers of phycobiliproteins used as fluorescent labeling reagents (Yen et al., 2013).

6.6.2 Mycosporines and mycosporine-like amino acids

Mycosporines and mycosporine-like amino acids are water-soluble molecules with low molecular weight that can absorb UV radiation (maximum absorbance 310–365 nm) (Oren and Gunde-Cimerman, 2007). From a structural perspective, mycosporines are composed of either an aminocycloheximine or an aminocyclohexenone ring, carrying nitrogen or imino alcohol substituents, and they are designated as mycosporine-like amino acids when substituted with amino acid residues (Fig. 6.2) (Carreto et al., 2011). They are typically accumulated by marine organisms, primarily cyanobacteria but also microalgae, yeasts, and fungi, and a variety of marine macroalgae, corals, and other marine life forms (Oren and Gunde-Cimerman, 2007). Their function as "microbial or natural sunscreens" is widely accepted and has been discussed in a number of review articles (Carreto et al., 2011; Shick and Dunlap, 2002; Sinha et al., 1998). Moreover, recent evidence indicates that mycosporine-like amino acids may serve additional functions as well, including, as antioxidant molecules, in reproductive and osmotic regulation and vision, or function as an accessory light-harvesting pigment in photosynthesis (Oren and Gunde-Cimerman, 2007; Shick and Dunlap, 2002; Wada et al., 2015). A few natural and synthetic analogs of mycosporine-like amino acids have already been commercialized. These products find applications as sun-protecting agents in skin care products as well as other nonbiological materials (i.e., in plastics, paints, and varnish as photostabilizing additives) (Cardozo et al., 2007). Examples of commercial products, used as a protection against UV-A-caused damage to the human skin, include Helionori which contains palythine, porphyra-334, and shinorine extracted from the *Porphyra umbilicalis* and Helioguard containing porphyra-334 and shinorine (Chrapusta et al., 2017).

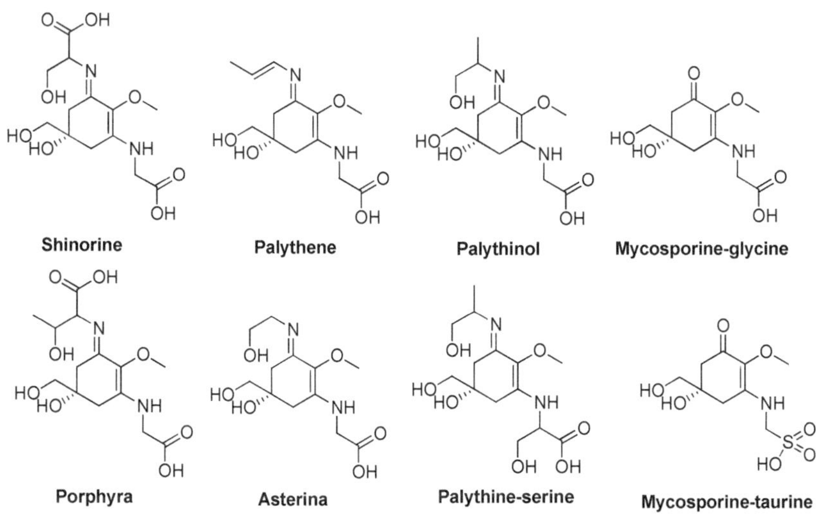

FIGURE 6.2 Chemical structures of mycosporine-like amino acids.

6.7 Algae-based polymers, blends, and composites

The increasing effect of nondegradable plastic wastes around the world is a growing concern. It is estimated that approximately 140 million tons of plastic are consumed on a global scale per year, which require processing of around 150 million tons of fossil fuels to be produced (Suriyamongkol et al., 2007). As an alternative to fossil-based plastics, there is a growing interest to utilize renewable sources for the production of biodegradable bioplastics. However, production costs for petroleum-derived polymers are significantly lower than most of the biodegradable alternatives, thus hindering commercial development and retail of these products. Among other microorganisms, certain cyanobacterial strains (e.g., *Chlamydomonas* sp., *Phormidium* sp., *Nostoc* sp.), and microalgae strains (i.e., *Ulva* sp., *Chlorella* sp., *Scenedesmus* sp.) have been suggested as potential candidates for bioplastic production (Khanra et al., 2018). Although, nowadays we are confident on the technical feasibility of algal bioplastics there are rare reports on the economic feasibility of such attempts. Nevertheless, algal biopolymers could be produced as byproducts of the biofuel production from algae with companies such as Petro Sun, Dow Chemicals, Cereplast, as well as others exploring the potential of algal-based biopolymer synthesis (Khanra et al., 2018).

6.7.1 Polyhydroxyalkanoates

The term *polyhydroxyalkanoates* (PHAs) is used for a group of polyhydroxy esters of 3-, 4-, 5-, and 6-hydroxy alkanoic acids that are synthesized by certain microbial species under nutrient-limiting conditions with excess of carbon (Philip et al., 2007). They are intracellularly accumulated as carbon and energy storage materials in discrete cytoplasmic inclusions within the microbial cells (Philip et al., 2007). Various types of PHAs exist; more than 150 different monomers could be combined within this group to produce materials with extremely different thermoplastic and elastomeric properties. The most important are poly(3-hydroxybutyrate) (PHB), the copolymer, poly(3-hydroxybutyrate-co-3-hydroxy valerate) (PHB/HV), and the elastomeric poly(3-hydroxy octanoate) (Sinha Ray and Sinha Ray, 2013). Currently, commercial PHAs production, which is restricted to a small market share, is achieved in fermentors by culturing mainly bacterial species with high concentrations of added organic carbon sources and salts, the addition of which constitutes approximately 50% of the total production cost (Costa et al., 2019). Although cyanobacterial biosynthesis of PHA has been known since the 60s, production from cyanobacteria was not pursued as the achieved yields were generally meager (Drosg et al., 2015). However, the advances in the field of genetic modification over the last 2 decades have led to significant strain improvements, achieved either by suppression of biochemical pathways or by gene amplification, resulting in a substantial increase of cyanobacterial PHA yields (Drosg et al., 2015). PHA content in cyanobacteria grown under various stress conditions or cultivation types were shown to range from 5% to 69% of the total dry weight with most of the studies indicating that nitrogen and phosphorus deficiencies can have pronounced effects on the ability of PHA synthesis (Costa et al., 2019). Although several cyanobacterial species are explored for their potential in PHAs production, *Synechocystis* sp. and *Nostoc* sp. grown under heterotrophic conditions are known to accumulate high (>30%) amounts of PHB or PHB/HV (Troschl et al., 2017). PHB is a naturally occurring intracellular storage compound produced by many prokaryotic organisms. This aliphatic polyester has shown interesting properties such as melting and glass transition temperatures, Young's modulus, and tensile strength, which are similar to polypropylene. Owing to their inherent biocompatibility, nonhazardous disintegration products, and controllable biodegradability, among other properties, both PHB and PHB/HV have found various medical applications ranging from the design of various in vivo implants to controlled release drug delivery systems (Singh et al., 2019). From a financial perspective, the global PHAs market reached US$73.6 million in 2016 with annual growth rate of approximately 5% and is expected to reach a value of $93.5 million by 2021 (Singh et al., 2019).

6.7.2 Polylactic acid

Polylactic acid (PLA) is a compostable, thermoplastic, aliphatic polyester with many properties, such as mechanical strength and stiffness, as well as gas permeability, which are comparable to those of fossil-based polymers (Hamad et al., 2018). Utilization of PLA-based materials though finds certain limitations for specific applications mainly due to the high production cost and slow biodegradation rate (Hamad et al., 2018). Although not much information exists on direct production of lactic acid by microalgae, it is known that starch containing microalgae like *Nannochlorum* sp. can convert starch into products like lactic acid, ethanol, acetic acid, and formic acid under dark and anaerobic conditions (Hirayama and Ueda, 2004). Nevertheless, microalgal biomass rich in carbohydrates could be utilized as a feedstock for microbial fermentation for production of lactic acid. Algal-based carbohydrates devoid of lignin have gained attention for their potential as suitable alternative substrates compared to lignocellulosic biomass (Abdel-Rahman et al., 2013). Microalgal species like *Hydrodictyon reticulum* (Nguyen et al., 2012), *Nannochloropsis salina* (Talukder et al., 2012), or a *Chlorella* sp. mutant (Lee et al., 2017), as well as macroalgae like *Ulva lactuca* (Wu et al., 2018) with high contents of reducing sugars, have been used as a carbohydrate substrate to produce L-lactic acid by various lactic acid bacteria.

6.7.3 Blends and composites

Algal biomass (both treated and untreated) or algal polysaccharides have shown a high potential for production of hybrid or composite materials for utilization in various industries and especially biomedical applications. Especially in the biomedical field, due to their biocompatibility and controlled biodegradation, algae-derived hybrid polyester scaffolds are widely applied for bone, cartilage, cardiac, and nerve tissue regeneration (Noreen et al., 2016). Moreover, capsules and spheres of algae-derived polyesters have been used for controlled release of pharmaceutical agents (Noreen et al., 2016). Typically, two main approaches in utilizing algae in composites have been reported in the literature. The first is using untreated or waste biomass as a filler with the primary aim to decrease the production cost and carbon footprint of polymers and second by applying, usually extracted or bleached (to increase cellulose content) biomass, as fibers reinforcement (Bulota and Budtova, 2015).

Examples of blends or composites include polyolefin/algal polysaccharide composites, which find applications in automotive industry and tissue engineering or seaweed residue—based polyolefin composites can be manufactured for industrial and biomedical applications (Zia et al., 2017). Moreover, biocomposites with PLA as matrix and various red, brown, or green macroalgae as a filler have been suggested for preparation of thermoplastic polymer composites (Bulota and Budtova, 2015, 2016). A number of materials could also be produced from a combination of alginate and polyurethanes; this combination has led to the creation of elastomers, nanocomposites, hydrogels, etc. with food and biomedical applications (Zia et al., 2015). Besides bioplastic production alginate could be used in composite materials for constructions and buildings, for example, as a bonding material in clay-based composites (Galán-Marín et al., 2010). Micro- and macroalgae have also been used as fillers in composites with poly-(ε-caprolactone), PHB, and polyvinyl alcohol (Noreen et al., 2016; Madera-Santana et al., 2011). Besides the abovementioned applications, algal-based materials have been suggested for bioremediation. Examples include a hybrid matrix of poly vinyl alcohol-sodium alginate embedded with *Jania rubens* for removal of lead from aqueous solutions (Kadimpati, 2017) or adsorption of uranium ions by functionalized hybrid biomass of *Arthrospira platensis* (Bayramoglu et al., 2015).

6.8 Algal products for agriculture

In the coming decades, a crucial challenge that societies will have to face would be meeting the raising food demands without causing further adverse environmental impact. The overuse of synthetic agrochemicals has resulted in detrimental ecological issues on a global scale, thus leading to the creation of ocean dead zones, eutrophication problems, soil infertility, and loss of biodiversity. For these reasons, modern agricultural practices would be required to reevaluate many of the currently applied agricultural practices, which comprise of using chemical fertilizers, pesticides, herbicides, fungicides, and insecticides and provide solutions in a sustainable manner.

6.8.1 Algae as biofertilizers and plant growth promoters

Biofertilizers are the products typically containing living or latent cells of bacteria, actinomycetes, fungi, or algae that are applied either individually or as a consortium of various microorganisms (Dineshkumar et al., 2018). In addition to the secretion of growth promoting substances, these products assist in nitrogen fixation and/or nutrient solubilization, thus enhancing crop growth and yield (Dineshkumar et al., 2018). Among the available biofertilizer types, formulations based on eukaryotic microalgae, anoxygenic phototrophs, or especially cyanobacteria are gaining importance (Renuka et al., 2018). Although the effect of specific products applied as biofertilizers are somewhat underexplored, data on cyanobacterial species such as *Anabaena variabilis*, *Nostoc muscorum*, *Aulosira fertissima*, and *Tolypothrix tenuis* indicate that they could be applied effectively as biofertilizers (Singh et al., 2016). The role of both symbiotic and free-living cyanobacteria in nitrogen fixation and in plant growth promotion is well documented (Hayat et al., 2010). Several studies showed that inoculation of cyanobacteria or cyanobacterial consortia results in a significant enhancement in the soil nitrogen content. The fixed nitrogen could be released, either by secretion or by microbial degradation after the cell death, in various forms such as ammonia, free amino acids, polypeptides, vitamins, and auxin-like substances (Subramanian and Shanmugasundaram, 1986). In rice agriculture, free-living cyanobacteria contribute on an average 20–30 kg of nitrogen per hectare, whereas the value for symbiotic complexes such as the *Azolla-Anabaena* system can reach 600 kg per hectare (Vaishampayan et al., 2001). It is perhaps characteristic that in an extensive analysis of rice soils from the Philippines, Malaysia, Portugal, and India heterocystous forms of cyanobacteria were present in all cases with *Nostoc*, *Anabaena*, and *Calothrix* being the dominant genera (Roger et al., 1987). Moreover in paddy fields in many Asian countries instead of nitrogen fertilizers various cyanobacteria are applied as biofertilizers (Singh et al., 2016). Microalgal bacterial flocs and a marine culture of *Nannochloropsis* sp. which were applied as an organic slow-release fertilizer resulted in tomatoes with increased carotenoid and sugar levels (Coppens et al., 2016). Supplementation of 0.5 g of the green microalgae *Tetraselmis* sp. in date palm led to higher plant growth rates, 100% survival rates, and a number of beneficial effects on roots, leaves, and stems as compared to conventional fertilizer (Saadaoui et al., 2019). Algae are also known as a rich source of plant growth-promoting substances/hormones (i.e., cytokinins, auxins, ethylene, betaine, gibberellins, abscisic acid, and polyamines) which can be found in cyanobacteria (Singh and Syiem, 2019) as well as green, brown, and red seaweeds or kelp species (Crouch and van Staden, 1993a). Several of these algal extracts have found commercial applications as growth stimulants in various agricultural crops and as amendments in crop production systems (Crouch and Van Staden, 1993b).

6.8.2 Algae as agricultural control agents

The massive application of chemical biocide agents against targeted pests and pathogenic microorganisms in agricultural practices is hazardous to sustainability and biodiversity of agroecosystems (Renuka et al., 2018). As a result, alternative sustainable approaches for pathogen control are required. A potential solution could derive from algal-based metabolites. In their natural environment algae are involved in a plethora of allelopathic interactions between competing organisms; this "battle" for survival and nutrients has equipped algae with tremendously diverse weaponry of active metabolites with biocontrol properties among which are antimicrobial, insecticidal, larvicidal, algicidal, and pesticidal properties. For example, algae are known to produce a variety of fatty acids (i.e., γ-linolenic acid from *Fischerella* sp.), macrolides, alkaloids (i.e., anhydrohapaloxindole A from *Hapalosiphon fontinalis*), aromatics, peptides (i.e., pahayokolide A from *Lyngbya* sp.). and terpenes (i.e., diterpenoids from *Nostoc* sp.) with a wide range of antimicrobial properties (Hernández-Carlos and Gamboa-Angulo, 2011). Additionally, compounds including several polyphenols, tocopherols, carbohydrates, proteins, and pigments, which can be helpful in pathogen control against soilborne diseases have been reported in algal extracts (Michalak and Chojnacka, 2015). Moreover, algal extracts have been reported to produce natural products with insecticidal activity: hapalindole L from *Fischerella* genus was active against the dipteran *Chironomus riparius* larvae, eremophilone from *Calothrix* sp. showed acute toxicity against insects, a number of products were reported on with effects in the development and survival of mosquito larvae (Hernández-Carlos and Gamboa-Angulo, 2011). Cyanobacteria and algae also produce toxins; however in spite of the existing reports on their occurrence and toxicity, the attempts to use algal toxins in allelopathy under field conditions are rather limited (Inderjit and Dakshini, 1994). Overall, although certain progress has been achieved in this sector, considering the extreme diversity and the number of underexplored species further knowledge of chemistry and biology of allelochemicals can help in their future use as biocontrol agents.

6.9 Algal biorefinery

In principle, biomass biorefineries are similar to crude oil refineries in terms of optimized manufacturing of a range of high value products from a single feedstock. Biorefineries are considered as an optimal strategy for a large-scale sustainable utilization of biomass in the bioeconomy, which should result in a cost-competitive coproduction of a wide range of products with optimal environmental and socioeconomic benefits (Hingsamer and Jungmeier, 2019). Due to economic hurdles as well as a market pull for green chemicals, it is likely that integrated algal-based biorefineries producing high-value bio-based products along with biofuels will be the future commercial template (Haznedaroglu et al., 2016). Research and development of algae-based biorefineries is therefore of major importance in this sector to explore mild cell disruption, and extraction and separation technologies on algal biomass (Wijffels et al., 2010). Fig. 6.3 shows a schematic presentation of the necessary steps along with the available methods for downstream processing of algal feedstocks. Products such as pigments, carbohydrates, ω-3-fatty acids, and proteins should maintain their functionality in this process and at the same time scalability, low energy costs, and ease of use also need to be taken into account (Wijffels et al., 2010). For the algal biorefinery concept to make sense, it would be essential to establish a proper connection between the various input and output streams of the products, but also the provided services by the participating industries (Trivedi et al., 2015). A top priority for the nascent algae-based bioeconomy would be the identification and addressing of critical improvements required in biomass, bioproduct, and biofuel productivity (Laurens et al., 2017). The resolution of these issues should be guided by economic and sustainability principles and will help to unravel the contentious water-food-energy-environment nexus that algae inhabit (Laurens et al., 2017).

6.10 Conclusions

Over the past years, there has been impressive progress both from the scientific and technological perspective on the potential uses and applications of algae as feedstocks of bio-based products. Reduction of greenhouse gas emissions, recycling, and production of bio-based products based on the biorefinery concept are within the environmental policy agenda and targets on a global level. Despite the major bottlenecks and challenges that algal biomass production faces, the nascent algal based sector is expected to grow rapidly and create novel products, markets, and jobs in the near future. Moreover, considering the enormous biodiversity and recent advances particularly in the methods of systems biology, genetic engineering, and biorefining the potential for algae bio-based products remains high.

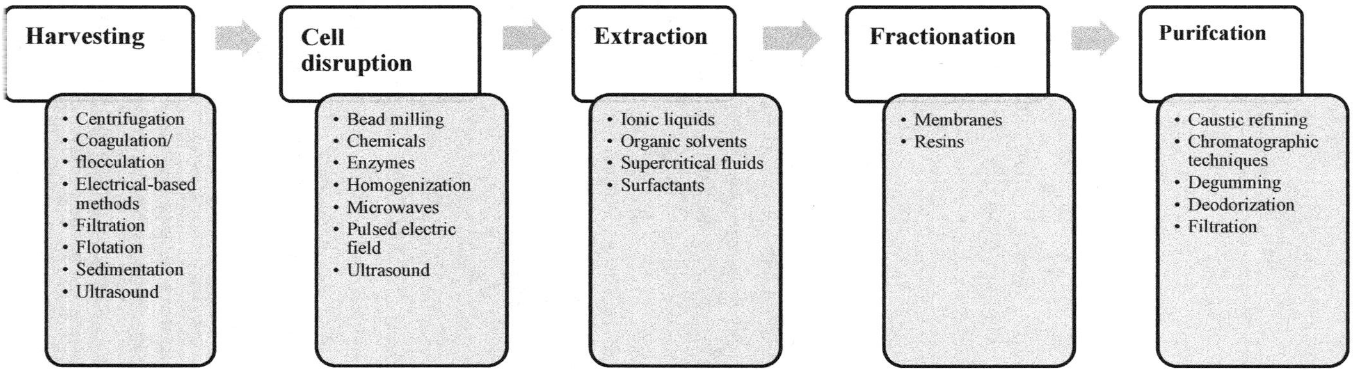

FIGURE 6.3 Downstream processing of algal feedstocks: Consecutive steps and available methods. *With information from Koutra, E., Economou, CN., Tsafrakidou, P., Kornaros M., 2018. Bio-based products from microalgae cultivated in digestates. Trends Biotechnol. 36, 819–833. https://doi.org/10.1016/J.TIBTECH.2018.02.015.*

References

Abdel-Aal, E.-S., Akhtar, H., Zaheer, K., Ali, R., 2013. Dietary sources of lutein and zeaxanthin carotenoids and their role in eye health. Nutrients 5, 1169–1185. https://doi.org/10.3390/nu5041169.

Abdel-Rahman, M.A., Tashiro, Y., Sonomoto, K., 2013. Recent advances in lactic acid production by microbial fermentation processes. Biotechnol. Adv. 31, 877–902. https://doi.org/10.1016/J.BIOTECHADV.2013.04.002.

Ahmad, A.L., Yasin, N.H.M., Derek, C.J.C., Lim, J.K., 2011. Microalgae as a sustainable energy source for biodiesel production: a review. Renew. Sustain. Energy Rev. 15, 584–593. https://doi.org/10.1016/J.RSER.2010.09.018.

Andersen, R.A., 1992. Diversity of eukaryotic algae. Biodivers. Conserv. 1, 267–292. https://doi.org/10.1007/bf00693765.

Banerjee, S., Mudliar, S., Sen, R., Giri, B., Satpute, D., Chakrabarti, T., et al., 2010. Commercializing lignocellulosic bioethanol: technology bottlenecks and possible remedies. Biofuel. Bioprod. Biorefin. 4, 77–93. https://doi.org/10.1002/bbb.188.

Bayramoglu, G., Akbulut, A., Arica, M.Y., 2015. Study of polyethyleneimine- and amidoxime-functionalized hybrid biomass of spirulina (*Arthrospira*) platensis for adsorption of uranium (VI) ion. Environ. Sci. Pollut. Res. 22, 17998–18010. https://doi.org/10.1007/s11356-015-4990-9.

Becker, E.W., 2007. Micro-algae as a source of protein. Biotechnol. Adv. 25, 207–210. https://doi.org/10.1016/J.BIOTECHADV.2006.11.002.

Begum, H., Yusoff, F.M., Banerjee, S., Khatoon, H., Shariff, M., 2016. Availability and utilization of pigments from microalgae. Crit. Rev. Food Sci. Nutr. 56, 2209–2222. https://doi.org/10.1080/10408398.2013.764841.

Bhatnagar, A., Chinnasamy, S., Singh, M., Das, K.C., 2011. Renewable biomass production by mixotrophic algae in the presence of various carbon sources and wastewaters. Appl. Energy 88, 3425–3431. https://doi.org/10.1016/J.APENERGY.2010.12.064.

Blatti, J.L., Michaud, J., Burkart, M.D., 2013. Engineering fatty acid biosynthesis in microalgae for sustainable biodiesel. Curr. Opin. Chem. Biol. 17, 496–505. https://doi.org/10.1016/J.CBPA.2013.04.007.

Borowitzka, M.A., 1999. Commercial production of microalgae: ponds, tanks, and fermenters. Prog. Ind. Microbiol. 35, 313–321. https://doi.org/10.1016/S0079-6352(99)80123-4.

Bozbas, K., 2008. Biodiesel as an alternative motor fuel: production and policies in the European Union. Renew. Sustain. Energy Rev. 12, 542–552. https://doi.org/10.1016/J.RSER.2005.06.001.

Brennan, L., Owende, P., 2010. Biofuels from microalgae—a review of technologies for production, processing, and extractions of biofuels and co-products. Renew. Sustain. Energy Rev. 14, 557–577. https://doi.org/10.1016/J.RSER.2009.10.009.

Bruton, T., Lyons, H., Lerat, Y., Stanley, M.S., Rasmussen, M.Bo., 2009. A Review of the Potential of Marine Algae as a Source of Biofuel in Ireland. Sustain Energy Ireland Technical Report, Dublin.

Bulota, M., Budtova, T., 2015. PLA/algae composites: morphology and mechanical properties. Compos. Part A Appl. Sci. Manuf. 73, 109–115. https://doi.org/10.1016/J.COMPOSITESA.2015.03.001.

Bulota, M., Budtova, T., 2016. Valorisation of macroalgae industrial by-product as filler in thermoplastic polymer composites. Compos. Part A Appl. Sci. Manuf. 90, 271–277. https://doi.org/10.1016/J.COMPOSITESA.2016.07.010.

Cardozo, K.H.M., Guaratini, T., Barros, M.P., Falcão, V.R., Tonon, A.P., Lopes, N.P., et al., 2007. Metabolites from algae with economical impact. Comp. Biochem. Physiol. C Toxicol. Pharmacol. 146, 60–78. https://doi.org/10.1016/J.CBPC.2006.05.007.

Carreto, J.I., Carignan, M.O., Carreto, J.I., Carignan, M.O., 2011. Mycosporine-like amino acids: relevant secondary metabolites. Chemical and ecological aspects. Mar. Drugs 9, 387–446. https://doi.org/10.3390/md9030387.

Chen, F., Li, H.-B., Wong, R.N.-S., Ji, B., Jiang, Y., 2005. Isolation and purification of the bioactive carotenoid zeaxanthin from the microalga *Microcystis aeruginosa* by high-speed counter-current chromatography. J. Chromatogr. A 1064, 183–186. https://doi.org/10.1016/J.CHROMA.2004.12.065.

Chen, C.-Y., Yeh, K.-L., Aisyah, R., Lee, D.-J., Chang, J.-S., 2011. Cultivation, photobioreactor design and harvesting of microalgae for biodiesel production: a critical review. Bioresour. Technol. 102, 71–81. https://doi.org/10.1016/j.biortech.2010.06.159.

Chen, H., Zhou, D., Luo, G., Zhang, S., Chen, J., 2015. Macroalgae for biofuels production: progress and perspectives. Renew. Sustain. Energy Rev. 47, 427–437. https://doi.org/10.1016/J.RSER.2015.03.086.

Chisti, Y., 2008. Biodiesel from microalgae beats bioethanol. Trends Biotechnol. 26, 126–131. https://doi.org/10.1016/j.tibtech.2007.12.002.

Chisti, Y., 2016. Large-Scale Production of Algal Biomass: Raceway Ponds. Springer, Cham, pp. 21–40. https://doi.org/10.1007/978-3-319-12334-9_2.

Chrapusta, E., Kaminski, A., Duchnik, K., Bober, B., Adamski, M., Bialczyk, J., 2017. Mycosporine-like amino acids: potential health and beauty ingredients. Mar. Drugs 15. https://doi.org/10.3390/md15100326.

Collet, P., Spinelli, D., Lardon, L., Hélias, A., Steyer, J.-P., Bernard, O., 2014. Life-cycle assessment of microalgal-based biofuels. Biofuels Algae 287–312. https://doi.org/10.1016/B978-0-444-59558-4.00013-9.

Cook, A.H., 1945. Algal pigments and their significance. Biol. Rev. 20, 115–132. https://doi.org/10.1111/j.1469-185X.1945.tb00445.x.

Coppens, J., Grunert, O., Van Den Hende, S., Vanhoutte, I., Boon, N., Haesaert, G., et al., 2016. The use of microalgae as a high-value organic slow-release fertilizer results in tomatoes with increased carotenoid and sugar levels. J. Appl. Phycol. 28, 2367–2377. https://doi.org/10.1007/s10811-015-0775-2.

Costa, S.S., Miranda, A.L., de Morais, M.G., Costa, J.A.V., Druzian, J.I., 2019. Microalgae as source of polyhydroxyalkanoates (PHAs) — a review. Int. J. Biol. Macromol. 131, 536–547. https://doi.org/10.1016/J.IJBIOMAC.2019.03.099.

Crouch, I.J., van Staden, J., 1993a. Evidence for the presence of plant growth regulators in commercial seaweed products. Plant Growth Regul. 13, 21–29. https://doi.org/10.1007/BF00207588

Crouch, I.J., Van Staden, J., 1993b. Commercial seaweed products as biostimulants in horticulture. J. Home Consum. Hortic. 1, 19–76. https://doi.org/10.1300/J280v01n01_03.

Das, D., Veziroğlu, T.N., 2001. Hydrogen production by biological processes: a survey of literature. Int. J. Hydrogen Energy 26, 13–28. https://doi.org/10.1016/S0360-3199(00)00058-6.

Debowski, M., Zieliński, M., Grala, A., Dudek, M., 2013. Algae biomass as an alternative substrate in biogas production technologies—Review. Renew. Sustain. Energy Rev. 27, 596–604. https://doi.org/10.1016/J.RSER.2013.07.029.

Dineshkumar, R., Kumaravel, R., Gopalsamy, J., Sikder, M.N.A., Sampathkumar, P., 2018. Microalgae as bio-fertilizers for rice growth and seed yield productivity. Waste Biomass Valori. 9, 793–800. https://doi.org/10.1007/s12649-017-9873-5.

Ding, L., Cheng, J., Lu, H., Yue, L., Zhou, J., Cen, K., 2017. Three-stage gaseous biofuel production combining dark hydrogen, photo hydrogen, and methane fermentation using wet *Arthrospira platensis* cultivated under high CO_2 and sodium stress. Energy Convers. Manag. 148, 394–404. https://doi.org/10.1016/J.ENCONMAN.2017.05.079.

Diprat, A.B., Menegol, T., Boelter, J.F., Zmozinski, A., Rodrigues Vale, M.G., Rodrigues, E., et al., 2017. Chemical composition of microalgae *Heterochlorella luteoviridis* and *Dunaliella tertiolecta* with emphasis on carotenoids. J. Sci. Food Agric. 97, 3463–3468. https://doi.org/10.1002/jsfa.8159.

Drosg, B., Fritz, I., Gattermayr, F., Silvestrini, L., 2015. Photo-autotrophic production of poly(-hydroxyalkanoates) in cyanobacteria. Chem. Biochem. Eng. Q. 29, 145–156. https://doi.org/10.15255/CABEQ.2014.2254.

Dumay, J., Morançais, M., 2016. Proteins and pigments. Seaweed Heal. Dis. Prev. 275–318. https://doi.org/10.1016/B978-0-12-802772-1.00009-9.

D'Orazio, N., Gemello, E., Gammone, M., de Girolamo, M., Ficoneri, C., Riccioni, G., et al., 2012. Fucoxantin: a treasure from the sea. Mar. Drugs 10, 604–616. https://doi.org/10.3390/md10030604.

Eriksen, N.T., 2008. The technology of microalgal culturing. Biotechnol. Lett. 30, 1525–1536. https://doi.org/10.1007/s10529-008-9740-3.

Eriksen, N.T., 2008. Production of phycocyanin – a pigment with applications in biology, biotechnology, foods and medicine. Appl. Microbiol. Biotechnol. 80, 1–14. https://doi.org/10.1007/s00253-008-1542-y.

Fernández-Rojas, B., Hernández-Juárez, J., Pedraza-Chaverri, J., Fernandez-Rojas, B., Hernandez-Juarez, J., Pedraza-Chaverri, J., 2014. Nutraceutical properties of phycocyanin. J. Funct. Foods 11, 375–392. https://doi.org/10.1016/j.jff.2014.10.011.

Fernandez-Sevilla, J.M., Fernandez, F.G.A., Grima, E.M., 2010. Biotechnological production of lutein and its applications. Appl. Microbiol. Biotechnol. 86, 27–40. https://doi.org/10.1007/s00253-009-2420-y.

Galán-Marín, C., Rivera-Gómez, C., Petric, J., 2010. Clay-based composite stabilized with natural polymer and fibre. Constr. Build. Mater. 24, 1462–1468. https://doi.org/10.1016/J.CONBUILDMAT.2010.01.008.

Galarza, J.I., Gimpel, J.A., Rojas, V., Arredondo-Vega, B.O., Henríquez, V., 2018. Overaccumulation of astaxanthin in *Haematococcus pluvialis* through chloroplast genetic engineering. Algal Res. 31, 291–297. https://doi.org/10.1016/J.ALGAL.2018.02.024.

Ghadiryanfar, M., Rosentrater, K.A., Keyhani, A., Omid, M., 2016. A review of macroalgae production, with potential applications in biofuels and bioenergy. Renew. Sustain. Energy Rev. 54, 473–481. https://doi.org/10.1016/J.RSER.2015.10.022.

Glazer, A.N., 1994. Phycobiliproteins — a family of valuable, widely used fluorophores. J. Appl. Phycol. 6, 105–112. https://doi.org/10.1007/BF02186064.

González-Fernández, C., Sialve, B., Bernet, N., Steyer, J.P., 2012. Thermal pretreatment to improve methane production of Scenedesmus biomass. Biomass Bioenergy 40, 105–111. https://doi.org/10.1016/J.BIOMBIOE.2012.02.008.

Granadolorencio, F., Herrerobarbudo, C., Acienfernandez, G., Molinagrima, E., Fernandezsevilla, J., Perezsacristan, B., et al., 2009. In vitro bioaccesibility of lutein and zeaxanthin from the microalgae *Scenedesmus almeriensis*. Food Chem. 114, 747–752. https://doi.org/10.1016/j.foodchem.2008.10.058.

Guedes, A.C., Amaro, H.M., Malcata, F.X., 2011. Microalgae as sources of carotenoids. Mar. Drugs 9, 625–644. https://doi.org/10.3390/md9040625.

Guerin, M., Huntley, M.E., Olaizola, M., 2003. *Haematococcus astaxanthin*: applications for human health and nutrition. Trends Biotechnol. 21, 210–216. https://doi.org/10.1016/S0167-7799(03)00078-7.

Guschina, I.A., Harwood, J.L., 2006. Lipids and lipid metabolism in eukaryotic algae. Prog. Lipid Res. 45, 160–186. https://doi.org/10.1016/J.PLIPRES.2006.01.001.

Hamad, K., Kaseem, M., Ayyoob, M., Joo, J., Deri, F., 2018. Polylactic acid blends: the future of green, light and tough. Prog. Polym. Sci. 85, 83–127. https://doi.org/10.1016/J.PROGPOLYMSCI.2018.07.001.

Hamed, I., 2016. The evolution and versatility of microalgal biotechnology: a review. Compr. Rev. Food Sci. Food Saf. 15, 1104–1123. https://doi.org/10.1111/1541-4337.12227.

Harun, R., Singh, M., Forde, G.M., Danquah, M.K., 2010. Bioprocess engineering of microalgae to produce a variety of consumer products. Renew. Sustain. Energy Rev. 14, 1037–1047. https://doi.org/10.1016/J.RSER.2009.11.004.

Harun, R., Singh, M., Forde, G.M., Danquah, M.K., 2010. Bioprocess engineering of microalgae to produce a variety of consumer products. Renew. Sustain. Energy Rev. 14, 1037–1047. https://doi.org/10.1016/j.rser.2009.11.004.

Harun, R., Yip, J.W.S., Thiruvenkadam, S., Ghani, W.A.W.A.K., Cherrington, T., Danquah, M.K., 2014. Algal biomass conversion to bioethanol – a step-by-step assessment. Biotechnol. J. 9, 73–86. https://doi.org/10.1002/biot.201200353.

Hayat, R., Ali, S., Amara, U., Khalid, R., Ahmed, I., 2010. Soil beneficial bacteria and their role in plant growth promotion: a review. Ann. Microbiol. 60, 579–598. https://doi.org/10.1007/s13213-010-0117-1.

Haznedaroglu, B.Z., Rismani-Yazdi, H., Allnutt, F.C.T., Reeves, D., Peccia, J., 2016. Algal biorefinery for high-value platform chemicals. Platf. Chem. Biorefinery 333–360. https://doi.org/10.1016/B978-0-12-802980-0.00018-3.

Hermanson, G.T., Hermanson, G.T., 2013. Fluorescent probes. Bioconjug. Tech 395–463. https://doi.org/10.1016/B978-0-12-382239-0.00010-8.

Hernández-Carlos, B., Gamboa-Angulo, M.M., 2011. Metabolites from freshwater aquatic microalgae and fungi as potential natural pesticides. Phytochem. Rev. 10, 261–286. https://doi.org/10.1007/s11101-010-9192-y.

Higuera-Ciapara, I., Felix-Valenzuela, L., Goycoolea, F.M., 2006. Astaxanthin: a review of its chemistry and applications. Crit. Rev. Food Sci. Nutr. 46, 185–196. https://doi.org/10.1080/10408690590957188.

Hingsamer, M., Jungmeier, G., 2019. Biorefineries. Role Bioenergy Bioecon. 179–222. https://doi.org/10.1016/B978-0-12-813056-8.00005-4.

Hirano, A., Ueda, R., Hirayama, S., Ogushi, Y., 1997. CO_2 fixation and ethanol production with microalgal photosynthesis and intracellular anaerobic fermentation. Energy 22, 137–142. https://doi.org/10.1016/S0360-5442(96)00123-5.

Hirayama, S., Ueda, R., 2004. Production of optically pure D-lactic acid by *Nannochlorum* sp. 26A4. Appl. Biochem. Biotechnol. Part A Enzyme Eng. Biotechnol. 119, 71–77. https://doi.org/10.1385/ABAB:119:1:71.

Hu, J.J., Nagarajan, D., Zhang, Q.G., Chang, J.S., Lee, D.J., 2018. Heterotrophic cultivation of microalgae for pigment production: a review. Biotechnol. Adv. 36, 54–67. https://doi.org/10.1016/j.biotechadv.2017.09.009.

Hughes, A.D., Kelly, M.S., Black, K.D., Stanley, M.S., 2012. Biogas from macroalgae: is it time to revisit the idea? Biotechnol. Biofuels 5, 86. https://doi.org/10.1186/1754-6834-5-86.

Inderjit, Dakshini, K.M.M., 1994. Algal allelopathy. Bot. Rev. 60, 182–196. https://doi.org/10.1007/BF02856576.

Inglesby, A.E., Fisher, A.C., 2012. Enhanced methane yields from anaerobic digestion of *Arthrospira* maxima biomass in an advanced flow-through reactor with an integrated recirculation loop microbial fuel cell. Energy Environ. Sci. 5, 7996. https://doi.org/10.1039/c2ee21659k.

Jambo, S.A., Abdulla, R., Mohd Azhar, S.H., Marbawi, H., Gansau, J.A., Ravindra, P., 2016. A review on third generation bioethanol feedstock. Renew. Sustain. Energy Rev. 65, 756–769. https://doi.org/10.1016/j.rser.2016.07.064.

Ji, S.-Q., Wang, B., Lu, M., Li, F.-L., 2016. Direct bioconversion of brown algae into ethanol by thermophilic bacterium *Defluviitalea phaphyphila*. Biotechnol. Biofuels 9, 81. https://doi.org/10.1186/s13068-016-0494-1.

John, R.P., Anisha, G.S., Nampoothiri, K.M., 2011. Micro and macroalgal biomass: a renewable source for bioethanol. Bioresour. Technol. 102, 186–193. https://doi.org/10.1016/J.BIORTECH.2010.06.139.

Johnson, E.J., 2014. Role of lutein and zeaxanthin in visual and cognitive function throughout the lifespan. Nutr. Rev. 72, 605–612. https://doi.org/10.1111/nure.12133.

Kadimpati, K.K., 2017. Design of hybrid PVA–CA– *Jania rubens* biomatrix for removal of lead. Int. J. Phytoremediation 19, 183–190. https://doi.org/10.1080/15226514.2016.1207603.

Kaliaperumal, N., Ramalingam, J.R., Kalimuthu, S., Ezhilvalavan, R., 2002. Seasonal Changes in Growth, Biochemical Constituents and Phycocolloid of Some Marine Algae of Mandapam Coast.

Khanra, S., Mondal, M., Halder, G., Tiwari, O.N., Gayen, K., Bhowmick, T.K., 2018. Downstream processing of microalgae for pigments, protein and carbohydrate in industrial application: a review. Food Bioprod. Process. 110, 60–84. https://doi.org/10.1016/J.FBP.2018.02.002.

Knothe, G., 2006. Analyzing biodiesel: standards and other methods. J. Am. Oil Chem. Soc. 83, 823–833. https://doi.org/10.1007/s11746-006-5033-y.

Koutra, E., Economou, C.N., Tsafrakidou, P., Kornaros, M., 2018. Bio-based products from microalgae cultivated in digestates. Trends Biotechnol. 36, 819–833. https://doi.org/10.1016/J.TIBTECH.2018.02.015.

Kuddus, M., Singh, P., Thomas, G., Al-Hazimi, A., 2013. Recent developments in production and biotechnological applications of C-phycocyanin. BioMed Res. Int. https://doi.org/10.1155/2013/742859.

Kumar, K., Mishra, S.K., Shrivastav, A., Park, M.S., Yang, J.-W., 2015. Recent trends in the mass cultivation of algae in raceway ponds. Renew. Sustain. Energy Rev. 51, 875–885. https://doi.org/10.1016/J.RSER.2015.06.033.

Lakaniemi, A.-M., Hulatt, C.J., Thomas, D.N., Tuovinen, O.H., Puhakka, J.A., 2011. Biogenic hydrogen and methane production from *Chlorella vulgaris* and *Dunaliella tertiolectabiomass*. Biotechnol. Biofuels 4, 34. https://doi.org/10.1186/1754-6834-4-34.

Laurens, L.M.L., Markham, J., Templeton, D.W., Christensen, E.D., Van Wychen, S., Vadelius, E.W., et al., 2017. Development of algae biorefinery concepts for biofuels and bioproducts; a perspective on process-compatible products and their impact on cost-reduction. Energy Environ. Sci. 10, 1716–1738. https://doi.org/10.1039/C7EE01306J.

Lee, Y.-K., 2001. Microalgal mass culture systems and methods: their limitation and potential. J. Appl. Phycol. 13, 307–315. https://doi.org/10.1023/A:1017560006941.

Lee, T.-M., Tseng, Y.-F., Cheng, C.-L., Chen, Y.-C., Lin, C.-S., Su, H.-Y., et al., 2017. Characterization of a heat-tolerant Chlorella sp. GD mutant with enhanced photosynthetic CO_2 fixation efficiency and its implication as lactic acid fermentation feedstock. Biotechnol. Biofuels 10, 214. https://doi.org/10.1186/s13068-017-0905-y.

Leite, G.B., Abdelaziz, A.E.M., Hallenbeck, P.C., 2013. Algal biofuels: challenges and opportunities. Bioresour. Technol. 145, 134–141. https://doi.org/10.1016/J.BIORTECH.2013.02.007.

Li, J., Zhu, D., Niu, J., Shen, S., Wang, G., 2011. An economic assessment of astaxanthin production by large scale cultivation of *Haematococcus pluvialis*. Biotechnol. Adv. 29, 568–574. https://doi.org/10.1016/J.BIOTECHADV.2011.04.001.

Li, W., Su, H.-N., Pu, Y., Chen, J., Liu, L.-N., Liu, Q., et al., 2019. Phycobiliproteins: molecular structure, production, applications, and prospects. Biotechnol. Adv. 37, 340–353. https://doi.org/10.1016/J.BIOTECHADV.2019.01.008.

Lin, J.-H., 2015. Lutein production from biomass: marigold flowers versus microalgae. Bioresour. Technol. 184, 421–428. https://doi.org/10.1016/J.BIORTECH.2014.09.099.

Lü, F., Ji, J., Shao, L., He, P., 2013. Bacterial bioaugmentation for improving methane and hydrogen production from microalgae. Biotechnol. Biofuels 6, 92. https://doi.org/10.1186/1754-6834-6-92.

Lüning, K., Pang, S., 2003. Mass cultivation of seaweeds: current aspects and approaches. J. Appl. Phycol. 15, 115–119. https://doi.org/10.1023/A:1023807503255.

Lüning, K., Yarish, C., Kirkman, H., 1990. Seaweeds: Their Environment, Biogeography, and Ecophysiology. Wiley.

Madera-Santana, T.J., Robledo, D., Freile-Pelegrín, Y., 2011. Physicochemical properties of biodegradable polyvinyl alcohol–agar films from the red algae *Hydropuntia cornea*. Mar. Biotechnol. 13, 793–800. https://doi.org/10.1007/s10126-010-9341-8.

Maity, J.P., Bundschuh, J., Chen, C.-Y., Bhattacharya, P., 2014. Microalgae for third generation biofuel production, mitigation of greenhouse gas emissions and wastewater treatment: present and future perspectives – a mini review. Energy 78, 104–113. https://doi.org/10.1016/J.ENERGY.2014.04.003.

Mata, T.M., Martins, A.A., Caetano, N.S., 2010. Microalgae for biodiesel production and other applications: a review. Renew. Sustain. Energy Rev. 14, 217–232. https://doi.org/10.1016/J.RSER.2009.07.020.

Mata-Alvarez, J., Macé, S., Llabrés, P., 2000. Anaerobic digestion of organic solid wastes. An overview of research achievements and perspectives. Bioresour. Technol. 74, 3–16. https://doi.org/10.1016/S0960-8524(00)00023-7.

Medipally, S.R., Yusoff, F.M., Banerjee, S., Shariff, M., 2015. Microalgae as sustainable renewable energy feedstock for biofuel production. BioMed Res. Int. 2015 https://doi.org/10.1155/2015/519513.

Mendez, L., Mahdy, A., Timmers, R.A., Ballesteros, M., González-Fernández, C., 2013. Enhancing methane production of *Chlorella vulgaris* via thermochemical pretreatments. Bioresour. Technol. 149, 136–141. https://doi.org/10.1016/J.BIORTECH.2013.08.136.

Metting, F.B., 1996. Biodiversity and application of microalgae. J. Ind. Microbiol. Biotechnol. 17, 477–489. https://doi.org/10.1007/BF01574779.

Michalak, I., Chojnacka, K., 2015. Algae as production systems of bioactive compounds. Eng. Life Sci. 15, 160–176. https://doi.org/10.1002/elsc.201400191.

Miyashita, K., Nishikawa, S., Beppu, F., Tsukui, T., Abe, M., Hosokawa, M., 2011. The allenic carotenoid fucoxanthin, a novel marine nutraceutical from brown seaweeds. J. Sci. Food Agric. 91, 1166–1174. https://doi.org/10.1002/jsfa.4353.

Monlau, F., Sambusiti, C., Barakat, A., Quéméneur, M., Trably, E., Steyer, J.-P., et al., 2014. Do furanic and phenolic compounds of lignocellulosic and algae biomass hydrolyzate inhibit anaerobic mixed cultures? A comprehensive review. Biotechnol. Adv. 32, 934–951. https://doi.org/10.1016/J.BIOTECHADV.2014.04.007.

Mooney-McAuley, K.M., Edwards, M.D., Champenois, J., Gorman, E., 2016. Best Practice Guidelines for Seaweed Cultivation and Analysis, Public Output report of the EnAlgae project, Swansea, p. 36. June 2016.

Mussgnug, J.H., Klassen, V., Schlüter, A., Kruse, O., 2010. Microalgae as substrates for fermentative biogas production in a combined biorefinery concept. J. Biotechnol. 150, 51–56. https://doi.org/10.1016/J.JBIOTEC.2010.07.030.

Nguyen, C.M., Kim, J.-S., Hwang, H.J., Park, M.S., Choi, G.J., Choi, Y.H., et al., 2012. Production of l-lactic acid from a green microalga, hydrodictyon reticulum, by Lactobacillus paracasei LA104 isolated from the traditional Korean food, makgeolli. Bioresour. Technol. 110, 552–559. https://doi.org/10.1016/J.BIORTECH.2012.01.079.

Noreen, A., Zia, K.M., Zuber, M., Ali, M., Mujahid, M., 2016. A critical review of algal biomass: a versatile platform of bio-based polyesters from renewable resources. Int. J. Biol. Macromol. 86, 937–949. https://doi.org/10.1016/J.IJBIOMAC.2016.01.067.

Olguin, E.J., 2012. Dual purpose microalgae-bacteria-based systems that treat wastewater and produce biodiesel and chemical products within a Biorefinery. Biotechnol. Adv. 30, 1031–1046. https://doi.org/10.1016/j.biotechadv.2012.05.001.

Oncel, S.S., 2013. Microalgae for a macroenergy world. Renew. Sustain. Energy Rev. 26, 241–264. https://doi.org/10.1016/J.RSER.2013.05.059.

Oren, A., Gunde-Cimerman, N., 2007. Mycosporines and mycosporine-like amino acids: UV protectants or multipurpose secondary metabolites? FEMS Microbiol. Lett. 269, 1–10. https://doi.org/10.1111/j.1574-6968.2007.00650.x.

Panis, G., Carreon, J.R., 2016. Commercial astaxanthin production derived by green alga *Haematococcus pluvialis*: a microalgae process model and a techno-economic assessment all through production line. Algal Res. 18, 175–190. https://doi.org/10.1016/J.ALGAL.2016.06.007.

Parmar, A., Singh, N.K., Pandey, A., Gnansounou, E., Madamwar, D., 2011. Cyanobacteria and microalgae: a positive prospect for biofuels. Bioresour. Technol. 102, 10163–10172. https://doi.org/10.1016/j.biortech.2011.08.030.

Passos, F., Uggetti, E., Carrère, H., Ferrer, I., 2014. Pretreatment of microalgae to improve biogas production: a review. Bioresour. Technol. 172, 403–412. https://doi.org/10.1016/j.biortech.2014.08.114.

Passos, F., Hom-Diaz, A., Blanquez, P., Vicent, T., Ferrer, I., 2016. Improving biogas production from microalgae by enzymatic pretreatment. Bioresour. Technol. 199, 347–351. https://doi.org/10.1016/j.biortech.2015.08.084.

Peng, J., Yuan, J.-P., Wu, C.-F., Wang, J.-H., 2011. Fucoxanthin, a marine carotenoid present in brown?seaweeds and diatoms: metabolism and bioactivities relevant to human health. Mar. Drugs 9, 1806–1828. https://doi.org/10.3390/md9101806.

Philip, S., Keshavarz, T., Roy, I., 2007. Polyhydroxyalkanoates: biodegradable polymers with a range of applications. J. Chem. Technol. Biotechnol. 82, 233–247. https://doi.org/10.1002/jctb.1667.

Raheem, A., Dupont, V., Channa, A.Q., Zhao, X., Vuppaladadiyam, A.K., Taufiq-Yap, Y.-H., et al., 2017. Parametric characterization of air gasification of *Chlorella vulgaris* biomass. Energy Fuel. 31, 2959–2969. https://doi.org/10.1021/acs.energyfuels.6b03468.

Raja, R., Hemaiswarya, S., Rengasamy, R., 2007. Exploitation of *Dunaliella* for β-carotene production. Appl. Microbiol. Biotechnol. 74, 517–523. https://doi.org/10.1007/s00253-006-0777-8.

Rammuni, M.N., Ariyadasa, T.U., Nimarshana, P.H.V., Attalage, R.A., 2019. Comparative assessment on the extraction of carotenoids from microalgal sources: astaxanthin from *H. pluvialis* and β-carotene from *D. salina*. Food Chem. 277, 128−134. https://doi.org/10.1016/J.FOODCHEM.2018.10.066.

Rashid, N., Rehman, M.S.U., Memon, S., Ur Rahman, Z., Lee, K., Han, J.-I., 2013. Current status, barriers and developments in biohydrogen production by microalgae. Renew. Sustain. Energy Rev. 22, 571−579. https://doi.org/10.1016/J.RSER.2013.01.051.

Rawat, I., Ranjith Kumar, R., Mutanda, T., Bux, F., 2013. Biodiesel from microalgae: a critical evaluation from laboratory to large scale production. Appl. Energy 103, 444−467. https://doi.org/10.1016/j.apenergy.2012.10.004.

Renuka, N., Guldhe, A., Prasanna, R., Singh, P., Bux, F., 2018. Microalgae as multi-functional options in modern agriculture: current trends, prospects and challenges. Biotechnol. Adv. 36, 1255−1273. https://doi.org/10.1016/J.BIOTECHADV.2018.04.004.

Rodrigues, D., Freitas, A.C., Pereira, L., Rocha-Santos, T.A.P., Vasconcelos, M.W., Roriz, M., et al., 2015. Chemical composition of red, brown and green macroalgae from Buarcos bay in Central West Coast of Portugal. Food Chem. 183, 197−207. https://doi.org/10.1016/J.FOODCHEM.2015.03.057.

Rodríguez, R., Espada, J.J., Moreno, J., Vicente, G., Bautista, L.F., Morales, V., et al., 2018. Environmental analysis of *Spirulina* cultivation and biogas production using experimental and simulation approach. Renew. Energy 129, 724−732. https://doi.org/10.1016/j.renene.2017.05.076.

Roger, P.A., Santiago-Ardales, S., Reddy, P.M., Watanabe, I., 1987. The abundance of heterocystous blue-green algae in rice soils and inocula used for application in rice fields. Biol. Fertil. Soils 5, 98−105. https://doi.org/10.1007/BF00257642.

Saadaoui, I., Sedky, R., Rasheed, R., Bounnit, T., Almahmoud, A., Elshekh, A., et al., 2019. Assessment of the algae-based biofertilizer influence on date palm (*Phoenix dactylifera* L.) cultivation. J. Appl. Phycol. 31, 457−463. https://doi.org/10.1007/s10811-018-1539-6.

Sahu, A., Pancha, I., Jain, D., Paliwal, C., Ghosh, T., Patidar, S., et al., 2013. Fatty acids as biomarkers of microalgae. Phytochemistry 89, 53−58. https://doi.org/10.1016/J.PHYTOCHEM.2013.02.001.

Sánchez, Ó.J., Cardona, C.A., 2008. Trends in biotechnological production of fuel ethanol from different feedstocks. Bioresour. Technol. 99, 5270−5295. https://doi.org/10.1016/J.BIORTECH.2007.11.013.

Sekar, S., Chandramohan, M., 2008. Phycobiliproteins as a commodity: trends in applied research, patents and commercialization. J. Appl. Phycol. 20, 113−136. https://doi.org/10.1007/s10811-007-9188-1.

Selvakumar, P., Umadevi, K., 2014. Enhanced lipid and fatty acid content under photoheterotrophic condition in the mass cultures of *Tetraselmis gracilis* and *Platymonas convolutae*. Algal Res. 6, 180−185. https://doi.org/10.1016/J.ALGAL.2014.10.002.

Shick, J.M., Dunlap, W.C., 2002. Mycosporine-like amino acids and related gadusols: biosynthesis, accumulation, and UV-protective functions in aquatic organisms. Annu. Rev. Physiol. 64, 223−262. https://doi.org/10.1146/annurev.physiol.64.081501.155802.

Shobana, S., Kumar, G., Bakonyi, P., Saratale, G.D., Al-Muhtaseb, A.H., Nemestóthy, N., et al., 2017. A review on the biomass pretreatment and inhibitor removal methods as key-steps towards efficient macroalgae-based biohydrogen production. Bioresour. Technol. 244, 1341−1348. https://doi.org/10.1016/J.BIORTECH.2017.05.172.

Sieburth, J.M., Smetacek, V., Lenz, J., 1978. Pelagic ecosystem structure: heterotrophic compartments of the plankton and their relationship to plankton size fractions. Limnol. Oceanogr. 23, 1256–1263. https://doi.org/10.4319/lo.1978.23.6.1256.

Sierra, E., Acién, F.G., Fernández, J.M., García, J.L., González, C., Molina, E., 2008. Characterization of a flat plate photobioreactor for the production of microalgae. Chem. Eng. J. 138, 136–147. https://doi.org/10.1016/J.CEJ.2007.06.004.

Singh, J., Gu, S., 2010. Commercialization potential of microalgae for biofuels production. Renew. Sustain. Energy Rev. 14, 2596–2610. https://doi.org/10.1016/j.rser.2010.06.014.

Singh, R.N., Sharma, S., 2012. Development of suitable photobioreactor for algae production – a review. Renew. Sustain. Energy Rev. 16, 2347–2353. https://doi.org/10.1016/J.RSER.2012.01.026.

Singh, A.K., Syiem, M.B., 2019. Plant growth-promoting abilities in cyanobacteria. Cyanobacteria 459–476. https://doi.org/10.1016/B978-0-12-814667-5.00023-4.

Singh, D., Puri, M., Wilkens, S., Mathur, A.S., Tuli, D.K., Barrow, C.J., 2013. Characterization of a new zeaxanthin producing strain of *Chlorella saccharophila* isolated from New Zealand marine waters. Bioresour. Technol. 143, 308–314. https://doi.org/10.1016/J.BIORTECH.2013.06.006.

Singh, J.S., Kumar, A., Rai, A.N., Singh, D.P., 2016. Cyanobacteria: a precious bio-resource in agriculture, ecosystem, and environmental sustainability. Front. Microbiol. 7 https://doi.org/10.3389/fmich.2016.00529.

Singh, A.K., Srivastava, J.K., Chandel, A.K., Sharma, L., Mallick, N., Singh, S.P., 2019. Biomedical applications of microbially engineered polyhydroxyalkanoates: an insight into recent advances, bottlenecks, and solutions. Appl. Microbiol. Biotechnol. 103, 2007–2032. https://doi.org/10.1007/s00253-018-09604-y.

Sinha, R.P., Klisch, M., Gröniger, A., Häder, D.-P., 1998. Ultraviolet-absorbing/screening substances in cyanobacteria, phytoplankton and macroalgae. J. Photochem. Photobiol. B Biol. 47, 83–94. https://doi.org/10.1016/S1011-1344(98)00198-5.

Sinha Ray, S., Sinha Ray, S., 2013. Environmentally friendly polymer nanocomposites using polymer matrices from renewable sources. Environ. Friendly Polym. Nanocompos. 89–156. https://doi.org/10.1533/9780857097828.1.89.

Sirajunnisa, A.R., Surendhiran, D., 2016. Algae – a quintessential and positive resource of bioethanol production: a comprehensive review. Renew. Sustain. Energy Rev. 66, 248–267. https://doi.org/10.1016/J.RSER.2016.07.024.

Soni, B., Kalavadia, B., Trivedi, U., Madamwar, D., 2006. Extraction, purification and characterization of phycocyanin from *Oscillatoria quadripunctulata* – isolated from the rocky shores of Bet-Dwarka, Gujarat, India. Process Biochem. 41, 2017–2023. https://doi.org/10.1016/j.procbio.2006.04.018.

Spiralg, 2019. https://www.spiralg.eu/.

Stahl, W., Sies, H., 2005. Bioactivity and protective effects of natural carotenoids. Biochim. Biophys. Acta 1740, 101–107. https://doi.org/10.1016/J.BBADIS.2004.12.006.

Stansell, G.R., Gray, V.M., Sym, S.D., 2012. Microalgal fatty acid composition: implications for biodiesel quality. J. Appl. Phycol. 24, 791–801. https://doi.org/10.1007/s10811-011-9696-x.

Subramanian, G., Shanmugasundaram, S., 1986. Uninduced ammonia release by the nitrogen-fixing cyanobacterium *Anabaena*. FEMS Microbiol. Lett. 37, 151–154. https://doi.org/10.1111/j.1574-6968.1986.tb01784.x.

Suganya, T., Varman, M., Masjuki, H.H., Renganathan, S., 2016. Macroalgae and microalgae as a potential source for commercial applications along with biofuels production: a biorefinery approach. Renew. Sustain. Energy Rev. 55, 909–941. https://doi.org/10.1016/J.RSER.2015.11.026.

Suh, I.S., Lee, C.-G., 2003. Photobioreactor engineering: design and performance. Biotechnol. Bioproc. Eng. 8, 313–321. https://doi.org/10.1007/BF02949274.

SundarRajan, P., Gopinath, K.P., Greetham, D., Antonysamy, A.J., 2019. A review on cleaner production of biofuel feedstock from integrated CO_2 sequestration and wastewater treatment system. J. Clean. Prod. 210, 445–458. https://doi.org/10.1016/J.JCLEPRO.2018.11.010.

Suriyamongkol, P., Weselake, R., Narine, S., Moloney, M., Shah, S., 2007. Biotechnological approaches for the production of polyhydroxyalkanoates in microorganisms and plants — a review. Biotechnol. Adv. 25, 148–175. https://doi.org/10.1016/J.BIOTECHADV.2006.11.007.

Suutari, M., Leskinen, E., Fagerstedt, K., Kuparinen, J., Kuuppo, P., Blomster, J., 2015. Macroalgae in biofuel production. Phycol. Res. 63, 1–18. https://doi.org/10.1111/pre.12078.

Takeda, H., Yoneyama, F., Kawai, S., Hashimoto, W., Murata, K., 2011. Bioethanol production from marine biomass alginate by metabolically engineered bacteria. Energy Environ. Sci. 4, 2575. https://doi.org/10.1039/c1ee01236c.

Talukder, M.M.R., Das, P., Wu, J.C., 2012. Microalgae (*Nannochloropsis salina*) biomass to lactic acid and lipid. Biochem. Eng. J. 68, 109–113. https://doi.org/10.1016/J.BEJ.2012.07.001.

Thomas, J.-B.E., Ramos, F.S., Gröndahl, F., 2019. Identifying suitable sites for macroalgae cultivation on the Swedish West Coast. Coast. Manag. 47, 88–106. https://doi.org/10.1080/08920753.2019.1540906.

Tibbetts, S.M., Milley, J.E., Lall, S.P., 2015. Chemical composition and nutritional properties of freshwater and marine microalgal biomass cultured in photobioreactors. J. Appl. Phycol. 27, 1109–1119. https://doi.org/10.1007/s10811-014-0428-x.

Trivedi, J., Aila, M., Bangwal, D.P., Kaul, S., Garg, M.O., 2015. Algae based biorefinery—how to make sense? Renew. Sustain. Energy Rev. 47, 295–307. https://doi.org/10.1016/J.RSER.2015.03.052.

Troschl, C., Meixner, K., Drosg, B., 2017. Cyanobacterial PHA Production-Review of Recent Advances and a Summary of Three Years' Working Experience Running a Pilot Plant, vol. 4. Bioeng, Basel, Switzerland. https://doi.org/10.3390/bioengineering4020026.

Tsoglin, L.N., Gabel, B.V., Fal'kovich, T.N., Semenenko, V.E., 1996. Closed photobioreactors for microalgal cultivation. Russ. J. Plant Physiol. 43, 131–136.

Uduman, N., Qi, Y., Danquah, M.K., Forde, G.M., Hoadley, A., 2010. Dewatering of microalgal cultures: a major bottleneck to algae-based fuels. J. Renew. Sustain. Energy 2, 012701. https://doi.org/10.1063/1.3294480.

Ugwu, C.U., Aoyagi, H., Uchiyama, H., 2008. Photobioreactors for mass cultivation of algae. Bioresour. Technol. 99, 4021–4028. https://doi.org/10.1016/J.BIORTECH.2007.01.046.

Vaishampayan, A., Sinha, R.P., Hader, D.-P., Dey, T., Gupta, A.K., Bhan, U., et al., 2001. Cyanobacterial biofertilizers in rice agriculture. Bot. Rev. 67, 453–516. https://doi.org/10.1007/BF02857893.

van Iersel, S., Gamba, L., Rossi, A., Alberici, S., Bart Dehue J van de, S., Flammini, A., 2009. ALGAE-BASED BIOFUELS: A Review of Challenges and Opportunities for Developing Countries. Rome.

Vasudevan, V., Stratton, R.W., Pearlson, M.N., Jersey, G.R., Beyene, A.G., Weissman, J.C., et al., 2012. Environmental performance of algal biofuel technology options. Environ. Sci. Technol. 46, 2451–2459. https://doi.org/10.1021/es2026399.

Vasumathi, K.K., Premalatha, M., Subramanian, P., 2012. Parameters influencing the design of photobioreactor for the growth of microalgae. Renew. Sustain. Energy Rev. 16, 5443–5450. https://doi.org/10.1016/J.RSER.2012.06.013.

Wada, N., Sakamoto, T., Matsugo, S., Wada, N., Sakamoto, T., Matsugo, S., 2015. Mycosporine-like amino acids and their derivatives as natural antioxidants. Antioxidants 4, 603–646. https://doi.org/10.3390/antiox4030603.

Wang, J., Yin, Y., 2018. Fermentative hydrogen production using pretreated microalgal biomass as feedstock. Microb. Cell Factories 17, 22. https://doi.org/10.1186/s12934-018-0871-5.

Wang, B., Lan, C.Q., Horsman, M., 2012. Closed photobioreactors for production of microalgal biomasses. Biotechnol. Adv. 30, 904–912. https://doi.org/10.1016/J.BIOTECHADV.2012.01.019.

Wang, M., Sahu, A.K., Rusten, B., Park, C., 2013. Anaerobic co-digestion of microalgae *Chlorella* sp. and waste activated sludge. Bioresour. Technol. 142, 585–590. https://doi.org/10.1016/j.biortech.2013.05.096.

Ward, A.J., Lewis, D.M., Green, F.B., 2014. Anaerobic digestion of algae biomass: a review. Algal Res. 5, 204–214. https://doi.org/10.1016/J.ALGAL.2014.02.001.

Wijffels, R.H., Barbosa, M.J., Eppink, M.H.M., 2010. Microalgae for the production of bulk chemicals and biofuels. Biofuels Bioprod. Biorefining 4, 287–295. https://doi.org/10.1002/bbb.215.

Wu, Z.-Z., Li, D.-Y., Cheng, Y.-S., 2018. Application of ensilage as a green approach for simultaneous preservation and pretreatment of macroalgae *Ulva lactuca* for fermentable sugar production. Clean Technol. Environ. Policy 20, 2057–2065. https://doi.org/10.1007/s10098-018-1574-7.

Xia, A., Jacob, A., Tabassum, M.R., Herrmann, C., Murphy, J.D., 2016. Production of hydrogen, ethanol and volatile fatty acids through co-fermentation of macro- and micro-algae. Bioresour. Technol. 205, 118–125. https://doi.org/10.1016/j.biortech.2016.01.025.

Yang, Z., Guo, R., Xu, X., Fan, X., Luo, S., 2011. Hydrogen and methane production from lipid-extracted microalgal biomass residues. Int. J. Hydrogen Energy 36, 3465–3470. https://doi.org/10.1016/J.IJHYDENE.2010.12.018.

Yen, H.-W., Hu, I.-C., Chen, C.-Y., Ho, S.-H., Lee, D.-J., 2013. Microalgae-based biorefinery – from biofuels to natural products. Bioresour. Technol. 135, 166–174. https://doi.org/10.1016/J.BIORTECH.2012.10.099.

Yuan, J.-P., Peng, J., Yin, K., Wang, J.-H., 2011. Potential health-promoting effects of astaxanthin: a high-value carotenoid mostly from microalgae. Mol. Nutr. Food Res. 55, 150–165. https://doi.org/10.1002/mnfr.201000414.

Zhan, J., Rong, J., Wang, Q., 2017. Mixotrophic cultivation, a preferable microalgae cultivation mode for biomass/bioenergy production, and bioremediation, advances and prospect. Int. J. Hydrogen Energy 42, 8505–8517. https://doi.org/10.1016/j.ijhydene.2016.12.021.

Zhang, Y., Liu, Z., Sun, J., Xue, C., Mao, X., 2018. Biotechnological production of zeaxanthin by microorganisms. Trends Food Sci. Technol. 71, 225–234. https://doi.org/10.1016/j.tifs.2017.11.006.

Zia, K.M., Zia, F., Zuber, M., Rehman, S., Ahmad, M.N., 2015. Alginate based polyurethanes: a review of recent advances and perspective. Int. J. Biol. Macromol. 79, 377–387. https://doi.org/10.1016/J.IJBIOMAC.2015.04.076.

Zia, K.M., Noreen, A., Bukhari, S.A., Aslam, N., Sultan, N., Jabeen, M., et al., 2017. Algae-based polyolefins. Algae based polym. blend. Compos 499–529. https://doi.org/10.1016/B978-0-12-812360-7.00013-6.

Chapter 7

Bio-based products from wood materials

Hamed Issaoui, Fatima Charrier - El Bouhtoury

CNRS/Univ. PAU & PAYS ADOUR/ E25 UPPA, Institute of Analytical Sciences and Physico-Chemistry for Environment and Materials (IPREM), UMR 5254, IUT des Pays de l'Adour, Mont de Marsan, France

7.1 Introduction

Due to the depletion of fossil resources and increasing of petroleum prices and environmental problems such as global warming and pollution, interest in green and renewable chemistry has increased in recent years. A strong focus is put on the biomass, especially wood, which plays an important role in the transition toward a bio-based and circular economy. Indeed, wood is an abundant bio-based and renewable material, which contributes to the reduction of the greenhouse effect by absorbing and storing durably of CO_2 gas during its growth, and its compounds represent a sustainable, abundant, and renewable source of products and molecules. Thus, wood chemistry, based on the valuation of the polymers and molecules that made up wood, is blossoming. Often overlooked, this sector is nevertheless considered as a path of the future for the many players in the wood industry. Lignin, which is the second major biopolymer of woody plants after cellulose, and tannins are abundant phenolic compounds. Despite the great availability of lignin, the major part of it is currently burnt and used as fuel and only about 2% is actually utilized for value-added applications (Younesi-Kordkheili et al., 2015); the current biorefinery processes mostly focus on the extraction and utilization of cellulose and hemicellulose. This can be partially explained by the variability in tannins and lignin compositions and structures, which depend on the feedstock but also on the extraction process (Smolarski, 2012). Consequently, research on cost-effective technologies for lignins and tannin isolation or conversion leading to easily further use in high-value applications are undergoing. This conversion will undoubtedly enhance overall biorefinery competitiveness; the good news is that there is already a range of value-added chemicals and materials including adhesives, carbon materials, etc., derived from lignin and/

Biobased Products and Industries. https://doi.org/10.1016/B978-0-12-818493-6.00007-5

or tannins and researches on their conversion into platform chemicals and alternative fuels are blossoming.

7.2 Wood lignins and tannins

Wood is a complex and multiscale structured composite material (Nilsson and Rowell, 2012). Whatever the species, wood is made of three high molecular components, which are cellulose, hemicelluloses and lignin, and various extractives in small quantities the presence of which varies by species (Fig. 7.1) (Van den Bosch et al., 2018). The extractives (aromatic and aliphatic

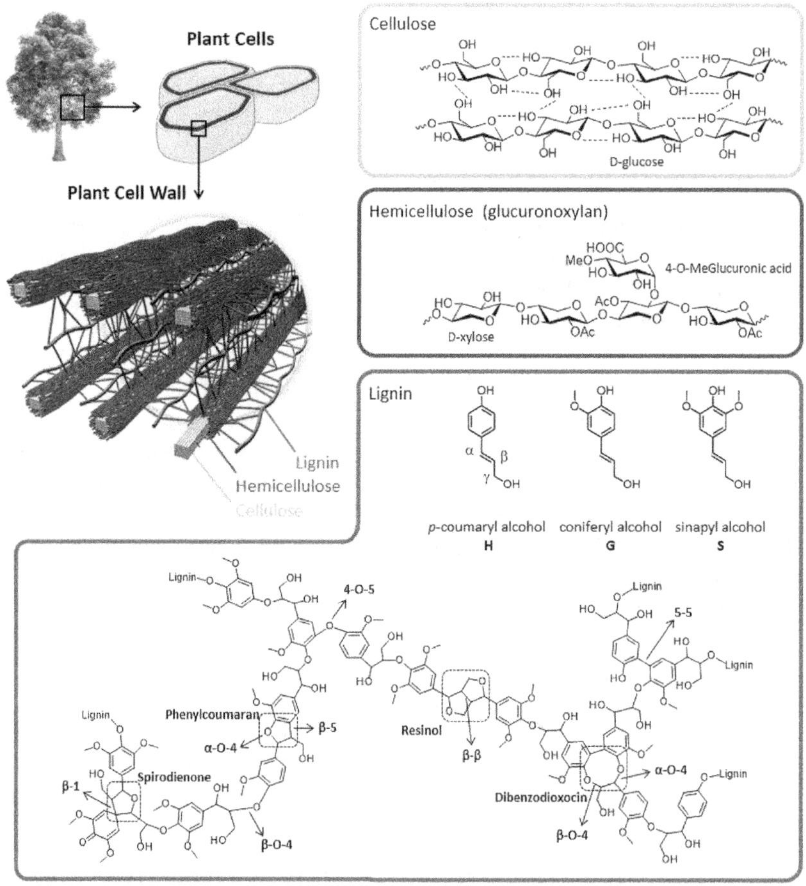

FIGURE 7.1 Schematic illustration of the lignocellulose matrix in plant cell walls, mainly composed of three biopolymers: cellulose, hemicellulose, and lignin. *Reprinted with permission from Van den Bosch, S., Koelewijn, S.F., Renders, T., Van den Bossche, G., Vangeel, T., Schutyser, W., Sels, B.F., 2018. Catalytic strategies towards lignin-derived chemicals. Top. Curr. Chem. 376, 36. https://doi.org/10.1007/s41061-018-0214-3*

chemical components of natural resins, dyes, lignans, glucosides, terpenes, alcohols, fats, waxes and tannins) represent between 1% and 10% by weight, are nonstructuring substances, and of low molecular weight. Wood is made up of 40–50 wt.% cellulose, 15–30 wt.% hemicellulose, 25–30 wt.% lignins, and 3–7 wt.% tannins (Poposcu, 2017).

7.2.1 Lignins: structures, isolation, and conversion

Lignin is a three-dimensional amorphous polymer, mainly found in the cell wall of woody tree species where it is cross-linked with cellulose and hemicellulose through covalent bonds and hydrogen bonds. The structure and the molecular mass of this native lignin varies according to the plant species. Chemically, lignin is a cross-linked heteropolymer, the highly branched structure of which is composed of partially methoxylated phenylpropane (C9) type units linked by specific ether or carbon-carbon linkages. Native lignin consists of three distinct types of phenylpropane units (hydroxyphenyl H, guaiacyl G, or syringyl S) distinguished by their degree of methoxylation (Van den Bosch et al., 2018) and the proportion of these three monolignol units in lignins varies according to the tree species. However, lignins of the conifers are mainly composed of guaiacyl units (95%) whereas the hardwood lignins consist of both guaiacyl and syringyl units.

The "fractionation" of lignocellulose (whether chemical, enzymatic, and/or mechanical) into its basic components alters the structure of native lignin, typically by selective disruption of certain bonds between phenylpropane units. The lignins thus obtained, designated under the name of "technical lignins," differ from the native lignins in terms of molecular mass, of nature of the bonds between the H units, G and S, by nature of functional groupings. Presently, there is no industrialized fractionation method allowing isolating lignin with high yield and high purity. The chemical isolation process, widely used for industrial production of lignin, can be divided into two main categories, which are sulfur containing lignin (kraft pulping process and sulfite pulping process) and non-sulfur biorefinery lignin (soda, organosolv, steam explosion, hydrolysis, diluted acid pyrolytic, high-pressure refining, ammonia-fiber-expansion lignin, etc.) (Hemmila et al., 2017). The Kraft lignin is obtained using an alkaline pretreatment while lignosulfonate entails an acidic pretreatment (Norgren and Edlund, 2014), both lignins are produced at industrial scale while organosolv, soda, and hydrolytic lignins are produced at industrial as well as at pilot scale. Organosolv method is a process where delignification is carried out using various organic solvents while soda lignins are obtained by a process that uses soda and anthraquinone (AQ) to depolymerize lignins.

A significant production of industrial lignins comes from the pulp and paper industry. Industrial lignin produced annually represents only 1.5–1.8 billion tons. In 2017, about 130 million tons of kraft pulp were produced worldwide, which alone extracts 50–70 million tons of lignin (Smolarski, 2012). Today, only 2% of the available lignin is exploited and significant amounts, of this byproduct, are currently burned at the source for heat and electricity production. Experts agree that the potential of lignin is underutilized with current trade of only US $ 300 million (Tribot et al., 2019). Therefore, lignin upgrading could be economically viable because of its low cost and abundant availability as a striking opportunity (Fig. 7.2) (Gillet et al., 2017). However, due to its complex structure, lignin has a low solubility in most of the usual solvents. Thus, lignins undergo important structural modifications to make them soluble (Kraft process) or transform them into soluble derivatives (bisulfite process). Organosolv methods permit to recover high quality lignin but remain expensive technology, while Kraft and sulfite processes provide lignin with impurities and soda extraction process has the lowest production costs as well as the lowest environmental impact. Nowadays, lignosulfonates are the only lignin derivatives currently present in quantity. Due to the presence of sulfonic acid groups, their lignin derivatives present amphiphilic properties and good solubility in water, but offer low benefit applications such as dispersing agents or adhesives, setting retarders in cements, additives in asphalt due to their antioxidant properties. In that respect, lignin structure, purity, properties and, therefore, cost largely depend on the feedstock but also on the delignification process.

Lignin valorization follows two main strategies; the first one is its functionalization for the formation of composites, copolymers, phenolic resins, reinforcements, or additives and the second one is its depolymerization which leads to the conversion of lignins into aliphatic and aromatic molecules

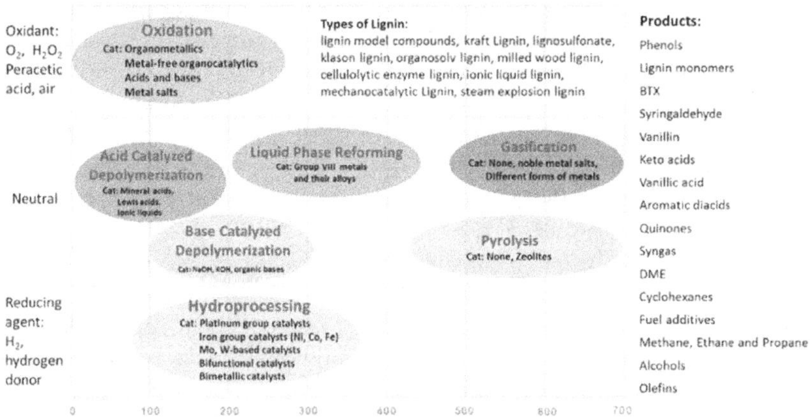

FIGURE 7.2 Summary of processes for the conversion of lignin. The abscissa represents the typical temperature range of the lignin conversion processes. *Reprinted with permission from Gillet, S., Aguedo, M., Petitjean, L., Morais, A.R.C., da Costa Lopes, A.M., Łukasik, R.M., Anastas, P.T., 2017. Lignin transformations for high value applications: towards targeted modifications using green chemistry. Crit. Rev. Green Chem. 19, 4200–4233.*

and functional synthons that can be used as bio-sourced monomers as an alternative to monomers from petrochemicals. Lignins depolymerization can be carried out by thermal treatments (pyrolysis/gasification), acid/base (neutral) reactions, reduction reactions, and oxidation reactions (Pandey and Kim, 2011). Laurichesse and Avérous (Laurichesse and Avérous, 2014) have described in detail the different chemical modifications of lignins and classified them into three groups: (1) lignin fragmentation into phenolic or other aromatic compounds for fine chemistry, (2) synthesis of new chemically active sites to impart new reactivity to lignin, and (3) functionalization of hydroxyl groups to enhance their reactivity. The summary of process for the conversion of lignins is given in Fig. 7.3 (Gillet et al., 2017).

7.2.2 Tannins: structure, isolation, and conversion

Tannins are secondary metabolites of plants where they play a role in their protection, their molar mass ranges between 300 and 3000 Da (gallic acid esters), which can in some cases go up to 30,000 Da for highly polymerized structures (proanthocyanidins). Tannins are astringent components, which have the ability to bind and precipitate proteins. From the standpoint of chemistry, tannins are polyhydroxylated aromatic compounds of great abundance in nature, which are able to form complexes with several kinds of macromolecules. This added to their antioxidant character makes them an

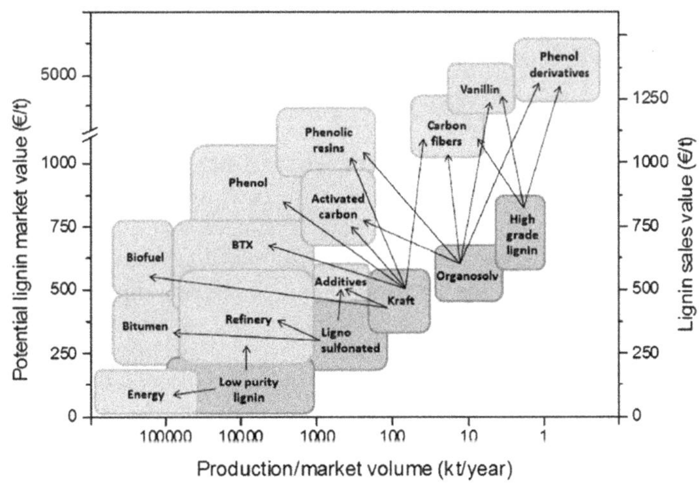

FIGURE 7.3 Potential market value for lignin-based products (orange [gray in print version]) and corresponding value-added chemicals (green [light gray in print version]). *Reprinted with permission from Gillet, S., Aguedo, M., Petitjean, L., Morais, A.R.C., da Costa Lopes, A.M., Łukasik, R.M., Anastas, P.T., 2017. Lignin transformations for high value applications: towards targeted modifications using green chemistry. Crit. Rev. Green Chem. 19, 4200–4233.*

attractive source of products. Indeed, they can be used in very different fields with a number of important applications. For example, tannins can be used as wine additives, pharmaceutical products, cement plasticizers, floatation agents, and as green renewable materials for novolac adhesives and resole resins (Peña et al., 2006, 2010). They are also studied to replace petro-based phenols in wood adhesive formulation (Ramires and Frollini, 2012). Based on their chemistry, tannins are classified in two major groups namely condensed and hydrolyzable tannins (Lochab et al., 2014). The hydrolyzable tannins are polyesters of carbohydrates and phenolic acid; they are easily hydrolyzable by acids and enzymes. Condensed tannins are the most common group of tannins. They have limited to no solubility in water, whereas in oligomeric form they are water-soluble and make up more than 90% of the annual industrial tannin production with 200,000 tons per year. Chemically, condensed tannins are polyflavonoids, their structure presents two aromatic rings, namely A and B rings, and they differ in number of hydroxyl groups. C4–C6 bonds can lead to branched polymers or oligomers, called proanthocyanidins (de Hoyos-Martínez et al., 2019). The structure of the most common flavan-3-ol monomers of condensed tannins are presented in Table 7.1.

Among the forest species richest in tannins, there are several species whose tannins are currently extracted on an industrial scale: Acacia (from which the tannins are extracted from the bark), Schinopsis or Quebracho (whose tannins are extracted from wood), Tsuga for which tannins are also extracted from bark and Rhus (sumac extracts), tannin extracts can also be produced from maritime pine bark (Chupin et al., 2015a).

As for lignins, the extraction process remains as the main hindrance for their valorization, due to their heterogeneous nature. In fact, the extraction yield and the purity of the obtained extracts relay on many factors such as the type of solvent, the extraction time, the temperature… de Hoyos-Martínez et al. (2019) carried out a literature review on tannin extraction method that highlights the main extraction parameters of each method and compares the different techniques between them. They concluded that, the new extraction methods, such as pressurized water extraction, microwave or ultrasound assisted extraction, are cleaner and more performant than the industrial traditional solid-liquid extraction which uses hot water as solvent in the presence of a base (carbonate, bicarbonate) and sodium sulfite or metabisulfite. Nonetheless, de Hoyos-Martínez et al. (2019) underlined that the new methods usually remain at the laboratory scale.

Tannins modification is led by the interest in obtaining new building blocks for polymer synthesis and their specific chemical structures allow this. Thus, Arbenz and Avérous (2015) have related the different chemical modification of condensed tannins structure, which can be divided into three major categories based on heterocycle reactivity, reactivity of nucleophilic sites, and hydroxyl group reactivity leading, respectively, to sulfonation, coupling reaction, and reaction with isocyanates.

TABLE 7.1 Structures of the most common flavan-3-ol monomers of condensed tannins.

A rings / B rings	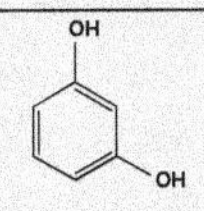 resorcinol ring	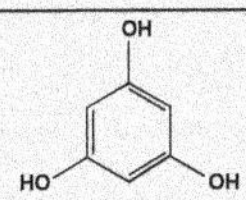 phloroglucinol ring
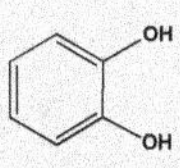 catechol ring	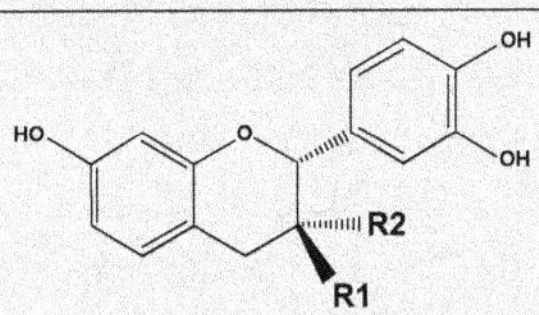 R1=OH R2=H fisetinidol R1=H R2=OH epifisetinidol	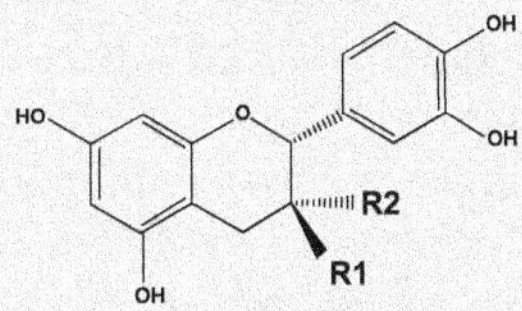 R1=OH R2=H catechin R1=H R2=OH epicatechin
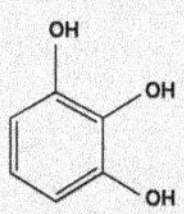 pyrogallol ring	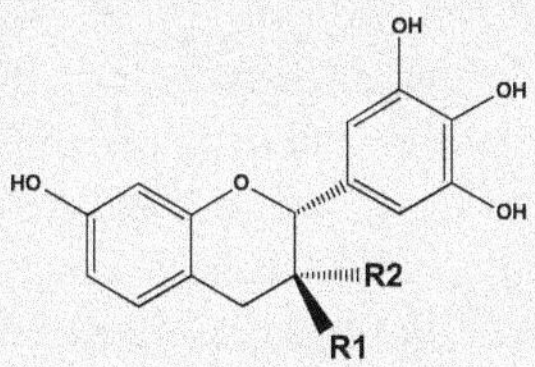 R1=OH R2=H robinetinidol R1=H R2=OH epirobinetinidol	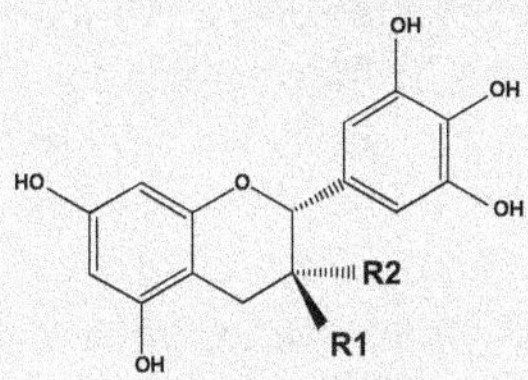 R1=OH R2=H gallocatechin R1=H R2=OH epigallocatechin

Reprinted with permission from de Hoyos-Martínez, P.L., Merle, J., Labidi, J., Charrier - El Bouhtoury, F., 2019. Tannins extraction: a key point for their valorization and cleaner production. J. Clean. Prod. 206, 1138–1155.

7.3 Lignin and tannins applications fields

The valorization of wood chemical components is a timely field of huge research. Currently, both lignin and tannins are open to new value-added recovery methods; inhere we focus on a part of this research which is extremely prolific.

7.3.1 Adhesives

The increasing need of environmentally friendly and emission free bonding processes has motivated the search for alternative adhesives. Tannins (Pizzi and Scharfetter, 1981) as well as lignins (Effendi et al., 2008) are the most abundant source of renewable resource phenols, have been subject to numerous studies to replace phenol in phenolic resins in wood composite adhesives. The raw materials used in industrial phenolic resin production (phenol and formaldehyde) are obtained from nonrenewable sources. For this reason, the substitution of these materials by others obtained from nonfossil source is desired and would provide economic and environmental advantages. The interest of substituting phenolic compounds in phenol formaldehyde resins is due to the molecular aromatic structure of tannins and lignin. Adhesives are a large variety of substances whose role is to allow the bonding of two items together.

Wood products are the largest users of adhesives, wood adhesives constituting more than 65% by volume of all the adhesives used in the world (Pizzi, 2016a). It is observed today a very strong international dynamic on the part of the academic world but more and more on the part of industrialists to identify and develop alternative solutions to urea-formaldehyde (UF) glues by the partial or total substitution of these glues by solutions bio-sourced techniques: soya-based glues, lignin or tannin glues, bio-based acrylic glues. Therefore, numerous research works developed on the properties of tannins since the mid-1970s have gradually confirmed their effectiveness in the replacement of synthetic phenolic adhesives. Condensed tannins are much interesting in adhesive chemistry because they lead to the same reactions types as phenols with hardeners, like the aldehydes in an acidic or basic environment. They are much more reactive than phenol because of the presence of resorcinol and the phloroglucinol nucleus found in condensed tannin structures. This increase of hydroxyl functions on the two aromatic nuclei unlike phenol generates a reactivity vis-a-vis formaldehyde, which can be greater than 10–50 times (Pizzi, 1978). Tannin glues have moisture resistance properties suitable for outdoor use. This property is related to the network formed by formaldehyde via methylene and methylene ether bridges during the polycondensation reaction. Research efforts are also deployed to manufacture other biosources such as urea-formaldehyde (Pizzi, 2016a) and to replace the use of formaldehyde, by a more environmentally friendly hardeners such as hexamine (Moubarik et al., 2009) and glyoxal (Navarrete et al., 2012a). Adhesives using hexamine as a hardener are already marketed.

Regarding lignin, the chemical structure of industrial one does not afford high reactivity and therefore restricts the commercial applications of lignins as adhesives. This because, even if lignins possess hydroxyl groups, either phenolic or aliphatic, their reactivity can be restricted by steric hindrance and only aliphatic OH groups of lignin are of high reactivity. To increase the reactivity to formaldehyde, lignins must be modified. Several works have studied the enhancement of lignin reactivity via different methods such as methylolation, hydroxymethylation (Wang and Ghen, 2014), or phenolation (Podschun et al., 2015). In general, lignins or lignosulfonates are mixed in low proportions with synthetic resins such as phenol-formaldehyde resins or urea-formaldehyde resins. In the case of using lignosulfonates, it is preferable to neutralize the HSO_3 group, which by its reactivity and its affinity for moisture, leads to a lower resistance of the resins. Vázquez et al. (1995), prepared plywood panels using lignin-phenol-formaldehyde adhesive. Lignin was obtained after acetosolv delignification of eucalyptus wood. The formaldehyde/phenol molar ratio (F/P) was between 1.5 and 2.5; the soda/phenol molar ratio (S/P) between 0.4 and 0.6, and the percentage of lignin relative to phenol varies between 20% and 40%. Their work showed that the reactivity of adhesives and the quality of plywood increase with the ratios F/P and S/P, but they decrease with the increase of the percentage of lignin. When the adhesive is prepared with methylolated or phenolated lignin, Vázquez et al. (1997) found that the manufactured plywood boards complied with European Standard EN 314-1:1993 and their performances where similar to those of boards manufactured with a commercial phenol formaldehyde resin. This last type of adhesives was also prepared using enzymatic hydrolyzed lignin (Jin et al., 2010) and steam explosion lignin (Zhao et al., 2016); in the latter case the authors reported that the substitution rate can reach 70% while satisfying Chinese standards for plywood panel bonding. In 2013, the company Weyerhaeuser Nr filed a patent (Winterowd, 2014) detailing the use of kraft lignin in the composition of phenol/formaldehyde liquid resins (LPF), used, for example, in the manufacture of particleboard, OSB board for wood construction, or furniture. Moreover, development of adhesives based on tannin and lignin mixture are now underway using industrial lignosulfonate (Chupin et al., 2015b). Aiming to manufacture wood adhesive with enhanced water-resistance, Faris et al. (2016) used a glyoxalated lignin polyol/tannin and polyethylenimine as raw materials. The resulting resins were used for plywood production; the obtained performances illustrate that the addition of polyethylenimine improves not only water-resistance but also the heat resistance of the prepared resins. Particleboard panels were prepared using mimosa tannin, hexamine as hardener, and kraft lignin or wheat straw glyoxalated lignin (Navarrete et al., 2012b) were classified as interior panel P2 according to the European Standard EN-312 and the formaldehyde emission was 75% lower than that of commercial panels glued with urea-formaldehyde or melamine-urea-formaldehyde. When different type of lignins (soda–anthraquinone flax lignin, kraft pine lignin, and ethanol–water wild tamarind lignin) are used to substitute phenol in phenolic resin, the obtained results demonstrated that kraft lignin is the best candidate because of its riches in free-ring positions, its higher Molecular weight (Mw), and greater thermal decomposition temperature in comparison to the other lignins tested (Tejado et al., 2007). Industrial tests for the panel industry were carried by the substitution of phenol with tannin or lignin. The results showed that the properties of the panels manufactured had the same properties as panels made with conventional phenol formaldehyde adhesives (Pizzi, 2016b), the mechanism of copolymerization, and hardening of glyoxalated lignin/tannin/hexamine-based adhesive was proposed by this author (Fig. 7.4). Formaldehyde emissions are close in terms of emissions than that of wood alone. In 2017, bio-based wood adhesives,

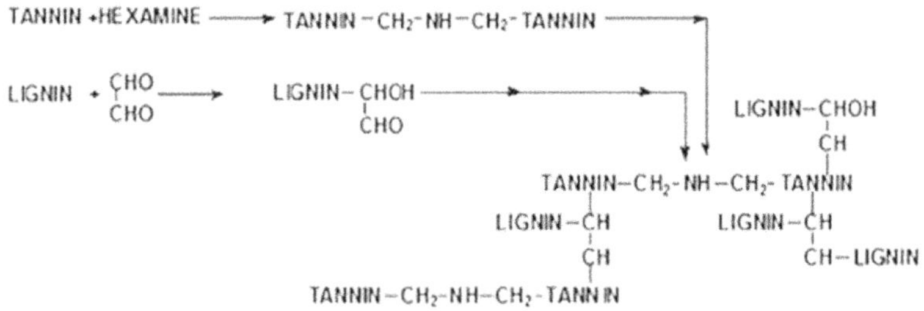

FIGURE 7.4 Schematic path of the simultaneous coreactions of glyoxalated lignin with the tannin/hexamine adhesive system. *Reprinted with permission from Pizzi, A., 2016. Wood products and green chemistry. Ann. For. Sci. 73, 185–203.*

including tannins and lignin-based adhesives, have been recently reviewed by He (2017).

In addition to work carried out on phenolic resins, there is a big interest in developing biosourced epoxy resins. Thus, resin-based 25 wt.% of commercial bisphenol A epoxy in which depolymerized kraft/organosolv lignin and epichlorohydrin were incorporated was manufactured (Ferdosian et al., 2016). Bio-based epoxy resins were also synthetized using multifunctionalized gallic acid (Aouf et al., 2013) or tara tannins as phenolic precursors (Aouf et al., 2014). However, it seems that the improvement of bioepoxy properties remains challenging.

7.3.2 Foams

Foams and more particularly polymeric foams are widely present in our daily life and their production is still highly petroleum-dependent. Foams can be defined as gas phase dispersed into a liquid or a solid matrix. Solid foam can be obtained after gelation or solidification of liquid phase containing polymerizable precursors. Foams can be open and/or closed cells and their proprieties such as density, size, and morphology influence their final application.

The opened cells are more suitable in acoustic insulation and depollution while the closed cells are more preferred in thermal insulation. Polymeric foams can be classified in two types: thermoplastic foams (polystyrene, polyethylene, polyvinyl chloride, etc.) and thermoset foams (polyurethane, epoxy foams, phenolic, etc.). The foaming of polymeric foams can be carried out by chemical, physical, or mechanical methods. In chemical process, gases released following an exothermic reaction increase the volume of the polymer. The blowing agents can be obtained after evaporation of added product, which have low-boiling temperature or can be chemically produced as a resulting of a reaction between the constituents of the liquid phase. In physical methods, a foaming agent such as pentane, N_2, or CO_2 injected in the polymer matrix is not chemically produced during the foaming process and does not react with the polymer. In mechanical foaming, the production of air is obtained by a strong mechanical agitation in the presence of surfactant, which stabilizes the foams. Less-conventional methods such as emulsion templating and microwave-assisted molding are developed to meet the current environmental challenges.

Due to their lower density compared to compact materials, high surface area, flexibility, or rigidity, polymeric foams have many useful applications such as acoustic and thermal insulation, packaging, transport, furniture, footwear, adsorption of pollutants.

The most important class of synthetic foams is phenolic and polyurethane foams; therefore, the development of greener foams to replace them is essential. Many renewal resources, such as cellulose, starch, chitosan, castor and soybean oil, tannins, and lignins have been used to produce green foam (El Bouhtoury, 2016). For example, from the adhesive resins previously commented foams can be derived. Thus, tannins are able to replace petro-sourced polyols in polyurethane foam production, since they contain polyphenols with both aliphatic and aromatic hydroxyl groups (Li et al., 2017a).

7.3.2.1 Bio-based phenolic foams

Synthetic phenolic foams are produced by reaction between phenol and formaldehyde, which can be catalyzed by an acid or a base. However, these two constituents are very toxic and classified as CMR agent. For this reason, several research works have been published aiming to reduce this toxicity by partially substituting or totally replacing some constituents with others that are less toxic or derived from renewable bioresources. Hexamine, glyoxal, and glutaraldehyde have been proposed in the literature to replace toxic formaldehyde. In addition, phenol, which is a product of cumene (petrochemical product), may be substituted or completely replaced by phenolic compounds from renewable resources such as tannins and lignins. Depolymerized hydrolysis lignins (molecular mass of 1500–2000 g/mol), obtained by a low-temperature (200–300°C) and low-pressure (1–5 MPa) process, have been used to replace 30 and 50 wt% of phenol in order to prepare bio-based phenol formaldehyde foams using polyetherpolysiloxane-copolymer, hexanes, and p-toluenesulfonic acid as surfactant, blowing agent, and catalyst, respectively (Li et al., 2019). The phenol substitution ratio influenced strongly on physical, mechanical, thermal, and morphological properties of obtained foams. With the increase in phenol substitution from 30%–50%, the apparent density, compressive strength, elastic modulus, and thermal conductivity were increased from 40–108 kg/m^3, 0.152–0.405 MPa, 2.16–7.56 MPa, and 0.033–0.04 W/m K, respectively. The use of kraft lignins, without any pre-treatment, as a substitute has allowed up to 50% reduction in the amounts of phenol and formaldehyde, which make the foams not only greener, but also more economical (Li et al., 2019).

The phenol substitution has also been achieved using tannins. In 1994, Meikleham and Pizzi developed formulations by stirring a homogeneous mixture of furfuryl alcohol, formaldehyde, water, and mimosa tannin; diethyl ether was used as a physical blowing agent. The properties of this kind of foams are close to those of phenolic foams in term of pore size and mechanical resistance to compression. It was possible to obtain low, standard, and high-density foams (respectively, 0.05, 0.08, and 0.12 g/cm^3) (Tondi and Pizzi, 2008). When foams were prepared from pine tannins, their thermal conductivity was lower than that of the mimosa tannins–based foams (Lacoste et al., 2012).

In 2014, Szczurek et al. proposed a new formulation without formaldehyde and chemical blowing agent. In this formulation, formaldehyde was replaced by hexamine; the foaming process was performed by strong mechanical agitation at a speed of 2000 rpm. A surfactant was added to the mixture before foaming to avoid destabilization of the foam. Among the used surfactants: Kolliphor ELP, Tegostab B8404 (silicon surfactant), Capstone 1470 (fluoro surfactant)... Then, the resulting liquid phases were placed in an oven at 85°C for 24 h. Finally, the materials obtained were dried at room temperature. Compared with conventional chemical foaming, this new method has similar characteristics. The thermal conductivity values are between 0.039 and 0.059 W/m K. For a thermal insulation application, the lowest thermal conductivity is required. However, this property is limited by the decrease in the mechanical properties of the foams. For this fact, Delgado-Sánchez et al. (2018) have studied the optimization of the formulation by modeling the mechanical and thermal properties according to the main ingredients: tannin-furfuryl alcohol-catalyst. The all prepared foams have a high porosity (95%–98%) and apparent densities ranging from 0.022 to 0.069 g/cm^3. The best thermomechanical properties of foams are reached when a moderate proportion of tannin, a high proportion of furfuryl alcohol, and a small amount of catalyst are used in the formulation. The thermal conductivity of the foams was less than 0.040 W/m K and their compressive strength was about 0.08 MPa. Condensed tannins can be replaced by lignins in the preparation of this type of foams, but, so far there are not enough studies in the literature. Furfuryl alcohol is manufactured industrially by reduction of a toxic product named furfural which is produced from hemicellulose of agricultural byproducts or sawdust. In a reactor under pressure and by acid catalysis the xylans of hemicellulose are hydrolyzed into xyloses and then dehydrated to form furfural. Replacing furfuryl alcohol with lignin results in more ecological and low-cost foams; the foams are made by valorizing of industrial waste.

Greener formulations of insulating foams, avoiding formaldehyde, and using mechanical frothing through high shear mixing in the presence of air were developed. The formulations in this case were based on a mixture of tannins and lignins, the use of hexamine as curing agent, glyoxal as hardener, and 80 as surfactant (Merle et al., 2016). All the obtained foams showed an open macroporous cellular morphology. The diameter of the cells varies from 96 to 115 μm. The porosity of the solid foams was about 90% and the thermal conductivity ranges from 0.035 to 0.044 W/m K.

The thermal conductivity of the most common insulating material varies from 0.02 to 0.05 W/m K; therefore, bio-based phenolic foams developed from wood products can be considered as good insulating materials. The tannin-based foams are particularly distinguished by their excellent fire resistance, which will favor their implementation in areas where fire safety is particularly important.

7.3.2.2 Bio-based polyurethane foams

Polyurethanes are copolymers synthesized by reacting polyols with a poly-isocyanate, and produced by chemical blowing, physical blowing, or a mixture of both process. In the chemical blowing, when water (the primary used as blowing agent) is added to the isocyanate, the reaction leads to a release of heat, as well as CO_2 gases, which allow the polymer to cross-link as well as in cell formation. In physical blowing an expanding agent is added to the formulation, such as pentane, which vaporizes due to the heat emission of the reaction between the polyisocyanate and the polyol. A mixture of the two processes may be used to control the density of the final material.

Tannins and lignins can be considered interesting sources of polyols in the production of polyurethane foams due to their aromatic and aliphatic hydroxyl groups, which are abundant in its structures. Mimosa condensed tannins have also been used, in 1998, to elaborate polyurethane targeting the partial replacement for synthetic polyol in the formulation of polyurethane, but the samples obtained were not stable at high temperature (Ge, 1998).

The development of lignin-based polyurethane foams is feasible but not easy due to their heterogeneous structure. Several approaches have been proposed in the literature for the synthesis of bio-based polyurethane foams with lignins: (i) polyols substitution by lignins without any prior modifications. (ii) Depolymerization of lignin in order to obtain new products, which have low molecular weight and higher hydroxyl groups. (iii) Liquefaction of lignins with other polyols such as glycerol, PEG, pulp fiber to increase the level of hydroxyl groups in the mixture. (iv) Chemical modification, such as esterification and etherification reactions to make hydroxyl functions more readily available to react with isocyanate.

Lignin, as a byproduct of the paper industry, has a complex structure with a very high molecular weight, making it difficult to access internal functional groups due to steric hindrance, which decreases its reactivity (Cateto and Rodrigues, 2008). Consequently, Mahmood et al. (2013) have proposed lignin depolymerization into polyols using alkaline hydrolysis with a ratio NaOH/lignin of 0.28 (w/w) at 250 and 300°C and reaction duration times ranging from 60 to 120 min. This study showed that the depolymerization has increased the total hydroxyl number, which passed from 275 mg KOH/g to 678–819 mg KOH/g. The number of aliphatic hydroxyl increased too, passing from 128 mg KOH/g measured for the nondepolymerized lignins to a number ranging between 236 and 352 mg KOH/g. The number of polyols hydroxyl required to manufacture rigid polyurethane, mentioned in the literature, is between 250 and 1000 mg KOH/g. In a recent study, depolymerized lignin by alkaline hydrolysis (Mw ~ 1700 g/mol, total hydroxyl number of 671 mg KOH/g, aliphatic hydroxyl number of 365 mg KOH/g) successfully replaces PPG400 and sucrose polyol with a high rate of 50 wt.% for the synthesis of rigid polyurethane foam (Mahmood et al., 2015).

Lignin liquefaction is another interesting approach proposed by many researchers for the production of polyurethane foams. It consists of mixing lignins with other polyols such as glycerol, crude glycerol, diethylene glycol, triethylene glycol, polyethylene glycol, polypropylene glycol triol, butanediol, etc., to improve the number of reactive OH groups. Muller et al. (2018) have developed rigid polyurethane foams by liquefaction of different lignins (hardwood calcium lignosulfonate, softwood kraft lignin, and organosolv lignin) in crude glycerol. The results show a significant difference between foams dependent of lignin used. With lignosulfonate, the resulting foam had highly irregular open cells and a density of 154 kg/m^3 higher than kraft lignin. Foam prepared with the later show more regular and smaller cell shapes, high compressive strength of 345 ± 63 KPa and low thermal conductivities of 0.039 ± 0.003 W/m K

Hatakeyama et al. (2013) prepared polyurethane foams using sodium lignosulfonate liquefaction with three types of polyols: diethylene glycol, triethylene glycol and polyethylene glycol (molecular mass of 200 g/mol). They found that the compressive strength of foams varied from 0.2 to 1 MPa which was increased with the lower molecular weight of polyol (diethylene glycol), and the glass transition temperature of foams can be ranged from 310 to 390K depending on the molecular weight of polyols, the amount of lignin and reaction time.

In a recent study, rigid polyurethane foams were developed using lignin liquefaction in crude glycerol and 1,4-butanediol at different temperatures (130–170°C) through microwave heating (Gosz et al., 2018). The higher hydroxyl number of 670 mg KOH/g was obtained with a yield of 93% using heating temperature of 150°C and times of 5 min. The foams were produced by replacing 0, 25 and 50 wt.% of reference polyols (Rokopol RF551) by liquefied lignin. The incorporation of liquefied lignin to the formulation increased the apparent density (83–150 kg/m^3) and compressive strength (627–855 kPa) and decreased the cells' size (236–168 μm), whose shapes became less regular and more inhomogeneous.

Cinelli et al. (2013) used liquefied lignin in glycerol/PEG400 in microwave that was added to polypropylene-glycol-triol or castor oil for the production of flexible polyurethane foams with water as the blowing agent. These authors suggest that their foams can be used in packaging furniture or in interior parts of car seats.

7.3.3 Carbon fibers

Carbon fibers are mainly applied for the development of composite materials that are used in many application fields such as aerospace, automotive, competitive sport, civil engineering, army, wind turbines, etc. (Push and Wohlmann, 2018).

Generally, the synthesis of the carbon fibers is according to the following steps: spinning and drawing of organic polymer, cross-linking of the chains between them by thermos-stabilization under oxidative atmosphere, carbonization, and graphitization under inert atmosphere. The most used spinning methods are melt-spinning, dry spinning, wet spinning, and electrospinning. Majorities of carbon fibers are made from a precursor of fossil origin, poly-acrylonitrile. The rest includes petroleum pitch or cellulosic precursors used in niche markets. With the rarefaction and rising cost of oil, the use of polyacrylonitrile will make carbon fibers very expensive, while its global demand will increase. Intense research work is carried out to replace precursors from fossil resources with biosourced precursors such as lignin.

The first patent on lignin conversion to carbon fibers dates from the year 1965. The company, Nippon Kayaku Co. Ltd. patented carbon fiber produced from different lignins (alkali-lignin, thiolignin, and ligninsulfonate) (Otani et al., 1969). Bayer AG described in a patent the preparation of carbon fiber from lignosulphonate in the presence of low amounts of plasticizers such as polyethylene oxide or acrylic acid–acrylamide copolymers (Mansmann et al., 1973). The mixture has the advantage of being easy to spin but has a carbon yield of 18–40wt. %. The mechanical stress properties were about 600–800 MPa and the modules were around 8–33 GPa. Around the same time, a patent published in 1974 by Gould describes the use of purified lignin as a precursor for carbon fiber. Lignins were extracted with alcohol followed by demethylation by hydrogen iodide. The stabilization takes place in an inert atmosphere at 140–240°C or via irradiation (beta or X-rays) and then carbonization ranges from 180 to 1800°C at a rate of 5°C/min. During the same period (1960–1970s), the development of high-performance carbon fiber using polyacrylonitrile as a precursor reduced the interest of lignin-based carbon fiber.

After two decades, Sudo and Shimizu (1994) described in a patent the preparation of carbon fiber from birch hardwood lignin exploded with steam. Lignin was phenolized by heating in a phenolic solvent such as phenol, creosole, creosote, methyl creosole, or xylenol in the presence of catalyst such as p-toluene sulfonic acid, sulfonyl chloride, amide salts, and hydroxyamine salts thereof. The polycondensation reaction leading to the densification of lignin is carried out at 180°C for 3–5 h with stirring. The phenolized lignin is then spun at a speed of 100 m/min at a temperature between 180 and 300°C from 30 min to 6 h, which converts it to infusible fiber. Subsequent carbonization in a nitrogen stream at a rate of 100°C/h up to 1000°C gave fibers with diameter of 21.4 ± 5.44 μm tensile strength of 518 MPa, elasticity modulus of 49 GPa, and a carbon yield of 28.8%.

In 2011, Baker et al. described in a patent the incorporation of carbon nanotubes (at a concentration of about 10 wt.% or less) into organosolv lignin, purified hardwood kraft ligninor softwood kraft lignin combined with the purified hardwood lignin using melt-spinning at about 150°C. The melt is extruded through the spinneret having apertures of diameter between 150 and 250 μm and is stretched and wound at a winding speed of 600–1200 m/min on a spool located 2 m from the spinneret. The diameter of obtained fiber may be between about 1 and 50 μm. Including of carbon nanotubes in lignin improved the spinning step via the increased heat capacity. This makes it possible to increase the cooling time of the fibers, limiting the rupture of filaments when quickly cooled and therefore they can be spun into more fine-diameter fibers. Carbon nanotubes also improved the mechanical resistance, since they appear mainly aligned along the axis of the fiber. In a later study, the same authors claim that the addition of carbon nanotubes improves by 20% of the tensile strength and by 50% of the modulus compared to the carbon fibers of the lignin alone (Baker and Rials, 2013). This poor increase in strength has been attributed to low interfacial adhesion between nonfunctionalized carbon nanotubes and lignin carbons.

More recently, Wang et al. (2016) grafted hardwood kraft lignin onto the surface of carboxyl multiwall carbon nanotubes. Then, these reinforcements (lignin-grafted carbon nanotubes) were mixed to hardwood kraft lignin in an internal mixer at 220°C at rates of 0, 0.5, 1, 2, and 3 wt.%. The addition of reinforcements seems to improve the melt spinnability of lignin but also improves the tensile strength of carbon fiber, which can be increased from 171.2 to 289.3 MPa when 0.5% reinforcement was incorporated, while the addition of 2 wt% nonfunctionalized carbon nanotubes lead to more efficient carbon fiber because of less porosity.

In 2012, Nordström et al. (2013) have reported on the manufacture of carbon fiber using softwood kraft lignin permeate or by the addition of hardwood kraft lignin permeate as a softening agent, which are obtained via ceramic membrane ultrafiltration of the black liquors. Once purified and dried, hardwood lignin changes from a Tg of 139−114°C, and a molar mass of 3500−1700 g/mol. and softwood lignin changes from a Tg of 150−146°C, and a molar mass of 6400−3300 g/mol. After stabilization and carbonization, carbon fibers containing about 93%−97% carbon atoms are produced but with a wide diameter ranging from 20 to 90 mm.

The complex structure of lignin makes its spinning difficult. Therefore, a plasticizing agent that gives it hold and flexibility during spinning step is needed. The polyethylene oxide seems to be a plasticizer of choice for different types of lignins. Kadla et al. (2002) used two types of lignin, an organosolv (Mw 3600 g/mol) and a purified kraft lignin (Mw = 4100 g/mol) in combination with polyethylene oxide at ratios of 95/5, 87.5/12.5, and 75/25. The addition of polyethylene oxide facilitated fiber spinning and carbon fibers produced had a carbon yield of 45% with a tensile and modulus ranging between 400 and 550 MPa and between 30 and 60 GPa, respectively.

Recently several research studies have been reported on the use of poly(ethylene oxide) as a plasticizer. Awal and Sain (2013) have mixed 5−20 wt% of PEO with soda hardwood lignin (Mw = 4000 g/mol, Tg = 126°C). Imel et al. (2016) have prepared carbon fiber by dissolving of kraft lignin (Tg = 153°C) in DMSO solvent and followed the addition of polyethylene oxide (17,000 g/mol) to the lignin solutions at 10 and 20 wt%.

Svinterikos and Zuburtikudis (2017) prepared nanofibers from a mixture of kraft lignin (Mw = 10,000 g/mol) and recycled polyethylene terephthalate using electrospinning method. The ratio of lignin in the polymer varied between 20, 35, and 50 wt.%. The study is focused on the optimization of process parameters in order to minimize the average diameter of fiber, which reached 191 ± 60 nm.

Another plasticizer potentially interesting for lignin is polylactic acid. It has a T_f toward 150°C and a T_g toward 60°C. The first activities with this type of blend for the manufacture of carbon fiber date from 2011 with the work of Thunga et al. (2014). The idea is to functionalize lignin kraft following butyration reaction, in order to optimize its affinity with PLA. Fine fibers were continuously spun from the blends with overall lignin ratios of 75 wt.%. SEM observations show a real heterogeneity of the compounds, where the PLA phase appears to be present in the form of fibrils in the blend with high ratios of lignin.

7.3.4 Adsorbents

Adsorption is a simple and widely used method in the treatment of water and air. It is a surface phenomenon based on the interactions of atoms, ions, or molecules (named adsorbates) with a solid surface (named adsorbent).

Depending on the nature of the surface interactions, adsorption can be classified into two main categories:

(i) Physical adsorption or physisorption involves low-binding energy type Van der Waals forces between the species adsorbed and the adsorbent. It is a reversible phenomenon.

(ii) Chemical adsorption or chemisorption involves high-binding energies, such as covalent, ionic, or metallic bonds, between the adsorbed species and the adsorbent. Generally, it is an irreversible phenomenon.

Many factors can affect the biosorption process such as biosorbent properties, nature of adsorbate, pH of a solution, biosorbent and adsorbate doses, contact time, temperature, reaction mechanism (complexation, electrostatic interactions, ion exchange) (Fomina and Gadd, 2014).

Several recent publications have reported the preparation of adsorbents from wood products. In particular, tannins and lignins are excellent candidates for developing biosorbents (Table 7.2). They are renewable, abundant, inexpensive, easy to obtain (waste from paper industries, bark, or sawdust), containing many organic functionalities such as carboxylic groups, ethers groups, and phenolic and aliphatic hydroxyl groups. Various adsorbents including tannins or lignins have been developed and extensively studied for the treatment of water and wastewater contaminated with heavy metals, dyes, pharmaceutical products, surfactant…, for air depollution (CO_2, SO_2, and H_2S) or for the recovery of noble and rare earth elements.

The phenolic groups present in the structure of lignins and tannins mark its anionic character, due to the possibility of forming phenoxide ions by deprotonation of phenolic groups and its resonance stabilization (Yin, 2010). Therefore, adsorbents-based lignin and tannin have been presented as promoter materials for the adsorption of cationic species by electrostatic attraction, and then an ion-exchange mechanism.

7.3.4.1 Tannin-based adsorbents

The water solubility of the tannins limits their use as an adsorbent. To overcome this problem, many researches have been investigated to make tannins in insoluble forms such as gelling, immobilization in supporting matrices, or producing tannin-based foams. Table 7.2 shows the comparison of various biosourced adsorbents studied for the removal of several adsorbates.

7.3.4.1.1 Tannin gels

Tannin gels were prepared following a polymerization reaction in the presence of an aldehyde such as formaldehyde using acidic or basic catalysis. In a typical experimental procedure, the tannin gels were prepared by dissolving the tannin powders in a solution of sodium hydroxide (Morisada et al., 2011; Gurung et al., 2013) or hydrogen chloride (Alvares Rodrigues et al., 2015).

TABLE 7.2 Adsorption capacities of various biosourced adsorbents studied for the removal of several adsorbates, Qm (maximum adsorption capacities). Self-developed table and not published elsewhere.

	Adsorbents	Adsorbates	Qm	References
Tannins	Tannin gel	Boron	24.3 mg/g	Morisada et al. (2011)
		Methylene blue Cetyltrimethylammonium bromide Zn(II)	1.35 mmol/g 2.12 mmol/g 1 mmol/g	Sánchez-Martín et al. (2011)
		Au(III) Pd(II) Pt(IV)	8.90 mmol/g 2.01 mmol/g 1.01 mmol/g	Gurung et al. (2013)
		Au(III)	7.7 mmol/g	Gurung et al. (2011)
		Cr(VI)	7.18 mmol/g	Inoue et al. (2010)
		Cr	488 mg/g	Alvares Rodrigues et al. (2015)
	Tannin-immobilized nanocellulose	Cu(II) Pb(II) Cr(VI)	51.846 mg/g 53.371 mg/g 104.592 mg/g	Xu et al. (2017)
	Tannin-immobilized gelatin/PVA nanofiber	U(VI)	170 mg/g	Menget al. (2019)
	Tannin-immobilized mesoporous silica	Au(III) Cr(III)	642.0 mg/g 1.30 mmol/g	Huang et al. (2010) Harmita et al. (2009)
	Tannin-immobilized collagen	Hg(II)	98.49 mg/g	Huang et al. (2009)
	Tannin r gid foam	Methylene blue Surfactant Trimethoprim	215 mg/g 65 mg/g 25 mg/g	Sánchez-Martín et al. (2013a)

Continued

TABLE 7.2 Adsorption capacities of various biosourced adsorbents studied for the removal of several adsorbates, Qm (maximum adsorption capacities). Self-developed table and not published elsewhere.—cont'd

Adsorbents		Adsorbates	Qm	References
Lignins	Kraft lignin	Cr(VI) Cd(II) Cu(II) Zn(II)	0.213 mmol/g 0.073 mmol/g 0.053 mmol/g 0.027 mmol/g	Šćiban et al. (2011)
	Organosolv and kraft lignin	Cu(II) Cd(II)	21.5–80.6 μmol/g 8.2–28.7 μmol/g	Harmita et al. (2009)
	Lignin-based activated carbon with physical activation	Methylene blue	92.51 mg/g	Fu et al. (2013)
	Lignin-based activated carbon with chemical activation	Ni(II)	14 mg/g	Gao et al. (2013)
		Methyl orange	300 mg/g	Mahmoudi et al. (2012)
		CO_2	8.6 mmol/g	Saha et al. (2017)
		CO_2	6.0 mmol/g	Hao et al. (2017)
		Benzene Toluene Xylene	~600 mg/g ~600 mg/g ~600 mg/g	Saha et al. (2018)
		SO_2	H_2SO_4	Rosas et al. (2017)

Lignin based activated carbon with microwave assistance	Cu(II)	1.54 mmol/g	Maldhure and Ekhe (2011b)
	Endosulfan	6.24 mg g	Maldhure and Ekhe (2011a)
Lignosulfonate sphere	Pb(II)	27.1 mg/g	Li et al. (2015c)
Sodium lignosulfonate/glucose	Pb(II) Cu(II) Cd(II) Ni(II) Cr(III)	194.5 mg/g 59.9 mg/g 48.8 mg/g 42.5 mg/g 41.8 mg/g	Liang et al. (2013a)
	Cr(VI)	57.68 mg/g	Liang et al. (2013b)
Lignin/methylamine/formaldehyde	Pb(II)	60.5 mg/g	Ge et al. (2015)
Lignin microspheres	Pb(II)	33.9 mg/g	Ge et al. (2016)
Aminated and sulfomethylated lignin	Cu(II) Pb(II)	0.71 mmol/g 0.26 mmol/g	Ge et al. (2014)
Lignin xanthate resin	Pb(II)	64.9 mg/g	Li et al. (2015a)
Surface functionalized porous lignin	Pb(II)	188 mg/g	Li et al. (2015b)
TiO_2/lignin TiO_2-SiO_2/lignin	Pb(II)	35.70 mg/g 59.93 mg/g	Klapiszewski et al. (2017)
Poly(ethylene imine) grafted lignin	Cu(II) Zn(II) Ni(II)	98 mg/g 78 mg/g 67 mg/g	Qin et al. (2017)
Lignin-grafted carbon nanotubes	Pb(II)	235 mg/g	Li et al. (2017b)

Addition of a certain amount of aldehyde like formaldehyde to start the gelling reaction, which was maintained at 353K for 8−12 h in alkaline condition (Morisada et al., 2011) (Gurung et al., 2013) and at 298K for 24 h in acidic condition (Alvares Rodrigues et al., 2015). Finally, the gels obtained were filtrated, washed, dried, and crushed into small particles.

Sánchez-Martín et al. (2011) have studied the removal of Zn(II), methylene blue, and cetyltrimethylammonium bromide in aqueous solution by adsorption onto four tannin extracts (Cypress, Quebracho, Pine, and Weibull black) which have been gelled with formaldehyde and acetaldehyde. The results show that formaldehyde has a strong gelling action compared to acetaldehyde, which gels only the tannin of Weibull black at acetaldehyde concentrations of 4.85 mmol/g of tannin. The best adsorbent has been attributed to the pine tannin gel made with concentrated formaldehyde whose adsorption capacity can reach 1, 1.35, and 2.12 mmol/g of Zn(II), methylene blue, and cetyltrimethylammonium bromide, respectively.

Alvares Rodrigues et al. (2015) studied the application of tannin gel as an adsorbent for the treatment of chromium-contaminated water. The tannin used in their study is extracted from black wattle, which has been cross-linked with formaldehyde in acidic condition. They have shown that the use of this tannin gel can reduced the Cr(IV) to less toxic Cr(III). The results showed that the tannin gel exhibits a high affinity for chromium with a maximum uptake capacity of 488 mg/g in the optimum condition of adsorbent dose, solution pH, and contact time of 0.1 g, 1, and 10−14 h respectively.

Morisada et al. (2011) have explored the influence of amine-modification of tannin gel on the adsorption of boron from an aqueous solution at different pH, temperature, and boron concentration. A wattle tannin gel was prepared in NaOH solution and gelled with formaldehyde. Its amine-modification was made in 10 wt.% aqueous ammonia at 50 g of gel/L. After modification, boron retention capacity showed a significant increase, which can reach 24.3 mg/g according to Langmuir model estimate. This is attributed to the stable coordination bond between boron and amino groups in the gels. In addition, the increase of temperature and pH above 7 has increased the boron adsorption.

Gurung et al. (2013) examined the effect of N-aminoguanidine modification of tannin gel for the selective adsorption and the recovery of precious metals: Au(III), Pd(II), and Pt(IV). In HCl solutions, N-aminoguanidine was used as a chelating ligand, which easily protonated and converted to positively charged sites, while precious metals are converted to chloro-anionic complexes and attracted onto the aminated tannin gel. The results showed that adsorption of precious metals was dependent on HCl concentration, and the maximum uptake capacity was evaluated as 1.01, 2.01 and 8.90 mmol/g for Pt(IV), Pd(II), and Au(III), respectively.

In another way, tannin gels were prepared without the use of aldehyde. It was prepared by mixing tannin with concentrated sulfuric acid, which cross-linked between a tannin polymer and polysaccharides present in the extract (Inoue et al., 2010; Gurung et al., 2011).

7.3.4.1.2 Tannin-immobilized matrix

In another approach, water-insoluble tannin has been successfully performed by its immobilization in a tridimensional matrix such as gelatin/PVA nanofibers, nanocellulose, mesoporous silica, collagen fiber.

More recently, Menget al. (2019) studied the immobilization of bayberry tannin on gelatin/PVA nanofibers and applied to uranium(VI) adsorptive removal from simulated seawater. The composite demonstrated their removal efficiency for uranium with a maximum uptake of 170 mg/g at the optimal pH of 5.5 when the initial concentration of uranium was 80 mg/L.

Xu et al. (2017) synthesized black wattle tannin−immobilized nanocellulose and employed it to extract Cu(II), Pb(II), and Cr(VI) from aqueous solutions. Firstly, a nanocellulose dialdehyde was prepared from bleached kraft pulp by sulfuric acid hydrolysis and sodium periodate oxidation. Then, black wattle tannin very reactive to aldehyde groups was covalently linked on nanocellulose matrix. It was found that the obtained nanocomposite was able to effectively remove Cr(VI) at pH 2 and Cu(II) and Pb(II) at pH 6. The equilibrium adsorption capacities of Cu(II), Pb(II), and Cr (VI) were estimated to 51.846, 53.371, and 104.592 mg/g, respectively.

Huang et al. (2010) have prepared a novel adsorbent by immobilization of bayberry tannin on mesoporous silica matrix and evaluated the recovery behavior of Au(III) from aqueous solutions. The obtained nanocomposite has a high recovery capacity of Au(III) from acid solutions whose original solution was concentrated about 18.0 times. The maximum adsorption capacity was estimated at 642.0 mg/g at 323K and was not affected by the presence of other coexisting metal ions.

In another study reported by Huang et al. (2009), bayberry tannin immobilized onto collagen fiber was prepared and used for Hg(II) adsorption from aqueous solution. The nanocomposite exhibited better adsorption capacity for Hg(II), up to 198.49 mg/g at pH 7 in 200 mg/L initial Hg concentration and 303K.

7.3.4.1.3 Tannin foams

Rigid foams based on tannin and furfuryl alcohol are widely studied for thermal insulation application. Their application as adsorbent is not much reported in the literature. Tannin foam was made as follows: firstly, a viscous mixture of tannin, furfuryl alcohol, formaldehyde, diethyl ether, and water was prepared. Then, p-toluenesulfonic acid as a catalyst was added to the mixture, which leads to an exothermic reaction due to the self-polymerization of furfuryl alcohol and its reaction with condensed tannin. At the same time, the generated heat leads to the evaporation of diethyl ether and thus to the formation of foam. Tannin foams have been shown to be excellent adsorbents for the heavy metals such as Cu(II) and Pb(II), dyes such as methylene blue, pharmaceuticals products such as trimethoprim, and surfactants such as sodium polyoxyethylene (3.5) lauryl ether sulfate (Tondi et al., 2009; Sánchez-Martín et al., 2013b).

7.3.4.2 Lignin-based adsorbents

Lignin is a highly aromatic low-value biomass residue, which can be utilized for chemicals, fuels, and materials production. In recent years, significant attention has focused on adsorbent materials from lignin. However, only 5% of available lignin is exploited worldwide; thus significant opportunities still exist for materials development. This review summarizes recent research advances in lignin-based adsorbents, with a particular emphasis on lignin, its modification and carbon materials derived from this abundant feedstock. Lignin derived activated carbons have been utilized for air pollutant adsorption (e.g., CO_2, SO_2 and H_2S), while modified lignin materials have been developed for the removal of organic dyes and organics (like methylene blue, Procion Blue MX-R and phenols), heavy metals (such as Cu, Zn, Pb and Cd), or recovery of noble metals (e.g., Pd, Au and Pt). Future perspectives highlight how green chemistry approaches for developing lignin adsorbents can generate added-value processes.

The biosorbents from lignins have been the subject of many research projects due to phenolic structure of lignin, its availability, and low cost. Absorbance can be achieved using lignin-derived activated carbon, non-modified (Ge and Li, 2018; Harmita et al., 2009) or modified lignin. The adsorbents based on nonmodified lignin have mostly poor adsorption capacity for heavy metals (Ge and Li, 2018) and both kraft lignin and lignosulfonates can lead to release of Sox on combustion and require the use of scrubbers to remove such pollutants from gas streams. Activated carbons have been employed for air pollutant adsorption (e.g., CO_2, SO_2, and H_2S), while modified lignin materials have been utilized for the removal of organic dyes and organics (such as methylene blue, phenols), heavy metals (e.g., Cu, Cd, Pb), or recovery of noble metals (like Au, Pt). In general, lignin adsorbents can be in different forms: activated carbon, composite with an organic or inorganic matrix, hydrogel.

7.3.4.2.1 Activated carbon

The improvement of the adsorption capacity was carried out according to different approaches and the most efficient one is the conversion of the lignin into porous activated carbon. The latter can be obtained by a physical (Fu et al., 2013), chemical (Gao et al., 2013; Rosas et al., 2017), or microwave irradiation (Maldhure and Ekhe, 2011a,b) activation process to increase the porosity of carbonized materials. The performance and properties of the obtained materials depend on the characteristics of the raw material and are influenced by the chosen activation process. In the physical process, the activation process can be performed using a gas such as CO_2, steam, or air. In the case of chemical activation, an activating agent such as KOH, $ZnCl_2$, or H_3PO_4 is mixed with the carbon precursor. Fu et al. synthesized activated carbon with black liquor lignin obtained in the pulp and paper industry using physical activation with steam (Fu et al., 2013). The activated carbon was prepared in two steps: carbonization in an oxygen-free atmosphere followed by steam activation. The highest surface area of activated carbon, with a value of 330.15 m^2/g, was obtained at a carbonization temperature of 450°C for 1 h and activation at 750°C for 40 min. This material has good potential for adsorption of methylene bromide with a retention capacity of 92.5 mg/g. Gao et al. prepared activated carbon from lignin of papermaking black liquor using chemical activation by KOH (Gao et al., 2013). The resulting biosorbent has a high specific surface area up to 2943 m^2/g. It was obtained with an activation temperature of 750°C and lignin:KOH ratio of 3:1. An adsorption capacity of Ni(II) up to 14 mg/g was obtained at a pH of 6.39 and a contact time of 8 h. Mahmoudi et al. investigated the preparation of activated carbon from lignin using $ZnCl_2$ (Mahmoudi et al., 2012). A surface area of activated lignin up to 587 m^2/g was obtained at activation temperature of 450°C for 2 h. The biosorbent was used for the uptake of methyl orange dye from aqueous solution and it showed a maximum adsorption capacity up to 300 mg/g.

Maldhure and Ekhe (2011a,b) prepared an activated carbon from black liquor lignin obtained from the kraft pulping process using $ZnCl_2$ and H_3PO_4 as an activating chemical agent with a microwave treatment and conventional heating. Microwave treatment showed several advantages compared to the conventional heating. The resulting activated carbon showed a maximum surface area of 1172 m^2/g, total pore volume of 0.640 cm^3/g, and a maximum adsorption capacity of copper of 1.541 mmol/g from microwave treatment at activation temperature 600°C. Whereas, the sample obtained with the conventional heating showed only 917.5 m^2/g, 0.506 cm^3/g, and 1.152 mmol/g, respectively, in the case of $ZnCl_2$ as activating agent. With H_3PO_4 activation, the sample obtained by microwave treatment revealed more oxygen surface functional groups, BET surface area, and pore volume than that derived under conventional heating. In addition, the maximum amount of adsorption of endosulfan was 6.4222 mg/g for activated carbon obtained with microwave treatment and higher than that with conventional heating which reached 3.9557 mg/g.

7.3.4.2.2 Modified lignins

Liang et al. (2013a) developed new ion-exchange resin from sodium lignosulfonate and glucose under acidic conditions. Under high acid concentration and high temperature, glucose and other sugars from raw lignin sulfonate were converted to 5-hydroxymethylfurfural, which is also converted to levulinic acid which then reacts with the lignin. These leads to the production of significant amount of carboxylic acids linked to lignin, therefore increasing oxygen-containing functional groups and enhancing the hydrophilicity of lignin. The resulting biosorbent showed a higher adsorption capacity of 194.553 mg/g for Pb(II), 59.998 mg/g for Cu(II), 48.80 mg/g for Cd(II), 42.450 mg/g for Ni(II) and 41.847 mg/g for Cr(III). In another study (Liang et al., 2013b), the authors reported the remove of Cr (VI) suggesting a 3-step elimination mechanism: electrostatic attraction of the chromate ion, reduction of Cr (VI) to Cr (III) and formation of Cr (III) bonds. The maximum adsorption capacity of 57.681 mg/g was obtained at a pH of 2, contact time of 5 h, temperature of 30°C, and biosorbent dose of 2 g/L. Biosorbent derived from alkaline lignin/methylamine/formaldehyde was also manufactured and its maximum adsorption capacity for Pb(II) was about 60.5 mg/g (Ge et al., 2015).

In the work led by Ge et al. (2016) an amino-functionalized lignin was prepared through Mannich reaction between alkaline lignin, methylamine, and formaldehyde. The obtained sample was used for lead removal from aqueous solution. The maximum adsorption capacity of Pb(II) was up to 60.5 mg/g on the biosorbent containing nitrogen groups which was 4 times higher than nonfunctionalized lignin (14.3 mg/g).

Ge et al. (2014) reported the preparation of bifunctionalized lignin with nitrogen and sulfur groups through Mannich reaction, with diethylenetriamine and sulfomethylation with Na_2SO_3. The bifunctionalized lignin has a pH at zero charge of 2.8 indicating that the biosorbent surface is negatively charged at pH greater than 2.8. Therefore, it can uptake the metal ions positively charged by electrostatic attractive forces. The maximum of the metal ions adsorbed at equilibrium reach to 0.7438 mmol/g for Cu(II) and 0.2884 mmol/g for Pb(II) according to Dubinin–Radushkevich model. The maximum adsorption capacity of a porous lignin xanthate resin for Pb^{2+} removal from aqueous solution was synthesized by Li et al. (2015a) was 64.9 mg/g. However, surface of such material is functionally enhanced to increase capacity of adsorption, which passes to 188 mg/g (Li et al., 2015b).

Li et al. (2015c) prepared a porous lignosulfonate-based sphere by gelation-solidification method. Lignin was mixed with sodium alginate solution and epichlorohydrin and after cross-linking reaction, the mixture was dripped in a $CaCl_2$ solution at 80°C for gelation. The obtained sphere showed a high porosity of 87.66%, a total pore volume of 0.416 cm^3/g and a maximum adsorption capacity for Pb(II) reached 27.1 mg/g at a dosage of 0.2 g of porous lignosulfonate-based sphere in 100 mL lead solutions, at pH of 5.0 and 30°C.

Klapiszewski et al. (2017) reported the preparation of functional hybrid materials TiO_2/lignin and TiO_2-SiO_2/lignin for adsorption of Pb(II) from the aqueous solutions. The adsorption capacity of 59.93 mg/g onto TiO_2-SiO_2/

lignin, higher than onto TiO_2/lignin of 35.70 mg/g, which indicates that the first hybrid materials are better for the uptake of Pb(II).

Nanocomposite based on lignin grafted carbon nanotubes has been used as adsorbent and reported by Li et al. (Qin et al., 2017). The nanocomposite showed a maximum adsorption capacity up to 235 mg/g for Pb(II) and a removal efficiency of 98.3% for oil droplets in water.

7.4 Conclusion and perspectives

The development of new materials from renewable resources is increasingly a necessity and the design of new materials not only respectful but also with multifunctional properties is a challenge. In future, platform chemicals will increasingly be produced from raw materials containing lignocellulose. Lignin and tannins are the options that could be taken to meet this challenge because they are highly valuable and their potential for value-added products is far from being exhausted. Developing bio-based products from lignin is a crucial part of an integrated biorefinery. Lignin and tannin extractions and characterizations methods are one of the major keys that not only could lead to a cleaner and more effective production, but also can promote and spread lignin and tannin utilization in several industrial fields. The work done so far on different foams, carbon fibers, adhesives... must continue to better compete and replace the materials and molecules of petroleum origin. Behind this development of materials and molecules, there are several major areas of interest such as potential economic impacts, environmental impacts.

References

Alvares Rodrigues, L., Koibuchi Sakane, K., Alves Nunes Simonetti, E., Patrocínio Thim, G., 2015. Cr total removal in aqueous solution by PHENOTAN AP based tannin gel (TFC). J. Environ. Chem. Eng. 3 (2), 725–733.

Aouf, C., Nouailhas, H., Fache, M., Caillol, S., Boutevin, B., Fulcrand, H., 2013. Multi-functionalization of galic acid. Synthesis of a novel bio-based epoxy resin. Eur. Polym. J. 55, 1185–1195.

Aouf, C., Benyahya, S., Esnouf, A., Caillol, S., Boutevin, B., Fulcrand, H., 2014. Tara tannins as phenolic precursors of thermosetting epoxy resins. Eur. Polym. J. 55, 186–198.

Arbenz, A., Avérous, L., 2015. Chemical modification of tannins to elaborate aromatic biobased macromolecular architectures. Green Chem. 17, 2626–2646.

Awal, A., Sain, M., 2013. Characterization of soda hardwood lignin and the formation of lignin fibers by melt spinning. J. Appl. Polym. Sci. 129 (5), 2765–2771.

Baker, D.A., Rials, T.G., 2013. Recent advances in low-cost carbon fiber manufacture from lignin. J. Appl. Polym. Sci. 130 (2), 713–728.

Baker, F.S., Baker, D.A., Menchhofer, P.A., November 24, 2011. Carbon Nanotube (CNT)-Enhanced Precursor for Carbon Fiber Production and Method of Making a CNT-Enhanced Continuous Lignin Fiber. US20110285049A1.

Cateto, C.A., Barreiro, M.F., Rodrigues, A.E., 2008. Monitoring of lignin-based polyurethane synthesis by FTIR-ATR. Ind. Crops Prod. 27 (2), 168–174.

Chupin, L., Maunu, S.L., Reynaud, S., Pizzi, A., Charrier, B., Charrier-EL Bouhtoury, F., 2015. Microwave assisted extraction of maritime pine (*Pinus pinaster*) bark: impact of particle size and characterization. Ind. Crops Prod. 65, 142–149.

Chupin, L., Charrier, B., Pizzi, A., Perdomo, A., El Bouhtoury, F.C., 2015. Study of thermal durability properties of tannin–lignosulfonate adhesives. J. Therm. Anal. Calorim. 119, 1577–1585.

Cinelli, P., Anguillesi, I., Lazzeri, A., 2013. Green synthesis of flexible polyurethane foams from liquefied lignin. Eur. Polym. J. 49 (6), 1174–1184.

Delgado-Sánchez, C., Santiago-Medina, F., Fierro, V., Pizzi, A., Celzard, A., 2018. Optimisation of "green" tannin-furanic foams for thermal insulation by experimental design. Mater. Des. 139, 7–15.

Effendi, A., Gerhauser, H., Bridgwater, A.V., 2008. Renew. Sustain. Energy Rev. 12 (8), 2092–2116.

El Bouhtoury, F.C., November 2016. Innovative green foams: properties and applications. In: Inamuddin, I. (Ed.), Green Polymer Composites Technology. CRC Press, ISBN 978-1-4987-1546-1, pp. 19–30. https://doi.org/10.1201/9781315371184.

Faris, A.H., Rahim, A.A., Ibrahim, M.N.M., Alkurdi, A.M., Shah, I., 2016. Combination of lignin polyol-tannin adhesives and polyethylenimine for preparation of green water-resistant adhesives. J. Appl. Polym. Sci. 133 (20), 43437. https://doi.org/10.1002/app.43437.

Ferdosian, F., Zhang, Y., Yuan, Z., Anderson, M., Xu, C., 2016. Curing kinetics and mechanical properties of bio-based epoxy composites comprising lignin based epoxy. Eur. Polym. J. 82, 153–165.

Fomina, M., Gadd, G.M., 2014. Biosorption: current perspectives on concept, definition and application. Bioresour. Technol. 160, 3–14.

Fu, K., Yue, Q., Gao, B., Sun, Y., Zhu, L., 2013. Preparation, characterization and application of lignin-based activated carbon from black liquor lignin by steam activation. Chem. Eng. J. 228, 1074–1082.

Gao, Y., Yue, Q., Gao, B., Sun, Y., Wang, W., Li, Q., Wang, Y., 2013. Preparation of high surface area-activated carbon from lignin of papermaking black liquor by KOH activation for Ni(II) adsorption. Chem. Eng. J. 217, 345–353.

Ge, J.J., 1998. Synthesis of biodegradable polyurethane foams from condensed tannin and bark of *Acacia mearnsii*. Bull. Kyushu Univ. For. 79, 21–85.

Ge, Y., Li, Z., 2018. Application of lignin and its derivatives in adsorption of heavy metal ions in water: a review. ACS Sustain. Chem. Eng. 6 (5), 7181–7192.

Ge, Y., Qin, L., Li, Z., 2016. Lignin microspheres: an effective and recyclable natural polymer-based adsorbent for lead ion removal. Mater. Des. 95, 141–147.

Ge, Y., Li, Z., Kong, Y., Song, Q., Wang, K., 2014. Heavy metal ions retention by bi-functionalized lignin: synthesis, applications, and adsorption mechanisms. J. Ind. Eng. Chem. 20 (6), 4429–4436.

Ge, Y., Song, Q., Li, Z., March 2015. A Mannich base biosorbent derived from alkaline lignin for lead removal from aqueous solution. J. Ind. Eng. Chem. 23, 228–234.

Gillet, S., Aguedo, M., Petitjean, L., Morais, A.R.C., da Costa Lopes, A.M., Łukasik, R.M., Anastas, P.T., 2017. Lignin transformations for high value applications: towards targeted modifications using green chemistry. Crit. Rev. Green Chem. 19, 4200–4233.

Gosz, K., Kosmela, P., Hejna, A., Gajowiec, G., Piszczyk, L., 2018. Biopolyols obtained via microwave-assisted liquefaction of lignin: structure, rheological, physical and thermal properties. Wood Sci. Technol. 52 (3), 599–617.

Gould, A.M., June 26, 1974. Manufacture of Carbon Fibre. Patent GB1358164 (A).

Gurung, M., Adhikari, B.B., Kawakita, H., Ohto, K., Inoue, K., Alam, S., 2011. Recovery of Au(III) by using low cost adsorbent prepared from persimmon tannin extract. Chem. Eng. J. 174 (2–3), 556–563.

Gurung, M., Adhikari, B.B., Morisada, S., Kawakita, H., Ohto, K., Inoue, K., Alam, S., 2013. N-aminoguanidine modified persimmon tannin: a new sustainable material for selective adsorption, preconcentration and recovery of precious metals from acidic chloride solution. Bioresour. Technol. 129, 108–117.

Hao, W., Björnerbäch, F., Trushkina, Y., Bengoechea, M.O., Salazar-Alvarez, G., 2017. High-performance magnetic activated carbon from solid waste from lignin conversion processes. 1. Their use as adsorbents for CO_2. ACS Sustain. Chem. Eng. 5 (4), 3087–3095.

Harmita, H., Karthikeyan, K.G., Pan, X., 2009. Copper and cadmium sorption onto kraft and organosolv lignins. Bioresour. Technol. 100 (24), 6183–6191.

Hatakeyama, H., Hirogaki, A., Matsumura, H., Hatakeyama, T., 2013. Glass transition temperature of polyurethane foams derived from lignin by controlled reaction rate. J. Therm. Anal. Calorim. 114 (3), 1075–1082.

He, Z., 2017. Bio-Based Wood Adhesives. Preparation, Characterization and Testing. CRC Press, Boca Raton, p. 336. ISBN 9781498740746 - CAT# K26705.

Hemmila, V., Adamopoulos, S., Karlsson, O., Kumar, A., 2017. Development of sustainable bio-adhesives for engineered wood panels – a review. RSC Adv. 7, 38604–38630.

Huang, X., Liao, X., Shi, B., 2009. Hg(II) removal from aqueous solution by bayberry tannin-immobilized collagen fiber. J. Hazard Mater. 170 (2–3), 1141–1148.

Huang, H., Wang, Y., Liao, X., Shi, B., 2010. Adsorptive recovery of Au^{3+} from aqueous solutions using bayberry tannin-immobilized mesoporous silica. J. Hazard Mater. 183 (1–3), 793–798.

Imel, A.E., Naskar, A.K., Dadmun, M.D., 2016. Understanding the impact of poly(ethylene oxide) on the assembly of lignin in solution toward improved carbon fiber production. ACS Appl. Mater. Interfaces 8 (5), 3200–3207.

Inoue, K., Paudyal, H., Nakagawa, H., Kawakita, H., Ohto, K., 2010. Selective adsorption of chromium(VI) from zinc(II) and other metal ions using persimmon waste gel. Hydrometallurgy 104 (2), 123–128.

Jin, Y., Cheng, X., Zheng, Z., 2010. Preparation and characterization of phenol–formaldehyde adhesives modified with enzymatic hydrolysis lignin. Bioresour. Technol. 101 (6), 2046–2048.

Kadla, J.F., Kubo, S., Venditti, R.A., Gilbert, R.D., Compere, A.L., Griffith, W., 2002. Lignin-based carbon fibers for composite fiber applications. Carbon 40 (15), 2913–2920.

Klapiszewski, L., Siwińska-Stefańska, K., Kołodyńska, D., 2017. Preparation and characterization of novel TiO_2/lignin and TiO_2-SiO_2/lignin hybrids and their use as functional biosorbents for Pb(II). Chem. Eng. J. 314, 169–181.

Lacoste, C., Basso, M.C., Pizzi, A., Laborie, M.P., Celzard, A., Fierro, V., 2012. Pine tannin-based rigid foams: mechanical and thermal properties. Ind. Crops Prod. 43, 245–250.

Laurichesse, S., Avérous, L., 2014. Chemical modification of lignins: towards biobased polymers. Prog. Polym. Sci. 39, 1266–1290.

Li, Z., Kong, Y., Ge, Y., 2015. Synthesis of porous lignin xanthate resin for Pb^{2+} removal from aqueous solution. Chem. Eng. J. 270, 229–234.

Li, Z., Xiao, D., Ge, Y., Koehler, S., 2015. Surface-functionalized porous lignin for fast and efficient lead removal from aqueous solution. ACS Appl. Mater. Interfaces 7 (27), 15000–15009.

Li, Z., Ge, Y., Wan, L., 2015. Fabrication of a green porous lignin-based sphere for the removal of lead ions from aqueous media. J. Hazard Mater. 285, 77–83.

Li, B., Wang, Y., Mahmood, N., Yuan, Y., Schmidt, J., Xu, C., 2017. Preparation of bio-based phenol formaldehyde foams using depolymerized hydrolysis lignin. Ind. Crops Prod. 97, 409–416.

Li, Z., Chen, J., Ge, Y., 2017. Removal of lead ion and oil droplet from aqueous solution by lignin-grafted carbon nanotubes. Chem. Eng. J. 308, 809–817.

Li, B., Yuan, Z., Schmidt, J., Xu, C., 2019. New foaming formulations for production of bio-phenol formaldehyde foams using raw kraft lignin. Eur. Polym. J. 111, 1–10.

Liang, F.B., Song, Y.L., Huang, C.P., Li, Y.X., Chen, B.H., 2013. Synthesis of novel lignin-based ion-exchange resin and its utilization in heavy metals removal. Ind. Eng. Chem. Res. 52 (3), 1267–1274.

Liang, F.B., Song, Y.L., Huang, C.P., Zhang, J., Chen, B.H., 2013. Adsorption of hexavalent chromium on a lignin-based resin: equilibrium, thermodynamics, and kinetics. J. Environ. Chem. Eng. 1 (4), 1301–1308.

Lochab, B., Shukla, S., Varma, I.K., 2014. Naturally occurring phenolic sources: monomers and polymers. RSC Adv. 4 (42), 21712–21752.

Mahmood, N., Yuan, Z., Schmidt, J., Xu, C., 2013. Production of polyols via direct hydrolysis of kraft lignin: effect of process parameters. Bioresour. Technol. 139, 13–20.

Mahmood, N., Yuan, Z., Schmidt, J., Xu, C., 2015. Preparation of bio-based rigid polyurethane foam using hydrolytically depolymerized Kraft lignin via direct replacement or oxy-propylation. Eur. Polym. J. 68, 1–9.

Mahmoudi, K., Hamdi, N., Kriaa, A., Srasra, E., 2012. Adsorption of methyl orange using activated carbon prepared from lignin by $ZnCl_2$ treatment. Russ. J. Phys. Chem. A 86 (8), 1294–1300.

Maldhure, A.V., Ekhe, J.D., 2011. Microwave treated activated carbon from industrial waste lignin for endosulfan adsorption. J. Chem. Technol. Biotechnol. 86 (8), 1074–1080.

Maldhure, A.V., Ekhe, J.D., 2011. Preparation and characterizations of microwave assisted activated carbons from industrial waste lignin for Cu(II) sorption. Chem. Eng. J. 168 (3), 1103–1111.

Mansmann, M., Winter, G., Pampus, G., Schon, N., March 27, 1973. Process for the Production of Carbon Fibers. US3723609A.

Meikleham, N.E., Pizzi, A., 1994. Acid- and alkali-catalyzed tannin-based rigid foams. J. Appl. Polym. Sci. 2 (11), 1547–1556.

Meng, J., et al., Apr. 2019. Preparation of tannin-immobilized gelatin/PVA nanofiber band for extraction of uranium (VI) from simulated seawater. Ecotoxicol. Environ. Saf. 170, 9–17.

Merle, J., Birot, M., Deleuze, H., Mitterer, C., Carré, H., El Bouhtoury, F.C., 2016. New biobased foams from wood byproducts. Mater. Des. 91, 186–192.

de Hoyos-Martínez, P.L., Merle, J., Labidi, J., Charrier - El Bouhtoury, F., 2019. Tannins extraction: a key point for their valorization and cleaner production. J. Clean. Prod. 206, 1138–1155.

Morisada, S., Rin, T., Ogata, T., Kim, Y.H., Nakano, Y., 2011. Adsorption removal of boron in aqueous solutions by amine-modified tannin gel. Water Res. 45 (13), 4028–4034.

Moubarik, A., Allal, A., Pizzi, A., Charrier, F., Charrier, B., 2009. Characterization of a formaldehyde-free cornstarch-tannin wood adhesive for interior plywood. Holz als Roh-Werkst. 68 (4), 427–433.

Muller, L.C., Marx, S., Vosloo, H.C., Fosso-Kankeu, E., Chiyanzu, I., Nov. 2018. Rigid polyurethane foams from unrefined crude glycerol and technical lignins. Polym. Renew. Resour. 9 (3–4), 111–132.

Navarrete, P., Pizzi, A., Tapin-Lingua, S., Benjelloun-Mlayah, B., Pasch, H., Rode, K., Delmotte, L., Rigolet, S., 2012. Low formaldehyde emittingbiobased wood adhesives manufactured from mixtures of tannin and glyoxalated lignin. J. Adhes. Sci. Technol. 26, 1667–1684.

Navarrete, P., Pizzi, A., Bernjelloun-Mlayah, B., Pasch, H., Rode, K., Delmonte, L., Rigolet, S., 2012. Low formaldehyde emitting biobased wood adhesives manufactured from mixture of tannin and glyoxaled lignin. J. Adhes. Sci. Technol. 26 (10–11), 1667–1684.

Nilsson, T., Rowell, R., 2012. Historical wood-structure-structure and properties. J. Cult. Herit. 13 (3), S5–S9. https://doi.org/10.1016/j.culher.2012.03.016.

Nordström, Y., Norberg, I., Sjöholm, E., Drougge, R., 2013. A new softening agent for melt spinning of softwood kraft lignin. J. Appl. Polym. Sci. 129 (3), 1274–1279.

Norgren, M., Edlund, H., 2014. Lignin: recent advances and emerging applications. Curr. Opin. Colloid Interface Sci. 19, 409–416.

Otani, S., Fukuoka, Y., Igarashi, B., Sasaki, K., August 12, 1969. Method for Producing Carbonized Lignin Fiber. US3461082A.

Pandey, M.P., Kim, C.S., 2011. Lignin depolymerization and conversion: a review of thermochemical methods. Chem. Eng. Technol. 34 (1), 29–41.

Peña, C., Larrañaga, M., Gabilondo, N., Tejado, A., Echeverria, J.M., Mondragon, I., 2006. Synthesis and characterization of phenolic novolacs modified by chestnut and mimosa tannin extracts. J. Appl. Polym. Sci. 100, 4412–4419.

Peña, C., de la Caba, K., Eceiza, A., Ruseckaite, R., Mondragon, I., 2010. Enhancing water repellence and mechanical properties of gelatin films by tannin addition. Bioresour. Technol. 101 (17), 6836–6842.

Pizzi, A., 1978. The chemistry and development of tannin-based adhesives for exterior plywood. J. Polym. Sci. 22 (8), 2397–2399.

Pizzi, A., 2016. Wood products and green chemistry. Ann. For. Sci. 73 (1), 185–203.

Pizzi, A., 2016. Wood products and green chemistry. Ann. For. Sci. 73, 185–203.

Pizzi, A., Scharfetter, H., 1981. Adhesives and techniques open new possibilities for the wood processing industry. Holz als Roh- Werkst. 39, 85–89.

Podschun, J., Saake, B., Lehnen, R., 2015. Reactivity enhancement of organosolv lignin by phenolation for improved bio-based thermosets. Eur. Polym. J. 67, 1–11.

Poposcu, C.M., 2017. Wood as bio-based building material. In: Jones, D., Brischke, C. (Eds.), Performance of Bio-Based Building Materials. Elsevier WP, pp. 21–96. https://doi.org/10.1016/B978-0-08-100982-6.00002-1.

Push, J., Wohlmann, B., 2018. Carbon fibers. In: Inorganic and Composite Fibers, first ed. Elsevier, ISBN 9780081022283, pp. 31–51.

Qin, L., Ge, Y., Deng, B., Li, Z., 2017. Poly (ethylene imine) anchored lignin composite for heavy metals capturing in water. J. Taiwan Inst. Chem. Eng. 71, 84–90.

Ramires, E.C., Frollini, E., 2012. Tannin – phenolic resins: synthesis, characterization, and application as matrix in biobased composites reinforced with sisal fibers. Compos. Part B 43, 2851–2860.

Rosas, J.M., Ruiz-Rosas, R., Rodríguez-Mirasol, J., Cordero, T., 2017. Kinetic study of SO_2 removal over lignin-based activated carbon. Chem. Eng. J. 307, 707–721.

Saha, D., Van Bramer, S.E., Orkoulas, G., Ho, H.-C., Chen, J., Henley, D.K., 2017. CO_2 capture in lignin-derived and nitrogen-doped hierarchical porous carbons. Carbon 121, 257–266.

Saha, D., Mirando, N., Levchenko, A., 2018. Liquid and vapor phase adsorption of BTX in lignin derived activated carbon: equilibrium and kinetics study. J. Clean. Prod. 182, 372–378.

Sánchez-Martín, J., Beltrán-Heredia, J., Gibello-Pérez, P., 2011. Adsorbent biopolymers from tannin extracts for water treatment. Chem. Eng. J. 168 (3), 1241–1247.

Sánchez-Martín, J., Beltrán-Heredia, J., Delgado-Regaña, A., Rodríguez-González, M.A., Rubio-Alonso, F., 2013. Adsorbent tannin foams: new and complementary applications in wastewater treatment. Chem. Eng. J. 228, 575–582.

Sánchez-Martín, J., Beltrán-Heredia, J., Delgado-Regaña, A., Rodríguez-González, M.A., Rubio-Alonso, F., 2013. Optimization of tannin rigid foam as adsorbents for wastewater treatment. Ind. Crops Prod. 49, 507–514.

Šćiban, M.B., Klašnja, M.T., Antov, M.G., 2011. Study of the biosorption of different heavy metal ions onto Kraft lignin. Ecol. Eng. 37 (12), 2092–2095.

Smolarski, N., 2012. High-Value Opportunities for Lignin: Unlocking its Potential. Frost & Sullivan. https://www.greenmaterials.fr/wp-content/uploads/2013/01/High-value-Opportunities-for-Lignin-Unlocking-its-Potential-Market-Insights.pdf.

Sudo, K., Shimizu, K., September 6, 1994. Method for Manufacturing Lignin for Carbon Fiber Spinning. US5344921A.

Svinterikos, E., Zuburtikudis, I., 2017. Tailor-made electrospun nanofibers of biowaste lignin/recycled poly(ethylene terephthalate). J. Polym. Environ. 25 (2), 465–478.

Szczurek, A., Fierron, V., Pizzi, A., Stauber, M., Celzard, A., 2014. A new method for preparing tannin-based foams. Ind. Crop. Prod. 54, 40–53.

Tejado, A., Peña, C., Labidi, J., Echeverria, J.M., Mondragon, I., 2007. Physico-chemical characterization of lignins from different sources for use in phenol-formaldehyde resin synthesis. Bioresour. Technol. 98, 1655–1663.

Thunga, M., Chen, K., Grewell, D., Kessler, M.R., 2014. Bio-renewable precursor fibers from lignin/polylactide blends for conversion to carbon fibers. Carbon 68, 159–166.

Tondi, G., Pizzi, A., 2008. Tannin-based rigid foams: characterization and modification. Ind. Crops Prod. 29, 356–363.

Tondi, G., Oo, C.W., Pizzi, A., Trosa, A., Thevenon, M.F., 2009. Metal adsorption of tannin based rigid foams. Ind. Crops Prod. 29 (2–3), 336–340.

Tribot, A., Amer, G., Abdou Alio, M., de Baynast, H., Delattre, C., Pons, A., Mathias, J.D., Callois, J.M., Vial, C., Michaud, P., Dussap, C.G., 2019. Wood-lignin: supply, extraction processes and use as bio-based material. Eur. Polym. J. 112, 228–240.

Van den Bosch, S., Koelewijn, S.F., Renders, T., Van den Bossche, G., Vangeel, T., Schutyser, W., Sels, B.F., 2018. Catalytic strategies towards lignin-derived chemicals. Top. Curr. Chem. 376, 36. https://doi.org/10.1007/s41061-018-0214-3.

Vázquez, G., Antorrena, G., González, J., Mayor, J., 1995. Lignin-phenol-formaldehyde adhesives for exterior grade plywoods. Bioresour. Technol. 51 (2–3), 187–192.

Vázquez, G., González, J., Freire, S., Antorrena, G., 1997. Effect of chemical modification of lignin on the gluebond performance of lignin-phenolic resins. Bioresour. Technol. 60 (3), 191–198.

Wang, G., Ghen, H., 2014. Carnbohydrate elimination of alkaline-extracted lignin liquor by steam explosion and its methylolation for phenolic adhesive. Ind. Crops Prod. 53, 93–101.

Wang, S., Zhou, Z., Xiang, H., Chen, W., Yin, E., Chang, T., Zhu, M., 2016. Reinforcement of lignin-based carbon fibers with functionalized carbon nanotubes. Compos. Sci. Technol. 128, 116–122.

Winterowd, J.G., October 2, 2014. Liquid Kraft Lignin Compositions. US20140296429A1. Weyerhaeuser NR Company. https://patents.google.com/patent/US20140296429.

Xu, Q., Wang, Y., Jin, L., Wang, Y., Qin, M., 2017. Adsorption of Cu (II), Pb (II) and Cr (VI) from aqueous solutions using black wattle tannin-immobilized nanocellulose. J. Hazard Mater. 339, 91–99.

Yin, C.Y., 2010. Emerging usage of plant-based coagulants for water and wastewater treatment. Process Biochem. 45 (9), 1437–1444.

Younesi-Kordkheili, H., Kazemi-Najafi, S., Eshkiki, R.B., Pizzi, A., 2015. Improving urea formaldehyde resin properties by glyoxalated soda bagasse lignin. Eur. J. Wood Wood Prod. 73 (1), 77–85.

Zhao, M., Jing, J., Zhu, Y., Yang, X., Wang, X., Wang, Z., 2016. Preparation and performance of lignin–phenol–formaldehyde adhesives. Int. J. Adhesion Adhes. 64, 163–167.

Chapter 8

Bio-based packaging materials

Aleksandra Nesic[1,2], Catalina Castillo[2], Patricia Castaño[2], Gustavo Cabrera-Barjas[2], Jesus Serrano[3]

[1]*University of Belgrade, Vinca Institute for Nuclear Sciences, Belgrade, Serbia;* [2]*University of Concepcion, Technological Development Unit, Concepcion, Chile;* [3]*Cipa Chile - Advanced Polymer Research Center, Concepcion, Chile*

8.1 Introduction

Since its discovery, plastic materials have facilitated human life and have contributed to the economic development of countries. However, plastic materials commonly used in industry such as polypropylene, polyethylene, polyethylene terephthalate, polystyrene come from nonrenewable resources—petroleum. It is known that the main drawback of using those plastics is related to the long time it could take to biodegrade (>100 years). Once they reach their end-life, they are disposed of in sanitary landfills, or they could reach undesired places such as oceans. In fact, the world oceans' microplastic contamination is a hot topic these days. Microplastics are affecting hundreds of marine species, and they could potentially affect also humans. This is because they could enter into the food chain, becoming a source of toxic compounds. According to statistics, at least one-third of the plastic waste generated each year in oceans and landfills comes from fossil-derived plastic used as food packaging (e.g., beverages, juices, wraps, bags, etc.). For this reason, it is important to develop novel food packaging bio-based materials that could progressively replace the use of conventional plastics.

The bio-based materials are formed by a group of biopolymer materials that could be obtained from different sources such as biomass (polysaccharides, proteins, waxes), microorganisms (biopolyesters), or synthesized from chemical building blocks of natural origin (biopolyesters). The use of biomass to develop novel bio-based materials for food packaging application would reduce the dependence on fossil-based plastics. In addition, products generated with these materials could be biodegradable or compostable at the end of their lifecycle, which could enable larger options for their final disposal, regarding petroleum-based products. It would also allow reducing the generation of microplastic pollutants that end into the oceans. In addition, the production of bio-based packaging should use sustainable methods, which

Biobased Products and Industries. https://doi.org/10.1016/B978-0-12-818493-6.00008-7

together with the lifecycle of the product, should contribute less to greenhouse gases emission than fossil-based polymers. For all these reasons, in recent years the market for bio-based food packaging materials has been raised. However, achieving a total replacement of conventional plastic as food packing is not an easy task due to high competence with the cost, material performance, and versatility. Therefore, the most effective strategy has been the gradual replacement of them by environmentally friendly materials. The growing perception of consumers along with government policy interventions have been the driving force behind the increasing use of these materials.

In order to be considered as a real alternative for conventional plastics, the bio-based plastics need to fulfill some technical expectations. A careful choice of bio-products or their blends for each industrial application is relevant, because these materials must display similar or better mechanical, barrier, and processing properties than petroleum-based plastics. In some cases, they showed remarkable biological activity such as fungicide or bactericide that could be exploited in several applications like food preservation and fruit shelf-life extension. Moreover, inside this family of materials, both polar (water soluble) and nonpolar (water-insoluble) biopolymers could be found, which allows preparing an interesting combination of biopolymers−natural bioactive compounds (plant aqueous extracts, essential oils, polyphenolic compounds, antimicrobials, antioxidants) with food packaging properties. Considering all these facts, the goal of this chapter is to provide an overview of the state-of-the-art of food packaging bio-based materials both from technical and market perspective.

8.2 Processing of bio-based food packaging materials

8.2.1 Extrusion

Extrusion is a continuous process that often constitutes the first key step in plastics processing, due to its capacity for converting resins from solid to molten phase. This transformation technique can homogenously compound different plastic components, additives, and eventually reinforcements (when composite materials are desired), into pellets, as well as convert solids into extruded profiles. Nowadays, machines are able to process solid materials with different shapes and geometric forms, including pellets, powder, scraps and, eventually, components in liquid phase. The extrusion process and equipment are increasingly sophisticated, and new technologies have allowed for significant advances in the field of bioplastics packaging.

During extrusion process, the raw materials are subjected to combinations of high friction (shear stress), pressure, and heat. The components of the machine feed convey, melt, mix, pump, degasify (vent), and shape the introduced materials, through the sections of the extruder (usually called feed, transition and metering zones). In general, an extruder machine is composed of the hopper and a feeding throat to dispense a constant flow of raw materials, the screw(s) that propels the materials and generates friction to melt them, the barrel that contains the screw(s) and provides heat through electrical resistance, the breaker plate that controls the flow of plastic entering the extrusion head as well as supports screen packs, and the extrusion head and die that define the shape of the extruded product.

Extruding thermoplastic is possible due to the combined effect of the screws with their specific designs that propel the plastics and other constituent elements, as they promote the melting of the materials, the external heating systems that provide extra heat, and the processing conditions set by the operators. Finally, the homogeneous materials generated are forced to pass through a die that defines the profile of the product. Several shapes can be produced by extrusion processes, including strands, films, sheets, tubes, rods, hoses, among others. Therefore, extrusion is a versatile and highly controlled method for compounding and manufacturing thermoplastic-based materials (Coles et al., 2003).

It is possible to find various types of extruding machines in the market, according to classifications based on their specific characteristics. A typical way to classify the extruders is based on the number of screws: single screw extruders and multiple-screw extruders (with twin-screw extruders being the most important). The former is usually a simpler and cheaper machine, while the latter is more complex, having two screws turning side by side in parallel or opposite directions, increasing friction and shear stress applied to the compound to reach a better mixing quality (Billmeyer, 1984).

Typically, problems with regular synthetic packaging are related to material-food interaction, environmental negative impact, and economic impact (disposal of materials) (Ahmed, 2018). Considering these issues, bioplastic packaging is a potential and promising solution. In order to generate better-quality bioplastic packaging, as well as extruded products with high added value, research and development efforts have focused on producing, analyzing, and studying bio-based thermoplastics in the following fields (Ahmed, 2018):

- Active biopackaging: packaging materials that are capable of delivering or releasing active components in a controlled rate.
- Smart or intelligent packaging: products with engineered functionality according to active, informative, ergonomic, and responsive packaging to optimize the efficiency and enhance the protection of the packaged product.
- Renewable-based materials: ecofriendly materials that can be obtained from renewable green resources.

Thermoplastic starches (TPSs), poly(lactic acid) (PLA), polyhydroxyalkanoates (PHAs), and cellulose are examples of some of the most common renewable-resource-based bioplastic materials used in packaging industry.

Although extruding machines were originally designed to be used in traditional processing techniques for synthetic materials, it has been widely demonstrated that they can work also for biomaterials and biopolymers processing. Recent extrusion studies in packaging include blends of thermoplastic starch and low density polyethylene with 40%–80% concentration of TPS, achieving cocontinuous morphology blends (Mazerolles et al., 2019); residual microalgae biomass plasticized with glycerol, water, and urea by extrusion processes with a twin screw extruding machine, demonstrating that microalgae-starch based materials could be used for the development of new extruded bioplastics (Mathiot et al., 2019); and the determination of nonvolatile components and compounds of a biodegradable food packaging material based on polyester and PLA and its migration to food simulants (Aznar et al., 2019); among others.

There are processing techniques derived from extrusion processes used for biomaterials to generate final products with different geometries and performances. Some of these techniques are blown film extrusion, cast film extrusion, and thermoforming. Since the blown extrusion process is the most commonly used technology, only this process will be described in following section.

8.2.2 Blown film extrusion

Among the different methods for producing films, blown film extrusion is the most common in the industry. The majority of the products obtained by this method constitute commodity plastics with low profit margins, which is directly related to the modern technologies for this technique that are able to produce high outputs of extruded materials with high dimensional stability and consistent mechanical properties. This process is widely used to produce bags and films in general for diverse applications (retail, agricultural, food industry, and others).

In blown film extrusion, the molten thermoplastics are forced through a vertical tubular die, forming a thin bubble by air blown through the die to its interior. The hollow tubular film is vertically extruded along a tower of guide rolls and collapsed on the top by frame and nip rolls. The bubble must be stabilized by optimizing a series of processing variables (profiles of temperature, screws speeds, nip and windup rollers speeds, air volumes and pressures, cooling rates) and using the guide rolls and tents that limit the mobility and dimensions of the extruded materials. The combination of materials and machine characteristics with the processing conditions will determine the final dimensions and performance of the extruded products. Changes in any processing variable may affect dimensions and performance of the films; thus, the product and its quality could widely fluctuate if the process is not properly controlled. The complexity of the blown film extrusion demands certain melt viscosity resistance from the materials to be able to withstand the process, as well as relatively high skills from the operator to obtain films with the size and thickness required for the specific desired applications (Billmeyer, 1984; Cantor, 2011).

The shelf life and quality of some aliments may require a set of properties that cannot be achieved by a single material. Typical examples of this include certain seafoods or bone-in meat that is packed by vacuum-seal packaging. In these cases, coextrusion processes are used to produce a multilayer blown film that combines the characteristics and properties of two or more materials, so the final packaging is able to properly preserve the food. In this method, the different components are extruded in separated extruders that are connected together at one single die, so that one layer covers other before coming out of the die and forming the bubble.

Although the blown film extrusion method increases the complexity and cost of production (due mainly to higher equipment cost and more difficult processing), it can produce a high quality film with layers that provide a better mechanical and chemical behavior (higher strength, improved puncture resistance, specific chemical resistances, etc.), a reduced permeability of vapors, good printability, enhanced sealability, and other properties of interest to the packaging industry. It is important to consider that due to the different nature of the plastic materials, the use of tie layers is usually needed so that the final film will not be easily delaminated and each layer provides its specific properties to the final product. Considering all the advantages, and despite the extra difficulties, the coextrusion processes open the market of plastic packaging to new opportunities and improved products (Cantor, 2011).

However, when it comes to bioplastic materials packaging, several disadvantages have been found by researchers. Some examples are listed below:

- Thermoplastic starch and cellulose: problems include low water vapor barrier, poor processability, brittleness, and susceptibility to degradation (Cyras et al., 2009; Gáspár et al., 2005; Joshi, 2008; Liu, 2006; Müller et al., 2011; Shen et al., 2009; Yu et al., 2006).
- Poly(lactic acid): some problems to be found were brittleness, thermal instability, low melt strength, low heat sealability, low oxygen and water vapor barrier (Feijoo et al., 2005; Jamshidian et al., 2010; Mensitieri et al., 2011; Rhim et al., 2009).
- Polyhydroxyalkanoates: some drawbacks were high brittleness and stiffness, poor impact resistance, and thermal instability (Cyras et al., 2009; Liu, 2006; Modi, 2010; Yu et al., 2006).

In this regard, recent attempts to improve the materials' functionality and extend the shelf life of the food include studies of the effect of electrospun coatings made from polycaprolactone PLA and polyhydroxybutyrate on oxygen and water vapor barrier properties of corn starch–based films (Fabra et al., 2016); analysis of different cross-linking agents used for starch/PHA blown films in order to study their effect on compatibility and thermal stability (finding high suitability when using citric acid and adipic acid) (Sun et al., 2018); thermoplastic starch/chitosan-based blown films where chitosan was able to increase processability of the TPS film and its strength, reduce its water absorption and surface stickiness (Dang and Yoksan, 2015), and lower the hydrophilicity and water vapor/oxygen permeability (Dang and Yoksan, 2016), among others. The recent advances in blown film extruding process for bioplastic materials are promising. This tendency to improve the materials and products encourages researchers to continue deepening their work in packaging fields, thus developing modern and innovative solutions to food industry.

8.3 Classification of bio-based materials

Biopolymers represent a special group of materials derived from renewable sources. Biopolymers are produced mainly from plant raw materials, with less energy consumption and minimal negative impact on the environment. Generally, biopolymers can be divided into three groups (Reddy et al., 2013; Siracusa et al., 2008):

(a) Polymers extracted/isolated directly from the biomass (mainly from plants but can also be of animal origin). This group of biopolymers includes polysaccharides (mostly applied in food packaging are starch and cellulose) and whey proteins, casein, collagen, etc. These materials have good barrier properties toward gases, but have a pronounced hydrophilic character.

(b) Polymers produced by chemical synthesis from bio-based monomers—monomers obtained from renewable sources. Poly(lactic acid) (PLA) is one of the most commonly used biopolymers for food packaging.

(c) Polymers obtained directly from natural or genetically modified organisms. This group of biopolymers includes Polyhydroxyalkanoates (PHAs) and bacterial cellulose. Their properties depend on the properties of monomers from which they are built, which allow a wide range of different biopolymers that can be synthesized by microbiological fermentation. The most commonly used is a polyhydroxybutyrate derivative, PHB.

Biopolymers have beneficial property in comparison to the synthetic polymers, such as biodegradation, which means that there is at least one enzyme that accelerates the degradation of the chemical chain of the given biopolymer. In addition to biodegradability and biocompatibility properties, it is very important that these materials can satisfy very demanding properties of food packaging materials, such as good mechanical resistance, thermal stability and high barrier properties toward gases and water vapor. In the following sections will be described the physicochemical properties, advantages, disadvantages and commercial potential of the main biopolymers used for food packaging application.

8.4 Biopolymers from bio-derived monomers—PLA

The environmental, economic, and safety challenges have provoked packaging scientists and producers to partially substitute petrochemical-based polymers with biodegradable ones (Farah et al., 2016). The poly(lactic acid) (PLA) is a thermoplastic aliphatic polyester that is generally derived from agricultural products. It is considered as a biodegradable and compostable material and is a sustainable alternative to conventional plastic products derived from petroleum. NatureWorks LLC is the major producer of PLA, with a capacity of 150,000 metric ton/year in its US manufacturing facility (in Blair, Nebraska) (Vink and Davies, 2015). PLA has several attractive properties such as biocompatibility, high strength, stiffness, excellent organoleptic characteristics, and thermoplasticity; however, it does have low impact strength (Elsawy et al., 2017). The properties that characterize PLA make it suitable to use in diverse applications, either the industrial packaging field, biocompatible/bioabsorbable medical device market, textile and leather production, or chemical industries (Garlotta, 2001; Madhavan Nampoothiri et al., 2010). PLA has been classified as GRAS (Generally Recognized as Safe) by the US Food and Drug Administration and authorized by the European Commission to be used in contact with food (Abdul Khalil et al., 2018; Langer et al., 2016). The degradation, physical, mechanical, and thermal properties of PLA, among others, influences the use of material on an industrial scale. The main properties of the PLA are summarized below.

8.4.1 Molecular structure and its production

Poly(lactic acid) or poly(lactide), PLA, is a partially crystalline thermoplastic polyester. PLA can be obtained by fermentation (from corn or sugar beets) or by chemical synthesis. Namely, the synthesis of PLA starts with lactic acid yielded from the fermentation of carbohydrates such as starch and cellulose. A large proportion is derived from the crops corn and cassava. Microorganism-based fermentation: the chemical reaction of the formation of lactide (cyclic lactic acid diphtheria) is the intermediate in the synthesis of PLA which in its chain can have two different optical stereoisomeric forms: L (−) - lactide (S, S); D (+) - lactide (R, R), and optically inactive mesolactide (R, S). By chemical synthesis, PLA is obtained by ring opening polymerization (ROP) reactions to obtain high molecular weight polymers. The racemic mixture of L and D-lactide is called D-L lactide, and L- and D-L lactide are used to prepare the polymer. The polymerization of an optically pure polymer leads to the formation of a stereoregular poly (L-lactide), PLLA and poly (D, L-lactide), PDLLA. The properties of PLA vary depending on the relationship and distribution of two stereoisomers or comonomers (Fig. 8.1).

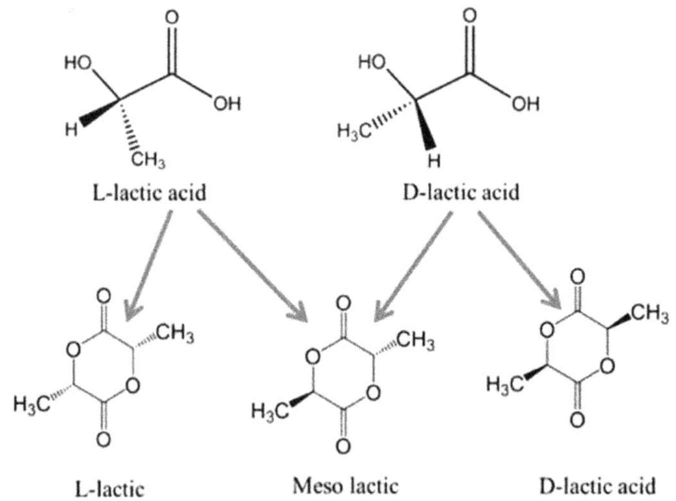

FIGURE 8.1 Steoreoforms of lactides (Madhavan Nampoothiri et al., 2010).

8.4.2 Physical-chemical properties of PLA

8.4.2.1 Mechanical properties

The tensile strength and elastic modulus of PLA are normally compared to PET, so its use in applications where is required a plastic deformation at higher stress levels (Rasal et al., 2010). The mechanical properties of PLA depend on crystallinity and optical purity between LA enantiomers in PLA chain. Namely, depending on the proportion of L- and D-enantiomers, PLA can crystallize in three forms—α the most stable with a melting point of 185°C, β with a slightly lower melting point of 175°C, and γ form (Jamshidian et al., 2010; Lim et al., 2008). The degree of crystallinity of PLA affects both the mechanical properties of the polymer, and the permeability of gases and thermal stability. PLA generally characterizes good mechanical properties in relation to other biopolymers, thus the values of the elastic modulus range between 3000 and 4000 MPa, and the tensile strength of 50–70 MPa. However, the values of elongation at break (ranging from 1% to 7%) greatly limit its widespread use (Auras et al. 2011; Raquez et al., 2013; Rasal et al., 2010). In Table 8.1 are presented the summarized mechanical properties of PLA, poly(L-lactide) (PLLA), and poly(D-lactide) (PDLLA) obtained according to ASTM standards.

8.4.2.2 Permeability

One of the most important properties of the material, when it comes to food packaging, is certainly the barrier properties, that is, the permeability to gases (O_2, N_2, CO_2, and water vapor) (Kerry, 2014; Peelman et al., 2014). Even

TABLE 8.1 Effects of stereochemistry and crystallinity on mechanical properties (ASTM D 638, 2003; ASTM D256, 2004; ASTM D790, 2017; ASTM D882-18, n.d.; Farah et al., 2016).

			Annealed	
Mechanical property	**ASTM method**	**PLA**	**PLLA**	**PDLLA**
Tensile strength (MPa)	D882	59	66	44
Elongation at break (%)	D882, D638	7	4	5.4
Elastic modulus (MPa)	D638	3750	4150	3900
Shear modulus (MPa)		1287		
Poisson's ratio (a.u)		0.36		
Yield strength (MPa)	D882, D638	70	70	53
Flexural strength (MPa)	D790	106	119	88
Unnotched izod impact (J/m)		195	350	150
Notched izod impact (J/m)	D256	26	66	18
Rockwell hardness (HR)		88	88	76
Heat deflection temp (°C)		55	61	50
Vicat penetration (°C)		59	165	52
Ultimate tensile strength (MPa)		73		
Percent of elongation (%)		11.3		
Young's modulus (MPa)	D882	1280		

small changes in relation to L and D enantiomers affect the barrier properties of PLA, primarily due to the amount of crystalline regions in the polymer matrix. In fact, diffusion of vapors gases occurs through the amorphous parts of the polymer, and the increase in the crystalline moiety results in a decrease in permeability (Auras et al., 2003). The permeability of gases and water vapor through PLA films has been widely investigated and reported in literature. It has been shown that permeability of CO_2 through PLA material is lower than through polystyrene, but higher than through poly(ethylene terephthalate). The same trend was obtained for permeability of oxygen through PLA films. Nevertheless, PLA materials showed beneficial oxygen barrier which was 20 times lower than for polystyrene. The barrier properties of PLA against water vapor, however, are not as good as those of PET and PS, although a decrease in the water permeability coefficient of PLA can be achieved by varying the film fabrication conditions (Colomines et al., 2010).

8.4.2.3 Thermal properties

The thermal properties of PLA are mainly obtained by differential scanning calorimetry (T_g, T_m, T_c), thermogravimetric analysis (degradation temperature), and dynamic mechanical analysis. The melting temperature (T_m) and the degree of crystallinity depend on the molar weight, thermal history of the material, and the purity of the polymer (López-De-Dicastillo et al., 2010). The crystallization process requires an optical purity between 72% and 75% (López-De-Dicastillo et al., 2010). PLA homopolymers prepared from optically pure L-LA or D-LA are semicrystalline polyesters with a T_m around 175°C and a glass transition temperature (T_g) of 55–60°C, whereas for a content in D-lactide below 93% the resulting PLA materials are amorphous with T_g value of 55–60°C and a fusion enthalpy between 40 and 50 J/g (Ni et al., 2009; Raquez et al., 2013). Properties of PLA depend on the component isomers, processing temperature, annealing time, and molecular weight. At atmospheric temperature, the PLA with high molecular weight may be amorphous or semicrystalline, depending on the concentration of L, D, and mesolactide in the structure. The concentration of L-lactide in PLA defines whether its structure is amorphous (50%–93%) or semicrystalline (>93%). In particular, PLLA has a crystallinity of around 37%, a glass transition temperature between 50 and 80°C and a melting temperature between 173 and 178°C. Because of the stereo regular chain microstructure, optically pure polylactides, PLLA and PDLLA, are semicrystalline. The crystallization ability of polylactides decreases with chain stereoregularity and below 43% optical purity crystallization is no longer possible (Madhavan Nampoothiri et al., 2010).

8.4.2.4 Other properties

In addition to all of these features, the PLA optical properties, such as color, transparency, and refractive index are of great importance as well (Auras et al., 2011). The refractive index, as a fundamental physical feature, is often used to identify or confirm the purity of the material. In polymeric materials this value depends on their structure, and although it would be expected that the PLA refraction index depends on the ratio of L- and D-lactides, it has been established that there is no statistical difference in refractive indexes at PLA with different L- and D-lactides (Hutchinson et al., 2006).

Taking into account the sensitivity of food products, the absorption and transmission of light in polymers is also of particular importance. Sensitive components of food products such as lipids, flavorings, vitamins, and pigments can be subjected to a degradation reaction if exposed to a light source. Spectrum, intensity of light source, exposure conditions, and percentage of light transmission of packaging materials can dramatically affect the quality of the packaged food. Thus, the packaging plays a key role in the prevention of photodegradation of some food components during storage. For example, with adequate selection of packaging materials, various reactions such as oxidation of grease and oils can be slowed down, the formation of sensory uncomfortable compounds (methional, aldehydes, and methyl ketones), loss of vitamins (riboflavin, β-carotene, vitamin C), degradation of free amino acids, as well as discoloration of pigments. Certainly in the selection of packaging material for certain foods, it is necessary to take into account the use of adequate additives and stabilizers in order to achieve better barrier properties for ultraviolet and visible (UV-VIS) radiation. PLA has a mild yellow color, and although it is transparent, the visible yellow color of the film can represent an obstacle in the end use of this material.

8.4.3 Biodegradation

Biodegradation is the conversion of organic polymers by fungi, algae, and bacteria in nature to carbon dioxide, methane, and other inorganic components (Moon et al., 2016). Degradation of PLA has been widely investigated in last two decades, particularly chemical hydrolysis and photodegradation. In general, PLA degradation takes place mainly through the scission of the ester bonds. In human body and animals, PLA degrade through hydrolysis process, where it comes to formation of soluble oligomers that are further metabolized by cells. On the other side, when PLA is disposed in nature (soil, compost, waters), hydrolysis leads to formation of low molecular weight oligomers, and biodegradation process is continued to the mineralization phase into CO_2 and H_2O by microorganisms presented in environment. It has been shown that PLA mostly degrade in presence of microorganisms that belong to the family of Pseudonocardiaceae (*Amycolatopsis*, *Lentzea*, *Kibdelosporangium*, *Streptoalloteichus,* and *Saccharothrix*) (Qi et al., 2017; Tokiwa and Calabia, 2006). Although PLA is completely biodegradable and has fast rate of biodegradation in comparison to the synthetic polymers, complete biodegradation can take place in environment for several years. For example, Ohkita et al. did not obtain any biodegradation of PLA sheets exposed in soil for 6 weeks (Ohkita and Lee, 2006), whereas several authors confirmed that biodegradation of PLA in soil or compost after 90 days of exposure is less than 5% (Akrami et al., 2016; Castillo et al., 2019; Vroman and Tighzert, 2009). It has been shown that degree of crystallinity, molecular weight, and stereoisomeric content of PLA influence the rate of biodegradation. Namely, PLA with higher degree of crystallinity and/or molecular weight have slower rate of biodegradation (Tokiwa and Calabia, 2006). In addition, higher rate of D-lactide in comparison to the L-lactide delays biodegradation process (Li and McCarthy, 1999). In order to accelerate biodegradation of PLA-materials, it has been combined with other biopolymers that have faster biodegradation rate, such as thermoplastic starch and cellulose (del Rosario Salazar-Sánchez et al., 2019; Luzi et al., 2015).

8.4.4 Ongoing research

Although there is a growing use of PLA in many industries, its use in food packaging is nevertheless limited to products with shorter shelf-life. In addition, its use at higher temperatures (or for foods that are packaged or used at higher temperatures) is not possible due to low T_g value. Moreover, moderate barrier properties limit the widespread use of PLA in food packaging. Restrictions on the use of PLA as packaging material can be overcome by the application of nanotechnology and the development of new hybrid materials (Raquez et al., 2013). For instance, it is found that different nanoclays such as Cloisite Na^+, Cloisite 20A, Cloisite 30B, and montmorillonite improve significantly water vapor barrier of PLA films (Molinaro et al., 2013; Rhim et al., 2009). Incorporation of metallic nanoparticles ZnO and TiO_2 provides higher mechanical, thermal, and barrier stability of PLA-based materials and adds one more beneficial function in terms of antimicrobial/antifungal activity toward foodborne pathogens (Pantani et al., 2013; Shankar et al., 2018; Xie and Hung, 2018). At the industrial level, there are extensive investigations of multilayer PLA films, usually containing silica and/or aluminum, in order to keep transparency of final product but increase the barrier properties (van den Oever et al., 2017).

In addition, PLA has been widely investigated as a matrix film for antimicrobial packaging, by incorporation of various natural bioactive components such as essential oil or other plant extracts (Agustin-Salazar et al., 2014; Arrieta et al., 2013; Kurek et al., 2017; Tawakkal et al., 2016). For example, Wang et al. showed that PLA films that are blended with thermoplastic starch and grapefruit seed extract possess high antimicrobial stability in in vitro studies, as well as in vivo studies by use of minced fish paste as a food preservation model (Wang and Rhim, 2016). Llana-Ruiz-Cabello et al. (2015) obtained prolonged shelf-life of ready-to-eat salads by use of PLA package that contains extract of *Allium* spp. Incorporation of cinnamaldehyde *Zataria multiflora* essential oil into PLA matrix provides longer shelf-life of vacuum-packed-cooked sausages (Rezaeigolestani et al., 2017), whereas *Origanum vulgare* L. essential oil influenced high antimicrobial activity of PLA packages toward *Staphylococcus aureus* and in in vivo studies with rainbow trout (Javidi et al., 2016).

8.4.5 Commercial products

The major supplier of PLA is the USA company, Nature Works LLC, with a production capacity of 100,000 tons/year. Beside Nature Works LLC, there are numerous big companies in Europe (Futerro, Belgium; Tate&Lyle, Netherlands), China (Hiusan Biosciences and Jiangsu Jiulding), and Japan that produce different grades of PLA suitable for automobile, medical device, and food packaging industrial sector. However, the average annual production of PLA in Europe, China, and Japan is less than 5000 tons.

One of the first companies that use PLA as a packaging material is French Danone. They use it to make lids for yogurts intended for the German market. Today PLA is used across Europe, America, and Japan, mainly for packaging of fresh foods with short shelf-life, such as fruits and vegetables. In addition, PLA thermoformed trays are available in the market, which are competitive with polystyrene trays for package of strawberries, mushrooms, or dairy products. On the other hand, due to poor water vapor barrier property of PLA films themself, these films/bags without additional layer or laminate are not able to be used for packaging of water-sensitive products that need to be stored for long-term periods such as cookies and breads. Moreover, PLA is not a suitable material for storage of gaseous and hot-fill drinks. In USA, PLA bottles are used only for storage of short-life fresh juices.

8.5 Biopolymers obtained from microorganisms—PHB, PHBV

Poly(hydroxyalkanoates) (PHAs) are a family of biodegradable aliphatic polyesters produced by microbial fermentation. Poly(3-hydroxybutyrate) (P(3HB) or PHB) and poly(3-hydroxybutyrate-co-3-hydroxyvalerate) (PHBV) are biopolymers belonging to PHAs. PHB was the first PHA to be identified in 1926 by Maurice Lemoigne in the bacterium *Bacillus megaterium* and to this day is the most widely studied PHA. On the other hand, PHBV is the most well-known copolymer derived from PHB (Anjum et al., 2016). Hereby, in this section special attention will be made to PHB and PHBV biopolymers.

8.5.1 Molecular structure

The general structural formula of PHAs is composed of hydroxyalkanoic acid as monomer units connected by an ester bond, which in turn is formed by connecting the carboxylic group of a monomer with the hydroxyl group of the adjacent one. In the general structural formula of PHAs, the number of repeating units "n" vary from 600 to 35,000.

Each PHA monomeric unit possesses a side chain (-R) group, which can differ from a hydrogen atom to methyl, ethyl, propyl, pentyl, nonyl, hexyl, or even to aromatic, branched, halogenated, epoxidized, and substituted alkyl groups. Both the side chain composition (-R) and "n" determine the identity of the monomeric unit. PHB is a homopolymer with a methyl group as side chain ($-CH_3$), while PHBV is a copolymer of PHB and polyhydroxyvalerate (PHV), with the last one having an ethyl group in the side chain ($-CH_2CH$) (Madison and Huisman, 1999). According to the number of carbon atoms in the monomers (hydroxyacids), PHAs are classified into three groups: short chain length (SCL) with 3–5 carbon atoms, medium chain length (MCL) with 6–14, and long chain length (LCL) with 15 or more than 15 carbon atoms (Singh et al., 2017). PHB and PHBV belong to the SCL group. Hybrid polymers comprising both SCL and MCL monomeric units; an example of this is the poly(3-hydroxybutyrate-co-3-hydroxyhexanoate) P(3HB-co-3HHx).

8.5.2 Sources and extraction

Nowadays, over 300 species of Gram-positive and Gram-negative bacteria are used to produce PHAs. Among the Gram-negative bacteria are those of the genus *Alcaligenes*, *Azotobacter*, *Bacillus,* and *Pseudomonas*, as well as Gram-positive are the *Rhodococcus*, *Nocardia,* and *Streptomyces* genus. In the sense of PHAs production on a large scale, the most used bacterium is the *Cupriavidus necator* (Gram-negative) to synthesize PHB/PHBV, which was previously mentioned in literature as *Ralstonia eutropha*, *Alcaligenes eutrophus*, or *Wautersia eutropha*. Also, *Bacillus sphaericus*, *Alcaligenes latus,* and recombinant *Escherichia coli* have been successfully used for getting PHB. Mutant *Burkholderia sacchari* (Mendonca et al., 2017) is a promising bacterium to produce P(3HB-co-3HHx), a biopolymer with similar characteristics to low density polyethylene (LLDPE), while for the MCL-PHAs production *Pseudomonas putida* are studied (Mozejko-Ciesielska and Kiewisz, 2016). In a general way, for obtaining PHAs, a bacteria culture is provided with a carbon feed source and nutrients, such as nitrogen, phosphorous, and oxygen, which will allow the growth of bacteria. Once the number of microorganisms is optimal, nutrients are reduced to put bacteria under stress conditions. Thus, bacteria begin to convert the extracellular carbon source in the form of polymeric granular inclusions of around 0.2 mm in diameter within their cell cytoplasm as a reserve energy source (Piergiovanni and Limbo, 2016). After this, the polymer mass reaches to a maximum level of cell dry weight percentage (% dcw), in which it is extracted by solvent extraction using sodium, hypochlorite, chloroform, etc. However, there exists another way to produce PHAs in which nutrients are not limited, the fermentation is made in two-stage, and the extraction is via enzymatic action (Varsha and Savitha, 2011). So far, different substrates (carbon sources) for these bacteria have been investigated. For example, pineapple and sugarcane have been used to obtain P(3HB) from *Bacillus* sp. *SV13* (Suwannasin et al., 2015), glucose and glycerol to obtain P(3HB) from *C. necator DSM 545* (Rodríguez-Contreras et al., 2015), or also wheat straw lignocellulosic hydrolysates (WSHs) to obtain P(3HB) from *B. sacchari DSM 17165* in (Suwannasin et al., 2015). Usually mix of more substrates allows getting copolymers such as PHBV.

The productivity of PHAs from bacteria is found to be expensive on a large scale. Some of the alternatives to decrease productivity costs are: using different fermentation forms as the fed-batch method (García et al., 2014; Shantini et al., 2015), implementing metabolic inhibitors to stimulate intracellular accumulation, suppressing and overexpressing of specific biosynthetic pathways by genetic engineering, producing PHAs extracellularly by genetically modified, developing transgenic crops to express PHA synthesis routes from bacteria, etc. However, economic analysis for PHAs production indicated that substrates constitute approx. 40%–50% of the overall production costs (Obruca et al., 2010). For that reason, researchers are evaluating PHAs biosynthesis from renewable feedstocks such as food and starch-based wastes, plant oils, organic industrial wastewater, and biodiesel industry residuals (Chee et al., 2010). It is also taking into account that producing strain must accumulate at least 60% of its cellular mass in polymer. As examples of some recent researches about it (Bhati and Mallick, 2015) used a combination of CO_2 and poultry litter as cultivation condition for *Nostoc muscorum Agardh Cyanobacteria*, getting from these 0.77 g/L (65% dcw) PHBV, during 7 incubation days. In the (García et al., 2013) work, obtained up 10.9 g/L (55.6 %dcw) PHBV from *C.necator*, using rapeseed meal hydrolysates, a byproduct from biodiesel industry as source, and a fed-batch fermentation method. Recently, PHAs obtained from cyanobacteria are considered, being that it is possible under photoautotrophic and chemoheterotrophic conditions. Despite this, productivity remains low at 0.016–0.0133 g/L h (Singh et al., 2017).

8.5.3 Physical-chemical properties

8.5.3.1 Molecular weight (M_w)

In general, the molecular weight of PHAs are in the typical range of 200,000–300,000 g/mol. This property will depend on the type of microbial species and its growth conditions such as pH, fermentation (batch, fed-batch, continuous), type, and concentration of the carbon source and extraction method. In the case of a homopolymer such as the PHB, increasing the molecular weight makes it possible to improve its physical properties. For a copolymer such as the PHBV, the strategy to change its properties is based on the variation of the mole percentage of 3-hydroxyvalerate (3HV mol%) monomer, which may range from 0 to 30 mol% (Ciesielski et al., 2015). The M_w analyses indicate that PHBV films have the highest M_w, followed by PHB, PLA, and PHBHB (Jost, 2018). Information about M_w is important to choose the appropriate processing method, for example, blow molding or injection molding are suitable for processing PHAs with low M_w, while thermoforming is suitable for PHAs with higher M_w (Bugnicourt et al., 2014). For miscibility studies, M_w is also important, for example, in PLA-PHB blending (Lai et al., 2017).

8.5.3.2 Mechanical properties

Mechanical properties as stiffness and flexibility are mainly affected by the biopolymer crystallinity degree. For example, SCL biopolymers as the P(3HB) that possess a high crystallinity degree (60%–80%) are stiff and brittle with a high Young's modulus, among 1–4 GPa and low elongation at break, typically below 15% (Anjum et al., 2016). Added to this, PHB suffers recrystallization of the amorphous phase during storage at room temperature, and this gets worse elongation at break. This amorphous phase part is commonly named rigid amorphous fraction (RAF) and is around 20%–30% of the overall sample mass (Di Lorenzo et al., 2012). This is why it is recommended to stock the final product in PHB at low temperature around 5°C. An interesting result was obtained by Lai et al. (2017), in which PLA/PHB were melt blended in a proportion of 80/20, without plasticizers and obtained 227% of elongation at break, and a yield strength of 46.7 MPa. Besides crystallinity, monomer composition also defines properties, so when PHBV has a 25 mol% 3HV, its Young's Modulus is 0.7 GPa, similar to that of HDPE (high density polyethylene), thus 0.4–1.0 GPa. While with 3 mol% 3HV has a Young's Modulus of 2.9 GPa in the order of Nylon—6.6 with 2.8 GPa. Elongation at break of PHBV is also typical less of 15% (Jost, 2018). This shows that increasing the 3HV mole percentage in PHVB decreases stiffness in the biopolymer, and that PHA copolymers and blends are efficient alternatives to improve mechanical properties by reducing the brittleness of PHAs-based plastics.

8.5.3.3 Permeability properties

Water vapor permeability (WVP, $Q_{100\,H_2O}$) and oxygen permeability (OP, $Q_{100\,O_2}$) are some of the main parameters studied in food packaging applications. Low WVP and OP of the polymer package wall could be beneficial for conserving quality and shelf-life of the product contained in the package. In general, PHAs have high WVP versus other non-bio-based polymers that are used for food packaging as the LDPE, HDPE, PP, or PET. The OPs of PHAs are usually lower to that of LDPE, HDPE, and PP, but higher than that of PET. That is why PHAs as packaging materials are promising for oxygen sensitive products, especially in PHBV. It is important to note that PHAs have better barrier properties than polylactic acid (PLA), being that this is one of the most widespread biopolymers used today, Jost (2018) made measurements of OP and WVP to different PHBV, thus PHBV3 (3 HV content: 3 mol%), PHBV7 (3 HV content: 7 mol%), and PHBV 11 (3 HV content: 11 mol%); as a result of this, he found that the permeability of PHBV7 and PHBV11 increased compared to PHBV3 but there was no significant difference in the permeability of PHBV7 and PHBV11. So, increasing the 3HV content led to increased WVP and OP, and this behavior was attributed to the decreased crystallinity. Besides, in all cases, PHBV had always less WVP and OP than PHB and PLA.

8.5.3.4 Thermal properties

Glass and melting transition temperatures (T_g and T_m) are important parameters in service applications of PHAs. SCL PHAs as PHB have a relative high T_g (approx. 2–5°C), versus PP (−10°C) (Mozejko-Ciesielska and Kiewisz, 2016). This high T_g explains why the elongation and impact strength of pure PHB are relative low. So, plasticizers are commonly used in PHB blends to get low T_g and improve these properties. Examples of plasticizers are: oxy-propylated glycerin, glycerol, 4-nonylphenol, acetyl tributyl citrate, salicylic ester, soybean oil, dioctyl phthalate, fatty alcohols with or without glycerol, polyethylene glycol (PEG), etc. (Bugnicourt et al., 2014). Because of its bimodal distribution of crystallite size, PHB hast two melting points (T_m), the highest one is in the order of the 177°C and thermally decomposes at 180°C just above its T_m. This low resistance to thermal degradation is the main problem for the PHB processing on an industrial scale by extrusion and injection molding. However, this outcome can be solved by use of lubricants or antioxidants that prevent chain scission, allowing the processing between 170 and 180°C. Another industrial processing problem, mainly in injection molding for PHAs is its low crystallization temperature (T_c) (e.g., 79°C for PHB). It leads to slow crystallization rates, poor efficiency for product fabrication, and high energy consumption, especially during injection. For solving this, nucleating agents could increase T_c of PHAs and speed up crystallization.

In the case of PHBV, its T_m is affected by the monomer content of valerate (HV). Increasing the HV content in the PHBV decreases T_m of the copolymer from 175.4 to 168.5°C, at 20 mol% of HV. Referring to the T_c, PHBV exhibits increased T_c as compared to the homopolymer PHB. In the work of Zhu et al. (2012) was found that PHBVs with 20% mol of HV with two nucleating agents of the heptane dicarboxylic derivatives have the best combination between T_c, T_m, and decomposition temperature (T_{decomp}) for processing it by injection molding, versus PHB and PHBV with a ratio of HV between 5% and 32.6%. This is why nucleating agents improve T_{decomp} and T_c, but decrease T_m.

8.5.3.5 Other properties

Transparency is an important aspect of packaging materials. In the PHB case, it presents an amber tonality and lower transparency in the visible region of the spectra (400–700 nm) (Arrieta et al., 2017). Therefore, a way to reduce the PHB yellowish trend is by blending with another biopolymer of lower crystallinity or with its plasticization. A striking property of PHB is that it acts as a better light barrier in the UV light region (250–400 nm) than PLA (Khosravi-Darani and Bucci, 2015). This is important for the elaboration of PLA-PHB transparent films with enhanced properties for food preservation. Surface wettability (SW) is another property to have in account for food packaging. Materials with low SW values are expected to protect food products from moisture and humidity during storage. This parameter is determined through the static water contact angle (WCA) measured. PHB usually have a WCA value higher than 65 degrees (Jordá-Vilaplana et al., 2014). However PP or LDPE have a WCA value higher than 100 degrees, indicating higher hydrophobic nature than PHB. This means that PHB has a higher SW than PP or LPDE.

8.5.4 Biodegradation

One of the most important characteristics of the PHAs is their biodegradability, since they could be degraded by a big population of microorganisms such as bacteria or fungi and in different environments. PHB can be biodegraded upon exposure to soil, compost, or marine sediment and in aerobic or anaerobic environments. The way in which degradation occurs is through the secretion of enzymes by the microorganism that break down the polymer into its hydroxyacid units. Under aerobic conditions, PHA biodegradation results in carbon dioxide and water, whereas in anaerobic conditions the degradation products are carbon dioxide and methane. Composting at high temperature for recycling PHB requires thermophilic microorganisms, example of these are the bacteria: *Streptomyces* sp. isolated from soil, *Actinomadura, Microbispora, Thermoactinomyces,* and *Saccharomonospora.* Other microorganisms are *Comamonas testosterone* and *Pseudomonas stutzeri* isolated from seawater and lake water (Bugnicourt et al., 2014).

8.5.5 Ongoing research deals

Various blends of PHAs have been developed to reduce the prices, enhance performance, or improve processing on an industrial scale of PHAs. Blends of PLA-PHB are among the most studied. PHB or PHBV have as foremost drawback a poor processability and formability, because of its low thermal stability that limits its wide industrial use. On the other hand, it is known that PLA limitations for food packaging are the sensitiveness to thermal degradation, poor barrier and mechanical performance. A solution to this is the PLA-PHB melt blending and the evaluation of plasticizers (acetyl tri-n-butyl citrate (ATBC), PEG, D-limonene, carvacrol, catechin, etc.), nanofillers (nanocellulose/OMMT), or nucleating agents (talc) in this system. The miscibility of both polymers is influenced by the molecular weight, processing temperature, and PLA:PHB proportion. It exits a consensus that PHB with a low M_w (approx. 9000 g/mol) has a good miscibility with PLA in melt blending (Lai et al., 2017). With respect to PLA:PHB proportion, different studies have proved that the optimal is 75:25 because of some trans-esterification reactions between PLA and PHB chains that take place during melting (Abdelwahab et al., 2012; Arrieta et al., 2015; Zhang and Thomas, 2011). Using this proportion and with addition of ATBC plasticizers in 15% has improved elongation at break reaching a value of 180% (Arrieta et al., 2014a,b). The presence of PHB increases the crystallinity of PHB-PLA blend, resulting in improved mechanical resistance and barrier performance (Zhang and Thomas, 2011). Thermal degradation of PLA-PHB blends occurs in a two-stage process, in the first at the lowest temperature, PHB degrades, while in the second PLA degrades. The benefit of this is that the degradation temperatures in PLA-PHB blends are higher than that of the neat PHB, while the resultant reduced PLA thermal stability is observed on the second degradation stage and with a minor decrement (Lai et al., 2017).

Other studies to improve the main downsides of PHAs are oriented to the polymeric nanocomposite elaboration, which are hybrid composites of nanomaterials dispersed into the polymer matrix. Nanomaterials already incorporated in PHB or PHBV are: organo modified nanoclay (OMMT, approx. 3%wt.), nanocellulose, and metal nanoparticles. The interaction between OMMT with PHB/PHBV results in the exfoliation or intercalation of nanoclay into the PHAs matrix, and usually improves PHB/PHBV processing, because it acts as a nucleating agent enhancing the matrix crystallization (Jandas et al., 2014; Puglia et al., 2013). OMMT could also improve barrier properties of PHAs or grant it antimicrobial properties. Natural fibers such as wood fibers (cellulose) to reinforce PHB/PHBV are of interest because of their relative cheapness and good relation between strength and weight. It could compete with conventional reinforcements as glass and aramid fibers. Metal nanoparticles of zinc, copper, silver, magnesium, gold, titanium, and platinum exhibit antimicrobial activity, so are mainly used for inhibiting or eliminating food pathogen microorganisms using antimicrobial packaging material (Abreu et al., 2015; Mlalila et al., 2017).

Moreover, bioplastics such as PHB/PHBV are recommended for applications in antimicrobial packaging because of their compatibility with a variety of antimicrobial agents and compounds (Nicosia et al., 2015). Some of the actual studied agents are mentioned below. Vanillin (4 hydroxy-3-methoxybenzaldehyde) from 10 to 200 μg/g has been incorporated in PHB to comprise *E.coli*, *S. flexneri*, *S. typhimurium*, *S. aureus*, and fungal growth (Xavier et al., 2015). Eugenol is also an antimicrobial agent compatible with PHB, but its effect is improved when is mixed with pediocin (Narayanan et al., 2013). Zinc nanoparticles (ZnNPs) have a strong antimicrobial effect against *E.coli* and *S. aureus* when is incorporated in PHB (Díez-pascual and Díez-vicente, 2014). On an industrial scale, processing antimicrobial materials could cause inactivation or destruction of the antimicrobial agents because of the high temperatures and mechanical efforts. Carvacrol is an antimicrobial agent that acts also as a plasticizer, and it has been successfully processed with PLA-PHB films by extrusion at 200°C, 6 min (Armentano et al., 2015). Also successfully, catechin could be incorporated to a PLA-PHB blend by melt blending (Arrieta et al., 2014a,b).

8.5.6 Commercial products

In the United States, in 1962, J. N. Baptist filed the first patent about PHAs production; however, the first industrial PHAs production occurred in the 80s. In 1982, the actually dissolved English company ICI marketed P(3HB) under the trade name Biopol. After, Biopol was managed by Monsanto (Zeneca Bioproducts, successor of ICI) and in 2001 it was sold to American Metabolix (USA), which worked on the production of P(3HB) and PHBV in genetically modified plants. In 2004, American Procter & Gamble (P&G) (USA) produced the copolymer poly(3-hydroxybutyrate-co-3-hydroxyhexanoate) namely Nodax, which was licensed to Kaneka Co.; and this Japanese company produced it on an industrial scale. Nowadays, PHAs industrial production is made by many companies, such as Bio on (Italy), PHB Industrial (Brazil), Tianan Biological Material Polyone (China), Tianjin Green Biomaterials (China), Mitshubishi Gas (Japan), Metabolix (USA), Procter & Gamble (USA), Goodfellow Cambridge Ltd. (UK), Tepha Inc (USA), Kaneka Corporation (Japan), Co., Ltd. Ningbo (China), ETH (Switzerland), Jiang Su Nan Tian (China), etc. (Albuquerque and Malafaia, 2018).

It is important to highlight that the commercial manufacturing processes of PHAs are still expensive. Obruca et al. (2010) calculated the theoretical price of PHAs produced in fed-batch mode and using waste materials in 3.51 Eur/kg, whereas synthetic alternatives like polypropylene and polyethylene cost 1.47 and 1.15 Eur/kg, respectively. In addition, increasing oil prices, shortage of the reserves, and growing environmental awareness may strengthen the PHA market. It is estimated that demand for PHAs will grow tenfold by 2020 (Aeschelmann and Carus, 2015). That is why producers consider that PHA market is still promising and still needs time to fully develop. Due to high price production, PHA packages are not found commonly in the market, however PHA-based foam cups are used in USA for short-term storage of food.

8.6 Biopolymers from agro-resources—starch

Starch, in addition to cellulose, is the most prevalent carbohydrate in nature and as such is one of the basic nutrients and sources of energy. Starch is found in all green plants. By photosynthesis, from carbon dioxide and water, is created glucose that produces starch by enzymatically catalyzed polymerization. The main plant sources of production are potato, corn, and rice. In all plants, the starch is produced in the form of granules, of different sizes and of somewhat different composition, depending on the plant. Starch is composed of two basic polysaccharides, amylose and amylopectin (Fig. 8.2). Amylose is a predominantly linear homopolymer of D-glucose in which the

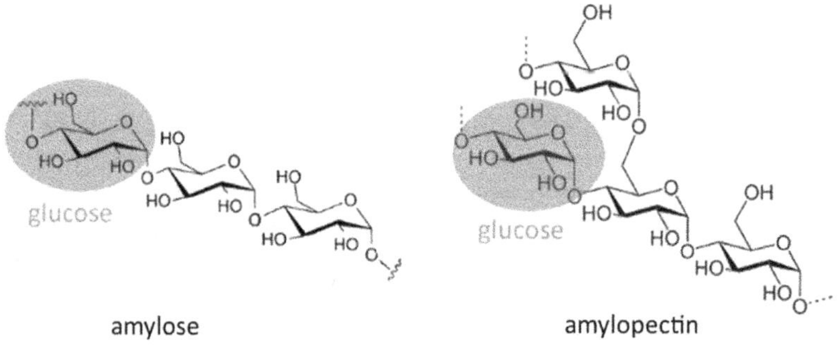

FIGURE 8.2 Molecular structure of amylose and amylopectin.

anhydroglucose units (AGU), in a pyranose form, are interconnected by α-1,4-glycosidic bonds. In each amylose molecule, there is one irregular and one reduction group. The molecular weight of amylose ranges from 10^5 to 10^6 g/mol. Amylopectin is a much larger molecule with branched structure and a pronounced (1 → 4) -α- (95%) and (1 → 6) -α-connections (5%). Generally, native starches contain about 70%–80% of amylopectin and 20%–30% of amylose (Jane, 2009).

There are three types of crystallinity in starch. Type "A" is found mainly in starch cereals, such as corn, wheat, and rice. Type "B" is a starch of the tuber (potato, sago); and type "C," the crystallinity found between crystalline "A" and "B" type, and is found in starch legumes and starch of other roots. Another type of crystallinity is the V-type, which is characteristic of the complex of amylose with fatty acids and monoglyceride.

The amylose and amylopectin molecules are organized in the form of semicrystalline starch granules. Although amylose and amylopectin can form the crystal structure, the crystal structure present in the native starch is mainly the result of double helices formed from lateral chains of amylopectin. Semicrystalline growing rings are composed alternately of amorphous and crystalline lamellas. Side chains of amylopectin form double helices, which are organized into clusters. Crystal lamella contains double helicase, while the amorphous lamella contains amylopectin branching sites that make them bind to double helices. The size of one crystal and one amorphous lamella is about 9–10 nm (Blazek and Copeland, 2008; Thompson, 2000).

8.6.1 Thermoplastic starch

One of the most famous modification of the natural starch is plasticization with the addition of plasticizers (water, glycerol, sorbitol, etc.) under high temperature and pressure, where thermoplastic starch (TPS) is obtained. Plasticization of starch causes molecular reorganization of chains and complete destruction of crystalline structure; thus it is obtained as a completely amorphous polymer.

At the industrial level, thermoplastic starch is produced by mixing of native starch with plasticizers in an extruder at specifically applied temperature (usually range of 90–180°C) and shear. Plasticizers have a role to interrupt inter/intramolecular hydrogen bonding in starch chains and increase chains mobility, which as a result lower the melting temperature of starch, that is, the processing temperature. During this process, first comes the fragmentation of starch granules, which is followed by splitting the hydrogen bond between the starch molecules, and finally resulting in a decrease in crystallinity and partial depolymerization of starch polymers. Mixing the compounds in extrusion allows complete homogenization of starch and plasticizers. The most commonly used plasticizers are polyglycols (glycerol, sorbitol), amines, amides, and water. Generally, water and glycerol have shown to be the most effective plasticizers, providing good mechanical properties of final TPS materials. However, the main drawback of TPS is its retrogradation—a phenomenon where natural increase in crystallinity over time is occurring. The retrogradation depends strongly on environmental conditions such as air humidity, and temperature, which can influence migration of plasticizers, and consequently change mechanical properties of materials during storage time, that is, increase the brittleness. Because of that, various new amine- and amide-based plasticizers have been investigated, in order to prevent retrogradation of TPS (Pushpadass et al., 2008; Zhang and Han, 2006).

8.6.2 Physical-chemical properties

The mechanical, barrier, and thermal properties of starch may vary greatly depending on type and concentration of plasticizers. Moreover, the tensile strength of TPS depends on the original sources of starch, glass-transition temperature, crystallinity, and ratio of amylose to amylopectin. In fact, it has been shown that higher content of amylose in starch gives materials with higher tensile strength (Lourdin et al., 1995). Generally, TPS materials have poor mechanical properties in comparison to other biopolymers and synthetic-based packages; thus, they are not applicable as single-layer material in food packaging. The tensile strength of TPS can vary between 1 and 4 MPa, elongation at break from 20% to 100%, and Young Modulus from 5 to 1000 MPa (Zhang et al., 2014). The Tg value can vary between −50 and 60°C, depending on the type of plasticizer. Due to high hydrophilic nature, TPS has very poor water vapor barrier properties, in the range of 1–30 g mm/m^2 h kPa. Generally, increased concentration of plasticizer promotes diffusion of water vapor through TPS material. Similar to water vapor permeability, oxygen permeability also increases with an increase of plasticizer content. However, TPS shows high oxygen barrier at low hydration conditions, estimated approximately in range of 1×10^{-2}–4×10^{-5} g mm/m^2 h kPa, which is for several order higher barrier than for LDPE (7.8×10^{-8} g mm/m^2 h kPa) and HDPE (1.8×10^{-8} g mm/m^2 h kPa) (Zhang et al., 2014).

8.6.3 Ongoing research

Generally, low water resistance and change in mechanical properties under humid conditions affect the applicative potential of thermoplastic starch. Due to high moisture absorption, the TPS material properties change according to the relative humidity of the air. The strong hydrophilic character, poor mechanical properties compared with conventional synthetic polymers, and retrogradation are the highest drawbacks of TPS that make it unsatisfactory for application as single layer in food packaging industry. There are a number of ways to mitigate these shortcomings. One of the solutions is starch modification through esterification reactions, which increases thermoplastic properties and thermal stability (Bergel et al., 2018b; Li et al., 2016). Another way of improving the functional properties of the starch films is starch blending with other biopolymers such as PHB, PCL, PLA, or processing of multilayer food packaging films (Bergel et al., 2018a; Garrido-Miranda et al., 2018; Mahieu et al., 2015).

8.6.4 Commercial products

Starch-based plastic found in market exists mainly in a form of thermoplastic starch/thermoplastic polyester blends. Generally, thermoplastic starch is blended with PLA, PHA, PBAT, PCL, or PBS, in order to decrease moisture sensitivity and to improve mechanical properties of final package. Moreover, the presence of TPS in the biopackage accelerates the biodegradation rate of thermoplastic polyesters. The percent of TPS in these packages is lower than 50%. Novamont (Italy) is the leading company in processing of starch-based products with annual production of 120 kt. Beside Novamont, other most important producers of starch-based products are Japan Corn Starch (Japan), Biotec (Germany), Rodenberg (Netherland), BIOP (Germany), Plantic (Australia), Wuhan Huali Environment Protection Sci. & Tech (China), and PSM (USA), that produce on annual level less than 15 kt. All of these companies produce various types of starch-based products using proprietary blend formulations. TPS/PBAT blends developed are used as biodegradable bags for potatoes and carrots, whereas TPS/PLA blends are used as candy multilayer wrapping of Mars. Moreover, TPS is used as laminate layer in cellophane for packaging of cereals and cookies. Paperfoam is one more starch-based product on market that represents thermoformed TPS/PLA package, commonly used as egg boxes.

8.7 Conclusions

Bio-based polymers demonstrate that they can be considered as a new class of beneficial biodegradable materials and meet the required features for sustainable technology in the framework of food packaging. The most investigated and exploited biopolymers intended for food packaging are PLA, PHAs, and TPS, showing the best performances in forms of their blends, in terms of improved biodegradation, mechanical and barrier stability. Moreover, these biopackages already found their route to the market for packaging of short shelf-life food products. However, higher price and lower water vapor barrier properties in comparison to synthetic packages still limits the widespread use of bio-based packages on market and their extended use in packaging the long shelf-life food products. Hence, further modifications and development by introduction of new bio-based plasticizers, cross-linkers, and additives is required, in order to completely meet food packaging criteria and promote sustainable future with lower environmental impact.

Acknowledgment

The authors gratefully acknowledge the project CONICYT PIA/CCTE AFB170007 for the financial support.

References

Abdelwahab, M.A., Flynn, A., Chiou, B., Imam, S., Orts, W., Chiellini, E., 2012. Thermal, mechanical and morphological characterization of plasticized PLA e PHB blends. Polym. Degrad. Stabil. 97, 1822–1828.

Abdul Khalil, H.P.S., Banerjee, A., Saurabh, C.K., Tye, Y.Y., Suriani, A.B., Mohamed, A., et al., 2018. Biodegradable films for fruits and vegetables packaging application: preparation and properties. Food Eng. Rev. 10 (3), 139–153. https://doi.org/10.1007/s12393-018-9180-3.

Abreu, A.S., Oliveira Rui, M., Miguel Rodrigues, M., Cerqueira António, A., Vicente, A., Machado, V., 2015. Antimicrobial nanostructured starch based films for packaging. Carbohydr. Polym. 129, 127–134. https://doi.org/10.1016/j.carbpol.2015.04.021.

Aeschelmann, F., Carus, M., 2015. Bio-based building blocks and polymers in the world. In: Capacities, Production and Applications : Status Quo and Trends towards 2020.

Agustin-Salazar, S., Gamez-Meza, N., Medina-Juàrez, L.À., Soto-Valdez, H., Cerruti, P., 2014. From nutraceutics to materials: effect of resveratrol on the stability of polylactide. ACS Sustain. Chem. Eng. 2 (6), 1534–1542. https://doi.org/10.1021/sc5002337.

Ahmed, S. (Ed.), 2018. Bio-Based Materials for Food Packaging. Springer, Singapore. https://doi.org/10.1007/978-981-13-1909-9.

Akrami, M., Ghasemi, I., Azizi, H., Karrabi, M., Seyedabadi, M., 2016. A new approach in compatibilization of the poly(lactic acid)/thermoplastic starch (PLA/TPS) blends. Carbohydr. Polym. 144, 254–262. https://doi.org/10.1016/j.carbpol.2016.02.035.

Albuquerque, P.B.S., Malafaia, C.B., 2018. Perspectives on the production, structural characteristics and potential applications of bioplastics derived from polyhydroxyalkanoates. Int. J. Biol. Macromol. 107, 615–625. https://doi.org/10.1016/j.ijbiomac.2017.09.026.

Anjum, A., Zuber, M., Zia, K.M., Noreen, A., Anjum, M.N., Tabasum, S., 2016. Microbial production of polyhydroxyalkanoates (PHAs) and its copolymers : a review of recent advancements. Int. J. Biol. Macromol. 89, 161–174. https://doi.org/10.1016/j.ijbiomac.2016.04.069.

Armentano, I., Fortunati, E., Burgos, N., Dominic, F., Luzi, F., Fiori, S., et al., 2015. Bio-based PLA_PHB plasticized blend films. Part I: processing and structural characterization. LWT Food Sci. Technol. 64 (2), 980–988. https://doi.org/10.1016/j.lwt.2015.06.032.

Arrieta, M.P., López, J., Ferrándiz, S., Peltzer, M.A., 2013. Characterization of PLA-limonene blends for food packaging applications. Polym. Test. 32 (4), 760–768. https://doi.org/10.1016/j.polymertesting.2013.03.016.

Arrieta, M.P., del Mar Castro-López, M., Rayón, E., Barral-Losada, L.F., López-Vilariño, J.M., González-Rodríguez, M.V., 2014a. Plasticized poly(lactic acid) – poly(hydroxybutyrate) (PLA – PHB) blends incorporated with catechin intended for active food- packaging applications. J. Agric. Food Chem. 62, 10170–10180.

Arrieta, M.P., Samper, M.D., López, J., Jiménez, A., 2014b. Combined effect of poly (hydroxybutyrate) and plasticizers on polylactic acid properties for film intended for food packaging. J. Polym. Environ. 22 (4), 460–470. https://doi.org/10.1007/s10924-014-0654-y.

Arrieta, M.P., López, J., López, D., Kenny, J.M., Peponi, L., 2015. Development of flexible materials based on plasticized electrospun PLA-PHB blends: structural, thermal, mechanical and disintegration properties. Eur. Polym. J. 73, 433–446. https://doi.org/10.1016/j.eurpolymj.2015.10.036

Arrieta, M.P., Samper, M.D., Aldas, M., López, J., 2017. On the use of PLA-PHB blends for sustainable food packaging applications. Materials 10, 1–26. https://doi.org/10.3390/ma10091008.

ASTM D256, 2004. Determining the Izod Pendulum Impact Resistance of Plastics. ASTM International, West Conshohocken, PA, 2018. https://doi.org/10.1520/D0256-10.

ASTM D790, 2017. Flexural Properties of Unreinforced and Reinforced Plastics and Electrical Insulating Materials. ASTM International, West Conshohocken, PA, 2018. https://doi.org/10.1520/D0790-03.

ASTM D882-18, n.d. Standard Test Method for Tensile Properties of Thin Plastic Sheeting. ASTM International, West Conshohocken, PA, 2018. https://doi.org/10.1520/D0882-09.

ASTM D 638, 2003. Standard Test Method for Tensile Properties of Plastics, vol. 8. ASTM International, West Conshohocken, PA, 2018. https://doi.org/10.1520/D0638-14.1.

Auras, R.A., Harte, B., Selke, S., Hernandez, R., 2003. Mechanical, physical, and barrier properties of poly(lactide) films. J. Plastic Film Sheeting 19 (2), 123–135. https://doi.org/10.1177/8756087903039702.

Auras, R.A., Lim, L.T., Selke, S.E., Tsuji, H. (Eds.), 2011. Poly(Lactic Acid): Synthesis, Structures, Properties, Processing, and Applications. John Wiley & Sons, Inc.

Aznar, M., Ubeda, S., Dreolin, N., Nerín, C., 2019. Determination of non-volatile components of a biodegradable food packaging material based on polyester and polylactic acid (PLA) and its migration to food simulants. J. Chromatogr. A 1583, 1–8. https://doi.org/https://doi.org/10.1016/j.chroma.2018.10.055.

Bergel, B.F., da Luz, L.M., Santana, R.M.C., 2018a. Effect of poly(lactic acid) coating on mechanical and physical properties of thermoplastic starch foams from potato starch. Prog. Org. Coating 118, 91–96. https://doi.org/10.1016/j.porgcoat.2018.01.029.

Bergel, B.F., Dias Osorio, S., da Luz, L.M., Santana, R.M.C., 2018b. Effects of hydrophobized starches on thermoplastic starch foams made from potato starch. Carbohydr. Polym. 200, 106–114. https://doi.org/10.1016/j.carbpol.2018.07.047.

Bhati, R., Mallick, N., 2015. Carbon dioxide and poultry waste utilization for production of polyhydroxyalkanoate biopolymers by Nostoc muscorum Agardh : a sustainable approach. J. Appl. Phycol. 28 (1), 161–168. https://doi.org/10.1007/s10811-015-0573-x.

Billmeyer, F.W., 1984. Textbook of Polymer Science, third ed. John Wiley & Sons, Inc., New York.

Blazek, J., Copeland, L., 2008. Pasting and swelling properties of wheat flour and starch in relation to amylose content. Carbohydr. Polym. 71 (3), 380–387. https://doi.org/10.1016/j.carbpol.2007.06.010.

Bugnicourt, E., Cinelli, P., Lazzeri, A., Alvarez, V., 2014. Polyhydroxyalkanoate (PHA): review of synthesis , characteristics , processing and potential applications in packaging. Express Polym. Lett. 8 (11), 791–808. https://doi.org/10.3144/expresspolymlett.2014.82.

Cantor, K., 2011. Blown Film Extrusion, second ed. Hanser Publications, Munich.

Castillo, C., Nesic, A., Urra, N., Maldonado, A., 2019. Influence of thermoplasticized starch on physical-chemical properties of new biodegradable carriers intended for forest industry. Int. J. Biol. Macromol. 122, 924–929. https://doi.org/10.1016/j.ijbiomac.2018.11.026.

Chee, J., Yoga, S., Lau, N., Ling, S., Abed, R.M.M., Sudesh, K., 2010. Bacterially produced polyhydroxyalkanoate (PHA): converting renewable resources into bioplastic bacterially produced polyhydroxyalkanoate (PHA): converting renewable resources into bioplastics. In: Méndez-Vilas, A. (Ed.), Current Research, Technology and Education Topics in Applied Microbiology and Microbial Biotechnology. Formatex Research Center, Spain, Badajoz, pp. 1394–1404.

Ciesielski, S., Mozejko, J., Pisutpaisal, N., 2015. Plant oils as promising substrates for polyhydroxyalkanoates production. J. Clean. Prod. 106, 408–421. https://doi.org/10.1016/j.jclepro.2014.09.040.

Coles, R., McDowell, D., Kirwan, M.J. (Eds.), 2003. Food Packaging Technology. Blackwell Publishing Ltd.

Colomines, G., Ducruet, V., Courgneau, C., Guinault, A., Domenek, S., 2010. Barrier properties of poly(lactic acid) and its morphological changes induced by aroma compound sorption. Polym. Int. n/a-n/a https://doi.org/10.1002/pi.2793.

Cyras, V.P., Soledad, C.M., Analía, V., 2009. Biocomposites based on renewable resource: acetylated and non acetylated cellulose cardboard coated with polyhydroxybutyrate. Polymer 50 (26), 6274–6280. https://doi.org/https://doi.org/10.1016/j.polymer.2009.10.065.

Dang, K.M., Yoksan, R., 2015. Development of thermoplastic starch blown film by incorporating plasticized chitosan. Carbohydr. Polym. 115, 575–581. https://doi.org/https://doi.org/10.1016/j.carbpol.2014.09.005.

Dang, K.M., Yoksan, R., 2016. Morphological characteristics and barrier properties of thermoplastic starch/chitosan blown film. Carbohydr. Polym. 150, 40–47. https://doi.org/https://doi.org/10.1016/j.carbpol.2016.04.113.

del Rosario Salazar-Sánchez, M., Campo-Erazo, S.D., Villada-Castillo, H.S., Solanilla-Duque, J.F., 2019. Structural changes of cassava starch and polylactic acid films submitted to biodegradation process. Int. J. Biol. Macromol. 129, 442–447. https://doi.org/10.1016/j.ijbiomac.2019.01.187.

Di Lorenzo, M.L., Gazzano, M., Righetti, M.C., 2012. The role of the rigid amorphous fraction on cold crystallization of poly(3-hydroxybutyrate). Macromolecules 45, 5684–5691.

Díez-pascual, A.M., Díez-vicente, A.L., 2014. Poly (3-hydroxybutyrate)/ZnO bionanocomposites with improved mechanical , barrier and antibacterial properties. Int. J. Mol. Sci. 15, 10950–10973. https://doi.org/10.3390/ijms150610950.

Elsawy, M.A., Kim, K.H., Park, J.W., Deep, A., 2017. Hydrolytic degradation of polylactic acid (PLA) and its composites. Renew. Sustain. Energy Rev. 79, 1346–1352. June 2016. https://doi.org/10.1016/j.rser.2017.05.143.

Fabra, M.J., López-Rubio, A., Cabedo, L., Lagaron, J.M., 2016. Tailoring barrier properties of thermoplastic corn starch-based films (TPCS) by means of a multilayer design. J. Colloid Interface Sci. 483, 84–92. https://doi.org/https://doi.org/10.1016/j.jcis.2016.08.021.

Farah, S., Anderson, D.G., Langer, R., 2016. Physical and mechanical properties of PLA, and their functions in widespread applications — a comprehensive review. Adv. Drug Deliv. Rev. 107, 367–392. https://doi.org/10.1016/j.addr.2016.06.012.

Feijoo, J., Cabedo, L., Pilar Villanueva, M., Lagaron, J.M., Gimémnez, E., 2005. Optimization of Biodegradable Nanocomposites Based on PLA/PCL Blends for Food Packaging Applications.

García, I.L., López, J.A., Dorado, M.P., Kopsahelis, N., Alexandri, M., Papanikolaou, S., et al., 2013. Bioresource technology evaluation of by-products from the biodiesel industry as fermentation feedstock for poly (3-hydroxybutyrate- co -3-hydroxyvalerate) production by *Cupriavidus necator*. Bioresour. Technol. 130, 16–22. https://doi.org/10.1016/j.biortech.2012.11.088.

García, A., Segura, D., Espín, G., Galindo, E., Castillo, T., Peña, C., 2014. High production of poly-B-hydroxybutyrate (PHB) by an Azotobacter vinelandii mutant altered in PHB regulation using a fed-batch fermentation process. Biochem. Eng. J. 82, 117–123. https://doi.org/10.1016/j.bej.2013.10.020.

Garlotta, D., 2001. A literature review of poly(lactic acid). Phys. Rev. Lett. 9 (2), 63–84. https://doi.org/10.1103/PhysRevLett.116.208301.

Garrido-Miranda, K.A., Rivas, B.L., Pérez -Rivera, M.A., Sanfuentes, E.A., Peña-Farfal, C., 2018. Antioxidant and antifungal effects of eugenol incorporated in bionanocomposites of poly(3-hydroxybutyrate)-thermoplastic starch. LWT 98, 260–267. https://doi.org/10.1016/j.lwt.2018.08.046.

Gáspár, M., Benkő, Z., Dogossy, G., Réczey, K., Czigány, T., 2005. Reducing water absorption in compostable starch-based plastics. Polym. Degrad. Stabil. 90 (3), 563–569. https://doi.org/https://doi.org/10.1016/j.polymdegradstab.2005.03.012.

Hutchinson, M.H., Dorgan, J.R., Knauss, D.M., Hait, S.B., 2006. Optical properties of polylactides. J. Polym. Environ. 14 (2), 119–124. https://doi.org/10.1007/s10924-006-0001-z.

Jamshidian, M., Tehrany, E.A., Imran, M., Jacquot, M., Desobry, S., 2010. Poly-lactic acid: production, applications, nanocomposites, and release studies. Compr. Rev. Food Sci. Food Saf. 9 (5), 552–571. https://doi.org/10.1111/j.1541-4337.2010.00126.x.

Jandas, P.J., Mohanty, S., Nayak, S.K., 2014. Morphology and thermal properties of renewable resource-based polymer blend nanocomposites influenced by a reactive compatibilizer. ACS Sustain. Chem. Eng. 2, 377–386.

Jane, J., 2009. Structural features of starch granules II. In: Starch. Elsevier, pp. 193–236. https://doi.org/10.1016/B978-0-12-746275-2.00006-9.

Javidi, Z., Hosseini, S.F., Rezaei, M., 2016. Development of flexible bactericidal films based on poly(lactic acid) and essential oil and its effectiveness to reduce microbial growth of refrigerated rainbow trout. LWT Food Sci. Technol. 72, 251–260. https://doi.org/10.1016/j.lwt.2016.04.052.

Jordá-Vilaplana, A., Fombuena, V., García-García, D., Samper, M.D., Sánchez-Nácher, L., 2014. Surface modification of polylactic acid (PLA) by air atmospheric plasma treatment. Eur. Polym. J. 58, 23–33.

Joshi, S., 2008. Can nanotechnology improve the sustainability of biobased products? J. Ind. Ecol. 12, 474–489. https://doi.org/10.1111/j.1530-9290.2008.00039.x.

Jost, V., 2018. Packaging related properties of commercially available biopolymers – an overview of the status quo. Express Polym. Lett. 12 (5), 429–435.

Kerry, J.P., 2014. New packaging technologies, materials and formats for fast-moving consumer products. In: Han, J. (Ed.), Innovations in Food Packaging. Academic Press, pp. 549–584.

Khosravi-Darani, K., Bucci, D.Z., 2015. Application of poly (hydroxyalkanoate) in food packaging : improvements by nanotechnology. Chem. Biochem. Eng. 29 (2), 275–285. https://doi.org/10.15255/CABEQ.2014.2260.

Kurek, M., Laridon, Y., Torrieri, E., Guillard, V., Pant, A., Stramm, C., et al., 2017. A mathematical model for tailoring antimicrobial packaging material containing encapsulated volatile compounds. Innov. Food Sci. Emerg. Technol. 42, 64–72. https://doi.org/10.1016/j.ifset.2017.05.014.

Lai, S.-M., Liu, Y.-H., Huang, C.-T., Don, T.-M., 2017. Miscibility and toughness improvement of poly (lactic acid)/poly (3-Hydroxybutyrate) blends using a melt-induced degradation approach. J. Polym. Res. 24, 1–12. https://doi.org/10.1007/s10965-017-1253-0.

Langer, R., Basu, A., Domb, A.J., 2016. Special issue: polylactide (PLA) based biopolymers. Adv. Drug Deliv. Rev. 107, 1–2. https://doi.org/10.1016/j.addr.2016.11.002.

Li, S., McCarthy, S., 1999. Influence of crystallinity and stereochemistry on the enzymatic degradation of poly(lactide)s. Macromolecules 32 (13), 4454–4456. https://doi.org/10.1021/ma990117b.

Li, X., He, Y., Huang, C., Zhu, J., Lin, A.H.-M., Chen, L., Li, L., 2016. Inhibition of plasticizer migration from packaging to foods during microwave heating by controlling the esterified starch film structure. Food Control 66, 130–136. https://doi.org/10.1016/j.foodcont.2016.01.046.

Lim, L.-T., Auras, R., Rubino, M., 2008. Processing technologies for poly(lactic acid). Prog. Polym. Sci. 33 (8), 820–852. https://doi.org/10.1016/j.progpolymsci.2008.05.004.

Liu, L., 2006. Bioplastics in Food Packaging: Innovative Technologies for Biodegradable Packaging.

Llana-Ruiz-Cabello, M., Pichardo, S., Baños, A., Núñez, C., Bermúdez, J.M., Guillamón, E., et al., 2015. Characterisation and evaluation of PLA films containing an extract of Allium spp. to be used in the packaging of ready-to-eat salads under controlled atmospheres. LWT Food Sci. Technol. 64 (2), 1354–1361. https://doi.org/10.1016/j.lwt.2015.07.057.

López-De-Dicastillo, C., Alonso, J.M., Catalá, R., Gavara, R., Hernández-Munoz, P., 2010. Improving the antioxidant protection of packaged food by incorporating natural flavonoids into ethylene-vinyl alcohol copolymer (EVOH) films. J. Agric. Food Chem. 58 (20), 10958–10964.

Lourdin, D., Della Valle, G., Colonna, P., 1995. Influence of amylose content on starch films and foams. Carbohydr. Polym. 27 (4), 261–270. https://doi.org/10.1016/0144-8617(95)00071-2.

Luzi, F., Fortunati, E., Puglia, D., Petrucci, R., Kenny, J.M., Torre, L., 2015. Study of disintegrability in compost and enzymatic degradation of PLA and PLA nanocomposites reinforced with cellulose nanocrystals extracted from *Posidonia oceanica*. Polym. Degrad. Stabil. 121, 105–115. https://doi.org/10.1016/j.polymdegradstab.2015.08.016.

Madhavan Nampoothiri, K., Nair, N.R., John, R.P., 2010. An overview of the recent developments in polylactide (PLA) research. Bioresour. Technol. 101 (22), 8493–8501. https://doi.org/10.1016/j.biortech.2010.05.092.

Madison, L.L., Huisman, G.W., 1999. Metabolic engineering of poly(3- hydroxyalkanoates): from DNA to plastic. Microbiol. Mol. Biol. Rev. 63, 21–53.

Mahieu, A., Terrié, C., Youssef, B., 2015. Thermoplastic starch films and thermoplastic starch/polycaprolactone blends with oxygen-scavenging properties: influence of water content. Ind. Crops Prod. 72, 192–199. https://doi.org/10.1016/j.indcrop.2014.11.037.

Mathiot, C., Ponge, P., Gallard, B., Sassi, J.-F., Delrue, F., Le Moigne, N., 2019. Microalgae starch-based bioplastics: screening of ten strains and plasticization of unfractionated microalgae by extrusion. Carbohydr. Polym. 208, 142–151. https://doi.org/https://doi.org/10.1016/j.carbpol.2018.12.057.

Mazerolles, T., Heuzey, M.C., Soliman, M., Martens, H., Kleppinger, R., Huneault, M.A., 2019. Development of co-continuous morphology in blends of thermoplastic starch and low-density polyethylene. Carbohydr. Polym. 206, 757–766. https://doi.org/https://doi.org/10.1016/j.carbpol.2018.11.038.

Mendonca, T.T., Tavares, R.R., Cespedes, L.G., Sánchez-rodriguez, R.J., Schripsema, J., Taciro, M.K., et al., 2017. Combining molecular and bioprocess techniques to produce poly (3-hydroxybutyrate- co -3-hydroxyhexanoate) with controlled monomer composition by *Burkholderia sacchari*. Int. J. Biol. Macromol. 98, 654–663.

Mensitieri, G., Di Maio, E., Buonocore, G.G., Nedi, I., Oliviero, M., Sansone, L., Iannace, S., 2011. Processing and shelf life issues of selected food packaging materials and structures from renewable resources. Trends Food Sci. Technol. 22 (2), 72–80. https://doi.org/https://doi.org/10.1016/j.tifs.2010.10.001.

Mlalila, N.G., Swai, H.S., Hilonga, A., Kadam, D.M., 2017. Antimicrobial dependence of silver nanoparticles on surface plasmon resonance bands against *Escherichia coli*. Nanotechnol. Sci. Appl. 10, 1–9.

Modi, S.K., 2010. Assessing the Feasibility of Poly-(3-Hydroxybutyrate-co-3-Hydroxyvalerate) (PHBV) and Poly-(lactic Acid) for Potential Food Packaging Applications. Ohio State University.

Molinaro, S., Cruz Romero, M., Boaro, M., Sensidoni, A., Lagazio, C., Morris, M., Kerry, J., 2013. Effect of nanoclay-type and PLA optical purity on the characteristics of PLA-based nanocomposite films. J. Food Eng. 117 (1), 113–123. https://doi.org/10.1016/j.jfoodeng.2013.01.021.

Moon, J., Kim, M.Y., Kim, B.M., Lee, J.C., Choi, M.C., Kim, J.R., 2016. Estimation of the microbial degradation of biodegradable polymer, poly(lactic acid) (PLA) with a specific gas production rate. Macromol. Res. 24 (5), 415–421. https://doi.org/10.1007/s13233-016-4060-2.

Mozejko-Ciesielska, J., Kiewisz, R., 2016. Bacterial polyhydroxyalkanoates : still fabulous ? Microb. Res. 192, 271–282. https://doi.org/10.1016/j.micres.2016.07.010.

Müller, C.M.O., Laurindo, J.B., Yamashita, F., 2011. Effect of nanoclay incorporation method on mechanical and water vapor barrier properties of starch-based films. Ind. Crops Prod. 33 (3), 605–610. https://doi.org/10.1016/j.indcrop.2010.12.021.

Narayanan, A., Neera, M., Ramana, K.V., 2013. Synergized antimicrobial activity of eugenol incorporated polyhydroxybutyrate films against food spoilage microorganisms in conjunction with pediocin. Appl. Biochem. Biotechnol. 170, 1379–1388. https://doi.org/10.1007/s12010-013-0267-2.

Ni, C., Luo, R., Xu, K., Chen, G.-Q., 2009. Thermal and crystallinity property studies of poly (L-lactic acid) Blended with oligomers of 3-hydroxybutyrate or dendrimers of hydroxyalkanoic acids. J. Appl. Polym. Sci. 4, 1720–1727. https://doi.org/10.1002/app.

Nicosia, A., Gieparda, W., Fosksowicz-Flaczyk, J., Walentowska, J., Wesolek, D., Vazquez, B., et al., 2015. Air filtration and antimicrobial capabilities of electrospun PLA/PHB containing ionic liquid. Separ. Purif. 154, 154–160.

Obruca, S., Marova, I., Snajdar, O., Mravcova, L., Svoboda, Z., 2010. Production of poly (3-hydroxybutyrate- co -3-hydroxyvalerate) by *Cupriavidus necator* from waste rapeseed oil using propanol as a precursor of 3-hydroxyvalerate. Biotechnol. Lett. 32, 1925–1932. https://doi.org/10.1007/s10529-010-0376-8.

Ohkita, T., Lee, S.-H., 2006. Thermal degradation and biodegradability of poly (lactic acid)/corn starch biocomposites. J. Appl. Polym. Sci. 100 (4), 3009–3017. https://doi.org/10.1002/app.23425.

Pantani, R., Gorrasi, G., Vigliotta, G., Murariu, M., Dubois, P., 2013. PLA-ZnO nanocomposite films: water vapor barrier properties and specific end-use characteristics. Eur. Polym. J. 49 (11), 3471–3482. https://doi.org/10.1016/j.eurpolymj.2013.08.005.

Peelman, N., Ragaert, P., Vandemoortele, A., Verguldt, E., De Meulenaer, B., Devlieghere, F., 2014. Use of biobased materials for modified atmosphere packaging of short and medium shelf-life food products. Innov. Food Sci. Emerg. Technol. 26, 319–329. https://doi.org/10.1016/j.ifset.2014.06.007.

Piergiovanni, L., Limbo, S., 2016. Plastic packaging materials. In: Food Packaging Materials. Springer International Publishing, Switzerland, pp. 33–49.

Puglia, D., Fortunati, E., Amico, D.A.D., Manfredi, L.B., Cyras, V.P., Kenny, J.M., 2013. In fluence of organically modified clays on the properties and disintegrability in compost of solution cast poly (3-hydroxybutyrate) films. Polym. Degrad. Stabil. 99, 127–135.

Pushpadass, H.A., Marx, D.B., Hanna, M.A., 2008. Effects of extrusion temperature and plasticizers on the physical and functional properties of starch films. Starch (Stärke) 60 (10), 527–538. https://doi.org/10.1002/star.200800713.

Qi, X., Ren, Y., Wang, X., 2017. New advances in the biodegradation of Poly(lactic) acid. Int. Biodeterior. Biodegrad. 117, 215–223. https://doi.org/10.1016/j.ibiod.2017.01.010.

Raquez, J.M., Habibi, Y., Murariu, M., Dubois, P., 2013. Polylactide (PLA)-based nanocomposites. Prog. Polym. Sci. 38 (10–11), 1504–1542. https://doi.org/10.1016/j.progpolymsci.2013.05.014.

Rasal, R.M., Janorkar, A.V., Hirt, D.E., 2010. Poly(lactic acid) modifications. Prog. Polym. Sci. (Oxf.) 35 (3), 338–356. https://doi.org/10.1016/j.progpolymsci.2009.12.003.

Reddy, M.M., Vivekanandhan, S., Misra, M., Bhatia, S.K., Mohanty, A.K., 2013. Biobased plastics and bionanocomposites: current status and future opportunities. Prog. Polym. Sci. 38 (10–11), 1653–1689. https://doi.org/10.1016/j.progpolymsci.2013.05.006.

Rezaeigolestani, M., Misaghi, A., Khanjari, A., Basti, A.A., Abdulkhani, A., Fayazfar, S., 2017. Antimicrobial evaluation of novel poly-lactic acid based nanocomposites incorporated with bioactive compounds in-vitro and in refrigerated vacuum-packed cooked sausages. Int. J. Food Microbiol. 260, 1–10. https://doi.org/10.1016/j.ijfoodmicro.2017.08.006.

Rhim, J.-W., Hong, S.-I., Ha, C.-S., 2009. Tensile, water vapor barrier and antimicrobial properties of PLA/nanoclay composite films. LWT Food Sci. Technol. 42, 612–617. https://doi.org/10.1016/j.lwt.2008.02.015.

Rodríguez-Contreras, A., Koller, M., Miranda De Soussa Dias, M., Calafell-Monforte, M., Braunegg, G., Marques-Calvo, M., 2015. Influence of glycerol on poly(3-hydroxybutyrate) production by *Cupriavidus necator* and *Burkholderia sacchari*. Biochem. Eng. J. 9450–9457.

Shankar, S., Wang, L.-F., Rhim, J.-W., 2018. Incorporation of zinc oxide nanoparticles improved the mechanical, water vapor barrier, UV-light barrier, and antibacterial properties of PLA-based nanocomposite films. Mater. Sci. Eng. C 93, 289–298. https://doi.org/10.1016/j.msec.2018.08.002.

Shantini, K., Yahya, A.R.M., Amirul, A.A., 2015. Influence of feeding and controlled dissolved oxygen level on the production of poly (3-hydroxybutyrate- co -3-hydroxyvalerate) copolymer by *Cupriavidus* sp . USMAA2-4 and its characterization. Appl. Biochem. Biotechnol. 176, 1315–1334. https://doi.org/10.1007/s12010-015-1648-5.

Shen, L., Haufe, J.I., Patel, M., 2009. Product Overview and Market Projection of Emerging Bio-Based Plastics-PRO-BIP 2009–Final Report.

Singh, A.K., Sharma, L., Mallick, N., Mala, J., 2017. Progress and challenges in producing polyhydroxyalkanoate biopolymers from cyanobacteria. J. Appl. Phycol. 29, 1213–1232. https://doi.org/10.1007/s10811-016-1006-1.

Siracusa, V., Rocculi, P., Romani, S., Rosa, M.D., 2008. Biodegradable polymers for food packaging: a review. Trends Food Sci. Technol. 19 (12), 634–643. https://doi.org/10.1016/j.tifs.2008.07.003.

Sun, S., Liu, P., Ji, N., Hou, H., Dong, H., 2018. Effects of various cross-linking agents on the physicochemical properties of starch/PHA composite films produced by extrusion blowing. Food Hydrocolloids 77, 964–975. https://doi.org/https://doi.org/10.1016/j.foodhyd.2017.11.046.

Suwannasin, W., Imai, T., Kaewkannetra, P., 2015. Potential utilization of pineapple waste streams for polyhydroxyalkanoates (PHAs) production via batch fermentation. J. Water Environ. Technol. 13, 335–347. July 2016. https://doi.org/10.2965/jwet.2015.335.

Tawakkal, I.S.M.A., Cran, M.J., Bigger, S.W., 2016. Release of thymol from poly(lactic acid)-based antimicrobial films containing kenaf fibres as natural filler. LWT Food Sci. Technol. 66, 629–637. https://doi.org/10.1016/j.lwt.2015.11.011.

Thompson, D.B., 2000. Strategies for the manufacture of resistant starch. Trends Food Sci. Technol. 11 (7), 245–253. https://doi.org/10.1016/S0924-2244(01)00005-X.

Tokiwa, Y., Calabia, B.P., 2006. Biodegradability and biodegradation of poly(lactide). Appl. Microbiol. Biotechnol. 72 (2), 244–251. https://doi.org/10.1007/s00253-006-0488-1.

van den Oever, M., Molenveld, K., van der Zee, M., Bos, H., 2017. Bio-Based and Biodegradable Plastics : Facts and Figures : Focus on Food Packaging in the Netherlands. https://doi.org/10.18174/408350.

Varsha, Y., Savitha, R., 2011. Microbial & biochemical technology overview on polyhydroxyalkanoates : a promising biopol. J. Microb. Biochem. Technol. 3 (5), 99–105. https://doi.org/10.4172/1948-5948.1000059.

Vink, E.T.H., Davies, S., 2015. Life cycle inventory and impact assessment data for 2014 ingeo™ polylactide production. Ind. Biotechnol. 11 (3), 167–180. https://doi.org/10.1089/ind.2015.0003.

Vroman, I., Tighzert, L., 2009. Biodegradable polymers. Materials 2 (2), 307–344. https://doi.org/10.3390/ma2020307.

Wang, L.-F., Rhim, J.-W., 2016. Grapefruit seed extract incorporated antimicrobial LDPE and PLA films: effect of type of polymer matrix. LWT 74, 338–345. https://doi.org/10.1016/j.lwt.2016.07.066.

Xavier, J.R., Babusha, S.T., George, J., Ramana, K.V., 2015. Material properties and antimicrobial activity of polyhydroxybutyrate (PHB) films incorporated with vanillin. Appl. Biochem. Biotechnol. 176 (5), 1498–1510. https://doi.org/10.1007/s12010-015-1660-9.

Xie, J., Hung, Y.-C., 2018. UV-A activated TiO_2 embedded biodegradable polymer film for antimicrobial food packaging application. LWT 96, 307–314. https://doi.org/10.1016/j.lwt.2018.05.050.

Yu, L., Dean, K., Li, L., 2006. Polymer blends and composites from renewable resources. Prog. Polym. Sci. 31 (6), 576–602. https://doi.org/https://doi.org/10.1016/j.progpolymsci.2006.03.002.

Zhang, Y., Han, J.H., 2006. Plasticization of pea starch films with monosaccharides and polyols. J. Food Sci. 71 (6), 253–261. https://doi.org/10.1111/j.1750-3841.2006.00075.x.

Zhang, M.I.N., Thomas, N.L., 2011. Blending polylactic acid with polyhydroxybutyrate : the effect on thermal, mechanical, and biodegradation properties. Adv. Polym. Technol. 30 (2), 67–79. https://doi.org/10.1002/adv.

Zhang, Y., Rempel, C., Liu, Q., 2014. Thermoplastic starch processing and characteristics—a review. Crit. Rev. Food Sci. Nutr. 54 (10), 1353–1370. https://doi.org/10.1080/10408398.2011.636156.

Zhu, C., Nomura, C.T., Perrotta, J.A., Stipanovic, A.J., Nakas, J.P., 2012. The effect of nucleating agents on physical properties of poly-3- hydroxybutyrate (PHB) and poly-3-hydroxybutyrate-co -3- hydroxyvalerate (PHB- co -HV) produced by Burkholderia cepacia ATCC 17759. Polym. Test. 31 (5), 579–585. https://doi.org/10.1016/j.polymertesting.2012.03.004.

Chapter 9

Bio-based electric devices

V. Hoffmann, M.P. Olszewski, K.M. Swiatek, B. Musa,
P. J. Arauzo Gimeno, C. Rodriguez Correa, A. Kruse
University of Hohenheim (UHOH), Conversion technologies of biobased resources (440f), Stuttgart-Hohenheim, Baden-Württemberg, Germany

9.1 Introduction

The depletion and increasing price of fossil fuels as well as the climate change and the need to decrease CO_2 emission are currently a major interest on a global scale. However, rapid economic growth and increasing populations result in high demand for modern electric devices. This requires continuous improvement of energy storage systems including faster charging, longer lifetime, and best conversion efficiency at the lowest price possible. To solve these challenges, one solution could be enhancing the use of biomass for more efficient production of electricity (via direct carbon fuel cells) and bio-based carbon materials for energy storage applications. According to recently published research, bio-based carbon materials achieve similar or even better properties than fossil equivalents. The bio-based materials are used as electrode materials in energy storage devices, including electrochemical double-layer capacitors (EDLCs, also known as supercapacitors), lithium-ion batteries (LIBs), sodium-ion batteries (SIBs), and as a fuel in direct carbon fuel cells (DCFCs) or Microbial Fuel Cells (MBFCs) (Paraknowitsch et al., 2009; Cao et al., 2007; WeiZi et al., 2016).

Rechargeable lithium-ion batteries are currently the most common in the market. They find application as energy backup sources on an industrial scale, for households and even as support for the energy transmission grid. They are also essential in hybrid and electric vehicles to store surplus electricity and to assist the conventional car engine, as well as in portable electronics for long power delivery. Supercapacitors, in turn, are used when large amounts of energy need to be stored or released in a relatively short time (high power densities) like in regenerative braking systems. Finally, direct carbon fuel cells have the huge advantage to be able to convert the energy chemically stored in the fuel directly into electricity which results in much higher efficiency than conventional combustion power plants, where the conversion takes place via various conversion steps.

Biobased Products and Industries. https://doi.org/10.1016/B978-0-12-818493-6.00009-9

The production of carbon materials from biomass mainly takes place in several stages such as pretreatment (washing, grinding, drying), thermal conversion (hydrothermal carbonization (HTC), pyrolysis), activation, and functionalization (Górka et al., 2016). The activation process can be divided into physical or chemical activation. During physical activation, steam or CO_2 is used as the activating agent whereas the chemical treatment involves bases and acids such as KOH, K_2CO_3, NaOH, $AlCl_3$, $ZnCl_2$, or H_3PO_4 (Ahmadpour and Do, 1996). The chemical activation leads to a higher surface area and pore volume compared to physical activation. But, it has to be considered that high amounts of wastewater are generated due to the posttreatment (washing and neutralizing) after chemical activation. The cost of chemicals and wastewater significantly increases the overall cost of chemical activation.

The thermal treatment (carbonization) takes place in a temperature range of 180°C to 1600°C. Organic precursors for the production of functional carbon materials are wood, agricultural and food production wastes (e.g., fruits peels and pomaces, nut shells, corn cobs), wood processing and paper production wastes (cellulose and lignin), among other. Carbon materials obtained from biomass and organic wastes are characterized by a partly amorphous structure and are often called "hard carbons." Biomass is characterized by high heterogeneity and is composed of biopolymers (hemicelluloses, crystalline cellulose, lignins, proteins, pectins, waxes, oils, and free sugars) and inorganic material (Emaga et al., 2011). Hard carbons have less organized structures (Fig. 9.1) than hexagonal planar carbon atoms in graphitic structure due to less order in the C-direction of the carbon lattice. The formation of graphite structures is limited even at high temperatures by dispersion of amorphous areas with graphite-like matrixes with strong cross-linkings. Crystalline cellulose could be graphitized in contrast to hemicellulose and lignin, which are unable to be graphitized due to strong cross-linking (Franklin, 1951).

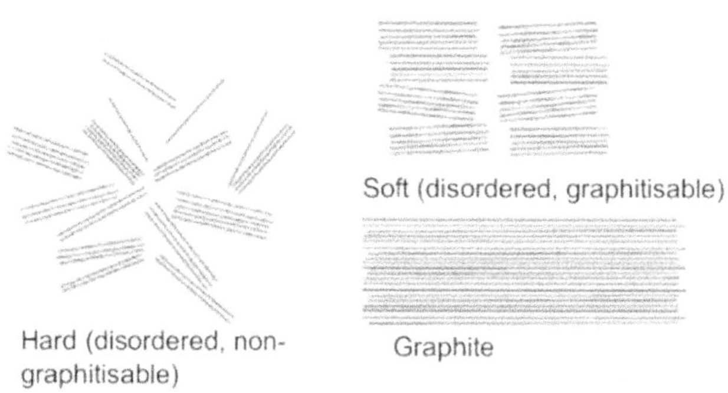

FIGURE 9.1 Schematic models of soft carbon, hard carbon, and graphite. *Adapted from Loeffler, N., Bresser, D., Passerini, S., 2015. Secondary lithium-ion battery anodes: from first commercial batteries to recent research activities. Johnson Matthey Technol. Rev. 59 (1), 34–44. https://doi.org/10.1595/205651314X685824*

The presence of certain surface functionalities (e.g., N- and O-groups) on the carbon, which is achieved via low-temperature carbonization such as HTC, oxidation reactions or doping, can significantly increase the capacities of the material due to pseudocapacitive effects. Furthermore, the structural properties of the carbon affect the suitability as electrode material or fuel in a direct carbon fuel cell. During pyrolysis, the biomass structure undergoes aromatization processes and extensive cross-linking. The carbonization process is accompanied by the release of volatile matter (a mixture of condensable hydrocarbons and noncondensable gases) (Górka et al., 2016; Dahn et al., 1995). Materials produced at lower pyrolysis temperatures are characterized by lower surface area and bigger pore sizes, which results in higher diffusions rates and more active sites. These properties improve rate capabilities and increase capacities (Han et al., 2014a). Consequently, increasing the pyrolysis temperature results in an increase of pore volume and surface area, but reduces the pore size. Additionally, higher pyrolysis temperature results in higher electric conductivity due to cross-linking and aromatization during carbonization process (Mochidzuki et al., 2003). The conversion of biomass provides an opportunity to design carbon materials tailored to the specific application, for example, for alternative batteries using cheaper sodium ion, which is larger than the commercially used lithium ion (Wen et al., 2014). However, the yield of the high-quality bio-based carbon material is below 10% (Yun et al., 2015). Thus, the challenge is to find optimal conditions for the production process of bio-based carbon materials for different energy storage and conversion applications as well as to increase the final yield of carbon material.

9.2 Energy storage devices (supercapacitors and batteries)

One application for bio-based carbon materials can be found in the large field of energy storage becoming increasingly important especially in times of renewable energy technologies which lack reliable storage systems or electric mobility. Supercapacitors or ultracapacitors are playing a key role in this context as they close the gap between batteries (see Section 9.2.1) and conventional capacitors by providing high specific capacities in a short time.

9.2.1 Supercapacitors

Supercapacitors are used in applications, where vast amounts of energy have to be stored or released in a very short time. Nowadays, supercapacitors are used primarily in hybrid electric vehicles, electric vehicles, and fuel cell vehicles like passenger cars, trains, or trolleybuses. Another area of application for supercapacitors is electronic devices such as uninterruptible power supplies and volatile memory backups in computers. The third area of use is energy harvesting systems, solar arrays, or wind turbines, where supercapacitors play a supplementary role next to conventional batteries (Wang et al., 2016; Miller, 2016; Zhang, 2017; Shukla et al., 2012).

Because of the wide range of applications of supercapacitors (beside electric or hybrid mobility, also in wind turbines or photovoltaic systems), they recently gained an increased interest and an enormous growth of the global demand for supercapacitors is expected in the next years. The global market volume reached a value of approximately €980 million in 2015 and is expected to increase to up to €4 billion in 2025. Approximately 20% of this increase is attributed to the demand that will arise from the growing market for electronic devices and integrated memories. However, the largest share is expected to come from the growing demand for supercapacitors from the mobility sector in connection with the development of emission-free or reduced drive technologies (e-mobility) (Yassine and Fabris, 2017).

Electrode materials for use in energy storage technologies such as supercapacitors must have certain properties to ensure high energy and power densities. The most important parameters in this context are (Hu et al., 2010; Zhao et al., 2010a; Kotz et al., 2000; Pandolfo and Hollenkamp, 2006; Falco et al., 2013; Ghosh and Hee, 2012):

- high surfaces, since the amount of energy stored correlates directly with the surface size (according to the equation (Eq. 9.1) for capacitance with ε_0 = vacuum permittivity, A = surface area, d = distance between charged layers)

$$C = \varepsilon_0 \times \frac{A}{d} \tag{9.1}$$

- certain pore size distributions with a high proportion of micropores
- high specific conductivity to ensure electron mobility and reduce ohmic losses
- surface functionalities to utilize pseudocapacitive effects

As shown in the RAGONE diagram (Fig. 9.2), batteries provide high energy densities, while pure electrolytic capacitors have very high power densities. The gap between these two technologies can be closed by supercapacitors with the goal of producing a double layer capacitor positioned in the upper right corner of the RAGONE diagram. At the same time, the supercapacitor should have the highest possible capacitance in the widest possible voltage range.

Electrostatic energy storage in supercapacitors is far superior to chemical energy storage (based on redox reactions) in batteries. Therefore, supercapacitors provide hundred to many thousand times higher power per volume compared to batteries. This means a capacitor storage system is often smaller

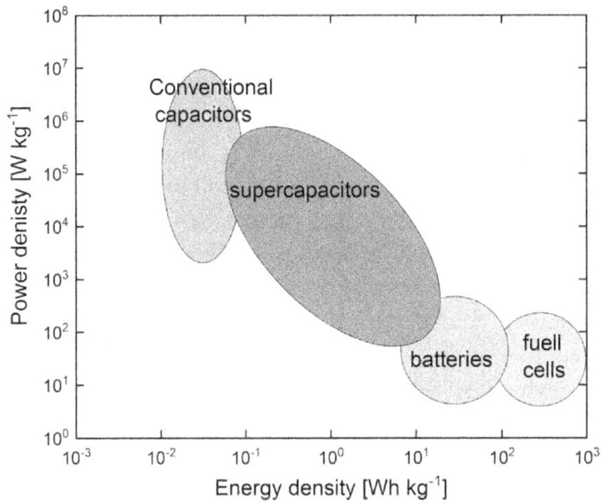

FIGURE 9.2 RAGONE plot. Power density as a function of energy density for different energy storage devices. *Own image based on Christen, T., Carlen, M.W., 2000. Theory of ragone plots. J. Power Sources 91 (2), 210–216. https://doi.org/https://doi.org/10.1016/S0378-7753(00)00474-2; Lee, S.C., Jung, W.Y., 2016. Analogical understanding of the ragone plot and a new categorization of energy devices. Energy Procedia 88, 526–530. https://doi.org/https://doi.org/10.1016/j.egypro.2016.06.073. (Copyright M. P. Olszewski).*

in size and lower in mass than a comparable battery system (Miller, 2017; Libich et al., 2018). Table 9.1 shows a comparison between batteries and supercapacitors in terms of key parameters such as energy storage mechanisms or power and cycle life limitations.

TABLE 9.1 Comparison between supercapacitors and batteries.

Parameter	Supercapacitor	Battery
Energy storage (limiting factor)	Limited (surface area)	High (bulk)
Storage mechanism	Mainly physical	Chemical
Power limitation	Electrolyte conductivity	Reaction kinetics/mass transport
Charge rate	High/= discharge rate	Kinetically limited
Cycle life limitations	Side reactions	Mechanical stability/chemical reversibility

Based on González, A., Goikolea, E., Barrena, J.A., Mysyk, R., 2016. Review on supercapacitors: technologies and materials. Renew. Sustain. Energy Rev. 58, 1189–1206. https://doi.org/https://doi.org/10.1016/j.rser.2015.12.249; Miller, J.R.; Simon, P., (2008). Electrochemical capacitors for energy management. Science 321 (5889), 651 LP-652.

Supercapacitors offer unlimited charge-discharge efficiency, excellent cycle life, long operational life, and exceptional power performance. These properties result from the mentioned physical energy storage mechanism in the electric double layer at the surface of an electrolyte interface. Electrode materials with high specific surface areas (SSA) and the small distance between opposed charges in the electric double layer lead to extremely high gravimetric and volumetric capacitances in supercapacitors (Miller, 2017; Libich et al., 2018; González et al., 2016). The electrostatic energy storage mechanisms also allow charging and discharging of the supercapacitor in (milli-) seconds and very long durability of about millions of cycles. However, the main disadvantage related to this mechanism is the operating voltage of the supercapacitor cell, which should be kept low in order to avoid the chemical decomposition of electrolytes (Miller, 2017).

In the following, the underlying energy storage mechanisms of supercapacitors are explained and an overview about the state-of-the art materials (electrolytes and electrode materials) is given before bio-based electrode materials are presented.

9.2.1.1 Operation principle: energy storage mechanisms

Supercapacitors combine two different energy storage mechanisms including electrostatic and pseudocapacitive energy storage. In order to understand these two energy storage principles, they will be explained separately by distinguishing three types: electrical double-layer capacitors (EDLC), pseudocapacitors, and hybrid capacitors (González et al., 2016) (see Fig. 9.3).

Electrical Double-Layer Capacitors (EDLC). The first type includes EDLCs which stores energy physically. EDLCs are the most common type of supercapacitors and represent the largest share of the commercial supercapacitor market. Unlike a traditional plate capacitor, an EDLC contains two separate charge layers at the interfaces of the electrolyte with the positive electrode and the negative electrode, respectively (Miller, 2017; Libich et al., 2018). A simple EDLC can be constructed by inserting two conductors in a beaker with an electrolyte. Fig. 9.3A shows a schematic representation of an EDLC. These supercapacitors use electrostatic processes to accumulate energy in double layers on the phase interface between the surfaces of the electrodes and the electrolyte. Charge separation occurs at each liquid-solid interface and potential is built up between the two rods. Solvated ions in the electrolyte are rapidly attracted to the solid surface by an equal but opposite charge located in the solid. In this type of supercapacitors, there is no electron exchange and no redox reaction. Extremely high capacity is the result of the high surface area of the electrodes and the layer thickness. Activated carbon obtained from fossil sources such as graphite or organic precursors such as coconut shell is widely used as an electrode material in industry due to the large surface areas of the materials (Miller, 2017; Libich et al., 2018; González et al., 2016; Chen et al., 2017; Wang et al., 2012).

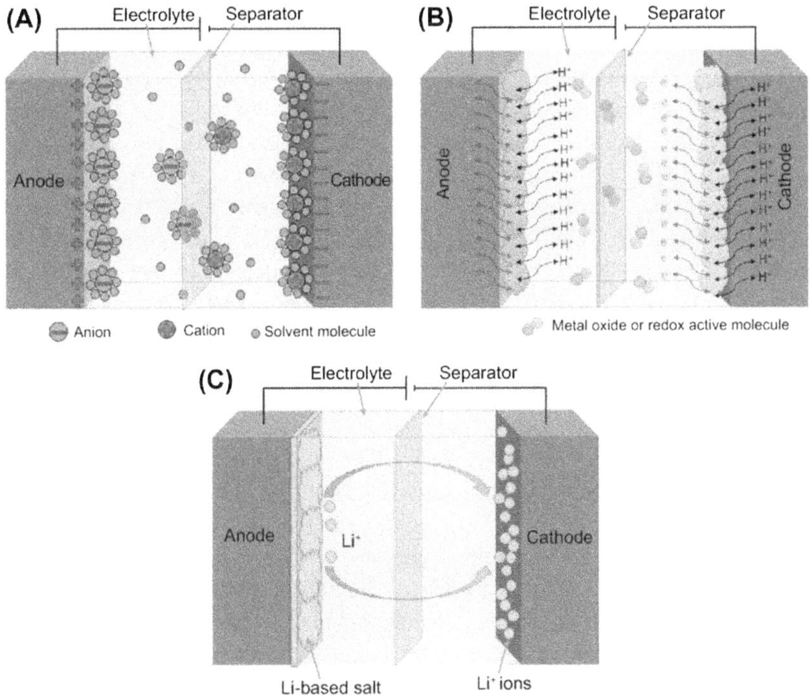

FIGURE 9.3 Schematic representation of (A) electrical double-layer capacitor (EDLC), (B) pseudocapacitor, (C) hybrid supercapacitor. *Adapted from Chen, X., Paul, R., Dai, L., 2017. Carbon-based supercapacitors for efficient energy storage. Natl. Sci. Rev. 4 (3), 453–489.*

Pseudocapacitors. The second type of capacitors is called pseudocapacitors, and they do not exist alone but form an essential part of a supercapacitor. In order to understand the basic charge storage principle, they are presented here as an independent type. A scheme of these pseudocapacitors is shown in Fig. 9.3B. By the operation principle, pseudocapacitors are more similar to batteries than to capacitors. The energy storage is based on a pseudocapacitance phenomenon, where electrode materials intermediate electron transfer and undergo redox reactions (Miller, 2017; González et al., 2016). The pseudocapacitance arises at the electrode surface, where faradaic reactions take place via reversible redox reactions. These redox reactions occur between a "desolvated" ion of the electrolyte (in contrast to solvation, which means the formation of a solvation shell of solvent molecules, for example, water molecules, around the dissolved molecules/ions), which is adsorbed to the electrode surface and makes a faradaic charge transfer between the adsorbed ion and the electrode material possible. This charge exchange includes the reversible exchange of only one electron between one adsorbed ion and the electrode material. The resulting pseudocapacitance gives rise to the total capacitance of a supercapacitor and can be several times higher than the capacitance caused by electrostatic energy storage. The disadvantage of these systems is the principle of operation due to charging and discharging; the electrodes are stressed and degrade faster (like in batteries), compared to the electrostatic storage principle of EDLCs. This is caused by the high internal resistance of supercapacitors. The stability and cyclability are lower than in the case of EDLC along with lower charging efficiency and lower discharge rate (Miller, 2017; Libich et al., 2018; González et al., 2016; Chen et al., 2017; Wang et al., 2012). The described energy storage mechanisms (for EDLCs and pseudocapacitors) at the positively charged electrode/cathode are shown schematically in Fig. 9.4. The double layer (Stern double layer) formed at the carbon surface via adsorption of ions to the electrode surface can be divided into a fixed double layer (also called Helmholtz double layer) and a diffuse layer in which the solvated ions are less closely packed due to the distance to the electrode surface. The Helmholtz double layer has a thickness of approximately one ion diameter and depends on the concentration of the respective electrolyte (e.g., 5–10 Å for concentrated electrolytes) (Kotz et al., 2000).

Hybrid capacitors. Hybrid capacitors are the newest type and currently have a technology readiness level between 2 and 4. This type is the most advanced among the available capacitors and combines previous types, the EDLC, and pseudocapacitor, on two different electrodes as shown in Fig. 9.3C. The main advantage is high volumetric and gravimetric energy density and the

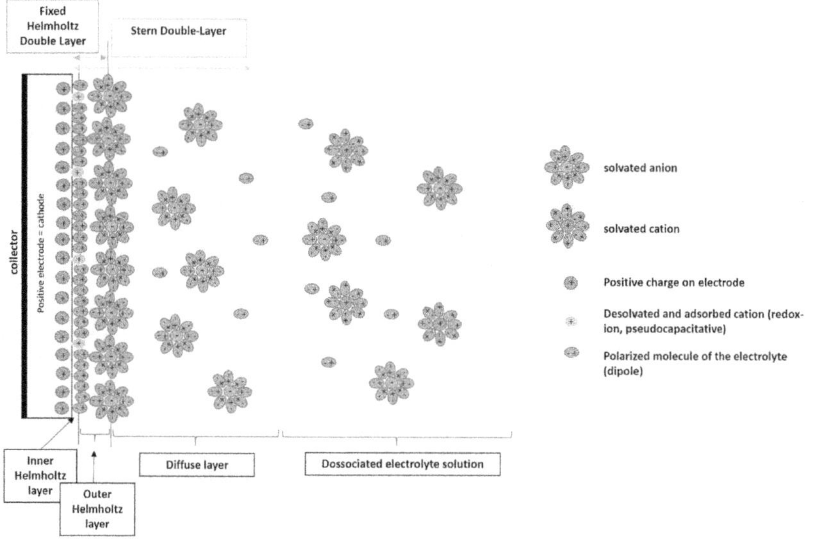

FIGURE 9.4 Schematic representation of the energy storage mechanisms at the cathode of a supercapacitor including electrostatic energy storage of an EDLC (in the double layers including Helmholtz- and Stern double layers) and pseudocapacitive energy storage in a PC (via ions, which have lost their solvation shell). *Own image (Copyright: Viola Hoffmann).*

capability to provide high currents. On the negative electrode, typically made from pseudocapacitive electrode material, occurs the faradaic reaction and therefore hybrid supercapacitors possess higher energy density. The positive electrode is typically made from activated carbon where stored energy is stored electrostatically in the double layer on the electrode surface. From the construction and operation point of view, hybrid supercapacitors are close to the lithium-ion batteries (Libich et al., 2018; González et al., 2016; Chen et al., 2017; Zhong et al., 2015).

Electrolytes. The electrolyte is one of the most important constituents of electrochemical supercapacitors due to its advantages of ionic conductivity and charge compensation on both electrodes of the cell, which is reflected in the supercapacitor performance. In general, electrolytes can be divided into two groups: aprotic and protic electrolytes. These electrolytes can further be classified as organic, aqueous, and solid-state polymer electrolytes, and ionic liquids (Wang et al., 2016; Libich et al., 2018).

The most important properties characterizing electrolytes are the temperature coefficient and the ionic conductivity. Besides, wide possible voltage window, high electrochemical stability, high ionic concentration, low solvated ionic radius, low viscosity, low volatility, low toxicity, low cost, and availability at high purity (Wang et al., 2012) are important. By mixing appropriate solvents, the specific conductivity of a solution can be modified. Besides, the potential range of electrochemical stability is critical, being higher for nonaqueous electrolytes than for aqueous solutions. Furthermore, the corrosion of electrodes and current collectors is important as well and depends on the nature of electrolytes, and for aqueous electrolytes on the pH (Wang et al., 2012).

Usually, aqueous electrolytes exhibit high conductivity compared to organic electrolytes. Moreover, aqueous electrolytes can offer a higher ionic concentration and contribute to a lower internal resistance due to the smaller ionic radius compared to organic electrolytes. Generally, aqueous electrolytes can be divided into acid, alkaline, and neutral solutions in which H_2SO_4, KOH, and Na_2SO_4 are representatives and also the most popular electrolytes. However, despite the advantages, this type of electrolytes shows a significant disadvantage in a much lower potential window than organic electrolytes, which is restricted by the electrochemical hydrogen and/or oxygen formation (Wang et al., 2016; Libich et al., 2018; González et al., 2016; Zhong et al., 2015).

The three types mentioned above of aqueous electrolytes could be applied in carbon-based EDLC, pseudocapacitors, and hybrid supercapacitors. However, it should be remembered that the electrolyte type significantly affects the pseudocapacitive properties of carbon-based electrode materials due to the different behavior of surface functionalities in different electrolytes. Besides, the operating temperature range of supercapacitors with aqueous electrolytes has also to be restricted to the freezing water point and below the boiling point (Wang et al., 2016; Zhong et al., 2015).

Supercapacitors based on organic electrolyte are currently dominating the commercial market owing to their high operation potential window. This allows obtaining higher energy and power density. However, organic electrolytes have a higher cost, a smaller specific capacitance, a lower conductivity, and less safety due to the flammability, volatility, and toxicity. Regarding industrial production, organic electrolytes are more expensive due to the purification from the water. The water content must be below 3–5 ppm; otherwise, the capacitor's voltage will be significantly reduced (Zhong et al., 2015; Kötz and Carlen, 2000).

The typical organic electrolytes for EDLC consist of conductive salts, for example, tetraethylammonium tetrafluoroborate (TEABF4)— dissolved in acetonitrile or propylene carbonate solvent. Furthermore, the additional advantage is that using organic electrolytes allows the use of cheaper materials—e.g., Al—for the current capacitors (Wang et al., 2016; Zhong et al., 2015).

Ionic liquids are generally defined as pure salts composed solely of ions, containing no solvents with melting points below 100°C. Ionic liquids, which are in the liquid state at ambient temperature, are of interest to supercapacitors because they are nonvolatile; poorly combustible; and have high thermal, chemical, and electrochemical stability. Furthermore, ionic liquids' physical and chemical properties can be highly tunable due to their large variety of combinations of cations and anions (Wang et al., 2016; Zhong et al., 2015; Galiński et al., 2006). The main drawback of ionic liquid electrolytes is their low electrical conductivity, particularly in comparison with aqueous electrolytes. The conductivity of ionic liquids is strongly correlated to their viscosity, with strong temperature dependence. The main ionic liquids investigated for supercapacitors are pyrrolidinium, imidazolium, or aliphatic quaternary ammonium salts coupled with anions (PF_6^-, BF_4^-, $TFSI^-$, or FSI^-) (Wang et al., 2016; Galiński et al., 2006).

Currently, ionic liquid electrolytes are considered to be an excellent solution to improve the performance of electrochemical capacitors. However, major impediments to the industrial use of ionic liquids as supercapacitor electrolyte are high cost, and ionic liquids require more stringent conditions for drying carbon materials in order to provide water-free environments in the cells. Another issue with these electrolytes is poor compatibility with microporous carbons (González et al., 2016).

9.2.1.2 Bio-based electrode materials for supercapacitors

The capacitance of supercapacitors directly depends on the electrode materials. When investigating new electrode materials, key properties include specific surface area, pore-size distribution, pore volume, morphology, surface modifications (surface functionalities), and material purity. Other essential electrode parameters include density, electric conductivity, and chemical homogeneity. In general, the electrode materials can be categorized into three

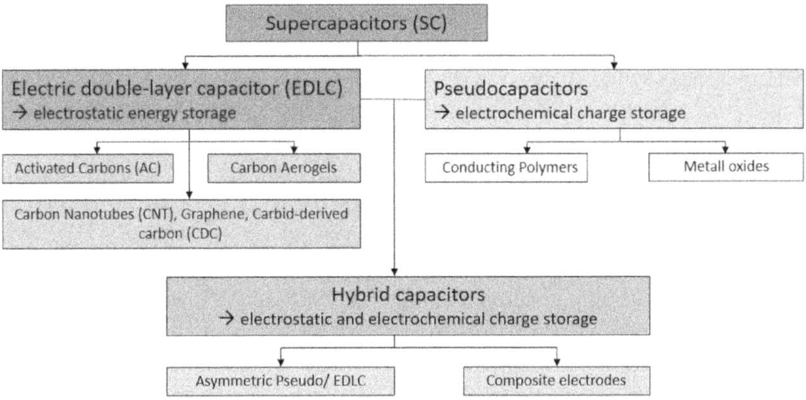

FIGURE 9.5 Overview of the different energy storage mechanisms and their respective suitable materials. *Own image based on Frackowiak, E, Béguin, F., 2013. Supercapacitors Materials, Systems and Application. WILEY-VCH Verlag GmbH & Co. KG, Weinheim. (Copyright: Viola Hoffmann).*

types depending on the underlying energy storage mechanism as shown in Fig. 9.5: carbon materials, conducting polymers, and metal oxides (González et al., 2016; Wang et al., 2012). Conducting polymers and metal oxides can also be classified as faradaic materials. Much research has been done in recent years to develop composite electrode materials which store energy electrostatically and allow the development of pseudocapacitance (Wang et al., 2012).

It has been shown that certain types of surface functionalities on the carbon surface are of particular interest for the application in energy storage as they enhance the pseudocapacitive effect. Therefore, the state of research regarding the production of bio-based, functionalized electrode materials from biomass is presented in this section and the gaps are pointed out.

While the surface size and the pore size distribution determine how many ions can be adsorbed and thus influence the energy storage capacity in the electrochemical double layer at the phase boundary between electrode and electrolyte, surface functionalities or foreign atoms enable the increase of the energy density by their pseudocapacitive properties. Oxygen-containing functional groups on the surface of activated carbon serve as electron acceptors due to their acidic character and nitrogen-containing surface functionalities serve as electron donors due to their basic character (Zhao et al., 2010a; Vagner et al., 2003). These properties lead to reversible redox reactions additionally to the electrostatic energy storage, which increases the total capacity of the supercapacitor.

At the same time, the importance of surface areas and pore size distributions of the respective carbon materials for energy and power density has to be taken into account. While large surface areas lead to increasing internal resistances and decreasing power densities, they are usually accompanied by a higher proportion of micropores (<2 nm), which in turn provide high energy densities because they have a pseudocapacitive effect. The latter is due to the fact that micropores are usually only accessible for ions without solvation shell (see Fig. 9.4). Large pores correlate with higher power densities (Bernstein, 2015).

Thus it becomes clear that due to the multitude and complexity of the parameters and the interactions between them, a finely tuned adaptation of the different material properties is required in order to identify the parameter combination that leads to both a high energy and power density.

State-of-the art, fossil-based carbon materials are characterized by low cost, easy processing, nontoxicity, high specific surface area, good electric conductivity, high chemical stability, and a wide operating temperature range. The most common carbon materials are activated carbon, carbon aerogels, carbon nanotubes, template porous carbons, and carbon nanofibers. However, most of the named materials are obtained from fossil sources (e.g., graphite). Besides, they also have high resistivity, because of the contact resistance between carbon particles giving rise to high internal series resistance and leading to reduced supercapacitor performance. Besides, the capacitive performance of carbon materials is often limited because not the whole surface area is accessible for electrolyte ions due to irregular pore-size distribution in the carbon materials (Chen et al., 2017; Wang et al., 2012). The production of biobased carbon materials with the same or even better properties would be not only more sustainable but also allow the production of functional carbon materials with tuned properties compared to the fossil-based materials. Before the process of biomass conversion for doing so is described, the fossil-based state-of-the-art materials are shortly presented in order to define the technological benchmark.

Conducting polymers, for example, polyacetylene, polypyrrole, polyaniline, have been considered as promising pseudocapacitive electrode materials due to their properties (Wang et al., 2016) such as high conductivity in a doped state, low cost, low environmental impact, high voltage window, high storage capacity, porosity, reversibility, and adjustable redox activity through chemical modification (Wang et al., 2012). They offer capacitive behavior through the redox process. During oxidation, ions are transferred to the polymer backbone, and when reduction takes place, the ions are released from this backbone into the electrolyte. Redox reactions occur not just on the surface, but also throughout the entire bulk (Wang et al., 2012, 2016). The redox processes are highly reversible because no structural changes, such as phase transformation, happen during the charging and discharging reactions. The most promising conducting polymers are polyaniline and polypyrrole because of their low cost, environmental stability, and facile synthesis (Wang et al., 2016). However, the disadvantages of these polymers are poor electric conductivity in the reduced state and low capacitance. For this reason, they are usually used as the positive electrode, and the negative electrode is made from another material such as hard carbons. The key factor for maintaining the supercapacitor performance is a selection of a suitable potential range. Conducting polymers work appropriately within a strict potential window and beyond this potential range, the material may be degraded at higher potentials. By contrast, if the potential is too negative, this material may be switched to an insulating state. Furthermore, conducting polymers may swell and shrink in the process of intercalation and deintercalation, which often results in low cycling stability. Polymers can be composed with the addition of other species, which improve their cycling stability (Wang et al., 2016).

Currently, commercial supercapacitors mainly use the described fossil-based carbon materials, except for activated carbon from coconut shells (e.g., KURARAY YP, NORIT Supra: 1990 $m^2\,g^{-1}$) (Bernstein, 2015; Kurzweil and Dietlmeier, 2015). Especially large surfaces and good capacitive properties are also achieved by the decomposition of fructose/glucose (TDA RESEARCH: 2260 $m^2\,g^{-1}$) or phenolic resins (KURARAY RP-15: 1350 $m^2\,g^{-1}$) (Kurzweil and Dietlmeier, 2015); however fructose/glucose should not be an option due to food competition aspects. Another promising fossil material for use as electrode material in supercapacitors is graphene, which can be produced from the mineral graphite. Graphene has high surface areas of up to 2630 $m^2\,g^{-1}$ and capacities of approx. 550 $F\,g^{-1}$ can be achieved (Wang et al., 2012; El-Kady et al., 2016). However, graphite is a fossil resource that was already counted among the 14 critical raw materials according to the EU resource ranking list in 2010, that is, a shortage of graphite is to be expected in the future (European Commission, 2010). The fact that the demand for graphite continuously increases in the course of the development in the battery sector, and only a few countries produce the mineral, further exacerbates the situation (European Commission, 2010; Feytis, 2010). This makes innovative and above all bio-based and environmentally friendly electrode materials as substitutes for fossil-based materials all the more urgently needed. Pseudocapacitive materials such as metal oxides are not bio-based but can be combined with bio-based carbons in order to improve their electrochemical performance. Metal oxides are characterized by high specific conductivity and capacitance, which is why these materials are suitable for electrode production focused on high energy and high power supercapacitors (González et al., 2016). In general, metal oxides have a higher energy density in supercapacitors than carbon materials and better electrochemical stability than conducting polymers. They both store energy like carbon materials and exhibit electrochemical faradaic reactions between electrode materials and ions within the appropriate potential range (Wang et al., 2012). There are several different metal oxide materials used for electrode production such as RuO_2, MnO_2, NiO, Co_2O_3, and others (Wang et al., 2012). Among the transition metal oxides, RuO_2 has been the most extensively studied electrode material due to its remarkably high specific capacitance among all pseudocapacitive materials. Besides, it has a wide potential window, highly reversible redox reactions, high proton conductivity, excellent thermal stability, long lifecycle, metallic-type conductivity, and high rate capability (Wang et al., 2012; Su et al., 2007). It also has three oxidation states accessible within 1.2 V voltage window. However, a disadvantage of this material is a very high cost, which reduces its applications (Miller and Simon, 2008). The pseudocapacitive behaviors of RuO_2 in acidic and alkaline environments involve different reactions (Wu et al., 2002). In acidic electrolyte solutions, according to Eq. (9.2), a fast reversible electron transfer and an electro-adsorption of protons on the surface of RuO_2 particles, where Ru oxidation states can change from Ru(II) to Ru(IV) (González et al., 2016; Wang et al., 2012) takes place.

$$RuO_2 + xH^+ + xe^- \leftrightarrow RuO_{2-x}(OH)_x \quad (9.2)$$

In an alkaline solution, the changes in the oxidation state of RuO_2 are quite different. It is assumed that when the carbon/ruthenium composite is charged the RuO_2 in the composite electrode will be oxidized to RuO_4^{2-}, RuO_4^-, and RuO_4 and is reduced back to RuO_2 on discharging (Su et al., 2007; Wu et al., 2002).

The behavior of the ruthenium oxides in the electrochemical process is also affected by specific surface area, the combined water in RuO_x, the crystallinity, the size of $RuO_2 \cdot H_2O$, and electrolytes used (Wang et al., 2012).

As an alternative material to RuO_2, MnO_2 has been proposed. It is cheaper than ruthenium oxides and an environmentally friendly material, which exhibits electrochemical behavior similar to RuO_2. The main pseudocapacitive energy storage mechanism in manganese oxides is attributed to reversible redox transitions involving the exchange of protons and/or cations with the electrolyte, as well as the transitions between Mn(III)/Mn(II), Mn(IV)/Mn(III), and Mn(VI)/Mn(IV) (Hu and Tsou, 2002; Messaoudi et al., 2001). The proposed mechanism is expressed in Eq. (9.3) (Wang et al., 2012; Hu and Tsou, 2002).

$$MnO_{\propto}(OC)_{\beta} + \delta C^+ + \delta e^+ \leftrightarrow MnO_{\alpha-\beta}(OC)_{\beta+\delta} \quad (9.3)$$

where C^+ denotes the protons and alkali metal cations (Li^+, Na^+, K^+) in the electrolyte, and $MnO_{\propto}(OC)_{\beta}$ and $MnO_{\alpha-\beta}(OC)_{\beta+\delta}$ are $MnO_2 \cdot nH_2O$ in high and low oxidation states, respectively.

Physical and chemical factors affect the pseudocapacitive behavior of manganese oxides (González et al., 2016). These include, among others: crystallinity, crystal structure, morphology, the thickness of the electrode layer, specific surface area, and pore structure (Wang et al., 2012).

The previously described carbon materials are currently mainly obtained from fossil sources. Fossil-based carbonaceous materials with their great range of allotropes (graphite, diamond, etc.), nanostructures (graphene, nanotube, fullerene) and good physicochemical properties (chemically and thermally stable) have been well proven in electrochemical applications regarding specific capacitance and cycling stability (Frackowiak and Béguin, 2001). Nevertheless, they are facing the problem of rising costs as fossil resources will be even scarcer in future.

A lot of research has been done in order to develop bio-based, highly porous materials which can substitute the conventional fossil-based electrode materials. As the charge-storage mechanism in supercapacitors is based on electrostatic adsorption and pseudocapacitive effects (including electro-sorption, redox-reactions, and intercalation processes on the surface of the conductive electrodes), the viability of a new electrode material highly depends on the respective porosity or specific surface area (Miller, 2011).

Bio-based carbonaceous materials characterized by a lower production cost (due to use of abundant and low-cost biomass) will be competitive as long as they reach the same electrochemical performance as state-of-the-art materials. Trials with activated hydochars (product from hydrothermal carbonization) and pyro-chars (product from pyrolysis) of different precursors have already been conducted and showed that under certain conditions (organic electrolyte, hydro-thermal carbonization of cellulose, starch, sawdust activated at 700 and 800°C with KOH/sample weight ratio of 4) the bio-based materials perform even better than the conventional one regarding the specific capacitance (= 236 F g^{-1} at 1 mV s^{-1} for bio-based electrodes) (Zhang et al., 2009a; Khairnar et al., 2008).

Furthermore, the electrochemical performance of activated hydrochars containing heteroatoms (e.g., nitrogen) or metal oxides (e.g., ruthenium oxide RuO_2) has been studied with the result that the specific capacitances of the functionalized or composite materials reached values up to 300 Fg^{-1} and 333 Fg^{-1}, respectively. It is assumed that the combination of capacitance and enhanced pseudocapacitance leads to this superior capacitance values (Su et al., 2007; Titirici et al., 2012; Li et al., 2007).

Nevertheless, it still remains unclear, which process leads to the most viable material for application in supercapacitors. As not only the specific surface area but also the pore size distribution (PSD), the electric conductivity, and the surface functionalities play a major role in this context; different nanocasting methods of the carbonaceous material have been tested (e.g., different activation methods; hydrothermal carbonization coating techniques with Mn, Fe, and Co; hydrothermal carbonization of silicon nanoparticles and glucose at 180°C) (Titirici and Antonietti, 2010). Also different biomasses have been used as precursor and the activation temperatures have been varied in order to understand which combination leads to which physicochemical properties of the activated carbon (Rodriguez Correa et al., 2016; Rodríguez Correa et al., 2017).

While untreated activated carbons have specific capacities of 100 F g^{-1} [37], oxygen-doped, porous carbon materials have achieved specific capacities of up to 220 F g^{-1} [56]. Nitrogen-doped activated carbons also show improved elec-trochemical properties (electrical conductivity, capacity) (Ania et al., 2007; Lota et al., 2005; Zhao et al., 2010b; Liu et al., 2016; Hulicova-Jurcakova et al., 2009). Up to 340 F g^{-1} were measured for microporous N-doped carbon ma-terials (Hulicova-Jurcakova et al., 2009). In addition, metal oxides such as RuO_2 or MnO_2 also provide higher capacities thanks to their pseudocapacitive properties (Zheng et al., 1995; Sassoye et al., 2009; Khomenko et al., 2006a; Wang et al., 2015). RuO_2 is the material with the most promising properties, since it alone has theoretical capacities between 1000 and 2000 F g^{-1} [63,66] and at the same time has a very good electrical conductivity with 105 S cm^{-1} [66]. In comparison, activated carbons only achieve electrical conductivities between 1250 and 3000 S m^{-1} [37]. Despite the excellent electrochemical properties of RuO_2, other metal oxides are preferable to RuO_2 from an economic point of view. Fe_3O_4 and MnO_2, for example, would be cheaper and environmentally friendly alternatives (Wang et al., 2015; Chen et al., 2013; Oh et al., 2015). However, the limiting factors here are the moderate electrical conductivity (e.g., 10^{-5} - 10^{-6} S cm^{-1} for MnO_2 (Bélanger et al., 2008)) and the relatively low capacitance values (e.g., 27 F g^{-1} for Fe_3O_4[69] and 160 F g^{-1} for MnO_2) (Khomenko et al., 2006b)).

From an environmental point of view, it would be even better to completely dispense with metal oxides or rare earths such as RuO_2. The incorporation of heteroatoms such as oxygen or nitrogen into the carbon lattice should therefore be investigated as a matter of priority. In principle, there are two possible methods: either the corresponding carbon material is subjected to an oxidative treatment with nitrogen/air or nitrogen- or oxygen-rich starting materials are carbonized so that the heteroatoms can already be integrated into the carbon material during carbonization (Titirici, 2013). The latter process is also referred to as in situ functionalization and offers the decisive advantage that two process steps (carbonization and functionalization) can be combined into one. Thus, in situ functionalization is not only more time-efficient but also more energy- and resource-efficient than the two-stage process.

The in-situ method for functionalization with nitrogen atoms has already been successfully applied on the basis of various nitrogen-rich starting materials such as melamine (Hulicova-Jurcakova et al., 2009), polyacrylonitrile (Béguin et al., 2005a), or polyvinylpyridine (Rufford et al., 2014). Oxygen-rich starting materials such as polysaccharide-based materials with an O content of 40–50 wt% (Zhao et al., 2010a) are suitable for functionalization with oxygen-rich surface groups. Apart from the starting materials mentioned above, which ultimately function as model biomasses, real biomasses have already been investigated in isolated cases. It could be shown that carbon materials from carbonized seaweed not only have a good conductivity, but also a better capacity (198 F g^{-1}) than commercial activated carbon (190 F g^{-1}), although the latter had a higher specific surface area. The good electrochemical performance was attributed to the presence of oxygen-containing surface functionalities and associated faradaic charge transfers (= pseudocapacity) (Raymundo-Piñero et al., 2006). Furthermore, carbon electrodes were produced from coffee grounds where capacities of up to 368 F g^{-1} and the presence of electrochemically active O- and N-surface groups could be determined (Titirici, 2013). However, partial pyrolysis (700°C) was used as the carbonization process, which resulted in a very low oxygen content of the coals (<5 wt.%) (Zhao et al., 2010a; Raymundo-Piñero et al., 2006; Béguin et al., 2005a). In contrast, the hydrothermal carbonization process (at 180–200°C) makes it easier to maintain the oxygen and nitrogen content of the starting material during carbonization and thus to produce carbons with a high proportion of N- and O-containing surface functionalities (Hu et al., 2010; Titirici and Antonietti, 2010). A further advantage of the hydrothermal carbonization process, taking into account resource and environmental aspects, is the low energy consumption due to the low process temperature.

It has therefore already been shown that the hydrothermal carbonization process is suitable for the production of bio-based, electrochemically active electrode materials from biomass. However, there are hardly any experiments with real biomass or organic residues. The latter would be of particular interest in the context of bioeconomy, since the conversion of regional organic waste materials into high-quality, functional materials for electromobility (supercapacitors) would be a resource-efficient and sustainable recycling method. If, in addition to the technological implementation, the positive, economic evaluation of the process in relation to the current benchmark were also successful, this could be an important step toward recycling management and bioeconomy. In view of the fact that in Germany alone 10.2 million tons of biowaste (2016) (Bundesamt, 2017) are produced per year and that the hydrothermal carbonization process is also suitable for the carbonization of wet biomass, the conversion of biowaste into high-quality carbon materials would be a highly interesting recycling option for organic residues in Germany and worldwide.

The current benchmark for commercial activated carbon for use in supercapacitors is currently about 28 $ kg^{-1} (for activated carbon with 100 F g^{-1}) (Yassine and Fabris, 2017). The production of bio-based activated carbons using hydrothermal carbonization and in situ functionalization should be possible at a lower price in order to make the desired substitution of commercial activated carbons possible.

In summary, the hydrothermal carbonization process, in combination with a functionalization process, is a promising path for the production of bio-based electrode materials for supercapacitors, as hydrothermal carbonization offers the following advantages over conventional processes:

- resource-saving: low-energy process, because of low process temperature and single-stage
- environmentally friendly: water as a reaction medium
- applicable to a wide range of biomasses (moist)

9.2.2 Batteries

Batteries are energy storage devices that have allowed different technologies to thrive: from smaller devices such as headsets, hearing aids, and smartphones to larger machines like electric cars. Furthermore, batteries and other energy storage systems are becoming a significant player in the transition toward renewable energies and fighting climate change. The greatest advantage of batteries is that they do not follow Carnot's principle as opposed to other thermal systems. Hence considerably higher energy conversion efficiencies can be achieved (Larcher and Tarascon, 2015).

A battery is an electrochemical cell that consists of two electrodes (usually metals) with different reduction potentials and an electrolyte. During the reduction and oxidation reactions an exchange of electrons and ions take place, and, in turn, a potential difference is generated that can be used to do *work*. Currently, several electronic devices work with rechargeable lithium-ion batteries.

9.2.2.1 Operation principle: Li-ion and Na-ions batteries

Li-ion batteries. As the name implies, these batteries exchange Li ions between the positive and the negative electrodes. The positive electrode consists of a current collector (typically copper), a binder to attach the active material to the current collector, and the active material, which is an inorganic material of the form LiM_xO_y, where M is a transition metal (e.g., Co, V, Ni, or Fe). These materials are organized in such a way (e.g., layers) that lithium can move in and out without disrupting the primary structure. The most common cathode material in commercially available lithium-ion batteries is $LiCoO_2$ Eq. (9.4); however, state-of-the-art materials have reached considerably higher specific energy densities and cyclabilities (Wu and Yushin, 2017). On the negative side of the electrochemical cell, the material is usually graphite. During charge, the lithium ion (Li^+) travels from the positive electrode toward the negative electrode and builds a lithiated carbon species formed by an intercalated array of graphene sheets and lithium (LiC_6). LiC_6 has a theoretical capacity of 372 Ah kg^{-1}, but in practice, the maximum capacity measured has been 335 Ah kg^{-1} for $Li_{0.9}C_6^{77}$. The anode volume should be reduced to improve the anode performance. Simultaneously, the amount of Li^+ per unit of volume should be increased. Ideally, the negative electrode should be pure metallic lithium (theoretical capacity of 3862 Ah kg^{-1}) (Winter et al., 1998). Moreover, this was one of the first configurations of lithium-ion batteries, and unwanted phenomena occurred during charge/discharge cycles. Including material deposition on sites with the same electrical potential as the anode (e.g., current collector), shape changes on the electrode due to the random deposition, like the formation of dendrites or filaments as well as energy losses due to exothermic reactions involving solid phase and morphology transformations, lead to cycling efficiencies lower than 99% (Winter et al., 1998; Huggins, 2010).

$$CoO_2 + Li^+ + e^- \underset{\text{charge}}{\overset{\text{discharge}}{\rightleftarrows}} LiCoO_2 \tag{9.4}$$

To overcome the issues with elemental lithium, carbon was chosen as the insertion material for anodes. Carbon showed higher specific charges and lower redox potentials than other metal oxides, chalcogenides, and polymers, as well as a better cycling performance (Winter et al., 1998). Graphite is a pure form of carbon that is chemically inert and thermally stable. Its crystalline structure of stacked graphene sheets allows lithium to be inserted in between the layers, changing the carbon stacking configuration from ABAB to an AAAA (Huggins, 2010). The lithium insertion in the carbon structure, also known as *staging* occurs during charging Eq. (9.5) and structures of the form Li_xC_6 are formed, where x can adopt values between 0 and 1 at room temperature. The value of x depends on the availability of lithium ions, the energy required to expand the graphene layers to insert the lithium, and the repulsive interactions between species occupying the interstice. Four main stages (at thermodynamic equilibrium) have been identified with x values corresponding to 0.22, 0.33, 0.5, and 1[80]. Theoretically, the extraction of lithium is an entirely reversible reaction (discharging); however, this is not the case in the praxis. During the first charge-discharge cycle, between 5% and 20% of the lithium stays irreversibly in the carbon structure. For the following cycles, the recovery rate is close to 100%. The loss of lithium during the first cycle is usually ascribed to the formation of a solid electrolyte interface, which is a thin protective film that protects the Li_xC_6 compounds from the electrolyte and it is only permeable for lithium ions (Winter et al., 1998).

$$Li_xC_6 \underset{\text{charge}}{\overset{\text{discharge}}{\rightleftarrows}} xLi^+ + xe^- + C_6 \tag{9.5}$$

Producing a highly ordered carbon material, also known as highly oriented pyrolytic graphite, is a very costly process (different production techniques include chemical vapor deposition, mechanical or chemical exfoliation, or the reduction of simple sugars) (Brownson et al., 2011). Hence, soft and hard carbons have become the preferred active materials for lithium-ion battery anodes. However, its electrochemical behavior strongly depends on the degree of order of the carbon (including crystal defects and heteroatoms). This is of major importance when dealing with hard carbons since their degree of order is considerably lower. The first notable aspect is the wide hysteresis formed in the charge-discharge curve. This indicates that higher voltages are needed to remove the lithium from the carbon. To compensate this, more cathode material is needed; hence, the weight and price of the cell increase (Buiel and Dahn, 1999). Another problem with hard carbons is that the turbostratic character hinders the shifting of the graphene layers to form the AAAA stacking pattern (Dahn et al., 1995). The primary challenge for hard carbons in batteries application is to reduce the irreversible capacity, which is a consequence of the SEI formation and the interaction of lithium with heteroatoms.

Many attempts to reduce the irreversible capacity have been related to controlling the interaction between lithium and heteroatoms; however, F. Beguin et al. found out that the heteroatoms are not the main problem but the active sites available for the chemisorption of heteroatoms (active surface area). Their results showed a direct correlation between the active surface area and the irreversible capacity. By decreasing the active surface, for example, by protecting the precursor during the carbonization, a less developed solid electrolyte interface was observed (Béguin et al., 2005b).

The work of developing hard carbons has continued since its production is less cost-intensive and renewable precursors like biomass can be employed. Hard carbons from biomass can be tailored depending on the application by changing the carbonization process. Depending on the precursor and heat treatment, hard carbons from biomass have shown reversible capacities as high as 1800 Ah kg^{-1} [84]. F. Zhang et al. produced hierarchical porous carbon from rice straw via pyrolysis and KOH (potassium hydroxide) activation with a surface area larger than 3300 $m^2\ g^{-1}$. They measured reversible capacities close to 1000 Ah kg^{-1}, but the irreversible capacity was close to 1000 Ah kg^{-1} as well (Zhang et al., 2009b). In the case of microporous carbons, lithium is located in the micropores, which can act either as lithium reservoirs, contributing to the reversible capacity or as traps, contributing to the irreversible capacity, depending on their size and shape (Wu et al., 1999). J. Hou et al. synthesized carbons from silk with a specific surface area of 2240 $m^2\ g^{-1}$ composed by large meso- and micropore volumes. They measured reversible capacities close to 1850 Ah kg^{-1}, which were a result not only of the carbon morphology but also due to the presence of N-heteroatoms. However, the irreversible capacity was almost as high as the reversible capacity, which was a result of the solid electrolyte interface formation. The solid electrolyte interface covers most of the carbon surface area; hence, the irreversible capacity increases the surface area (Hou et al., 2015). Another critical parameter concerning hard carbons is the H-content. According to several studies, the presence of H-heteroatoms increases the reversible capacity since the C-Li binding energy is comparable to that of the C-H bond (Dahn et al., 1995). It was shown by S. Han et al. after studying the effect of pyrolysis temperature on the properties of carbons from tea leaves. They observed that the carbons obtained at lower pyrolysis temperatures, that is, those with the highest H/C ratios, showed higher reversible capacities at higher currents (Han et al., 2014b).

Another way to improve the capacity of a lithium-ion battery is to change the oxidation partner. From the electrochemical point of view, Li-F would be the best choice as a redox pair. The formation of metal fluorides is characterized by a vast free enthalpy change and, consequently, high potential differences (5.88 V). However, fluorine is a gas and has a high chemical reactivity, which makes it difficult to incorporate it as an electrode element. Furthermore, the reversibility of the fluorine-metal reactions has not been demonstrated yet (Anji Reddy and Fichtner, 2011). Two alternatives to fluorine are oxygen (Li-air battery) and sulfur (Li-S battery). The Li-S battery has been gaining increasing attention due to its high theoretical specific capacity (1675 Ah kg^{-1}) and power density (2600 Wh kg^{-1}) (Zhang, 2013). These values were calculated assuming pure lithium as the negative electrode; however, a lithium anode in Li-S systems suffers also from dendrite formation and low coulombic efficiencies due to the interaction with polysulfides, an intermediate product of the sulfur reduction reaction. Consequently, Li-S batteries have usually a short cycle life, low charging efficiencies, poor safety, and a high self-discharge rate (Zhang et al., 2018). A possible solution to this problem is to use carbon materials due to their relatively high electrical conductivity, mechanical stability, and porous structure. At the moment, much effort has been dedicated to studying graphene and graphene derivates in this context, which leaves an open window to explore hard carbons from biomass (Zhang et al., 2018; Brückner et al., 2014; Schuster et al., 2012; Abdul Razzaq et al., 2019; Jin et al., 2018).

Na-ion batteries. Lithium has significant advantages over other materials: (i) it is the lightest metal available, (ii) it has one of the lowest reduction potentials ($E°_{(Li+/Li)} = -3.04$ V versus standard hydrogen electrode), and (iii) it has one of the fastest ion mobility in solids (Ellis and Nazar, 2012). Therefore, it has been the preferred material in the development of batteries. Nevertheless, lithium is a relatively rare metal, and the largest reserves are in America (Bolivia, Chile, Argentina, and the United States), China, and Australia. On the other hand, most of the processing takes place in Asia. Consequently, the production costs of lithium-ion batteries are remarkably high and the CO_2-print can be compared to that of the combustion of 35 L of gasoline (Larcher and Tarascon, 2015). Additionally, other metals used in the electrodes such as cobalt, are also limited. Another downside of lithium-ion batteries is recycling. This task is still in its infancy due to the high toxicity of the compounds and the difficulty in separating the active material from the current collector (Zeng et al., 2014). Hence, despite the prevalence, versatility, and formidable specific energy and power of lithium-ion batteries, there is currently a great deal of research involved in improving the environmental character of lithium-ion batteries. There are several alternatives to achieve this; for example, developing other technologies involving lithium, like Li-S (Zhang, 2013) or Li-Air (Girishkumar et al., 2010) batteries, or by replacing lithium with other active materials, like sodium. Sodium batteries were the prelude to lithium-ion batteries. In the late 1960s, engineers from the Ford Motor co. laboratories developed a solid material called sodium beta alumina with interesting ionic transport properties (Kummer and Weber, 1968). This material was used as an electrolyte in a battery with a liquid electrode and a solid electrolyte. Due to its lower weight and higher voltage, analogous systems with lithium were conducted, introducing the development of lithium-ion batteries (Huggins, 2010).

Sodium, being also an alkali metal, has similar properties to lithium. Sodium has a higher standard potential ($E°_{(Na+/Na)} = -2.71$ V versus standard hydrogen electrode) than lithium, and its theoretical capacity is 1165 Ah/kg. However, the most significant advantage of sodium is its availability since it can be found everywhere (Dahbi and Komaba, 2015). On the other hand, it is a heavier metal; the cation radius is 30% larger than that of lithium which reduces the specific capacity. Hence, it does not form stable intercalation compounds with graphite (electrochemical reduction experiments under vacuum or helium have demonstrated that the compound NaC_{64} can be formed, but the insertion capacity is considerably lower than that of Li^+ and even K^+) (GE, 1988).

The electrochemistry of Na-insertion in the carbon structure has been studied to a much lesser extent than lithium. M. Doeff et al. studied the insertion mechanism of Na in petroleum coke and graphite using organic and polymer electrolytes. They identified that the maximum composition was NaC_{15} for graphite and NaC_{24} for petroleum coke (Doeff et al., 1993). In 2000, D. Stevens and J. Dahn showed that Na^+ ions could be inserted in hard carbons made of pyrolyzed glucose, and they measured reversible capacities of approximately 300 Ah/kg[102]. This study was followed by P. Thomas and D. Billaud who replicated the experiments using different types of hard carbons, obtaining similar results. They proposed that the sodium uptake in hard carbons follows a mechanism that consists of adsorption on the graphene layers and insertion in the porosity (Thomas and Billaud, 2002). R. Alcántara et al. compared three different hard carbons in lithium and sodium cells. The carbon with the lowest surface area showed the highest reversible capacity and lowest irreversible capacity for both ions. Based on this finding, they proposed that sodium ions are more likely to be inserted in pores and spaces between bent graphene layers (Alcántara et al., 2001). S. Wenzel et al. also found a relation between the textural properties of carbon and Na-insertion. They compared six samples with different surface areas and pore size distributions and found that the carbon with a more developed and controlled macro- and mesoporosity (template carbon) showed the highest reversible capacity (ca. 130 Ah/kg) (Wenzel et al., 2011). Also, R. Alcántara et al. produced carbon microspheres with a surface area of 3 m^2/g using a mixture of resorcinol and formaldehyde, which leads to reversible capacities of 285 Ah/kg[106].

9.2.2.2 Bio-based materials in batteries

Until now, the positive results have been obtained mainly with hard carbons produced from fossil sources. However, hard carbons with different morphologies and promising performances can be synthesized parting from renewable sources, like biomass, as well. N. Sun et al. produced hard carbons with surface areas between 25 and 83 m^2/g parting from shaddock (pomelo) peel. The measured reversible capacities were between 260 and 430 Ah/kg, which they ascribe to the honeycomblike morphology and the large interlayer distance obtained with increasing pyrolysis temperature (Sun et al., 2015). I. Izanzar et al. produced hard carbons from date palm seeds and pulp carbonized between 800 and 1400°C, which presented reversible capacities between 280 and 300 Ah/kg and very high coulombic efficiencies of approx. 88% (Izanzar et al., 2018). It has also been shown, that by doping biomass-based hard carbons with N-heteroatoms improves the electrochemical performance of the cell by providing active sites for Na^+ ion storage (Zhang et al., 2016).

9.3 Energy conversion devices (fuel cells)

The chemical energy contained in the fuel can be converted directly into electricity using an electrochemical device called fuel cell. There are many different types of fuel cells available. They can be divided according to the type of used electrolyte, for example, proton exchange membrane (PEM), molten carbonate fuel cells (MCFC), and solid oxide fuel cells (SOFC). These fuel cells used polymer membrane, molten carbonates, and ceramic materials, respectively, for conducting ions between electrodes. On the other hand, it may be distinguished by the type of used fuel, for example, direct methanol fuel cell (DMFC) and direct carbon fuel cell (DCFC). In the fuel cells usually, gaseous (hydrogen, carbon monoxide, and directly reformed methane) or liquid (methanol) fuels are used (Cao et al., 2007; Giddey et al., 2012).

A typical fuel cell (Fig. 9.6) is built of three main elements including an anode, electrolyte, and cathode. The electrolyte is designed considering two

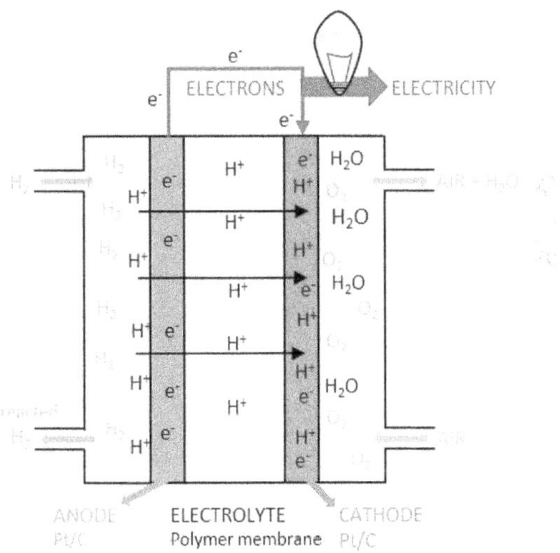

FIGURE 9.6 A simplified model of a proton exchange membrane fuel cell. *Own image. (no permission needed, Copyright M. P. Olszewski).*

crucial functions. First, it must be an electrical insulator that separates two electrodes (anode and cathode). Second, it must conduct the ions between electrodes, to chemical reactions may occur. On the other hand, the cathode and the anode should be good conductors, to allow the flow of electrons. The fuel (hydrogen) is supplied to the anode where hydrogen ions and electrons are created as shown in Eq. (9.7). Then, the hydrogen ion goes through the electrolyte to the cathode. The electrons flow through the electric circuit from anode to cathode and generate electric current. The oxidizing agent (air or oxygen) is delivered to the cathode, where electrons are consumed. Oxidizing of the fuel occurs according to Eq. (9.6). The process results in production of electricity as well as water formation as a byproduct (Cao et al., 2007; Giddey et al., 2012). The overall reaction for a fuel cell is described by Eq. (9.8) (Daud et al., 2017). The electrode potentials (E°) are presented versus the standard hydrogen electrode.

$$\text{Cathode reaction}: \frac{1}{2}O_2 + 2H^+ + 2e^- \rightarrow H_2O \quad E^{\circ} = 1.23V \tag{9.6}$$

$$\text{Anode reaction}: H_2 \rightarrow 2H^+ + 2e^- \quad E^{\circ} = 0V \tag{9.7}$$

$$\text{Overal reaction}: H_2 + \frac{1}{2}O_2 \rightarrow H_2O \quad E^{\circ} = 1.23V \tag{9.8}$$

The factor affecting the efficiency of the process is the conversion of the fuel in the cell. The value of the factor depends on the type of fuel cell, operating temperature, and used fuel. Generally, for gaseous and liquid fuels the conversion is limited up to 85%. It is caused by not fully reacted fuel and separation of byproducts. The net electrical efficiency for fuel cell used gaseous, and liquids fuels are in the range of 40%–60%. However, part of the energy contained in the fuel is converted into heat. The heat can be recovered in the cogeneration system and leads to an increase in the overall electrical efficiency to ~80% (Giddey et al., 2012; Cooper and Selman, 2009; Cherepy et al., 2005).

9.3.1 Direct carbon fuel cell (DCFC)

One of the first direct carbon fuel cell (DCFC) was patented in 1891 by Thomas A. Edison (Thomas, 1891). A DCFC uses a solid carbon-rich material (biomass, coal, charcoal) as a fuel. It distinguishes a DCFC compare to other types of fuel cells, for example, the proton exchange membrane and the direct methanol fuel cell, which use hydrogen and methanol as a fuel. The construction of DCFC (Fig. 9.7) is very similar to the other fuel cells (Fig. 9.7); it also consists of an anode, electrolyte, and cathode. However, the operation principle is closer to solid oxide fuel cell and molten carbonate fuel cell than to proton exchange membrane fuel cell. Here, the air is also supplied to the cathode although oxygen is reduced

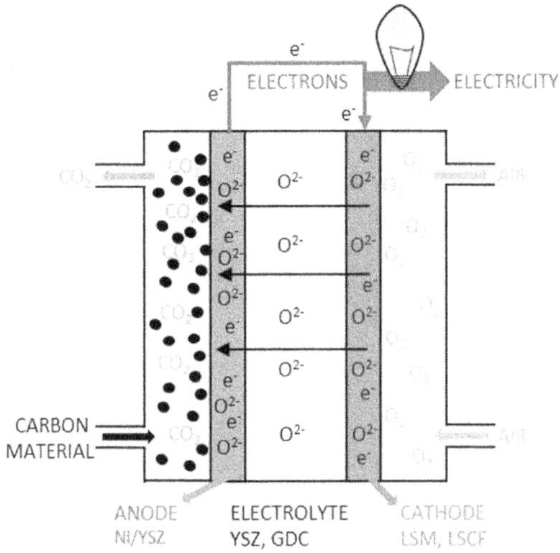

FIGURE 9.7 A simplified model of direct carbon fuel cell with solid oxide electrolyte. *Own image. (no permission needed, Copyright M. P. Olszewski).*

according to Eq. (9.9) consuming electrons and producing oxygen ion. Then the oxygen ions are transported by the electrolyte to the anode. The carbon material is used as fuel and must be supplied to the anode container during the process. The fuel is oxidized on the anode according to Eq. (9.10) by consuming oxygen ions and producing electrons and carbon dioxide as a byproduct. Electrons flow from the anode to the cathode and generate electrical power from the cell (Cao et al., 2007; Giddey et al., 2012). The overall reaction taking place in the cell may be represented as Eq. (9.11). In this type of fuel cells, the fuel, and any byproduct of the reaction remain on the anode side. However, they occur in gas (CO_2) and solid state. It results in easy separation of almost pure carbon dioxide which can be used for other purposes (for example, oil enhancement and feed to the greenhouses). Also, it keeps the chemical potentials of reactant and product at a constant level, regardless of the conversion of the fuel. It results in the theoretically full utilization of the fuel (close to 100%) in the single cell and maintaining theoretical voltage at $E^{\circ} = 1.02$ V. Theoretical efficiency of the direct carbon fuel cell due to the thermodynamics of the process is also close to 100% (Cao et al., 2007; Giddey et al., 2012; Cherepy et al., 2005; Zecevic et al., 2005). However, due to operating at high temperatures carbon dioxide can react with a solid carbon material in reverse Boudouard reaction Eq. (9.12). The operating temperature of the fuel cell has a significant effect on the kinetics of the reaction. Temperature above 705°C shifts the reaction to the right resulting in the formation of the product (CO). It causes carbon losses without generating electricity. In practice, the direct electrical efficiency obtained from DCFC stacks are up to 80%. However, the use of cogeneration can increase total efficiency up to 90% (Cao et al., 2007; Giddey et al., 2012).

$$\text{Cathodereaction}: O_2 + 4e^- \rightarrow 2O^{2-} \quad E^\circ = 0V \tag{9.9}$$

$$\text{Anodereaction}: C + 2O^{2-} \rightarrow CO_2 + 4e^- \quad E^\circ = 1.02V \tag{9.10}$$

$$\text{Overalreaction}: C + O_2 \rightarrow CO_2 \quad E^\circ = 1.02V \tag{9.11}$$

$$\text{Boudouardreaction}: C + CO_2 \leftrightarrow 2CO \quad E^\circ = 0V \tag{9.12}$$

Direct carbon fuel cells are characterized by higher electrical conversion efficiency $\eta_{el} \sim 80\%$ compared to conventional (thermal) electricity production which is thermodynamically limited by Carnot's theorem for heat engine cycles (Rankine cycle for coal, biomass, and nuclear power plants) where practical efficiency η_{el} is up to 45%. Electricity can also be generated using gas turbines, which are limited by the same theorem applying the Brayton cycle. When natural gas is used to produce electricity, gas turbines can be combined with steam turbines, which results in higher efficiency up to 64% (achieved by General Electric, 2017). Higher electrical efficiency for fuel cells is caused by direct electricity generation in electrochemical reactions which take place in the cell (Cooper and Selman, 2009; Cherepy et al., 2005). Moreover, coupling fuel cells in the trigeneration system (electricity, heating, and cooling generation) increases the efficiency of the system up to 90% (Giddey et al., 2012). Additionally, the utilization of charcoals in fuel cells leads to a reduction of flue gas volume compared to conventional combustion. It is due to the avoidance of mixing the fuel directly with the air, the electrooxidation of the fuel occurs on the anode side. It results in the production of relatively pure carbon dioxide which is not mixed with nitrogen contained in the air (Cao et al., 2007; Cooper and Selman, 2009). The high efficiency of the direct carbon fuel cell as well as the production of CO_2 which can be separated by a relatively low cost (due to high concentration) may lead to a future reduction of CO_2 emission.

Fuel cells are characterized by modular construction. To build a fuel cell stack, it is necessary to connect a large number of individual cells depending on the nominal power output. This creates a technical concern when constructing stacks with power in mega-watt (MW) scale, for example, managing the thermal load of the stack and issues related with replacing of damaged or defected cells (Chen et al., 2006; Faghri and Guo, 2005; Mueller et al., 2007). On the other hand, due to modular structure fuel cell stacks are accessible for downsizing (1–10 kW) without the efficiency losses. For comparison, a gas turbine with 1 kW electricity output has 50% less efficiency compared to hundreds of MWs industrial systems (Giddey et al., 2012). For these reasons, fuel cell stacks are suitable for decentralized electricity production in small scale. It will also ensure the reduction of losses caused by the transmission of electrical energy.

9.3.1.1 Operation principle: types of DCFCs

The direct carbon fuel cells operate at high temperatures from 500 to 1000°C. The need for high temperatures results from the very slow kinetics of electrochemical oxidation of carbonaceous materials (Cherepy et al., 2005; Vutetakis and Byker, 1987; Peelen et al., 2000; Thonstad, 1970). It results in the use of special materials for an electrolyte which will be thermally stable and able to transport ions at high temperatures. This introduces the division of direct carbon fuel cells due to the type of used electrolyte. Generally, three main types can be distinguished: ceramic, molten carbonate, and molten hydroxide (Cao et al., 2007; Giddey et al., 2012).

Ceramic. A direct carbon fuel cell using a ceramic electrolyte is called direct carbon solid oxide fuel cell. The solid oxide fuel cell operates at the highest temperatures (700 −1000°C) compared to other DCFC. It is due to properties of an electrolyte. The oxygen ions (O^{2-}) conductivity of ceramic electrolytes increases with temperature (Rady et al., 2012). Also, the use of high temperature may intensify the corrosion of cell components, limit the used materials (due to mechanical stress and thermal expansion), as well as increase the start-up time of the fuel cells stack. Also, it allows using a cheaper catalyst in comparison to platinum which is susceptible to CO poisoning (Rady et al., 2012). The simplified model of direct carbon solid oxide fuel cell is shown in Fig. 9.7. The working principle is described in Section 3.1 (direct carbon fuel cell).

As mentioned above, the electrolyte must conduct the oxygen ions and be an electric insulator to limit the loss of current in the cell. For this reason, the most commonly used materials for solid oxide fuel cell electrolytes are porous ceramic composites such as: yttrium oxide (Y_2O_3) in zirconium dioxide (ZrO_2) called Yttria-stabilized zirconia, scandium oxide (Sc_2O_3) in zirconium dioxide (ZrO_2) called scandia-stabilized zirconia, and Gandolinium in cerium oxide (CeO_2) called gandolinium-doped ceria (Rady et al., 2012). The operating temperatures depend on the ions conductivity of used electrolytes and are in the range of 800−1000°C and ~700°C for Yttria-stabilized zirconia and gandolinium-doped ceria, respectively.

The next structural element of the fuel cell is the cathode. In the solid oxide fuel cell, the cathode is made of thin sheet of porous ceramic which has to conduct electrons and oxygen ions. The typically used material for the cathode is lanthanum-strontium-manganite with the formula $La_{1-x}Sr_xMnO_3$, the final composition depends on the doping level with lanthanum and strontium. The lanthanum-strontium-manganite has similar thermal expansion factor to yttria-stabilized zirconia electrolyte which results in proper mechanical fitting and longer lifetime of the cell. A new promising cathode material is lanthanum-strontium-cobalt-ferrite. However, it is still under research for an industrial application in fuel cells (Rady et al., 2012).

The properties of the anode must be similar to the cathode; good conductivity of electrons as well as oxygen ions is required for proper operation of the fuel cell. The typical material used is a composite made of ceramic and metal called cermet. Yttria-stabilized zirconia is often chosen as a ceramic material while as a metal nickel (Ni) is used. It creates a functional configuration in combination with yttria-stabilized zirconia electrolyte due to similar thermal expansion factors. From the technical point of view, at the anode oxidation (Eq. 9.10) takes places which are related to the generation of CO_2 which involves the release of heat. For this reason, the anode needs cooling to avoid overheating of the cell materials (Rady et al., 2012). The solid oxide fuel cell reaches the power density in the range of 400−500 mW cm^{-2}, but the overall cost is high due to corrosive environment and application of rare-earth elements. The current research is focused on developing new materials (electrodes and electrolyte) which allow decreasing the operating temperature and as results decrease corrosion and broader a range of construction materials (Cao et al., 2007; Giddey et al., 2012; Rady et al., 2012).

Molten carbonate. The direct carbon fuel cell may also use molten carbonate salt as the electrolyte. This type of electrolyte is formed by a mixture of lithium carbonate (Li_2CO_3), sodium carbonate (Na_2CO_3), and potassium carbonate (K_2CO_3) and supported in a porous ceramic material (zirconia or alumina). The electrolyte can contain ternary eutectics of $(Li/Na/K)_2CO_3$ or binary $(Li/Na)_2CO_3$, or $(Li/K)_2CO_3$ (Cao et al., 2007; Giddey et al., 2012; Rady et al., 2012). These molten salts have to conduct the carbonate ions (CO_3^{2-}) as well as can work in a CO_2 environment. The operating temperature for fuel cell based on molten carbonate electrolyte is in the range of 600 −800°C to achieve optimal ions conductivity. The high temperature and the presence of molten salts limit the useable materials for the construction of the fuel cell due to corrosive conditions (Cao et al., 2007; Rady et al., 2012; Desclaux et al., 2011; Dicks, 2006). However, the temperature used in MCFC (600 −800°C) is lower compared to SOFC (800 −1000°C). It shifts the reverse Boudouard reaction (Eq. 9.12) to the left side and decreases the amount of produced CO in the exhaust gases, and rather clean CO_2 is produced (Cherepy et al., 2005; Rady et al., 2012).

The schematic model of the direct carbon molten carbonate fuel cell (DCMCFC) is shown in Fig. 9.8. The carbon material is delivered to the anode container. The oxidization of the fuel takes place on the anode according to Eq. (9.14). Carbon material reacts with carbonate ion (CO_3^{2-}) producing carbon dioxide (CO_2) and electrons. Electrons flow through the electric circuit from anode to cathode and generate electric current. Electrons are consumed together with oxygen and carbon dioxide on the cathode according to Eq. (9.13). It results in the formation of carbonate ions. The carbonate ions are transported by the electrolyte from the cathode to the anode. The electrolyte (carbonate ions) is consumed on the anode side and produced at the cathode side. It results in the consumption of CO_2 on the cathode side; to keep the

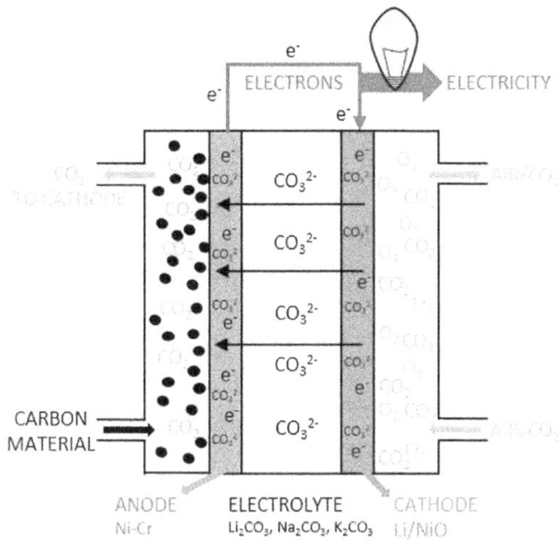

FIGURE 9.8 A simplified model of direct carbon fuel cell with molten carbonate electrolyte. *Own Image. (no permission needed, Copyright M. P. Olszewski).*

balance it is necessary to recirculate produced CO_2 at the anode side to the cathode container. Due to the presence of CO_2 in cathode reaction, the partial pressure of carbon dioxide affects the cell voltage, in practice voltage is ~0.8 V. The overall cell reaction can be expressed as Eq. (9.15) (Cao et al., 2007; Giddey et al., 2012; Rady et al., 2012).

$$\text{Cathode reaction}: 2CO_2 + O_2 + 4e^- \rightarrow 2CO_3^{2-} \quad E^{\circ} = 0V \tag{9.13}$$

$$\text{Anode reaction}: C + 2CO_3^{2-} \rightarrow 3CO_2 + 4e^- \quad E \sim 0.80V \tag{9.14}$$

$$\text{Overal reaction}: C + O_2 \rightarrow CO_2 \quad E \sim 0.80V \tag{9.15}$$

The most commonly used electrode materials are porous nickel oxide (NiO) doped with lithium (Li), and porous Ni/Cr alloy for cathode and anode, respectively. In the early stages of carbon fuel cell development, a carbon material was used as an anode (for example, graphite rod). However, it caused many problems among others poor mechanical stability of carbon electrodes and leakage of ash and other impurities directly to the electrolyte. Generally, MCFC offers power density at 120–160 mW cm^{-2} but lower temperatures allow using cheaper materials in the construction of the cells compared to SOFC (Giddey et al., 2012; Rady et al., 2012).

Molten hydroxide. The last category of carbon fuel cells uses molten hydroxide (MHFC) as an electrolyte. This fuel cell has an identical structure to molten carbonate fuel cell. However, the electrolyte instead of sodium and potassium carbonates is built from molten sodium hydroxide (NaOH) or potassium hydroxide (KOH) (Cao et al., 2007; Giddey et al., 2012; Rady et al., 2012). The use of molten hydroxides has several advantages over other types of fuel cells. It has a higher conductivity of the ions, faster carbon oxidation kinetics, and lower melting point. These features allow decreasing the operating temperature of molten hydroxide fuel cell to 600°C. It shifts the equilibrium of the Boudouard reaction (Eq. 9.12) to the left and decrease loses of carbon in the side reaction and increases the overall efficiency of the process (Cao et al., 2007). Additionally, the lower temperatures slightly decrease the corrosion of the cell and allow using cheaper construction materials, for example, stainless steel. These factors significantly reduce the price of the fuel cells stack production (Rady et al., 2012).

Fig. 9.9 shows a model of a direct carbon fuel cell with molten hydroxide electrolyte. The fuel in the form of the carbon material is delivered to the anode, where the electrooxidation reaction takes place according to Eq. (9.17). Carbon particles react with hydroxide ions (OH^-) and produce carbon dioxide (CO_2), water, as well as electrons. The electrons flow from the anode to the cathode through the electric circuit and generate electric current. The air is supplied to a cathode chamber; then the oxygen contained in the air reacts (Eq. 9.16) with the water molecule and consumes electrons delivered from the anode. The overall cell reaction is expressed as Eq. (9.18) (Giddey et al., 2012).

As the cell component materials graphite rods was initially used as the anode and fuel, whereas the cathode was made of cast iron. Such solution was patented in 1896 by William W. Jacques. He constructed a stack which was built from 100 singe fuel cells and achieved a total electric power of 1 kW[125]. Currently, the direct application of carbon material as an anode and as fuel has ceased to be used due to technical problems such as leakage at the anode side and the entry

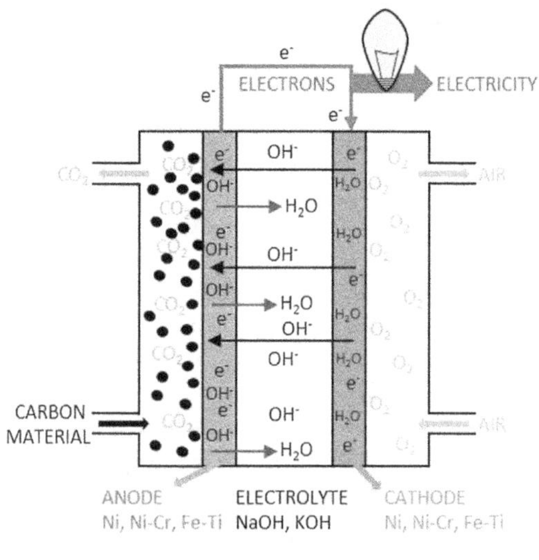

FIGURE 9.9 A simplified model of direct carbon fuel cell with molten hydroxide electrolyte. *Own Image. (no permission needed, Copyright M. P. Olszewski).*

of oxygen from the atmosphere that decreases the efficiency. Additionally, the impurities contained in the carbon material go directly to the electrolyte which caused contamination (Giddey et al., 2012). Nowadays both electrodes in this type of fuel cell are made of metal alloys, for example: (Ni mesh, Ni foam, Ni-Cr alloy, Fe-Ti alloy) (Zecevic et al., 2005; Guo et al., 2013).

Although this type of fuel cell has a drawback associated with the side reactions occurring in the cell, these reactions cause the formation of carbonate ions (CO_3^{2-}) and can be divided into two stages. The first stage is described by Eq. (9.19), and it is a fast chemical reaction. The second stage is an electrochemical reaction (Eq. 9.20) which is rate-limited. It results in reduced efficiency of the fuel cell due to the lower ionic conductivity of carbonate ions compared to hydroxide ions (OH^-) (Cao et al., 2007; Giddey et al., 2012; Zecevic et al., 2005; Rady et al., 2012). The power densities produced by direct carbon molten hydroxide fuel cell are in the range of 40–60 mW cm^{-2}. For these reasons, these cells are not widely used on an industrial scale (Cao et al., 2007; Rady et al., 2012; Guo et al., 2013).

$$\text{Cathode reaction: } O_2 + 2H_2O + 4e^- \rightarrow 4OH^- \quad (9.16)$$

$$\text{Anode reaction: } C + 4OH^- \rightarrow CO_2 + 2H_2O + 4e^- \quad (9.17)$$

$$\text{Overal reaction: } C + O_2 \rightarrow CO_2 \quad (9.18)$$

$$\text{Side reaction 1: } 6OH^- \leftrightarrow 3O^{2-} + 3H_2O \quad (9.19)$$

$$\text{Side reaction 2: } C + 3O^{2-} \rightarrow CO_3^{2-} + 4e^- \quad (9.20)$$

9.3.1.2 Bio-based carbon materials for DCFCs

As described above in the early stages of development of direct carbon fuel cells in the late 19th century, graphite rod was used as a fuel and as an anode due to its high electrical conductivity. During the operation of the cell, the anode was consumed; it was dissolved into the electrolyte ("like an icicle in warm water"). However, graphite is a very stable allotropic form of carbon, and therefore graphite has low reactivity. Also, graphite is a very expensive material to use for electricity generation. This introduced the need to use another carbon material, for example, coal or charcoal. It caused technical problems with the cell's operation due to impurities contained in the coals and charcoals (mainly mineral substances and sulfur) (Cao et al., 2007; Giddey et al., 2012; Rady et al., 2012). The direct use of coal as the anode leads to the penetration of a mineral substance (ash) and unreacted particles of coal into the electrolyte. It resulted in contamination of the electrolyte and unstable cell operation. In addition, consumption of the anode rod leads to leakage of oxygen from the atmosphere which resulted in side oxidation due to high operating temperatures of the cell. Moreover, coals and chars have much lower electrical conductivity compared to graphite. This resulted in the introduction of a separate anode in direct carbon fuel cells; on the anode's surface occurs electrooxidation of carbon particles as well as a collection of the electric charge. For this purpose, good contact of fuel with the surface of the anode is needed (Cao et al., 2007; Giddey et al., 2012). To increase contact between solid fuel particles and the anode in the direct carbon fuel cells operating in a flow mode is required (for example, a fluidized bed).

The properties of carbon materials used as a fuel in direct carbon fuel cells depend on the type of fuel cell used. Generally, the fuel particles should be a small size to minimize the mass transfer limitations. Sulfur contained in the fuel must be removed to avoid corrosion problems at high temperatures. Also, other impurities should be reduced to possibly lowest level to extend the lifetime of the fuel cell. The production of solid carbonaceous fuels from biomass and agricultural wastes in pyrolysis or either hydrothermal carbonization process is cheaper and requires less energy than production of hydrogen or hydrogen-rich fuels (Cao et al., 2007; Giddey et al., 2012; Dicks, 2006).

9.3.2 Microbial fuel cells (MBFCs)

In general, a *fuel cell* refers to an electrochemical device which produces electricity from chemical energy contained in a fuel by separating oxidation and reduction reactions. Oxygen or other exogenous electron acceptors like carbonates, sulfates, and nitrates are commonly employed as the terminal electron acceptors at the cathode forming water, methane, hydrogen sulfide, or ammonia, respectively, depending on the one used. This therefore suggests that a microbial fuel cell (MBFC) is a bioelectrochemical system that generates electrical energy by harnessing the natural metabolism of microbes.

9.3.2.1 Operation principle

By principle, microbes within the anode chamber of the MBFC consume the nutrients, usually organic matter such as that which is contained in wastewater and convert a portion of the chemical energy contained in the carbon-source into electricity. Theoretically, most microorganisms are potential biocatalysts in microbial fuel cells and the earliest MBFC concept was demonstrated by Potter in 1910. It was later in 1912; Potter reported that viable cultures of *Saccharomyces* and *Escherichia coli* produced electrical power after using platinum electrodes in the culture media. However, this new area of research did not attract the attention of most researchers until 1980s, when addition of electron mediators was found to greatly enhance power output and the current density (Du et al., 2007). In general, MBFC comprise of a negatively charged electrode called cathode, a positively charged electrode called anode, an ion-selective membrane for exchange of protons between the chambers, and an electrical circuit. The bacteria (typically anaerobic or facultatively anaerobic) live in the anode (Du et al., 2007) due to both internal and external porosity and convert a substrate such as glucose or acetate into carbon dioxide (CO_2), protons, and electrons. Under aerobic conditions, bacteria use oxygen (O_2) or nitrate as the terminal electron acceptor to produce H_2O or ammonia. However, in the anode of an MBFC, no oxygen is present (anoxygenic) and the bacteria need to switch from their natural electron acceptor to an insoluble alternative terminal electron acceptor such as the MBFC anode made from pyrolyzed biomass. Interestingly, certain chemical compounds are used as electron mediators to accelerate the electron transfer processes as reported by (Davis and Higson, 2007).

An appropriate design and operation of an MBFC greatly determine its performance (Du et al., 2007). An MBFC can be either double or single-chambered depending on the need for a cathode chamber, which is completely absent (exposure to air instead) in the latter. Strikingly, in double-chambered MBFC a close proximity between the anode and the cathode over a large surface area of proton exchange membrane results in a low internal resistance, leading to ultimate increase in voltage efficiency of the MBFC (He and Angenent, 2006).

Applications, prospects, and current challenges. Although, since the conception of its idea, MBFC technology had a reputation only of a laboratory scale curiosity rather than a practical technology (Du et al., 2007), it however still bears numerous potential applications in different fields; these include but is not limited to the following;

(a) *Generation of Electricity*: This involves microbial conversion of chemical energy stored in chemical compounds present in a biomass into electricity.

(b) *Production of Biohydrogen*: Biohydrogen can be produced instead of electricity by simple modification of the MBFC. In this case, the need for oxygen in the cathode chamber is eliminated leading to an improved MBFC efficiency since oxygen leak into the anode is no more a problem. MBFCs can therefore provide a renewable source of hydrogen that can significantly contribute to the overall hydrogen demand in a hydrogen economy (Du et al., 2007).

(c) *Treatment of Wastewater*: Since municipal wastewater contains a large amount of various organic compounds, it can be used as a fuel for MFCs operation. Consequently, the power generated from the MBFC during treatment of wastewater can potentially reduce the amount of electrical required for aeration of conventional activated sludge system. Similarly, the residual solid to be disposed is reduced by nearly 90% (Du et al., 2007). Interestingly, MBFC application in wastewater treatment has good operational stabilities, since the growth of the bioelectrochemically active microorganisms are enhanced during the process.

(d) *Biosensing*: MBFC technology is also used as a sensor in analyzing pollutants as well as in situ process monitoring and control (Chang et al., 2005). MBFC is a possible sensor for biological oxygen demand (BOD), because of a strong positive correlation between the wastewater strength and the coulombic yield of the MBFC. Thus, excellent operational stability as well as good reproducibility and accuracy are some of the advantages of MBFC-type BOD sensor over other types of BOD sensors.

Current Challenges in MBFC Technology. The scaleup of MBFC application remains a big challenge despite considerable improvements in reaching the primary power target (Pant et al., 2010). Additionally, the amount of power generated as well as the practical application of MBFCs are limited by certain factors such as the high cost of proton-exchange membrane as well as the potential biofouling effect of the electrode and associated high internal resistance as reported by Hu (2008). Another major challenge associated with MBFCs is the startup time, which may take from 4 to 103 days depending on the inoculum, electrode materials, reactor design, operating conditions such as temperature, as well as the type of substrate used in the system (Wang et al., 2009). Similarly, scaling-up of MBFCs in wastewater treatment is highly impeded by lack of buffering capacity of electrolytes (You et al., 2009).

Prospects of MBFCs. MBFCs are capable of converting biomass at low temperatures and low substrate concentrations; both of these are impossible for their methanogenic competitors (Pham et al., 2006). However, it is likely that MBFC technology will coexist in future with the methanogenic anaerobic digestion technology. This is because anaerobic digesters eliminate the need for biofilms in mediatorless electron transport, by efficiently reusing the microbial consortium without cell immobilization as opposed to the MBFCs that completely rely on biofilm formation in the absence of a mediator (Pham et al., 2006). Additionally, a breakthrough is highly needed to develop cheap electrodes with large surface area and porosity but resistant to biofouling; this is especially useful when MBFCs are employed in the treatment of wastewater (Du et al., 2007). Similarly, bio-based materials stand the possibility of replacing synthetic materials in microbial fuel cells in not so distant future.

9.3.2.2 Bio-based materials in microbial fuel cells

With increased technological advancements, several products from bio-based materials such as hydrochars and pyrochars have recently found application in new areas; these include the microbial fuel cell technology, which can bear a double-edged advantage of wastewater treatment and concomitant electricity generation. The electrode materials and the feedstock for the operation of MBFCs can be completely of bio-based origin.

Bio-based Electrodes. In Microbial Fuel Cells (MBFC), different carbon materials are used to prepare the anode electrode which serves as the alternative solid electron acceptor used by the electrogenic bacteria in the anode chamber.

These include but is not limited to graphite, carbon paper, carbon cloth, carbon fiber brush, carbon mesh, and more recently carbon nanotube (Cho et al., 2008; Zhou et al., 2012; Feng et al., 2010; Logan et al., 2007). However, most of these carbon materials commonly used as electrodes have an array of drawbacks ranging from high cost, low surface area, limited porosity, fouling effect (clogged by the electrogenic bacteria) to low electrical conductivity. Thus, the anode electrode plays a significant role in predicting the cost and performance of the MBFC. High electrical conductivity, high surface area and porosity, nonfouling, and cheapness characterize an ideal anode material (Logan et al., 2007). These greatly affect the power output of MBFCs, simply because of the more suitability of the conditions for the electrogenic bacteria to form biofilms and transfer electrons onto the anode electrode. Although bio-based materials such as biochars have relatively lower conductivity, yet they might replace the existing electrodes in the near future since they have high surface area, high porosity, and are resistant to biofouling.

Bio-based Feedstock in MBFC. Agricultural residues of lignocellulosic origin are a promising feedstock for cost-effective energy production due to their renewable nature and relative abundance (Huang and Angelidaki, 2008). However, most microbes cannot directly utilize this lignocellulosic biomass in MBFCs for electricity generation, and therefore need to be hydrolyzed to monosaccharides or other low-molecular-weight compounds as reported by Ren et al. (2007). Interestingly, all monosaccharides generated directly from hydrolysis of the lignocellulosic biomass are good substrates for electricity generation in MBFCs and could be used directly as they require no pretreatment. However, when cellulose is used as the substrate, electricity generation requires a microbial consortium possessing a good cellulolytic machinery and exoelectrogenic activities (Rezaei et al., 2009).

Several methods of substrate pretreatment have so far been tested ranging from hydrothermal and chemical treatments to mechanical fractionation. In an experiment by Zuo et al. (2006), waste biomass of corn stover treated with acid was used for electricity generation in MBFC. Similarly, the use of raw corn stover as a substrate for electricity generation in a single-chambered MBFC was also demonstrated by Wang et al. (2009), though the power output was much less than that obtained with glucose as substrate. Most bacteria are not known to be effective in the fermentation of pentoses (one of the main components in lignocellulose hydrolysates), thus rendering a large fraction of residual plant materials unsuitable for electricity generation. For instance, at a concentration of 10 mM, xylose, which is a typical pentose, yielded a power density of 69 mW m^{-2}, which was less than 97 mW m^{-2}, generated with glucose at the same concentration. This therefore suggests that xylose is more difficult to utilize for power generation than glucose (Huang and Angelidaki, 2008).

9.4 Conclusion

Carbon materials have a wide range of applications including simple combustion to generate heat through advanced carbon materials, the use as electrode materials in electric devices for energy storage (supercapacitors) as well as the use as fuel in highly efficient devices for the production of electrical energy from electrochemical reactions (DCFC). Power density and energy density characterize the energy storage devices and the fuel cells. Individual devices can be used for different purposes depending on the application; often they are used in combined systems, for example, in electric cars (battery plus supercapacitor).

Nowadays, bio-based carbon materials have similar or even better properties compared to their fossil-based counterparts. In supercapacitors, the charge is stored physically on the surface of carbon electrodes (e.g., activated carbons, carbons aerogels, carbon nanotubes). For this purpose, a high specific surface area, a certain pore-size distribution, and good electric conductivity are required—characteristics, which bio-based carbon materials do possess. In Li-ion batteries, the charge is stored chemically. Currently, for the production of batteries graphite is used as the anode and lithium ions in combination with graphite create an array of graphene sheets and lithium (LiC_6). In the case of microporous carbons produced from biomass, lithium is located in the micropores. An alternative solution for Li-ion batteries are Na-ion batteries, which avoid using a rare and expensive element such as lithium. However, this type of energy storage is still under investigation.

Direct carbon fuel cells (DCFC) are a promising technology for the production of electricity using solid fuels. Their advantage is the production of electricity directly from the electrochemical reactions. It results in a much higher electrical efficiency (80%) for direct carbon fuel cells compared to coal combustion plant (up to 45%). Historically in this type of fuel cell, the carbon material was used both as anode and fuel. However, due to technical problems such as the penetration of impurities from the carbonaceous anode to the electrolyte, a separate anode is currently used, and the carbon material is used as fuel. To increase the contact between the fuel and the anode, the anode compartment is operated in flow mode, for example, fluidized bed. For this reason, the fuel should have a fine grain size. Direct carbon fuel cells operate at high temperatures (600°C −1000°C), where corrosion of cell forming materials is severe. Therefore, it is necessary to find fuels with low sulfur and ash content, which is given in the case of bio-based carbon materials.

Since huge amounts of waste biomasses are available worldwide and carbonization processes such as hydrothermal carbonization or pyrolysis make it possible to convert these organic precursors into value-added, functional advanced carbon materials, biomass is an attractive and promising feedstock for the application in energy storage and conversion devices.

References

Abdul Razzaq, A., Yao, Y., Shah, R., Qi, P., Miao, L., Chen, M., Zhao, X., Peng, Y., Deng, Z., 2019. High-performance lithium sulfur batteries enabled by a synergy between sulfur and carbon nanotubes. Energy Storage Mater. 16, 194–202. https://doi.org/10.1016/j.ensm.2018.05.006.

Ahmadpour, A., Do, D.D., 1996. The preparation of active carbons from coal by chemical and physical activation. Carbon N. Y. 34 (4), 471–479. https://doi.org/https://doi.org/10.1016/0008-6223(95)00204-9.

Alcántara, R., Jiménez-Mateos, J.M., Lavela, P., Tirado, J.L., 2001. Carbon black: a promising electrode material for sodium-ion batteries. Electrochem. Commun. 3 (11), 639–642. https://doi.org/10.1016/S1388-2481(01)00244-2.

Ania, C.O., Khomenko, V., Raymundo-Piñero, E., Parra, J.B., Béguin, F., 2007. The large electrochemical capacitance of microporous doped carbon obtained by using a zeolite template. Adv. Funct. Mater. 17 (11), 1828–1836. https://doi.org/10.1002/adfm.200600961.

Anji Reddy, M., Fichtner, M., 2011. Batteries based on fluoride shuttle. J. Mater. Chem. 21 (43), 17059. https://doi.org/10.1039/c1jm13535j.

Béguin, F., Szostak, K., Lota, G., Frackowiak, E., 2005. A self-supporting electrode for supercapacitors prepared by one-step pyrolysis of carbon nanotube/polyacrylonitrile blends. Adv. Mater. 17 (19), 2380–2384. https://doi.org/10.1002/adma.200402103.

Béguin, F., Chevallier, F., Vix-Guterl, C., Saadallah, S., Bertagna, V., Rouzaud, J.N., Frackowiak, E., 2005. Correlation of the irreversible lithium capacity with the active surface area of modified carbons. Carbon N. Y. 43 (10), 2160–2167. https://doi.org/10.1016/J.CARBON.2005.03.041.

Bélanger, D., Brousse, T., Long, J.W., 2008. Manganese oxides: battery materials make the leap to electrochemical capacitors. Electrochem. Soc. Interface 17 (1), 49–52.

Bernstein, H., 2015. Bauelemente Der Elektronik. Walter de Gruyter GmbH, Berlin, München, Boston, p. 273. https://doi.org/10.1007/978-3-7091-7946-8_1.

Brownson, D.A.C., Kampouris, D.K., Banks, C.E., 2011. An overview of graphene in energy production and storage applications. J. Power Sources 196 (11), 4873–4885. https://doi.org/10.1016/j.jpowsour.2011.02.022.

Brückner, J., Thieme, S., Böttger-Hiller, F., Bauer, I., Grossmann, H.T., Strubel, P., Althues, H., Spange, S., Kaskel, S., 2014. Carbon-based anodes for lithium sulfur full cells with high cycle stability. Adv. Funct. Mater. 24 (9), 1284–1289. https://doi.org/10.1002/adfm.201302169.

Buiel, E., Dahn, J.R., 1999. Li-insertion in hard carbon anode materials for Li-ion batteries. Electrochim. Acta 45 (1), 121–130. https://doi.org/10.1016/S0013-4686(99)00198-X.

Bundesamt, S., 2017. Pressemitteilungen - 6 Kilogramm Pro Kopf Mehr Bioabfälle Im Jahr 2016 Gegenüber Vorjahr - Statistisches Bundesamt (Destatis).

Cao, D., Sun, Y., Wang, G., 2007. Direct carbon fuel Cell : fundamentals and recent developments. J. Power Sources 167, 250–257. https://doi.org/10.1016/j.jpowsour.2007.02.034.

Chang, I.S., Moon, H., Jang, J.K., Kim, B.H., 2005. Improvement of a microbial fuel cell performance as a BOD sensor using respiratory inhibitors. Biosens. Bioelectron. 20 (9), 1856–1859. https://doi.org/10.1016/j.bios.2004.06.003.

Chen, Q., Weng, Y., Zhu, X., Weng, S., 2006. Design and partial load performance of a hybrid system based on a molten carbonate fuel cell and a gas turbine. Fuel Cells 6 (6), 460–465. https://doi.org/10.1002/fuce.200500228.

Chen, L.Y., Hou, Y., Kang, J.L., Hirata, A., Fujita, T., Chen, M.W., 2013. Toward the theoretical capacitance of RuO 2 reinforced by highly conductive nanoporous gold. Adv. Energy Mater. 851–856. https://doi.org/10.1002/aenm.201300024.

Chen, X., Paul, R., Dai, L., 2017. Carbon-based supercapacitors for efficient energy storage. Natl. Sci. Rev. 4 (3), 453–489.

Cherepy, N.J., Krueger, R., Fiet, K.J., Jankowski, a F., Cooper, J.F., 2005. Direct conversion of carbon fuels in a molten carbonate fuel cell. J. Electrochem. Soc. 152, A80–A87. https://doi.org/Doi 10.1149/1.1836129.

Cho, Y.K., Donohue, T.J., Tejedor, I., Anderson, M.A., McMahon, K.D., Noguera, D.R., 2008. Development of a solar-powered microbial fuel cell. J. Appl. Microbiol. 104 (3), 640–650. https://doi.org/10.1111/j.1365-2672.2007.03580.x.

Christen, T., Carlen, M.W., 2000. Theory of ragone plots. J. Power Sources 91 (2), 210–216. https://doi.org/https://doi.org/10.1016/S0378-7753(00)00474-2.

Cooper, J.F., Selman, J., 2009. Electrochemical oxidation of carbon for electric power generation: a review. ECS Trans. 19, 15–25. May 2012.

Dahbi, M., Komaba, S., 2015. Fluorine chemistry for negative electrode in sodium and lithium ion batteries. In: Advanced Fluoride-Based Materials for Energy Conversion, pp. 387–414. https://doi.org/10.1016/B978-0-12-800679-5.00016-6.

Dahn, J.R., Zheng, T., Liu, Y., Xue, J.S., 1995. Mechanisms for lithium insertion in carbonaceous materials, 80 Science 270 (5236), 590 LP-593.

Daud, W.R.W., Rosli, R.E., Majlan, E.H., Hamid, S.A.A., Mohamed, R., Husaini, T., 2017. PEM fuel cell system control: a review. Renew. Energy 113, 620–638. https://doi.org/https://doi.org/10.1016/j.renene.2017.06.027.

Davis, F., Higson, S.P.J., 2007. Biofuel cells-recent advances and applications. Biosens. Bioelectron. 22 (7), 1224–1235. https://doi.org/10.1016/j.bios.2006.04.029.

Desclaux, P., Nürnberger, S., Rzepka, M., Stimming, U., 2011. Investigation of direct carbon conversion at the surface of a YSZ electrolyte in a SOFC. Int. J. Hydrogen Energy 36 (16), 10278–10281. https://doi.org/https://doi.org/10.1016/j.ijhydene.2010.10.027.

Dicks, A.L., 2006. The role of carbon in fuel cells. J. Power Sources 156 (2), 128–141. https://doi.org/https://doi.org/10.1016/j.jpowsour.2006.02.054.

Doeff, M.M., Ma, Y., Visco, S.J., Jonghe, L. C. De, 1993. Electrochemical insertion of sodium into carbon. J. Electrochem. Soc. 140 (12), 169–170. https://doi.org/10.1149/1.2221153.

Du, Z., Li, H., Gu, T., 2007. A state of the art review on microbial fuel cells: a promising technology for wastewater treatment and bioenergy. Biotechnol. Adv. 25 (5), 464–482. https://doi.org/10.1016/j.biotechadv.2007.05.004.

El-Kady, M.F., Shao, Y., Kaner, R.B., 2016. Graphene for batteries, supercapacitors and beyond. Nat. Rev. Mater. 1 (7), 1–14. https://doi.org/10.1038/natrevmats.2016.33.

Ellis, B.L., Nazar, L.F., 2012. Sodium and sodium-ion energy storage batteries. Curr. Opin. Solid State Mater. Sci. 16 (4), 168–177. https://doi.org/10.1016/j.cossms.2012.04.002.

Emaga, T.H., Bindelle, J., Agneesens, R., Buldgen, A., Wathelet, B., Paquot, M., 2011. Ripening influences banana and plantain peels composition and energy content. Trop. Anim. Health Prod. 43 (1), 171–177. https://doi.org/10.1007/s11250-010-9671-6.

European Commission, 2010. Report Forecasts Shortages of 14 Critical Mineral Raw Materials; Germany.

Faghri, A., Guo, Z., 2005. Challenges and opportunities of thermal management issues related to fuel cell technology and modeling. Int. J. Heat Mass Transf. 48 (19), 3891–3920. https://doi.org/https://doi.org/10.1016/j.ijheatmasstransfer.2005.04.014.

Falco, C., Marco-Lozar, J.P., Salinas-Torres, D., Morallón, E., Cazorla-Amorós, D., Titirici, M.M., Lozano-Castelló, D., 2013. Tailoring the porosity of chemically activated hydrothermal carbons: influence of the precursor and hydrothermal carbonization temperature. Carbon 62, 346–355. https://doi.org/10.1016/j.carbon.2013.06.017.

Feng, Y., Yang, Q., Wang, X., Logan, B.E., 2010. Treatment of carbon fiber brush anodes for improving power generation in air-cathode microbial fuel cells. J. Power Sources 195 (7), 1841–1844. https://doi.org/10.1016/j.jpowsour.2009.10.030.

Feytis, A., 2010. The Bright Side of Wrong.

Frackowiak, E., Béguin, F., 2001. Carbon materials for the electrochemical storage of energy in capacitors. Carbon N. Y. 39 (6), 937–950. https://doi.org/10.1016/S0008-6223(00)00183-4.

Frackowiak, E., Béguin, F., 2013. Supercapacitors Materials, Systems and Application. WILEY-VCH Verlag GmbH & Co. KG, Weinheim.

Franklin, R.E., 1951. Crystallite growth in graphitizing and non-graphitizing carbons. Proc. R. Soc. London. Ser. A. Math. Phys. Sci. 209 (1097), 196 LP-218.

Galiński, M., Lewandowski, A., Stepniak, I., 2006. Ionic liquids as electrolytes. Electrochim. Acta 51 (26), 5567–5580. https://doi.org/https://doi.org/10.1016/j.electacta.2006.03.016.

GE, P., 1988. Electrochemical intercalation of sodium in graphite. Solid State Ionics 28–30 (PART 2), 1172–1175. https://doi.org/10.1016/0167-2738(88)90351-7.

Ghosh, A., Hee, Y., 2012. Carbon-based electrochemical capacitors. ChemSusChem 5, 480–499. https://doi.org/10.1002/cssc.201100645.

Giddey, S., Badwal, S.P.S., Kulkarni, A., Munnings, C., 2012. A comprehensive review of direct carbon fuel cell technology. Prog. Energy Combust. Sci. 38 (3), 360–399. https://doi.org/https://doi.org/10.1016/j.pecs.2012.01.003.

Girishkumar, G., McCloskey, B., Luntz, A.C., Swanson, S., Wilcke, W., 2010. Lithium–Air battery: promise and challenges. J. Phys. Chem. Lett. 1 (14), 2193–2203. https://doi.org/10.1021/jz1005384.

González, A., Goikolea, E., Barrena, J.A., Mysyk, R., 2016. Review on supercapacitors: technologies and materials. Renew. Sustain. Energy Rev. 58, 1189–1206. https://doi.org/https://doi.org/10.1016/j.rser.2015.12.249.

Górka, J., Vix-Guterl, C., Matei Ghimbeu, C., 2016. Recent progress in design of biomass-derived hard carbons for sodium ion batteries. J. Carbon Res. https://doi.org/10.3390/c2040024.

Guo, L., Calo, J.M., DiCocco, E., Bain, E.J., 2013. Development of a low temperature, molten hydroxide direct carbon fuel cell. Energy Fuel. 27 (3), 1712–1719. https://doi.org/10.1021/ef302100h.

Han, S.-W., Jung, D.-W., Jeong, J.-H., Oh, E.-S., 2014. Effect of pyrolysis temperature on carbon obtained from green tea biomass for superior lithium ion battery anodes. Chem. Eng. J. 254, 597–604. https://doi.org/https://doi.org/10.1016/j.cej.2014.06.021.

Han, S.W., Jung, D.W., Jeong, J.H., Oh, E.S., 2014. Effect of pyrolysis temperature on carbon obtained from green tea biomass for superior lithium ion battery anodes. Chem. Eng. J. 254, 597–604. https://doi.org/10.1016/j.cej.2014.06.021.

He, Z., Angenent, L.T., 2006. Application of bacterial biocathodes in microbial fuel cells. Electroanalysis 18 (19–20), 2009–2015. https://doi.org/10.1002/elan.200603628.

Hou, J., Cao, C., Idrees, F., Ma, X., 2015. Hierarchical porous nitrogen-doped carbon nanosheets derived from silk for ultrahigh-capacity battery anodes and supercapacitors. ACS Nano 9 (3), 2556–2564. https://doi.org/10.1021/nn506394r

Hu, Z., 2008. Electricity generation by a baffle-chamber membraneless microbial fuel cell. J. Power Sources 179 (1), 27–33. https://doi.org/10.1016/j.jpowsour.2007.12.094.

Hu, C.C., Tsou, T.W., 2002. Capacitive and textural characteristics of hydrous manganese oxide prepared by anodic deposition. Electrochim. Acta 47 (21), 3523–3532. https://doi.org/10.1016/S0013-4686(02)00321-3.

Hu, B., Wang, K., Wu, L., Yu, S.H., Antonietti, M., Titirici, M.M., 2010. Engineering carbon materials from the hydrothermal carbonization process of biomass. Adv. Mater. 22 (7), 813–828. https://doi.org/10.1002/adma.200902812.

Huang, L., Angelidaki, I., 2008. Effect of humic acids on electricity generation integrated with xylose degradation in microbial fuel cells. Biotechnol. Bioeng. 100 (3), 413–422. https://doi.org/10.1002/bit.21786.

Huggins, R.A., 2010. Energy Storage. Springer US, Boston, MA. https://doi.org/10.1007/978-1-4419-1024-0.

Hulicova-Jurcakova, D., Kodama, M., Shiraishi, S., Hatori, H., Zhu, Z.H., Lu, G.Q., 2009. Nitrogen-enriched nonporous carbon electrodes with extraordinary supercapacitance. Adv. Funct. Mater. 19 (11), 1800–1809. https://doi.org/10.1002/adfm.200801100.

Izanzar, I., Dahbi, M., Kiso, M., Doubaji, S., Komaba, S., Saadoune, I., 2018. Hard carbons issued from date palm as efficient anode materials for sodium-ion batteries. Carbon N. Y. 137, 165–173. https://doi.org/10.1016/j.carbon.2018.05.032.

Jin, C., Sheng, O., Zhang, W., Luo, J., Yuan, H., Yang, T., Huang, H., Gan, Y., Xia, Y., Liang, C., et al., 2018. Sustainable, inexpensive, naturally multi-functionalized biomass carbon for both Li metal anode and sulfur cathode. Energy Storage Mater 15, 218–225. https://doi.org/10.1016/j.ensm.2018.04.001.

Khairnar, V., Jaybhaye, S., Hu, C.-C., Afre, R., Soga, T., Sharon, M., Sharon, M., 2008. Development of supercapacitors using porous carbon materials synthesized from plant derived precursors. Carbon Lett. 9 (3), 188–194. https://doi.org/10.5714/CL.2008.9.3.188.

Khomenko, V., Raymundo-Piñero, E., Frackowiak, E., Béguin, F., 2006. High-voltage asymmetric supercapacitors operating in aqueous electrolyte. Appl. Phys. Mater. Sci. Process 82 (4), 567–573. SPEC. ISS. https://doi.org/10.1007/s00339-005-3397-8.

Khomenko, V., Raymundo-Piñero, E., Béguin, F., 2006. Optimisation of an asymmetric manganese oxide/activated carbon capacitor working at 2 v in aqueous medium. J. Power Sources 153 (1), 183–190. https://doi.org/10.1016/j.jpowsour.2005.03.210.

Kötz, R., Carlen, M., 2000. Principles and applications of electrochemical capacitors. Electrochim. Acta 45 (15), 2483–2498. https://doi.org/https://doi.org/10.1016/S0013-4686(00)00354-6.

Kotz, R., Carlen, M., Kötz, R., Kötz, R., Carlen, M., Carlen, M., 2000. Principles and applications of electrochemical capacitors. Electrochim. Acta 45, 2483–2498. https://doi.org/10.1016/S0013-4686(00)00354-6.

Kummer, J.T., Weber, N., 1968. A sodium-sulfur secondary battery. SAE Intl. 1003–1007. https://doi.org/10.2307/44564986.

Kurzweil, P., Dietlmeier, O.K., 2015. Elektrochemische Speicher - Superkondensatoren, Batterien, Elektrolyse-Wasserstoff, REchtliche Grundlagen. Springer.

Larcher, D., Tarascon, J.-M., 2015. Towards greener and more sustainable batteries for electrical energy storage. Nat. Chem. 7 (1), 19–29. https://doi.org/10.1038/nchem.2085.

Lee, S.C., Jung, W.Y., 2016. Analogical understanding of the ragone plot and a new categorization of energy devices. Energy Procedia 88, 526–530. https://doi.org/https://doi.org/10.1016/j.egypro.2016.06.073.

Li, W., Chen, D., Li, Z., Shi, Y., Wan, Y., Wang, G., Jiang, Z., Zhao, D., 2007. Nitrogen-containing carbon spheres with very large uniform mesopores: the superior electrode materials for EDLC in organic electrolyte. Carbon N. Y. 45 (9), 1757–1763. https://doi.org/10.1016/j.carbon.2007.05.004.

Libich, J., Máca, J., Vondrák, J., Čech, O., Sedlaříková, M., 2018. Supercapacitors: properties and applications. J. Energy Storage 17, 224–227. https://doi.org/https://doi.org/10.1016/j.est.2018.03.012.

Liu, B., Zhang, L., Qi, P., Zhu, M., Wang, G., Ma, Y., Guo, X., Chen, H., Zhang, B., Zhao, Z., et al., 2016. Nitrogen-doped banana peel–derived porous carbon foam as binder-free electrode for supercapacitors. Nanomaterials 6 (1), 18. https://doi.org/10.3390/nano6010018.

Loeffler, N., Bresser, D., Passerini, S., 2015. Secondary lithium-ion battery anodes: from first commercial batteries to recent research activities. Johnson Matthey Technol. Rev. 59 (1), 34–44. https://doi.org/10.1595/205651314X685824.

Logan, B., Cheng, S., Watson, V., Estadt, G., 2007. Graphite fiber brush anodes for increased power production in air-cathode microbial fuel cells. Environ. Sci. Technol. 41 (9), 3341–3346. https://doi.org/10.1021/es062644y.

Lota, G., Grzyb, B., Machnikowska, H., Machnikowski, J., Frackowiak, E., 2005. Effect of nitrogen in carbon electrode on the supercapacitor performance. Chem. Phys. Lett. 404 (1–3), 53–58. https://doi.org/10.1016/j.cplett.2005.01.074.

Messaoudi, B., Joiret, S., Keddam, M., Takenouti, H., 2001. Anodic behaviour of manganese in alkaline medium. Electrochim. Acta 46 (16), 2487–2498. https://doi.org/10.1016/S0013-4686(01)00449-2.

Miller, J.M., 2011. Ultracapacitor Applications. IET Power and Energy Series, London.

Miller, J.R., 2016. Engineering electrochemical capacitor applications. J. Power Sources 326, 726–735. https://doi.org/https://doi.org/10.1016/j.jpowsour.2016.04.020.

Miller, J.R., 2017. Perspective on electrochemical capacitor energy storage. Appl. Surf. Sci. 460. https://doi.org/https://doi.org/10.1016/j.apsusc.2017.10.018.

Miller, J.R., Simon, P., 2008. Electrochemical capacitors for energy management. Science 321 (5889), 651 LP-652.

Mochidzuki, K., Soutric, F., Tadokoro, K., Antal, M.J., Tóth, M., Zelei, B., Várhegyi, G., 2003. Electrical and physical properties of carbonized charcoals. Ind. Eng. Chem. Res. 42 (21), 5140–5151. https://doi.org/10.1021/ie030358e.

Mueller, F., Jabbari, F., Gaynor, R., Brouwer, J., 2007. Novel solid oxide fuel cell system controller for rapid load following. J. Power Sources 172 (1), 308–323. https://doi.org/https://doi.org/10.1016/j.jpowsour.2007.05.092.

Oh, I., Kim, M., Kim, J., 2015. Deposition of Fe3O4 on oxidized activated carbon by hydrazine reducing method for high performance supercapacitor. Microelectron. Reliab. 55 (1), 114–122. https://doi.org/10.1016/j.microrel.2014.09.014.

Pandolfo, A.G., Hollenkamp, A.F., 2006. Carbon properties and their role in supercapacitors. J. Power Sources 157 (1), 11–27. https://doi.org/10.1016/j.jpowsour.2006.02.065.

Pant, D., Van Bogaert, G., Diels, L., Vanbroekhoven, K., 2010. A review of the substrates used in microbial fuel cells (MFCs) for sustainable energy production. Bioresour. Technol. 101 (6), 1533–1543. https://doi.org/10.1016/j.biortech.2009.10.017.

Paraknowitsch, J.P., Thomas, A., Antonietti, M., 2009. Carbon colloids prepared by hydrothermal carbonization as efficient fuel for indirect carbon fuel cells. Chem. Mater. 21 (7), 1170–1172. https://doi.org/10.1021/cm801586c.

Peelen, W.H.A., Olivry, M., Au, S.F., Fehribach, J.D., Hemmes, K., 2000. Electrochemical oxidation of carbon in a 62/38 mol % Li/K carbonate melt. J. Appl. Electrochem. 30 (12), 1389–1395. https://doi.org/10.1023/A:1026586915244.

Pham, T.H., Rabaey, K., Aelterman, P., Clauwaert, P., De Schamphelaire, L., Boon, N., Verstraete, W., 2006. Microbial fuel cells in relation to conventional anaerobic digestion technology. Eng. Life Sci. 6 (3), 285–292. https://doi.org/10.1002/elsc.200620121.

Rady, A.C., Giddey, S., Badwal, S.P.S., Ladewig, B.P., Bhattacharya, S., 2012. Review of fuels for direct carbon fuel cells. Energy Fuel. 26 (3), 1471–1488. https://doi.org/10.1021/ef201694y.

Raymundo-Piñero, E., Leroux, F., Béguin, F., 2006. A high-performance carbon for supercapacitors obtained by carbonization of a seaweed biopolymer. Adv. Mater. 18 (14), 1877−1882. https://doi.org/10.1002/adma.200501905.

Ren, Z., Ward, T.E., Regan, J.M., 2007. Electricity production from cellulose in a microbial fuel cell using a defined binary culture. Environ. Sci. Technol. 41 (13), 4781−4786. https://doi.org/10.1021/es070577h.

Rezaei, F., Xing, D., Wagner, R., Regan, J.M., Richard, T.L., Logan, B.E., 2009. Simultaneous cellulose degradation and electricity production by Enterobacter cloacae in a microbial fuel cell. Appl. Environ. Microbiol. 75 (11), 3673−3678. https://doi.org/10.1128/AEM.02600-08.

Rodriguez Correa, C., Otto, T., Kruse, A., 2016. Influence of the biomass components on the pore formation of activated carbon. Biomass Bioenergy 97, 53−64. https://doi.org/10.1016/j.biombioe.2016.12.017.

Rodríguez Correa, C., Stollovsky, M., Hehr, T., Rauscher, Y., Rolli, B., Kruse, A., 2017. Influence of the carbonization process on activated carbon properties from lignin and lignin-rich biomasses. ACS Sustain. Chem. Eng. 5 (9) acssuschemeng.7b01895. https://doi.org/10.1021/acssuschemeng.7b01895.

Rufford, T.E., Zhu, J., Hulicova-Jurcakova, D., 2014. Green Carbon Materials: Advances and Applications. Taylor&Francis Group, Boca Raton, Florida, p. 304. https://doi.org/10.4032/9789814411141.

Sassoye, C., Laberty, C., Khanh, H. Le, Cassaignon, S., Boissière, C., Antonietti, M., Sanchez, C., 2009. Block-copolymer-templated synthesis of electroactive ruo2-based mesoporous thin films. Adv. Funct. Mater. 19 (12), 1922−1929. https://doi.org/10.1002/adfm.200801831.

Schuster, J., He, G., Mandlmeier, B., Yim, T., Lee, K.T., Bein, T., Nazar, L.F., 2012. Spherical ordered mesoporous carbon nanoparticles with high porosity for lithium-sulfur batteries. Angew. Chem. Int. Ed. 51 (15), 3591−3595. https://doi.org/10.1002/anie.201107817.

Shukla, A.K., Banerjee, A., Ravikumar, M.K., Jalajakshi, A., 2012. Electrochemical capacitors: technical challenges and prognosis for future markets. Electrochim. Acta 84, 165−173. https://doi.org/https://doi.org/10.1016/j.electacta.2012.03.059.

Su, Y., Wu, F., Bao, L., Yang, Z., 2007. RuO2/Activated carbon composites as a positive electrode in an alkaline electrochemical capacitor. N. Carbon Mater. 22 (1), 4−9.

Sun, N., Liu, H., Xu, B., 2015. Facile synthesis of high performance hard carbon anode materials for sodium ion batteries. J. Mater. Chem. 3 (41), 20560−20566. https://doi.org/10.1039/C5TA05118E.

Thomas, A., 1891. Edison. Process of and Apparatus for Generating Electricity. US patent, 460122A.

Thomas, P., Billaud, D., 2002. Electrochemical insertion of sodium into hard carbons. Electrochim. Acta 47 (20), 3303−3307. https://doi.org/10.1016/S0013-4686(02)00250-5.

Thonstad, J., 1970. The electrode reaction on the C, CO2 electrode in cryolite-alumina melts-I. Steady state measurements. Electrochim. Acta 15 (10), 1569−1580. https://doi.org/10.1016/0013-4686(70)80079-2.

Titirici, M.-M., 2013. Sustainable Carbon Materials from Hydrothermal Process. John Wiley and Sons, West Sussex.

Titirici, M.-M., Antonietti, M., 2010. Chemistry and materials options of sustainable carbon materials made by hydrothermal carbonization. Chem. Soc. Rev. 39 (1), 103−116. https://doi.org/10.1039/b819318p.

Titirici, M.-M., White, R., Falco, C., Sevilla, M., 2012. Black perspectives for a green Future : hydrothermal carbons for environment protection and energy storage. Energy Environ. Sci. 5 (6796), 6796−6822. https://doi.org/10.1039/c2ee21166a.

Vagner, C., Finqueneisel, G., Zimny, T., Burg, P., Grzyb, B., Machnikowski, J., Weber, J.V., 2003. Characterization of the surface properties of nitrogen-enriched carbons by inverse gas chromatography methods. Carbon N. Y. 41 (14), 2847–2853. https://doi.org/10.1016/S0008-6223(03)00412-3.

Vutetakis, D.G., Byker, H.J., 1987. Electrochemical oxidation of molten carbonate-coal slurries. J. Electrochem. Soc. 134 (12), 3027–3035. https://doi.org/10.1149/1.2100334.

Wang, X., Feng, Y., Wang, H., Qu, Y., Yu, Y., Ren, N., Li, N., Wang, E., Lee, H., Logan, B.E., 2009. Bioaugmentation for electricity generation from corn stover biomass using microbial fuel cells. Environ. Sci. Technol. 43 (15), 6088–6093. https://doi.org/10.1021/es900391b.

Wang, G., Zhang, L., Zhang, J., 2012. A review of electrode materials for electrochemical supercapacitors. Chem. Soc. Rev. 41 (2), 797–828. https://doi.org/10.1039/C1CS15060J.

Wang, J.-W., Chen, Y., Chen, B.-Z., 2015. A synthesis method of MnO2/activated carbon composite for electrochemical supercapacitors. J. Electrochem. Soc. 162 (8), A1654–A1661. https://doi.org/10.1149/2.0031509jes.

Wang, Y., Song, Y., Xia, Y., 2016. Electrochemical capacitors: mechanism, materials, systems, characterization and applications. Chem. Soc. Rev. 45 (21), 5925–5950. https://doi.org/10.1039/C5CS00580A.

WeiZi, C., Qian, Z., YongMin, X., Jiang, L., GuoHui, L., Shuang, C., MeiLin, L., 2016. A direct carbon solid oxide fuel cell operated on a plant derived biofuel with natural catalyst. Appl. Energy 179, 1232–1241.

Wen, Y., He, K., Zhu, Y., Han, F., Xu, Y., Matsuda, I., Ishii, Y., Cumings, J., Wang, C., 2014. Expanded graphite as superior anode for sodium-ion batteries. Nat. Commun. 5. https://doi.org/10.1038/ncomms5033.

Wenzel, S., Hara, T., Janek, J., Adelhelm, P., 2011. Room-temperature sodium-ion batteries: improving the rate capability of carbon anode materials by templating strategies. Energy Environ. Sci. 4 (9), 3342. https://doi.org/10.1039/c1ee01744f.

Winter, M., Besenhard, J.O., Spahr, M.E., Novák, P., 1998. Insertion electrode materials for rechargeable lithium batteries. Adv. Mater. 10 (10), 725–763. https://doi.org/10.1002/(SICI)1521-4095(199807)10:10<725::AID-ADMA725>3.0.CO;2-Z.

Wu, F., Yushin, G., 2017. Conversion cathodes for rechargeable lithium and lithium-ion batteries. Energy Environ. Sci. 10 (2), 435–459. https://doi.org/10.1039/C6EE02326F.

Wu, Y.-P., Wan, C.-R., Jiang, C.-Y., Fang, S.-B., Jiang, Y.-Y., 1999. Mechanism of lithium storage in low temperature carbon. Carbon N. Y. 37 (12), 1901–1908. https://doi.org/10.1016/S0008-6223(99)00067-6.

Wu, N.L., Kuo, S.L., Lee, M.H., 2002. Preparation and optimization of RuO2-impregnated SnO2xerogel supercapacitor. J. Power Sources 104 (1), 62–65. https://doi.org/10.1016/S0378-7753(01)00873-4.

Yassine, M., Fabris, D., 2017. Performance of commercially available supercapacitors. Energies 10 (9). https://doi.org/10.3390/en10091340.

You, S.-J., Ren, N.-Q., Zhao, Q.-L., Kiely, P.D., Wang, J.-Y., Yang, F.-L., Fu, L., Peng, L., 2009. Improving phosphate buffer-free cathode performance of microbial fuel cell based on biological nitrification. Biosens. Bioelectron. 24 (12), 3698–3701. https://doi.org/10.1016/j.bios.2009.05.015.

Yun, Y.S., Park, K.-Y., Lee, B., Cho, S.Y., Park, Y.-U., Hong, S.J., Kim, B.H., Gwon, H., Kim, H., Lee, S., et al., 2015. Sodium-ion storage in pyroprotein-based carbon nanoplates. Adv. Mater. 27 (43), 6914–6921. https://doi.org/10.1002/adma.201502303.

Zecevic, S., Patton, E.M., Parhami, P., 2005. Direct electrochemical power generation from carbon in fuel cells with molten hydroxide electrolyte. Chem. Eng. Commun. 192 (12), 1655–1670. https://doi.org/10.1080/009864490896241.

Zeng, X., Li, J., Singh, N., 2014. Recycling of spent lithium-ion battery: a critical review. Crit. Rev. Environ. Sci. Technol. 44 (10), 1129–1165. https://doi.org/10.1080/10643389.2013.763578.

Zhang, S.S., 2013. Liquid electrolyte lithium/sulfur battery: fundamental chemistry, problems, and solutions. J. Power Sources 231, 153–162. https://doi.org/10.1016/J.JPOWSOUR.2012.12.102.

Zhang, S.S., 2017. Eliminating pre-lithiation step for making high energy density hybrid Li-ion capacitor. J. Power Sources 343, 322–328. https://doi.org/https://doi.org/10.1016/j.jpowsour.2017.01.061.

Zhang, L.L., Zhou, R., Zhao, X.S., 2009. Carbon-based materials as supercapacitor electrodes. J. Mater. Chem. 38 (29), 2520–2531. https://doi.org/10.1039/c000417k.

Zhang, F., Wang, K.-X., Li, G.-D., Chen, J.-S., 2009. Hierarchical porous carbon derived from rice straw for lithium ion batteries with high-rate performance. Electrochem. Commun. 11 (1), 130–133. https://doi.org/10.1016/J.ELECOM.2008.10.041.

Zhang, Y., Chen, L., Meng, Y., Xie, J., Guo, Y., Xiao, D., 2016. Lithium and sodium storage in highly ordered mesoporous nitrogen-doped carbons derived from honey. J. Power Sources 335, 20–30. https://doi.org/10.1016/J.JPOWSOUR.2016.08.096.

Zhang, Y., Gao, Z., Song, N., He, J., Li, X., 2018. Graphene and its derivatives in lithium–sulfur batteries. Mater. Today Energy 9, 319–335. https://doi.org/10.1016/j.mtener.2018.06.001.

Zhao, L., Fan, L.Z., Zhou, M.Q., Guan, H., Qiao, S., Antonietti, M., Titirici, M.M., 2010. Nitrogen-containing hydrothermal carbons with superior performance in supercapacitors. Adv. Mater. 22 (45), 5202–5206. https://doi.org/10.1002/adma.201002647.

Zhao, L., Crombez, R., Pérez, F., Antonietti, M., Texter, J., Titirici, M., 2010. Sustainable nitrogen-doped carbon latexes with high electrical and thermal conductivity. Polymer 51 (20), 4540–4546. https://doi.org/10.1016/j.polymer.2010.07.044.

Zheng, J.P., Cygan, P.J., Jow, T.R., 1995. Hydrous ruthenium oxide as an electrode material for electrochemical capacitors. J. Electrochem. Soc. 142 (8), 2699–2703.

Zhong, C., Deng, Y., Hu, W., Qiao, J., Zhang, L., Zhang, J., 2015. A review of electrolyte materials and compositions for electrochemical supercapacitors. Chem. Soc. Rev. 44 (21), 7484–7539. https://doi.org/10.1039/C5CS00303B.

Zhou, M., Chi, M., Wang, H., Jin, T., 2012. Anode modification by electrochemical oxidation: a new practical method to improve the performance of microbial fuel cells. Biochem. Eng. J. 60, 151–155. https://doi.org/10.1016/j.bej.2011.10.014.

Zuo, Y., Maness, P.C., Logan, B.E., 2006. Electricity production from steam-exploded corn stover biomass. Energy Fuel. 20 (4), 1716–1721. https://doi.org/10.1021/ef060033l.

Further reading

Alcantara, R., Lavela, P., Ortiz, G.F., Tirado, J.L., 2005. Carbon microspheres obtained from resorcinol-formaldehyde as high-capacity electrodes for sodium-ion batteries. Electrochem. Solid State Lett. 8 (4), A222. https://doi.org/10.1149/1.1870612.

Dahn, J.R., 1991. Phase diagram of LixC6. Phys. Rev. B 44 (17), 9170–9177. https://doi.org/10.1103/PhysRevB.44.9170.

Du, X., Wang, C., Chen, M., Jiao, Y., Wang, J., 2009. Electrochemical performances of nanoparticle Fe 3 O 4/activated carbon supercapacitor using KOH electrolyte solution. J. Phys. Chem. C (113), 2643–2646. https://doi.org/10.1021/jp8088269.

Moseley, P.T., Garche, J., 2014. Electrochemical Energy Storage for Renewable Sources and Grid Balancing. https://doi.org/10.1016/C2012-0-01253-7.

Potter, M.C., 1911. Electrical effects accompanying the decomposition of organic compounds. Proc. R. Soc. B Biol. Sci. 84 (571), 260–276. https://doi.org/10.1098/rspb.1911.0073.

Stevens, D.A., Dahn, J.R., 2000. High capacity anode materials for rechargeable sodium-ion batteries. J. Electrochem. Soc. 147 (4), 1271. https://doi.org/10.1149/1.1393348.

William, W., 1896. Jacques. Method of Converting Potential Energy of Carbon into Electrical Energy. US patent, 555511A.

Chapter 10

Bio-based textile coatings and composites

David De Smet, Frederik Goethals, Brecht Demedts, Willem Uyttendaele, Myriam Vanneste
Centexbel, Textile Competence Centre, Ghent, Belgium

10.1 Introduction

Generally there is a move toward renewable, natural, green, and environmental friendly products due to the increasing consciousness of consumers, the influence of environmental organizations (e.g., Greenpeace), and the impact of legislation (e.g., REACH). Furthermore, current global municipal waste in 2016 amounted to approximately 2.01 billion tonnes and is expected to increase to 3.4 billion tonnes by 2050 (World Bank, 2019). About 45.5% of waste were landfilled and 37.8% were recycled in the EU in 2016 (EUROSTAT, 2019). The European Coatings supply chain is increasing its efforts to ensure that a much higher proportion of its products are derived from renewable sources. The stimulus is the climate change. The European Union has set a target of 30% of existing petrochemical-derived chemicals and materials being replaced by bio-based and biodegradable alternatives by 2030. This objective is in line with the long-term aim of the European Union of establishing low-carbon, circular economies with substantial reductions in CO_2 emissions and greater reliance on its own raw materials with much more recycling and reuse of materials to achieve zero waste.

At the same time there is a widespread desire to move away from fossil fuels to reduce air pollution. There is also, however, a gradually urgent need for action to deal with frequent deficiencies of petrochemical-derived raw materials (Coatings World, 2018). The textile market covers a broad range of materials. Consequently, the textile industry has a significant impact on the environment. Therefore, extensive research in the field of bio-based, biodegradable, or even recyclable textiles is being done to reduce the environmental impact of textiles. CO_2-neutral renewable materials are being developed due to the depletion of fossil fuels and the impact of CO_2-emission on the global climate. Some fabrics are already made from renewable resources (e.g., cotton,

Biobased Products and Industries. https://doi.org/10.1016/B978-0-12-818493-6.00010-5

flax, etc.) but others are synthetic and oil-derived, such as polyester and polyamide. One response to this need has been the development of synthetic bio-based fibers from renewable raw materials. Many bio-based polymers have been developed, including polyethylene terephthalate (PET), polyethylene (PE), polylactic acid (PLA), starch blends, biodegradable polyesters such as polybutylene succinate (PBS) and poly(butylene adipate-co-terephthalate) (PBAT) and cellulose acetate. The development and production of bio-based fibers for textiles is widely reported and will not be discussed in this chapter. Textile fabrics can be used as reinforcement in composites or can be finished or coated to introduce new functionalities (e.g., flame retardancy, antimicrobial activity, etc.). The elaboration of novel bio-based finishing and coating formulations remains a challenge.

The use of textiles as reinforcement in bio-based composites and the development of bio-based coatings for textiles will be further discussed in the following paragraphs. In the following paragraphs an overview of some work done at Centexbel and other institutes in the area of bio-based textiles is presented as well as bio-based products commercially available for the textile industry (renewable textile-based composites and bio-based coating and finishing formulations).

10.2 Bio-based composites

Composite materials offer outstanding properties, such as lightweight, high specific strength, and corrosion resistance, which make them excellent for applications in aerospace, automotive, wind energy, sporting, transport, and construction sector. They are composed of reinforcing fibers (mostly glass, carbon, or aramid) and a bed matrix. The matrix can be thermosetting polymer (e.g., epoxy, vinyl ester, polyester) or thermoplastic (e.g., PP, PA, PEEK).

Bio-based composites cover composites that are made entirely or largely of biomass. As this section is part of the bio-based textiles chapter we will only focus on composites which contain bio-based textile reinforcements and bio-based matrices.

The market of fiber reinforced polymers are still dominated by nonnatural fibers (i.e., glass, carbon, aramid), but there is a growing interest in bio-based fiber reinforced materials. The main reason for this interest is the lower carbon footprint of these materials compared to the traditional composites made from glass and carbon fibers. Other interesting features are the low weight, interesting weight/performance balance, the special looks, and touch (Good market developments for bio-based composites—Bio-based News, 2018).

10.2.1 Bio-based fibers

Bio-based fibers can be derived from plants and animals of which the plant-based fibers are the most interesting one for composite applications because they have in general a higher stiffness and strength compared to animal fibers. Furthermore, from an economic and ecological point of view plant fibers can grow in many countries, are a renewable resource, and are considered as CO_2 neutral. To be competitive with synthetic fibers the bio-based fibers should have comparable mechanical properties.

A comprehensive overview of the mechanical properties of natural and synthetic fibers was prepared by Pickering et al. and is shown in Table 10.1 (Pickering et al., 2016).

From the table, it can be seen that the highest strength and stiffness are obtained for the cellulose-based fibers ramie, flax, and hemp. When compared to the e-glass fiber, which is the most used fiber in composites, it can be observed that glass fibers have still a higher strength in absolute values, but when looking at the specific tensile strength comparable values can be obtained with natural fibers and the specific young's moduli can even be higher.

Further, there is also a huge variation in mechanical properties within one fiber family. One reason is that it is a natural fiber, making it difficult to obtain always the same properties (Eve et al., 2009). The SEM image of flax fibers (Fig. 10.1) clearly shows that the fibers differ in diameter, have kink bands, and some of them are damaged.

On the other hand, the SEM image of glass fibers (Fig. 10.2) reveals that the fibers are uniform and very smooth making it possible to obtain always the same mechanical properties.

Other reasons for the huge variation in mechanical properties are that fibers were damaged during processing; different characterization methods were used such as testing single fibers instead of fiber bundles; some misalignment of fibers; or different testing conditions are applied such as difference in speed, gauge length, moisture content, or temperature (Pickering et al., 2016).

In Europe flax is one of the most used natural fiber for composite applications because the climatology conditions are beneficial to grow flax especially in the countries of France, Belgium, and the Netherlands (about 85% −90% of the worldwide flax production). Also, flax is a crop that requires less fertilizers and crop protection; no irrigation is necessary and it is a zero waste plant making it a very sustainable fiber source (Gardetti and Muthu, 2015). All parts of the plant can be used: long fibers for textile and composites; short fibers for textiles, insulation, car components, and paper industry; linseed for nutrition, animal feed, health care, and industrial use; shives for litter, boards, and building industry (ABV-Algemeen Belgisch Vlasverbond, 2019).

In order to find the best mechanical properties of flax fibers, the influence of several parameters was investigated by many research groups. Martin et al. (2013) investigated the influence of the degree of retting of flax fibers on the tensile properties. Retting is a process to remove pectin and thus releasing the fiber material. From an economic point of view most of the flax is dew retted on the field and is thus dependent on the weather conditions. The researchers collected samples with retting times of 1, 5, 9, 14, 16, and 19 days. They found

TABLE 10.1 Mechanical properties of natural and glass fibers (Pickering et al., 2016).

Fiber	Density (g/cm^3)	Length (mm)	Failure strain (%)	Tensile strength (MPa)	Young's modulus (GPa)	Specific tensile strength ($MPa/g\ cm^3$)	Specific Young's modulus ($GPa/g\ cm^3$)
Ramie	1.5	900–1200	2.0–3.8	400–938	44–128	270–620	29–85
Flax	1.5	5–900	1.2–3.2	345–1830	27–80	230–1220	18–53
Hemp	1.5	5–55	1.6	550–1110	58–70	370–740	39–47
Jute	1.3–1.5	1.5-120	1.5-1.8	393–800	10–55	300–610	7.1–39
Harakeke	1.3	4–5	4.2–5.8	440–990	14–33	338–761	11–25
Sisal	1.3–1.5	900	2.0–2.5	507–855	9.4-28	362–610	6.7–20
Alfa	1.4	350	1.5–2.4	188–308	18–25	134–220	13–18
Cotton	1.5–1.6	10–60	3.0–10	287–800	5.5-1.3	190–530	3.7–8.4
Coir	1.2	20–150	15–30	131–220	4–6	110–180	3.3–5
Silk	1.3	Continuous	15–60	100–1500	5–25	100–1500	4–20
Feather	0.9	10–30	6.9	100–203	3–10	112–226	3.3-11
Wool	1.3	38–152	13.2–35	50–315	2.3-5	38–242	1.8–3.8
E-glass	2.5	Continuous	2.5	2000–3000	70	800–1400	29

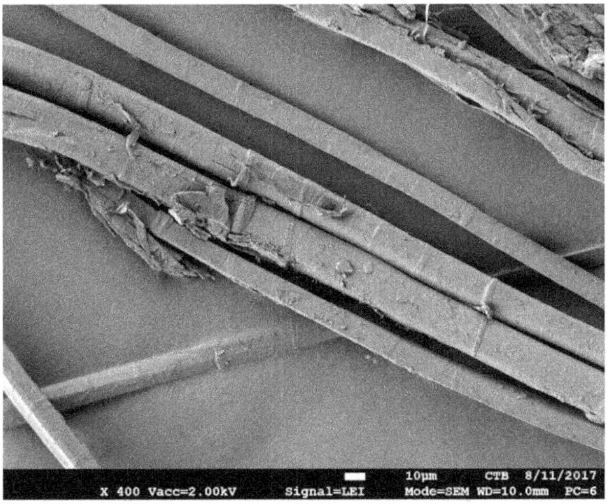

FIGURE 10.1 SEM image of flax fibers.

FIGURE 10.2 SEM image of glass fibers.

that the water uptake is higher for low retted fibers, which is unfavorable for composite applications. Also, the tensile properties of single fiber increased with the degree of retting. Another factor that influences the fiber property is the growing condition (year effect). It was found that composites from the flax grown in 2013 had higher stiffness than those with flax grown in 2012 (Haag et al., 2017). The same research group also identified flax varieties which are showing higher or lower mechanical properties in composites. In the Interreg project Biocompal it is also found that the flax variety can have an influence on the mechanical properties of flax (Pisupati et al., 2019). The tensile strength and modulus of the fibers were determined by the impregnated fiber bundle test (based on ISO 10618) instead of testing the individual fibers. This in situ behavior testing of the technical fiber is much more realistic and relevant for composite applications than its behavior when tested as an individual technical fiber (Bensadoun et al., 2017).

The results of the tensile modulus (Fig. 10.3) and strength (Fig. 10.4) of different flax varieties from 1 test field and 1 year were compared with a commercially available flax type. The results show that almost all the tested varieties possess higher mechanical properties than the reference. An explanation for this could be that the reference fibers were more damaged due to processing (hackling) than the fibers that were only scutched and combed (Bos et al., 2002). Methods to minimize the degradation of the fibers during processing are treating them with ammonia instead of retting or skipping the hackling step during preform manufacturing (Zeng et al., 2015).

Although natural fibers have intrinsic good mechanical properties and are considered as durable, they are most of time incinerated after use. For recycling applications thermoplastic bio-based fibers are an interesting approach. Although mostly known as a bio-based matrix system (see bio-based matrices), poly(lactic acid) (PLA) fibers possess interesting properties to be utilized as reinforcement in composites. PLA is not only made from 100% renewable sources such as corn, but is also compostable. It is a linear aliphatic thermoplastic polyester which has mechanical properties similar to conventional PET, but with lower melting and softening temperatures. The density of the fibers is 1.25 g/cm^3 which is lower than natural fibers and they have a low moisture regain (Farrington et al., 2005). The Young's modulus of PLA fibers derived by melt-spinning ranges from 5.2 to 9.2 GPA and the strength varies from 280 to 870 MPA depending on the initial weight and process parameters (draw ratio and temperature) (Goutianos et al., 2019).

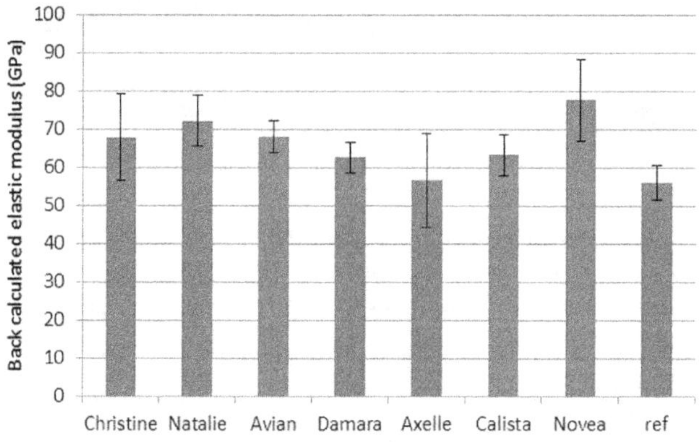

FIGURE 10.3 Tensile modulus of different flax varieties and a commercially available reference sample determined by the impregnated fiber bundle test.

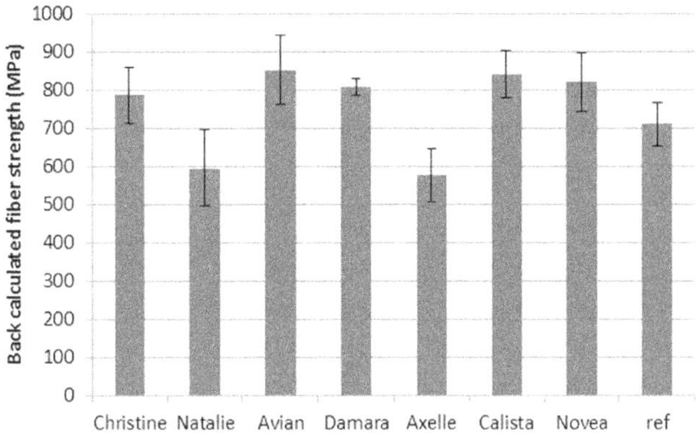

FIGURE 10.4 Tensile strength of different flax varieties and a commercially available reference sample determined by the impregnated fiber bundle test.

10.2.2 Bio-based matrix

In composite materials, the matrix binds the fibers and transfers the load between the fibers. The concentration of the matrix can vary between 30% and 80%, meaning that a large part of the composite consists of matrix polymers. Therefore, to obtain fully bio-based composites also the matrix should be derived from natural sources.

The most used matrix resins in conventional composites are unsaturated polyester resins, because they have acceptable mechanical, electrical, and chemical properties and are quite cheap. Because they have a high share in the composite market, the replacement of these resins by bio-based resins can have a huge impact on the sustainability of composite materials. Feedstocks for these resins are soybean oil and corn oil. These oils cannot be used as such and other functional groups such as epoxy or acrylate groups have to be introduced to convert these oils into thermoset resins (The matrix: CompositesWorld, 2016).

To replace petroleum-based unsaturated polyester resin acrylated epoxidized soybean oil (AESO) and maleinated epoxidized oil (MAESO) are proposed. However these resins need similar to standard resins styrene as reactive diluent in a concentration of about 35%. Besides that, styrene is non-bio-based, it also has other drawbacks of which one is that it is reported as a potential human carcinogen. So the development of bio-based diluents that can replace styrene are gaining more and more interest. Zhang et al. (2018) determined that methacrylated eugenol is the most promising, sustainable reactive diluent to replace styrene for MAESO resin.

The first commercially available resins are the Envirez resins from Ashland (Ashland, 2019). These resins meet the same performance and processing requirements as 100% petroleum-based unsaturated polyester resins. The biocontent of these resins varies between 10% and 20%. Another commercially available bio-based resin is ENVIROLITE from Reichhold (Reichhold, 2019). This resin is based on soy oil and has a green content of 25%. The resin can be used as a stand-alone resin or in combination with other unsaturated polyester resins. The resulting laminates have similar mechanical performance compared to laminates with standard resins. AOC resins offer bio-based unsaturated polyester (Eko Tek) resins with a biocontent up to 30% (AOC).

When higher mechanical properties of composite are required, epoxies are selected as matrix material. Epoxies are cured with a harder element, which are in most cases amines or anhydrides. Bio-based alternatives are derived from vegetable oils, lignin, rosin, tannin, sugar, cardanol or itaconic acid (Ramon et al., 2018). Bio-based hardeners can be derived from cashew nut shell liquid (Benega et al., 2017). Companies that already offer bio-based epoxies or hardeners are, for example, Sicomin (Sicomin, 2014), Cardolite (Cardolite, 2017), Anacarda (Anacarda, 2019), and Entropy Resins (Entropy Resins, 2019).

When composites need to be flame retardant, phenolic resins are often used due to their superior fire, smoke, and toxicity properties compared to other polymeric matrices. Good bio-based alternatives are furan resins. They are produced from agriculture waste products such as sugarcane bagasse. Hemicellulose is turned into furfural via steam digestion and dehydration. Next furfuryl alcohol is formed by hydrogenation of furfural which is the precursor to synthesize furan prepolymers. Furan resins possess interesting properties such as a high stiffness, fire resistance, and chemical resistance to organic and inorganic acids. One of the pioneers in furfural chemicals is the company TransFurans Chemicals and has a yearly manufacturing capability of more than 40,000 MT (TransFurans Chemicals, 2019).

For challenging applications benzoxazine resins are interesting resins. They almost do not shrink and do not release volatiles during curing. Other properties of these resins are high stiffness, excellent thermal resistance with glass transition temperatures that can go beyond 200°C, and similar flammability, smoke, and toxicity performance of phenolic resins. Bio-based benzoxazine alternatives can be derived from natural phenolic compounds such as cardanol, vanillin, and eugenol (Trejo-Machin et al., 2018). Further Dumas et al. (2016) synthesized fully bio-based benzoxazines using hydroquinone and resorcinol as phenolic compounds resulting in glass transition temperatures higher than 280°C. The same researchers also reported that adding carbon nanotubes (CNTs) leads to a significant improvement of the thermomechanical behavior (about 50°C) (Dumas et al., 2014).

Besides thermoset matrices, thermoplastic polymers are also used in composite manufacturing mainly because they can be recycled or reshaped by just melting the polymer. One of the most studied bio-based thermoplastic matrices for composites is PLA. PLA was already mentioned as fiber, but it is more investigated as matrix in composite materials. The glass transition temperature of PLA is around 55–60°C and has melting points of 130–230°C. Similar to the melting range, the mechanical properties can be varied from soft and elastic plastics to stiff and high strengths depending on the molecular weight, crystallinity, and stereometry of the molecule. A drawback of PLA is poor toughness which limits the amount of applications (Murariu and Dubois, 2016).

However, research is going to reduce the brittleness of the material by adding plasticizers of which low molecular weight polyethylene glycol and and oligomeric lactic acid are the most common (Farah et al., 2016). Two main suppliers of PLA are NatureWorks (NatureWorks, 2019) and Corbion (Corbion, 2019). Other bio-based and biodegradable polymers are polyhydroxyalkanoates (PHAs). They are polyesters and are derived from bacteria which use them as energy storage medium. The most commercialized PHAs are P3HB, P(3HB-co-3HV), and P(3HB-co-3-hydroxyhexanoate) (P(3HB-co-3HH)). The t_g of these polymers is quite low namely 4°C for P3HB and −35°C for P(3HB-co-3HH). The melting temperature is 177 and 61°C, respectively. The tensile strength is 40 MPa for P3HB and 9 MPa for P(3HB-co-3HH) (Nakajima et al., 2017). Some examples of suppliers are Biomer (BIOMER and PHB, 2019), BIO-ON, (BIO-ON, 2019) and Kaneka (Kaneka, 2019).

Polyamide (PA) is a high-performance engineering polymer with high strength and good fatigue resistance making it an attractive polymer to use as matrix in composite materials. Unlike other bio-based polymers, biopolyamides are high performance polymers as they are long-lived, durable, and do not give rise to technical weaknesses. Evonik (2019) supplies these bio-based polyamides under the name Vestamid Terra of which even a 100% bio-based material is available (Vestamid Terra DS (PA 1010)). Other suppliers of bio-based polyamides are Arkema (Arkema, 2019) and DSM (DSM, 2019).

10.2.3 Interface fiber–matrix

To achieve the best composite performance, a strong interaction between fiber and matrix is critical. A bad adhesion between the fiber and matrix leads to rapid failure of the composite as the load cannot be efficiently distributed between the fibers. Natural fibers are hydrophilic due to the presence of hydroxyl groups from cellulose and lignin. On the other hand, the matrix is most of the time hydrophobic making the adhesion very difficult. In order to overcome this mismatch, the surface of natural fibers can be modified by chemical treatment to increase the compatibility with the resin (Sood and Dwivedi, 2018).

Possible treatments are alkali treatment, isocyanate treatment, peroxide treatment, vinyl grafting, acetylation, and treatment with coupling agents. Alkali treatment removes fiber constituents such as lignin, wax, oils, and hemicellulose and modifies the cellulose surface via following reaction:

$$\text{Cellulose-OH} + \text{NaOH} \rightarrow \text{Cellulose-O}^-\text{Na}^+ + H_2O$$

The effect on the fiber is an increased roughness what results in a better interlocking and an increase of the amount of cellulose at the fiber surface meaning that more reaction sites are available (Li et al., 2007). An increase of

30% in tensile properties is possible for alkali treated flax/epoxy composites (de Weyenberg et al., 2003). Also, Mishra et al. (2003) reported that a 5% sodium hydroxide (NaOH) concentration results in better strength than when the fibers are treated with a 10% solution. The higher amount of NaOH seems to damage the fiber.

Isocyanates can react with the hydroxyl groups of the natural fiber to form a urethane binding. Strong covalent bonds are formed and improve the compatibility between fiber and matrix (Mohanty et al., 2001). Peroxides form radicals rapidly, which can react with the hydrogen atoms of the hydroxyl groups of the cellulose fibers. Examples of peroxides that are used are benzoyl peroxide and dicumyl peroxide. By treating sisal fibers with benzoyl peroxide and dicumyl peroxide, the tensile strength of sisal LDPE composites increases with increase in concentration of the peroxide up to 4% for dicumyl peroxide and up to 6% of benzoyl peroxide (Joseph et al., 1996).

A vinyl grafting was done by Mohanty et al. by subjecting jute fabrics to a graft graft-copolymerization with methyl methacrylate and acrylonitrile using $CuSO_4-KIO_4$ combination as initiator in the aqueous medium. Among the methyl methacrylate and acrylonitrile grafted samples, a 10% AN-grafted material shows the highest improvement of mechanical properties (Mohanty et al., 2000).

Acetylation is an esterification of the cellulose fibers.

$$\text{Cellulose-OH} + CH_3-C(=O)-O-C(=O)-CH_3 \text{ - > Cellulose-}OCOCH_3 + CH_3COOH$$

Hydrophilic OH groups are replaced by acetyl groups making the fiber more hydrophobic and this improves the dimensional stability of the composites (Bledzki et al., 2008).

Coupling agents are frequently used to improve the compatibility between natural fibers and organic matrices. To be compatible with PP, fibers can be treated with maleic anhydride grafted polypropylene (MAPP) providing covalent bonds at the interface. After this treatment, an increase of about 72% in flexural strength can be achieved compared to nontreated jute/PP composites (Mohanty et al., 2004). Silane coupling agents are used in the glass fiber industry to improve the adhesion with organic matrices. Silane coupling agent forms in water silanol groups which can react with the OH groups of the glass fibers. The organic functional group can interact with the matrix. As natural fibers contain OH groups the same process can be applied:

$$\text{RSi(OR')3} + \text{H2O - > } RSi(OH)_3 + 3\text{ R'OH (with R'} = \text{methyl, ethyl)}$$

$$\text{Cellulose-OH} + \text{RSiOH3 - > Cellulose-OSiR} + \text{H2O}$$

The R group can contain different functional groups and depending on the functionality they can couple different matrices. An overview is given in the Table 10.2 below:

From the table, it is clear that silane can be used to couple a lot of organic resins, but it is important to select the right silane to achieve a good interaction.

TABLE 10.2 Compatibility between functionalized silanes and organic resin. ++ means a very good compatibility (Shin-Etsu, 2017).

	Thermoplastic										Thermoset					
Resin funtional group	PE	PP	PS	Acrylic	PVC	PC	PA	PU	PET	ABS	Phenolic	Epoxy	PU	Polyimide	Unsaturated polyester	Furan
Vinyl	++	++													+	
Epoxy	+	+	+	+	+	+	+	+	+	+	+	+	+	+	+	+
Styryl			+	+												
Methacryloxy	++	++	++	+		+		+		++					++	
Acryloxy	+	+	+	+		+		+		++					++	
Amino	+	+	++	++	++	+	++	+	+	+	++	++	+	+		++
Ureide							++				+		+	+		
Mercapto	+	+	+		+			+		+	+	+	+			
Isocyanate						+	+	++	+	+	+	+	++	+		+

Xie et al. wrote a comprehensive review about silane coupling agents used for natural fiber/polymer composites. In the report, it is mentioned that a proper treatment of fibers with silanes can increase the adhesion between fiber and matrix and improves the mechanical and outdoor performance of the resulting composites (Xie et al., 2010).

10.2.4 Bio-based composites

Bio-based composites can be prepared via traditional composite manufacturing techniques such as:

- Injection molding: a fast, high-volume closed molding process suitable for short fibers and thermoplastic matrices.
- Hand lay-up: this is a simple processing method, but needs skilled operators to obtain good-quality composites. A gel coat is first applied to the mold and afterward fiber reinforcements are placed in the mold. The thermoset resin is poured on the reinforcement and distributed by brushing or using paint rollers or squeezes. This process is repeated till the desired thickness is obtained. Curing occurs usually at room temperature.
- Resin transfer molding (RTM): the fiber reinforcement is placed in a closed mold and the liquid resin is injected by pressure. This method is suitable to produce complex parts with smooth finishes.
- Vacuum infusion or vacuum assisted (VA) RTM: fiber reinforcements are placed in a mold and are covered with a vacuum foil. The resin is driven into the laminate by vacuum force. This method is normally used to produce large structures.
- Pultrusion: this is a continuous process in which continuous fibers, rovings, or mats are impregnated with resin and pulled through a steel die.
- Compression molding: fiber reinforcements and matrix are already mixed in the right concentrations and are placed in a heated mold. The resin, if thermoset, is cured or reshaped when thermoplastic, under pressure.

However, when using natural fibers, the temperature should not be higher than 180°C to avoid degradation of the fibers and a predrying step is recommended to remove the absorbed moisture as this negatively influences the interaction between fiber and matrix (Venkateshwaran et al., 2013).

The fibers can be used as such or arranged in a network forming nonwovens (Fig. 10.5), unidirectional preforms (Fig. 10.6), or woven structures such as twill (Fig. 10.7) or plain weaves (Fig. 10.8).

Although there exist already several publications on bio-based composites with natural fibers (see reviews of Elanchezhian et al., 2018; Ku et al., 2011; Li et al., 2007; Mohammed et al., 2015; Pickering et al., 2016; Sood and Dwivedi, 2018), reports of composites consisting of a bio-based fiber and resin are rather rare probably due to the immaturity and limited availability of high-quality matrices.

FIGURE 10.5 Jute nonwoven.

FIGURE 10.6 UD flax tape.

FIGURE 10.7 Flax twill weave.

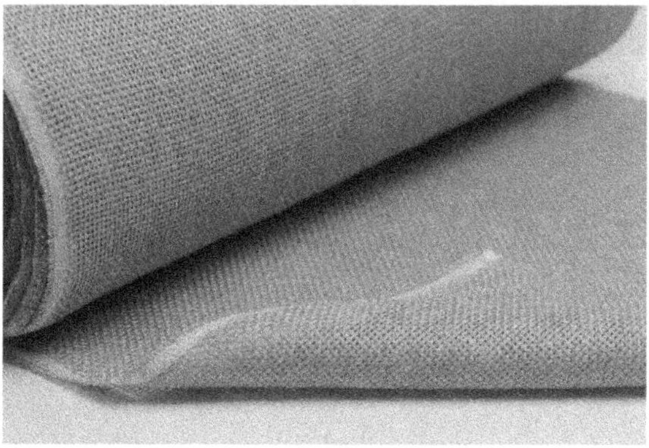

FIGURE 10.8 Flax plain (basket) weave.

Table 10.3 shows the different manufacturing methods and mechanical properties of different composite materials with bio-based fibers and matrices. Also, the tensile properties of two non-bio-based reference composites (i.e., glass and carbon fiber) are presented to compare their properties with the bio-based composites.

The results from the table show that there are some differences in the tensile properties of the bio-based composites not only due do the use of different fibers or matrices, but also by differences in manufacturing method and fiber volume content. As fiber volume content increases, the tensile properties should also increase until there is not enough resin to distribute the load to the fibers or there must be a mismatch between fiber and matrix. When comparing the tensile properties of the bio-based composites with glass nonwoven composites prepared by hand lay-up better result can be obtained and thus biocomposites can offer a more sustainable alternative. However, the UD glass example shows that much higher strengths can be obtained with glass composites, meaning that bio-based fibers can only match with the stiffness of glass fibers which was already clear from Table 10.1. Compared to carbon composites, the current biocomposites cannot offer the same performance regarding the tensile properties.

10.2.5 Applications

The green image of bio-based composites (CO_2 emission reduction) makes, rather than their technical performance or price, them to be used in applications like automotive, construction, packaging, and consumer goods. For automotive, the main motivations are weight reduction (about 10%–30%), day-to-day fuel efficiency, and CO_2 emission reduction. Parts that are made are door and instrument panels, body panels, underbody panels, roof tops, and front-end fenders. Manufacturers that incorporated bio-based composites are,

TABLE 10.3 Tensile properties of different bio-based fibers and three non-bio-based references.

Bio-based reinforcement	Bio-based resin	Treatment	Fiber volume fraction	Manufacturing method	Tensile strength (MPa)	Tensile modulus (GPa)	References
Flax unidirectional	Epoxy	/	32.1	VARTM	/	18	Schuster et al. (2014)
Flax unidirectional	Polyester	/	41.1	VARTM	/	13	Schuster et al. (2014)
Flax UD	Epoxy	/	30	Resin infusion	150	16	Michelena et al. (2015)
Hemp plain weave	Epoxy	/	40	RTM	63	5.8	Di Landro and Janszen (2014)
Flax woven	Epoxy	/	/	VARTM	80	1.2	Bertomeu Perelló et al. (2012)
Flax UD (cross- ply lay up [0/90]S)	PFA	/	/	Compression molding	64	8.5	Giannis et al. (2008)
Flax UD + nonwoven	Epoxy	/	/	Compression molding	185	14	Brighton et al. (2015)
Flax UD + nonwoven	Epoxy	Silane (1%)	/	Compression molding	178	12	Brighton et al. (2015)
Flax UD + nonwoven	Epoxy	Alkali (5%)	/	Compression molding	175	11	Brighton et al. (2015)
Flax fiber bundles	PPolyester	/	40	VARTM	140	6	Haag et al. (2017)
Kenaf fiber bundle	PLA	/	70	Compression molding	223	23	Ochi (2008)

Continued

TABLE 10.3 Tensile properties of different bio-based fibers and three non-bio-based references.—cont'd

Bio-based reinforcement	Bio-based resin	Treatment	Fiber volume fraction	Manufacturing method	Tensile strength (MPa)	Tensile modulus (GPa)	References
Hemp short fiber	PLA	/	30	Compression molding	41	5.6	Hu and Lim (2007)
Hemp short fiber	PLA	/	40	Compression molding	45	7.4	Hu and Lim (2007)
Hemp short fiber	PLA	/	50	Compression molding	44	7	Hu and Lim (2007)
Hemp short fiber	PLA	Alkali (6%)	30	Compression molding	39	7.6	Hu and Lim (2007)
Hemp short fiber	PLA	Alkali (6%)	40	Compression molding	54	8.5	Hu and Lim (2007)
Hemp short fiber	PLA	Alkali (6%)	50	Compression molding	42	7	Hu and Lim (2007)
Flax twill	PLA	/	40	Compression molding	110	14	Biotex (2018)
Flax nonwoven	PHB	/	20	Compression molding	30	4	Barkoula et al. (2010)
Flax nonwoven	PHB	/	30	Compression molding	30	7	Barkoula et al. (2010)
Flax nonwoven	PHB	/	40	Compression molding	40	8.5	Barkoula et al. (2010)

Kenaf nonwoven	PLA	/	70	Wet impregnation	60	6.4	Nishino et al. (2003)
Flax short fibers	PLA	/	28	Compression molding	53	8.3	Oksman et al. (2003)
Kenaf carded	PLA	/	35	Compression molding	53	7.1	Graupner and Müssig (2011)
Hemp carded	PLA	/	35	Compression molding	58	8.1	Graupner and Müssig (2011)
PLA UD	PLA	/	/	Compression molding	107	6.2	Goutianos et al. (2019)
Cellulose fibers	PA	/	/	Pultrusion	120	5	Feldmann and Bledzki (2014)
Ramie plain woven	PLA	Cyclic loading	/	Compression molding	77	11	Zhou et al. (2013)
Glass nonwoven	Polyester	/	19	Hand lay-up	99	4	Varga et al. (2010)
Glass UD	Epoxy	/	30	Hand lay-up	432	18	Ekşi and Genel (2017)
Carbon UD	Epoxy	/	48	Vacuum bagging	1059	80	Rahmani et al. (2014)

for example, Ford, Opel, Daimler Chrysler, BMW, Fiat, Audi, Peugeot, Renault, Mercedes Benz, and Volvo (Campaner et al., 2010).

In the construction sector, bio-based composites can be used for decking, railing systems, window frames, fencing, and panels (Ariadurai, 2012). Further, researchers of the TU Eindhoven developed a pedestrian bridge made of a PLA core and skins made of flax and hemp combined with a bio-based epoxy matrix (Lepelaar et al., 2016).

Besides the green image, the good demping properties and looks of natural fiber composites make them attractive for designers to create consumer and sporting goods. This includes tennis rackets, surfboards, skis and ski pole, bicycles, archers, woofer cones, helmets, scooters, guitars, trays, and chairs (Pil et al., 2016).

10.2.6 Tendency: looking for sustainable bio-based carbon fibers

It is clear that natural fibers and PLA fibers can compete with glass fibers or thermoplastic fibers as more sustainable alternative to make composites, but their mechanical properties are far below that of carbon fiber (tensile modulus: 230–600 GPa, tensile strength 3000–6000 MPa) meaning that it is not possible to use them in high performance composite materials.

On the other hand, the most commonly applied technique to produce carbon fibers starts from polyacrylonitrile (PAN), a polymer synthesized from crude oil. Through a wet spinning process, this polymer is converted into a precursor fiber. The precursor fiber is then subjected to three processes, all under controlled conditions: stabilization, carbonization, and graphitization. The stabilization process is performed in an oxygen atmosphere at 200–300°C. This process converts thermoplastic PAN to a nonplastic cyclic or a ladder compound. After stabilization, the fibers are carbonized in an inert atmosphere up to 1000°C. To further enhance the orientation of the crystallites, the fiber can undergo a graphitization process at 1500–3000°C (Rahaman et al., 2007).

From an environmental point of view, this production process could be improved in several ways: from the natural resources (crude oil) to the wet-spinning process (based on solvents) and the energy-consuming stabilization, carbonization, and graphitization steps.

Therefore, research is going on to reduce the cost and the dependence on fossil sources by looking at biomass-derived precursors. One of these research projects is the H2020 project LIBRE (LIBRE – Lignin Based Carbon Fibers for Composites, 2019).

First aspect targeted by LIBRE is the natural resource: instead of fossil-based PAN, the project uses blends from modified lignin and biopolymers. Lignin is a constituent of wood, forming the glue that holds cellulose and the hemicellulose together. Hence, it is also a valuable byproduct of the paper and pulp industry, produced in quantities exceeding 50 million tons annually. For now, the majority of this byproduct disposed as low-cost fuel. However, it has a large potential for value-added products, for example, carbon fibers. Next aspect in LIBRE is the spinning process: instead of solvent-based wet-spinning, the lignin blends are converted into precursor fibers through melt-spinning. Finally, the project also targets the energy-consuming heat treatment steps: using microwave and radio frequency (MW/RF) heating technologies, it is possible to reduce energy use and greenhouse gas emissions during the manufacturing process.

Other bio-based carbon fiber precursors, besides lignin, that are investigated are glycerol and lignocellulosic sugars (Milbrandt and Booth, 2016). However, a technical report from Milbrandt and Booth concluded that currently no biomass-based carbon fiber has been developed with the required properties to be used in aerospace, wind, and automotive applications (Milbrandt and Booth, 2016). So, further research is necessary to find bio-based carbon fibers that can compete with the non-bio-based carbon fibers qua performance and cost.

10.3 Coating and finishing

In order to impart the required functional properties to a fabric, the material is subjected to physical and chemical treatments. Coating and finishing refers to the processes performed after dyeing the fabric. Coating can be defined as applying a polymer or elastomer (in viscous or solid form) directly to the surface of a fabric and drying and curing it, if necessary. Several coating layers may be applied, resulting in better performance (Smith, 2018). A process related to coating is finishing, in which a chemical or polymer covers only the yarns and not the whole fabric. In fabric coating the small holes in between the individual yarns are covered (Fung, 2002). Some coating and finishing technologies will be shortly discussed hereafter.

Fabric finishing is usually performed by a padding or impregnation process. This process step is commonly used in textile industry to impart among others water and oil repellency, softening, or easy-care properties to a textile. During padding (also called foulard) a textile is impregnated with functional additives and a binder via immersion in a solution (often water-based). The solution contains a specific concentration of additives and binder. The concentration of these products depends of the desired effect and the wet pick-up of the textile. After the textile is immersed in the solution, the textile is squeezed between two rolls to remove the excess of solution. Fig. 10.9 represents the foulard process.

Liquids can also be sprayed directly onto a textile. Spraying is very well suited to apply low concentrations and for functionalizing delicate fabrics which cannot be padded (e.g., carpet). A concern of spraying is the difficulty to obtain uniform coatings and therefore the spray parameters need to be governed and controlled well (Joshi and Butola, 2013).

Coating is essentially spreading a polymer (in the form of a solid material, viscous dispersion, or viscous solution) onto a fabric to form a continuous layer. The material needs to be viscous so that it does not penetrate through the fabric. Thickening agents need to be added sometimes to increase the viscosity

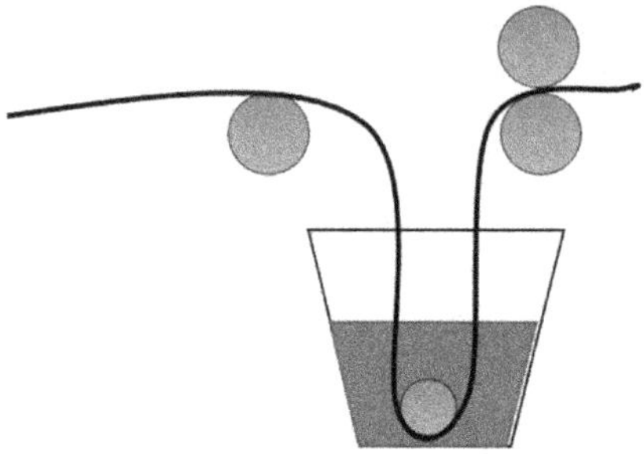

FIGURE 10.9 Padding or impregnation process.

of the coating formulation prior to application. A coating formulation is composed of binder (polymer) and (functional) additives which determine the final characteristics of the coating.

In knife coating, an excess of coating formulation is applied to the textile and the amount of coating applied (layer thickness) is controlled by a blade. In floating knife (knife-over-air), the knife is fixed above the textile and touches the textile (Fig. 10.11). This technique is usually done to apply low amounts of coating. The applied amount of coating is controlled by fabric tension and the distance between the fabric and the knife blade (Shim, 2018; Conway, 2016). The sharpness and the angle of the blade also determine the amount of coating applied. Using knife over air coating a higher penetration of the coating material into the textile is obtained, but this results in a decrease of flexibility.

Knife-over-roll is the most common knife-coating process (Fig. 10.10). In this technique, the knife blade is above a roller and doesn't touch the fabric. The height of the blade above the textile controls the applied amount of coating (layer thickness). A coating layer of 1 μm results in a coating weight of 1 g/m^2 if the coating material is 100% solid with a density of 1 g/cm^3. The aperture between the roller and the knife blade can be set accurately to control precisely the thickness (Shim, 2018; Conway, 2016).

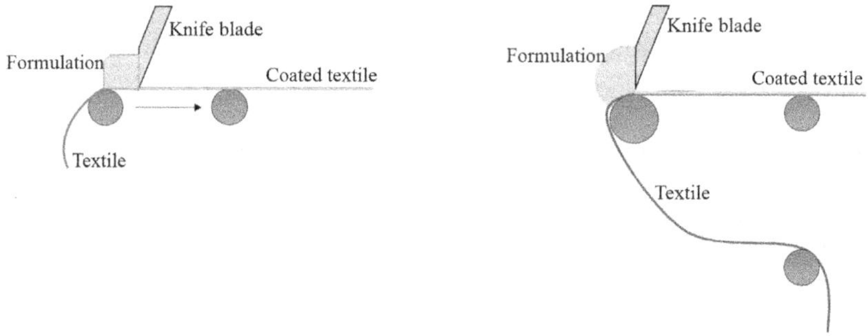

FIGURE 10.10 Knife coating: knife-over-air (left) and knife-over-roll (right) coating.

Transfer coating is an indirect coating technique, since the coating material is applied by a knife blade on a release paper and dried. The base coat is then applied over the first layer using a second knife blade and the coated release paper passes through the laminating rollers with the textile. The coated paper and textile on it goes into an oven, to dry and cross-link the two together. Afterward the coated fabric is peeled off the release paper, which can be reused. An advantage of transfer coating is the low penetration of the coating material, which decreases stiffness and promotes flexibility (Shim, 2018; Conway, 2016; Fung, 2002). For this reason, this technique is often used for coating knitted fabrics.

Another coating technique involves coating rollers. Kiss roll coating is the simplest and most commonly used roll coating, using a single roller (Fig. 10.11). The coating is applied by a roller which is with its bottom half immersed in a coating bath and as it rotates, the coating material forms a film on the roller surface. The film is partly transferred to the textile, which is in contact with the roller. The rotating speed of the roller, the speed of the textile, and the coating fluid properties determine the amount of coating material to be applied on to the textile.

Hot melt or slot die coating is used for thermoplastic polymers (100% solid) such as polyolefins and polyurethane. A plasticized compound (thermoplastic resin or reactive adhesives) is pressed through a die directly to the textile (Whiteman, 1993; Zickler, 1978). In the conventional slot die coating, the die lip is in contact with the textile and backed by the rollers, putting pressure on the textile. Some slot die systems allow a gap between the die lip and the textile (Anon, 2002; Glawe et al., 2003; Shim, 2010). Cooling down of the melt onto the textile results in a coating film. Adhesion between the melt/coating and textile is determined by the compatibility of the melt and the textile, but it is also affected by other parameters such as textile line speed, film thickness, air gap, and melt temperature (Mamish, 1990). Slot die coating can also be used as hot melt lamination technique using hot melt adhesives such as the bonding agent. After applying the hot melt on a substrate, the coated substrate is subsequently bonded with another substrate to form a laminate. Hot melt adhesives are 100% solid and replace water- and solvent-based adhesives as environmentally friendly alternative technique. Hot melt adhesives can be distinguished into thermoplastic and reactive hot melt adhesives. Thermoplastics melt and solidify solely according to temperature (Glawe, 2003). Reactive hot melt adhesives are

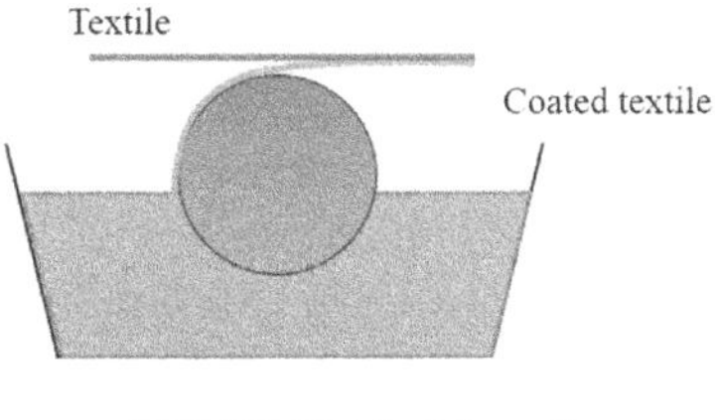

FIGURE 10.11 Kiss roll.

fully cross-linked by the reaction with moisture after being applied on the textile. Due to the cross-linking step, reactive hot melt adhesives are not sensitive to heat or washing; and they are more durable compared to thermoplastics (Mansfield, 2003; Wodruff, 2002).

For all discussed coating techniques, coating products and adhesives are complex mixtures composed of multiple components to deliver the desirable properties in the final textile product and to provide good adhesion properties (increase compatibility with the textile). Most coating mixtures are composed of a binder, medium, pigments, fillers (such as chalk to reduce the cost), and process functional additives. The exact composition is dependent of the coating process. The medium is the carrier of the coating components. Depending on the medium used, coating materials are classified as solvent-based, water-based, high solids, and 100% solid (Wicks et al., 1999; Shim, 2018). Solvent-based coating was the most popular; however, its use is substantially decreased because of environmental and toxic concerns and governmental regulations regarding volatile organic compound emission and (environmental) toxicity. Bio-based coating components will be discussed hereafter.

10.3.1 Bio-based binder

The binder is the film-forming element of a coating or adhesive. It provides adhesion to a textile, bonding among other components and the textile, and determines important mechanical properties such as durability and flexibility. The binders used in textile coating and lamination are often organic polymers (e.g., acrylate, polyurethane, polyvinyl chloride, ethyl vinyl acetate, …) or organic monomers or oligomers to be polymerized after the coating has been applied (Wicks et al., 1999; Shim, 2018).

Although a wide variety of bio-based polymers are available or in development, most often they do not comply to the requirements (e.g., cost, durability, and mechanical properties) for fabric finishing and coating. Regarding the application, bio-based polymers can be classified into two groups:

- Drop-in bio-based polymers: polymers chemically identical to the current petroleum-based polymers but made out of renewable resources (e.g., bio-based polyurethane).
- New bio-based polymer chemistry such as polylactic acid and polyhydroxy alkanoate.

Much research is performed in the field of drop-in bio-based polymers toward bio-based polyurethane, based on renewable polyester polyols. Lubrizol markets bio-based thermoplastic polyurethanes originally developed by Merquinsa. Pearlbond ECO 590 has a renewable content of 67% and can be added to hot melt polyurethane formulations to improve crystallization speed and is also used for hot melt adhesives in heat sealable fabrics (Lubrizol, 2019). Uni-Rez, originally developed by Arizona Chemicals, are hot melt polyamides from Kraton with a renewable content ranging from 75% to 95%. Uni-Rez are pine-based chemicals (IMCD, 2019). With respect to hot melts, bio-based and biodegradable types are available on the market, developed primarily for paper and cardboard (Guo et al., 2010). Toyobo developed amorphous polylactic acid for coating applications, mainly in the field of packaging. The hot melt coating was too brittle; however, the urethane modified polylactic acid dispersions are applicable for textile coating if plasticizer is added. Fig. 10.12 demonstrates scanning electron microscope pictures of urethane modified polylactic textile coatings with and without plasticizer. No defects were observed in the plasticized coating after 8000 Crumple flex cycles (EN ISO 7854-C); nonetheless, the polylactic acid showed no wash durability (De Smet et al., 2015).

Vercet (2017), a division of NatureWorks, produces and markets special PLA grades (PLA of low molecular weight with lower crystallinity) for adhesive and coating applications on different substrates such as textile, wood, paper, etc.

Among organic coatings, polyurethane coatings have excellent abrasion resistance, high flexibility, hydrolysis, and chemical resistance and therefore used in many textile coating applications. Bio-based polyurethanes are commonly synthesized by reacting bio-based polyols with diisocyanates. Analysts forecast the global bio-based polyurethane market to grow at a compounded annual growth rate of 6.75% during the period 2016–20 (Business Wire, 2019). Polyurethanes are synthesized from vegetable oils obtained from various plant seeds such as neem, castor, rapeseed, jatropha, palm, soybean, etc. (Noreen et al., 2016). Covestro, a major supplier of polyurethane dispersions, has marketed a polyol made from CO_2 for the production of polyurethanes. Covestro is also interested to establish bio-based value chains in chemicals as well. It has developed a technology for making the polyurethane precursor aniline from vegetable raw materials (Coating Worlds, 2018). Covestro developed and launched bio-based waterborne

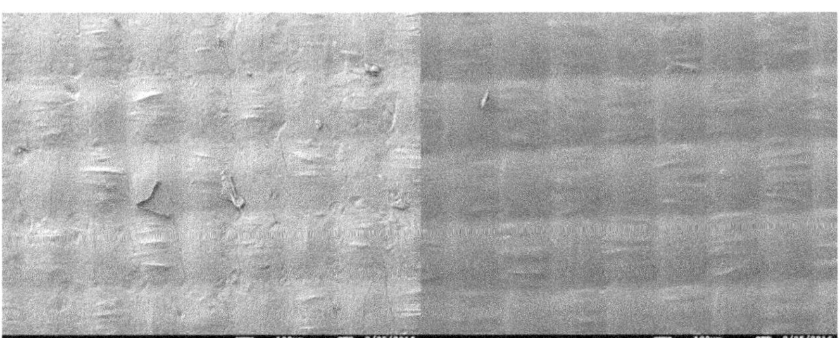

FIGURE 10.12 Scanning electron microscopy analysis revealed microcracking in the urethane modified polylactic textile coatings without plasticizer while no defects are observed in the coating with plasticizer (right).

polyurethane dispersions in April 2015 under the trade name Impranil Eco. The bio-based polyurethane dispersions have a high content of renewable raw materials (up to 65%). The polyurethanes are made from bio-based succinic acid (from BioAmber) instead of adipic acid. Bio-based succinic acid is derived from corn starch that is not produced in competition with the food chain. The Impranil Eco range offers drop-ins of existing Impranil polyurethane dispersions retaining the same performance standards with little reformulation required. Products which are part of the Impranil Eco portfolio are Impranil Eco DLS, Impranil Eco DL-519, and Impranil Eco DLP-R. The product range covers tie coatings, intermediate coatings, and top coatings (Covestro (a); Bittner, 2017). Coatings (Knife-over-roll and transfer) based on Impranil Eco DL-519 exhibited a good abrasion resistance and wash fastness (minimum 20 cycles at 40°C according to ISO 6330) if cross-linker is added. The coating was flexible. No compatibility issues were observed in formulating Impranil Eco DL-519 in a coating mixture.

Next to examining and evaluating commercially available waterborne bio-based polyurethane dispersions, Centexbel is developing and assessing two-component bio-based polyurethane (2K Bio PU) for textile coating. Bio-based polyol and bio-based polyisocyanate are mixed in presence of a catalyst. The ratio of polyol to isocyanate is calculated according to the hydroxy equivalent weight of the polyol and the isocyanate equivalent weight of the polyisocyanate. Bio-based polyols are widely commercially available, contrary to polyisocyanates. Nonetheless, commercially available renewable polyols are polyester based and not polyether or polycarbonate based. Mitsui Chemicals markets Stabio D-370N and Stabio D-376N with a bio-based content of 70% and 67%, respectively (Mitsui Chemicals, 2019). Both products are applicable in applications for which hexamethylene diisocyanates are used. Vencorex chemicals (2019) developed Tolonate X FLO 100, a partially (25%) bio-based low viscosity aliphatic isocyanate polymer. Covestro produces Desmodur eco N 7300 for solventborne and solvent free polyurethane coatings. Desmodur eco N 7300 is a bio-based aliphatic polyisocyanate (pentamethylene diisocyanate-trimer) with a renewable content of 70% (Covestro (b)). Desmodur Eco N 7300 was mixed with several bio-based polyols in presence of a catalyst and coated on textile via knife-over-roll coating. The mixture exhibited a long pot life ranging from 30 min to several hours depending on the type of polyol. The coating was cured at 155°C for 2 min. The bio-based 2K PU coating was flexible, showed good wash fastness (minimum 20 cycles at 40°C according to ISO 6330), good to excellent abrasion resistance, and excellent resistance to UV and water penetration (hydrostatic head of minimum 10 m according to EN 20811) even after 10 washing cycles at 40°C (ISO 6330). However, the coating is not weldable and showed higher penetration in the fabric compared to waterborne polyurethane coatings.

OC-BioBinder is a binder solution for nonwoven and textiles that is made from bio-based, renewable, and biodegradable compounds such as modified carbohydrates, proteins, and organic acids. OrganoClick has developed this versatile binder solution to improve the mechanical properties such as dimensional stability, optimizing stiffness and softness, prevent fraying and linting, and enhance dry and wet strength (OrganoClick (a)).

Next to polyurethanes and polylactic acid, renewable acrylates also are getting commercially available (e.g., Sarbio from Sartomer), but as in the case of polylactic acid, the renewable acrylates are not developed for textile coating and therefore at this moment not suited for this application. Since acrylates are commonly used in textile coating and finishing, the development of bio-based drop-in acrylates for textile coating and finishing would have a significant impact in lowering the environmental footprint of textile industry.

10.3.2 Bio-based additives

Additives and pigments render the properties required from the coating or finish such as functional properties (e.g., flame retardancy, antimicrobial activity, etc.), color, opacity, increased abrasion resistance, and processability. Viscosity modifiers, emulsifiers, dispersing agents, flow modifiers, catalysts, and wetting agents are the most common process additives. These processing aids can also be bio-based. As examples polysaccharides can be used to increase the viscosity of coating formulations and fatty acid esters are ideal emulsifiers to prepare homogeneous water-in-oil or oil-in-water emulsions. Thickening agents (viscosity modifiers) are used to increase the viscosity of a coating formulation to ensure the correct balance between penetration, which is required for adhesion, and surface properties. Centexbel has scanned a range of bio-based thickening agents and Arbocel P3500 performed better in water-based dispersions than the currently used viscosity modifiers due to the fast response time.

Selecting (functional) additives for coating and finish applications depends on the following factors next to the required functionality efficiency of the additive:

- Compatibility with the binder or matrix of the coating; the additives should mix and dissolve or disperse well in the coating material.
- Minimal effect on coating or finishing process; the functional additives should not degrade during the application process and have no significant effect on the rheology of the mixture.
- Minimal effect on the overall textile product properties including mechanical properties, haptics, and aesthetics.

Additives need to be uniformly distributed on a coated or finished textile (Wicks et al., 1999; Horrocks, 2018; Shim, 2018). Specific examples of bio-based additives and pigments will be discussed below.

10.3.3 Flame retardant

Flame retardants (FR) are necessary in many products to inhibit a rapid flame propagation to save lives in case of a fire. FR are applied in textiles as an impregnation, back-coating, or by adding to the polymer melt. FRs are a group of anthropogenic compounds used at relatively high concentrations in many applications. Currently, the largest market group of FR is brominated FR, of which some are considered toxic, persistent, and/or bioaccumulative (Eriksson et al. 2001; Lunder and Sharp 2003).

Currently used FRs are not limited to a specific type of chemical; they comprise a whole range of chemically diverse products possessing FR properties. Major classes within FR are the following: halogenated (mostly brominated compounds or chlorinated paraffins), nitrogen/phosphorus based, inorganics (minerals, ceramics), and others like, for example, nanomaterials, graphite, borates.

Halogenated FRs are often used, as they are showing good FR properties in very diverse applications. The halogenated FRs are usually combined with antimony trioxide as they show a synergistic effect. But its use is placed under scrutiny as the inhalation of dust originating from antimony trioxide may cause cancer. Furthermore, several of these components, which were intensively used in the past, are listed under REACH like the decabromodiphenyl oxide due to its environmental fate (persistent and bioaccumulative). Most companies switched to decabromodiphenyl ethane, but its future as FR is uncertain. The use of FR based on nitrogen/phosphorus compounds is vastly growing. Often, they are well-performing alternatives to the restricted halogenated ones but overall, environmentally friendly FR textile finishing presents a lack of performance for the most demanding applications and low to moderated durability. So far, not many bio-based FRs have made it to the market and are being applied in the textile sector.

Research is being performed in order to find potent FRs originating from natural sources. Components containing sufficient amounts of nitrogen and/or phosphorus can be regarded as possible candidates. Examples are DNA, keratin, phytic acid, or lignin. Preliminary results indicated that DNA can be applied as renewable FR for cotton fabrics. The flammability of DNA treated cotton fabrics was evaluated. Above a 10 wt% add-on, the DNA coating extinguished cotton as soon as the flame was removed (Alongi et al., 2013). Whey proteins were also examined as FR, coating enables to slow down the burning rate and hence to increase the burning time (Bosco et al., 2013). Costes et al. investigated the FR effect of various metallic phytates in PLA (Costes et al., 2015). Phytates promote formation of charred layer during combustion. Zhou et al. examined a new FR for viscose fiber using cyclotriphosphazenekeratin from keratin, which significantly improved the flame retardancy (Zhou et al., 2015).

Structures with numerous alcohol moieties (polyols) can also serve as FR precursors. These molecules can be modified, for example, phosphorylated, to enhance the FR effect. Examples are chitosan and different lignin types. Lignin is extensively studied as single additive to improve the flame retardancy (Costes et al., 2016 and Ferry et al., 2014). Chitosan also raised interest in FR formulation, but mainly for layer-by-layer technique which is less suitable for functionalizing textiles (Alongi et al., 2012 and Jimenez et al., 2016).

Centexbel developed together with Maes Dyeing and Finishing a bio-based FR formulation with a renewable content of 92%. The FR formulation is developed for mattress ticking and complies to flame retardancy test EN597-1 and EN-597-2. Maes Dyeing and Finishing implemented the bio-based FR formulation in their industrial finishing of mattress tickings.

10.3.4 Antimicrobial

Antimicrobials currently applied in textile finishing and coating are often silver or quaternary ammonia-based products (Gao and Cranston, 2008; Joshi et al., 2009). The extensive use of antimicrobials resulted in severe legislation concerning biocides. The application of chitosan in textile finishing has been widely examined and reported. Chitosan is obtained via the de-acetylation of chitin, which is mainly found in the exoskeletons of invertebrates. Chitosan finished polyester and cotton textiles exhibited a significant antibacterial efficacy after 1-h contact time (ASTM E2149, shake flask test) against *Klebsiella pneumoniae* and *Staphylococcus aureus* (De Smet et al., 2015). Modified chitosan has also been studied as antimicrobial agent for textiles. Cheng et al. synthesized a new N-halamine chitosan derivative, which was applied on cotton fabric using 1.2,3,4-butanetetracarboxylic acid as a cross-linking agent. The coated cotton swatches could load 0.25% oxidative chlorine upon exposure to dilute household bleach and exhibited very strong antimicrobial activity against *Escherichia Coli* and *Staphylococcus aureus* (Cheng et al., 2014). Fu et al. studied chitosan derivatives bearing double functional groups. These modified chitosan types were applied on cotton fabrics using citric acid as cross-linker. The finished fabrics had strong antibacterial efficiency against *E. coli* and *S. aureus* (reduction of 96% and 99%, respectively) (Fu et al., 2011). Natural herbal products also show antimicrobial activity (Joshi et al., 2009). Thymol and carvacrol are found in essential oils (e.g., thyme oil). The antimicrobial activity and the synergistic effect is described (Nostro et al., 2007). Cotton knitted fabrics were finished with a mixture of thymol and carvacrol (1:1) and showed antibacterial activity against *E. coli* and *S. aureus* (De Smet et al., 2015). Shahidi et al. reported that plasma treatment resulted in increased durability due to higher uptake of thymol (Shahidi et al., 2014). Malpani reported the antibacterial effect of cotton fabrics finished with neem oil, tulsi and aloe vera (Malpani, 2013). Thyme oil applied to linen-cotton blended fabric showed high antibacterial (AATCC 147) and antifungal (EN 14119) activity against a wide range of bacteria and fungi (Walentowska and Foksowicz-Flaczyk, 2013). Centexbel studied different biobased antibacterial products. Tannic acids are naturally occurring polyphenols of which the antibacterial activity against *S. aureus* and *K. pneumoniae* is strongly dependent on the type of fiber and the type and weight of the fabric. In case of cotton fabric finished with tannic acid a good activity against *S. aureus* was observed. The antibacterial efficacy of tannic acid finished polyester fabrics against *S. aureus* and *K. pneumoniae* varied from no effect to strong effect depending on the structure and fabric weight. Monolaurin (glycerol monolaurate) is an ester from glycerol and lauric acid and is found in coconut oil. Cotton and polyester fabrics finished with monolaurin exhibited good antibacterial activity against *S. aureus* and *K. pneumoniae* (Fig. 10.13).

10.3.5 Repellents

Fluorocarbons are today still the most efficient water, oil, and dirt repellents due to their low surface tension. Environmental problems associated with C8 fluorocarbons resulted in increasingly restrictive worldwide legislation and pressure from environmental organizations to ban these products. As for example, perfluoro octanoic acid is put on the Substances of Very High Concern list of REACH limiting use of long chain fluorocarbons. ECHA has recently added perfluorohexane-1-sulphonic acid and its salts to the REACH candidate list because of very persistent and very bioaccumulative properties. The German Environment Agency is currently investigating short-chain per- and polyfluoroalkyl substances. New fluor-free repellents are being developed and commercialized. Less toxic alternative water repellents mimic nature (Eadie and Gosh, 2011). However, although fluor-free repellents exhibit water repellency, they do not provide any solvent and oil repellency, which is needed for many textile applications such as protective clothing and work wear. Opwis and Gutmann published surface modification of textiles with hydrophobins. Hydrophobins are small proteins consisting of a hydrophilic and a hydrophobic part. These proteins will adhere to

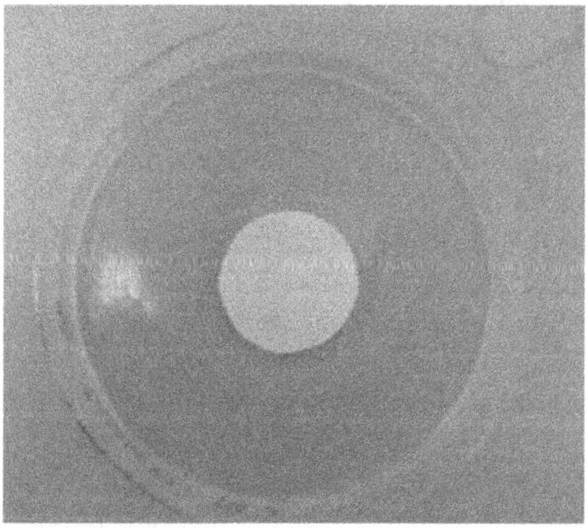

FIGURE 10.13 Monolaurin finished cotton fabric exhibiting antibacterial activity against *S. aureus*.

hydrophilic substrates with their hydrophilic counterpart, creating a hydrophobic surface. Vice versa hydrophobins will adhere to hydrophobic substrates with their hydrophobic side, resulting in a hydrophilic surface (Opwis and Gutmann, 2011).

Organoclick developed a range of OC-aquasil TEX hydrophobic products. OC-aquasil Tex is a product range of water repellents for technical textiles and nonwoven that confers water repellency, while being free from both fluorocarbons (PFCs), VOC, and isocyanates. The OrganoTex technology is based on nature's own chemistry, using plant-based catalysts to bind water repellent hydrophobic polymers directly to the textile fibers. The plant-based catalysts create a 3D-structured fabric surface of hydrophobic molecules, resulting in a water repellent effect (OrganoClick (b)). Schoeller Technologies launched the water repellent Ecorepel Bio, which is based on renewable primary products; agricultural products not used for foodstuffs or animal feed. The plant cuticle is the very outer layer of leaves which protects them from uncontrolled evaporation. This waxy film enveloping the leaves and stems is responsible for water droplets running off the plant. This effect is the exact inspiration for Ecorepel Bio. Ecorepel Bio imitates plants' natural protection with the aid of a finish. It is PFC-free and is obtained entirely from renewable primary products. The finish envelopes the fibers of the fabric in a thin hydrophobic film, providing the repellent effect (Schoeller Technologies, 2019). Zelan R3 is a renewably sourced, plant-based, non-fluorinated fabric treatment for durable water repellency and is manufactured with 63% renewably sourced raw materials, carefully selected to be from non-genetically modified and non-food-source feedstock. The product is based on a new technology from Chemours and marketed by Huntsman. The chemical structure of the product is alkyl urethane. Durability to home laundering and dry cleaning can be improved by adding extenders (Huntsman, 2019). Centexbel evaluated the performance of Zelan R3. Polyester fabric was finished with a formulation based on Zelan R3. The water repellency was assessed according to the spray method (ISO 4920: determination of resistance to surface wetting). The spray rating is determined by comparing the appearance of the specimen with descriptive standards and photographs. The spray rating scale is defined as follows:

0 —Complete wetting of the entire face of the specimen
1 —Complete wetting of the entire specimen face beyond the spray points;
2 —Partial wetting of the specimen face beyond the spray points
3 —Wetting of specimen face at spray points
4 —Slight random sticking or wetting of the specimen face
5 —No sticking or wetting of the specimen face.

Table 10.4 exhibits the performance of polyester fabric finished with Zelan R3 before and after five washing cycles (ISO 6330 at 40°C) followed by line drying or tumble drying. The fabric samples showed no wetting of the surface when the spray test was performed before washing the samples. Addition of blocked isocyanate improved the water repellent performance after washing

TABLE 10.4 Spray test results of polyester fabric finished with Zelan R3.

Formulation	Spray after five washing cycles – line drying	Spray after five washing cycles – tumble drying	Spray
5% Zelan R3	2	5	5
5% Zelan R3 + 0.625% blocked isocyanate	4	5	5

and line drying (spray rating changes from two to four by adding blocked isocyanate). No difference was observed in spray rating between the Zelan R3 formulation with and without extender in case washing was followed by tumble drying due to the higher dry temperature (compared to line drying) which reactivates the repellent product after washing.

10.3.6 Pigments

There are several bio-based dyes and pigments available on a commercial scale, but the use of them is usually limited to food or cosmetic applications. Although natural dyes have been used on textiles for centuries, the use of them is nearly completely replaced by synthetic alternatives. These usually perform better in terms of color fastness and a much broader range of bright colors can be achieved. However, their stability and toxicity is having a harmful effect on the environment due to the release of heavy metals and persistent organic compounds (Bhatt and Rani, 2013). Natural dyes often offer the advantage of being biodegradable and are therefore less accumulative and harmful for the environment. This implicates that waste water can often be treated by natural processes. Biodegradability however has as disadvantage that many natural dyes are less stable and often display a notably poor resistance toward UV. Another disadvantage of natural dyes is the lack of variety in colors. Especially blue or bright shades are difficult to find. Table 10.5 underneath gives an overview of natural colorants that are available at industrial scale. A majority of natural dyes are plant derived and belong to either the terpenes or phenylpropanoids. The more stable colors are usually less vivid and often brownish shades. Some of the dyes display different colors, depending on the used pH (e.g., anthocyanins) or mordants (e.g., tannins).

Most of the bio-based dyes are small molecules that depend on their chemical properties and can be applied to different textile substrates. In textiles, a dye is usually classified as "acid dye" when it bonds to the textile using ionic interactions (typical for polyamide, silk, and wool), "reactive dye" when it makes covalent bonds with the substrate (typically cellulosic substrates), or "disperse dye" when

TABLE 10.5 Overview of natural colorants available at industrial scale.

Name	Type of molecule	Color
Annatto	Terpene	Orange, red
Anthocyanin	Phenylpropanoid	Red, purple
Bio indigo	Indigo	Blue
Brazilwood	Brazilin	Red/pink
Caramel	Sugar/fatty acid	Brown
Carbon	Carbon black	Black
Carmin	Anthracene	Purple
Carminic acid	Anthracene	Red
Carotene	Terpene	Orange
Chlorophyllin	Porphyrin	Green
Curcumin	Phenylpropanoid	Yellow
Lac	Laccaic acid	Red
Logwood	Hematoxylin	Purple/pink
Lutein	Terpene	Orange
Lycopene	Terpene	Red
Madder Lake	Anthraquinone	Red
Norbixin	Terpene	Orange, red
Phycocyanin	Protein	Blue
Riboflavin	Vitamin B2	Yellow
Tannin	Phenylpropanoid	Brown, red, purple

the dye migrates into the synthetic fiber at high temperatures (typically polyester or PLA). Although PLA can be dyed at high temperatures using disperse dyes, it is prone to shrinkage. Therefore, PLA is usually colored during melt-spinning using masterbatches. In general, the field of natural dyes and pigments is considered a niche market for textiles and there are very few companies that focus on it. There are a few suppliers of masterbatches with natural dyes (e.g., Lyondellbasell, Polyone), where biodegradable additives are desired in, for example, PLA-based agrotextiles. For dyes there is a range of natural dyes provided by Archroma under the brand name "EarthColors" (Archroma, 2019). These dyes have been designed specifically as reactive dyes for cellulosic fibers as an alternative for sulfur dyes and should equal them in color fastness. Rubia Natural Colors also provides several natural dyes for broader substrates (Rubia Natural Colors, 2017).

A particular case of dye is phycocyanin, which is a bright blue protein (phycocyanin) derived from certain species of *Arthrospira*, commonly referred to as Spirulina. In a recent research project, a method was developed in which the water-soluble phycocyanin is bound to cellulosic substrates as a water insoluble precipitate. The dye is applied by a simple dipping process and requires a textile which is pretreated with tannin. Although the textiles display good color fastness toward rubbing, temperature, and sweat, the colorfastness toward UV remained poor, illustrating that UV-resistance remains a challenge for many natural dyes (Demedts, 2018).

It should also be noted that a biosourced dye is not always ecologically the better choice. When extraction and purification processes require solvents, distillation or energy-intensive drying steps, the biosourced dye may be ecologically worse compared to a synthetically produced one. LCA studies can provide an objective view on the ecological impact of a dye, but are difficult to find. One such example is indigo, which originates from nature, but is nearly exclusively produced by synthetic means from fossil oil (Saling et al., 2002). There are also biotechnological approaches to produce indigo by genetically modified bacteria and fungi. This has as advantage that the synthesis occurs at low temperatures, but it still requires energy intensive purification steps. These approaches have proven successful so far, and given some improvements in efficiency might in the future be an economically and environmentally alternative to synthetically produced indigo.

Another promising bio-based dye is curcumin which is present in Yellowroot at a concentration of 2%–6% (Sachan and Kapoor, 2007). The powder has a vibrant yellow color and Centexbel is investigating its possibilities as a textile dye. Curcumin was evaluated as dye for woven PLA by varying the pH, concentration, and liquor ratio. The process yielded good results at a pH of 4.5 and minimal coloration at a pH of 8.5, with no noticeable effect of the used concentration and liquor ratios. The color intensity was then further improved by adding bio-based dispersants with a fabric that is up to 4% brighter and 6% more yellow according to the CIE L*a*b color space. The light fastness was determined for all samples but these were extremely poor.

10.3.7 Plasticizers

Plasticizers are used to improve the properties of hard, brittle polymer materials by increasing flexibility, plasticity, process ability, and elongation (Jia et al., 2018). This effect is achieved because plasticizers act as a high boiling solvent for the polymers which break interactions between the chains resulting in increased mobility and decreased crystallinity (Bleys, 2015). Secondary properties such as optical clarity, fire behavior, and conductivity are also influenced.

Due to these improvements plasticizers are used in a wide range of polymers with a total annual production that is estimated to be around five million ton

(Lowell Center for Sustainable Production at the University of Massachusetts Lowell, 2011). Most plasticizers include ester functionalities, such as adipates, citrates, and phthalates. Phthalates are most commonly used and account for 85% of the plasticizers, which are mainly used in combination with PVC. Plasticized PVC is used in a wide range of applications such as flooring, hoses, blood bags, and textile coatings. This means that PVC industries use 90% of the global plasticizer production (Bleys, 2015). Plasticizers in textile industry are mainly used in PVC coatings. Plasticizers, especially low molecular weight, tend to migrate to the surface of the coating and diffuse in the environment, causing toxicity problems over time (Carbonell-Verdu et al., 2016). Since plasticizers are used on large scale in many applications, they have been extensively tested for possible health and environmental effects. Exposure to phthalates has been proven to lower testosterone levels, decrease lung function, and induce genital failure and deficiencies in the metabolism. A relation between phthalate concentrations and asthma and diabetes has also been observed (Hosney et al., 2018).

Nine low molecular weight plasticizers, classified as category 1B reproductive agents, are listed as Substances of Very High Concern and placed on the REACH Candidate List in Europe. Four of them (dibutyl phthalate, di-(2-ethylhexyl) phthalate, benzyl butyl phthalate, and diisobutyl phthalate) are on the Authorization List (European Plasticisers, 2018). In addition, the US Environmental Protection Agency is concerned about phthalates because of their toxicity and the evidence of pervasive human and environmental exposure to them. They intend to initiate action to address the manufacturing, processing, distribution, and use of eight low molecular weight phthalates. The classified phthalates are also restricted for toys and childcare products in Europe and the United States (U.S. Environmental Protection Agency, 2012). Due to the toxicity of some plasticizers and the legislation ban, researchers and industry have widely studied and developed bio-based and non-bio-based alternatives with a lower toxicity profile. Renewable plasticizers originate from vegetable oils, cardanol, vegetable fatty acids, glycerol, and citric acid (Jia et al., 2018). The majority of epoxidized fatty acid esters are compatible with PVC and the efficiency decreases with increasing molecular weight (Chaudhary et al., 2015). Ester-amides of ricinoleic acid are evaluated as plasticizer in PVC. The glass transition temperature of PVC blends containing up to 40 wt% of the plasticizer decreased down to −13.5°C. Viscosity of PVC decreased with increased addition of ester-amide plasticizer (Savvashe et al., 2015). Epoxidized castor oil–based diglycidyl ester improved the flexibility and thermal stability of PVC. The elongation at break amounted to 332.9% (Chen et al., 2018). The plasticizing effect of epoxidized vegetable oil polyol esters on PVC films was studied. Results indicated that these molecules present good plasticizing effect and can be used as alternative for dioctyl phthalate in PVC films (Jia et al. (2015) and Jia et al. (2016a)). Palm oil–based polyester plasticizer exhibited lower plasticizing effect than dioctyl phthalate for PVC (Jia et al. (2016b)). Benzyl ester (synthesized out of dehydrated castor oil fatty acid and benzyl alcohol) was studied as coplasticizer in combination with soy bean oil

(Mehta et al., 2014). Cardanol, produced form cashew nut liquid shell, presents chemical and physical characteristics close to those of diisononyl phthalate or di-(2-ethylhexyl) phthalate; hence, many studies focused on the synthesis of cardanol-based plasticizers (Caillol, 2018). Esterified cardanols are studied as secondary plasticizers in presence of di-(2-ethylhexyl) phthalate. Results indicated that esterification of cardanol yields a partial miscibility with PVC, whereas esterification and subsequent epoxidation yield a complete miscibility with PVC (Greco et al., 2010). Epoxidation of cardanol derivatives resulted in higher plasticizing efficiency. Mechanical properties of soft PVC plasticized by cardanol epoxy derivatives were comparable to those of PVC with commercial plasticizers. Aging tests showed that higher yield of epoxidation causes a reduction of plasticizer leaching (Greco et al., 2016). Epoxidized cardanol butyl ether as plasticizer for PVC instead of phthalates enhanced the thermal and mechanical properties (Li et al., 2017). Citric acid—based plasticizers are being produced on large scale. Citric acid ester plasticizers are environment friendly, nontoxic, and renewable. They are approved in the European Union and the United States to be used in products in close contact with the human body. Acetyl tributyl citrate and tributyl citrate are the most common citric acid—based plasticizers. Tributyl citrate is very compatible with PVC. The main drawback of citrate plasticizers is the relatively high cost price (Jia et al., 2018). Glycerol-based plasticizers are also assessed as alternative phthalate-free plasticizers. Low molecular weight glycerol triester plasticizers are synthesized from butanoic, propanoic, isobutanoic, isopentanoic, and benzoic acids. Results demonstrate that these molecules can be used as poly(vinyl chloride) plasticizers in some applications (Palacios et al., 2014).

Jungbunzlauer produces renewable phthalate-free citrate ester plasticizers (Citrofol). Their product range comprises tributyl citrate (Citrofol BI), tributyl o-acetylcitrate (Citrofol BII), and tris (2-ethylhexyl) o-acetyl citrate (Citrofol AHII). Citrofol BI and Citrofol AHII are used in coatings and artificial leather (Jungbunzlauer, 2019). Danisco markets Grindsted Soft-N-Safe, a renewable and biodegradable acetylated monoglyceride PVC plasticizer, based on hydrogenated castor oil, glycerine, and acetic acid. Fig. 10.14 represents the two main components of Grindsted Soft-N-Safe: acetylated glycerol monoester on 12-hydroxystearic acid (85%) and acetylated glycerol monostearate (10%) (Danisco, 2019).

Oleon manufactures bio-based chemicals out of vegetable oil and fats, of which plasticizers is one of the applications. Radiamuls Acetem 2130 and

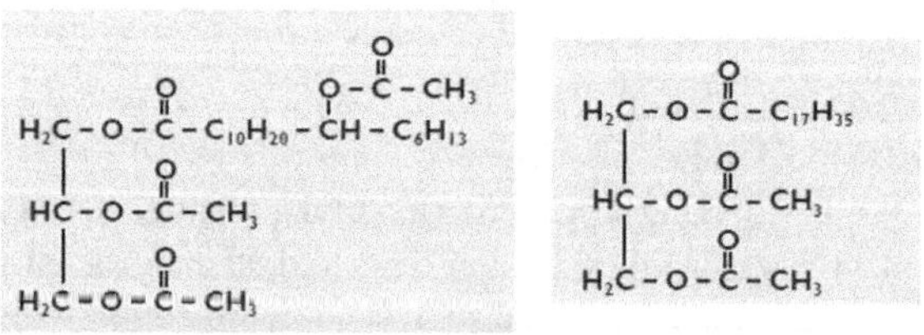

FIGURE 10.14 The two main components of Grindstef Soft-N-Safe (Danisco, 2019).

Radiamuls Acetem 2134 (acetylated monoglyceride) can be used as bio-based phthalate free (secondary) plasticizers for PVC (Oleon, 2019).

10.3.8 Other bio-based additives

Other functional additives used in textile coating and finishing include softening agents, abrasion resistant additives, and odor managing agents. Abrasion resistance is an important parameter determining the lifetime of textile. Additives improving the abrasion resistance of textiles include waxes, nano and microparticles, and surface active or lubricating agents to reduce the friction between individual fibers. BYK Additives & Instruments developed Ceraflour 1000, a bio-based and biodegradable micronized polymer with waxlike properties to improve the abrasion resistance of surfaces (BYK Additives & Instruments, 2019). Ceraflour 1000 can be used in textile coating to improve the abrasion resistance but also for matting the surface.

Softening agents can also be applied onto textiles to improve the abrasion resistance by smoothing the surface of the fibers. Softening agents confer soft, supple, and smooth fabric handle and better drape. Polysiloxanes are commonly used as softening agents, but some companies also commercialize bio-based softening agents. Rudolf Group markets Rucofin Avo New (silicone softener with avocado oil) and Rucofin Lan New (silicone softener with lanolin) for all type of fibers (Rudolf Group, 2019).

Cyclodextrins are cyclic oligosaccharides produced from starch by enzymatic conversion. The three major cyclodextrins are α-, β-, and γ-cyclodextrin. Cyclodextrins can form inclusion complexes with a variety of (hydrophobic) molecules. Thus textiles finished with cyclodextrins can be used to remove unpleasant odors since hydrophobic odor molecules are trapped in the cyclodextrin cavities (Bhaskara-Amrit et al., 2011; Dehabadi et al., 2014; Grigoriu et al., 2008). This process is reversible, that is, in contact with water (laundry process), the odor molecules are released and the cyclodextrins are free to absorb a new load. Several processes to attach cyclodextrins to various textile fibers have been developed and different industrial applications in clothing, medical textiles, filtration, etc., are known. Since cyclodextrins can host a wide range of hydrophobic molecules, they can also be used to complex and slowly release substances. Citronella oil was complexed with β-cyclodextrin on cotton and polyester fabrics. The release of citronellal oil shifted from seconds to hours because of the complexation with β-cyclodextrin (Lis et al., 2018). Cotton fabrics functionalized with β-cyclodextrin can be used as filter to capture organic micropollutants (e.g., bisphenol A) out of water or volatile organic compounds. The functionalized cottons can be regenerated multiple times (Alzate-Sanchez et al., 2016).

10.3.9 Tendency

Textile and chemical industry is focusing on the synthesis and application of sustainable nontoxic bio-based materials. Although bio-based materials are already produced and used by the textile industry, a strong shift toward nontoxic products originating from renewable resources is expected in the near future. The driving forces are the increasing consciousness of consumers, the influence of environmental organizations, legislation, and the impact of brands. Much research has been done and will be still done in developing new bio-based and nonhazardous chemicals for textile industry. One of the research topics interesting for textile coating and finishing is the development of bio-based binders, such as bio-based acrylates and bio-based isocyanate-free technologies. Research on appropriate technologies for PU coating is necessary due to increasingly severe legislative restrictions, toxicity concerns, and consumer demands. Accordingly, new technologies adapted by (textile) coating companies need to comply with various regulations (e.g., REACH in the European Union). The European Union has introduced measures to encourage improvements in the safety and health of workers working with diisocyanates ("Framework Directive"). Germany submitted a restriction proposal (~REACH; EC 927-229-7) for diisocyanates and products containing them, such as polyurethane. The committees for risk assessment and socioeconomic analysis agreed to restrict the use of diisocyanates at the workplace, following the proposal by Germany, in the European Union. The main goal is to prevent new cases of occupational asthma from exposure to diisocyanates among industrial workers and professionals (ECHA, 2017). Next to this, coating industry faced a PU shortage and raw material price increase due to a tightening in the supply of isocyanates in 2017. Being able to use PU free of isocyanate would solve the previously discussed problem. Providing an overall solution by developing environment friendly solvent-free bio-based nonisocyanate polyurethane technologies for (textile) coating industry is the next challenging step for the future. Changing to these techniques, companies will improve the safety for the employees working with this type of coatings since emissions of hazardous substances will be decreased. Nonisocyanate polyurethanes are recently developed as a new class of polyurethane polymers to mitigate health concerns. A nonisocyanate route to polyurethane—however not developed to a technical process—is based on the reaction of an activated carbonic acid ($X{-}OCO{-}R^2\text{-}OCO\text{-}X$; with X = chlorine or phenyl) with diamines ($H_2N\text{-}R^1\text{-}NH_2$). An alternative isocyanate- and phosgene free route which leads to hydroxyl-polyurethanes starts with di-, tri-, or multifunctional ethylene carbonates and di-, tri-, or multifunctional amines. By ring opening of the ethylene carbonate rings with amines, urethane- and hydroxyl groups are formed in a polyaddition reaction (Ubaghs et al., 2004; Prömpers et al., 2005; Prömpers et al., 2006; Hahn et al., 2012; Keul et al., 2013).

10.4 Perspectives

The general trend in textile industry is not only applying renewable and less harmful chemicals, but also using less energy and water and implementing recycled or recyclable materials to lower the impact of textile industry on the environment. The solvent and water-free process techniques will gain importance for the textile industry in the near future. Currently many of these techniques (UV coating, hotmelt, digital finishing, and plasma coating) are being assessed and once the feasibility is demonstrated, these techniques will be implemented at large scale. Infrared and UV curing are being evaluated and partly implemented in the textile coating industry instead of thermal curing at high temperature in an oven, resulting in a significant decrease in energy consumption.

Nonetheless, novel techniques are being assessed; the last decade has experienced a strong increase in technological developments for improved material functionality. Recently innovation in textile industry is primarily focusing on new environmentally friendly chemicals with lower environmental impact. Therefore, textile industry *depends on the developments done by chemical producers. The search for renewable chemicals will be continued by the chemical industry, exploring new synthesis* and conversion routes and new renewable resources (e.g., marine resources such as algae). As renewable chemicals with the required functionalities (such as FR, antimicrobial, or water-repellent properties) will be widely available in the future, these will be commonly used by the textile industry.

As the use of renewable chemicals is one route to lower the impact of the textile industry, another option is recycling materials. Aquafil produces ECONYL regenerated nylon by recovering nylon waste (e.g., carpets destined for landfill, end-of-life fishing nets, etc.) and turning it back into virgin quality nylon (Aquafil, 2019). The ECONYL regenerated nylon is processed into carpet yarn and textile yarn for the fashion and interior textile industry (Econyl, 2019). Centexbel, FILK, and Hochschule Niederrhein are examining the implementation of recycled polyvinyl butyral, which is used in laminated safety glass, as coating binder in textile applications. Research is also done in enabling circular use of coated textiles. Coated textiles need to be designed to enable easy and complete separation of coating and textile. In the project EcoMeTex Centexbel developed an adhesive layer which can be easily separated from the textile by an external stimulus. VinyLoop, patented by Solvay, is a technology to recycle polyvinyl chloride waste in a recycled polyvinyl chloride compound, by selective dissolution and filtration.

Textile structures are being increasingly used in multifunctional lightweight composite structures as reduction of mass is one of the major drivers to develop new materials and concepts. With the increasing availability of bio-based resins and fibers, the application of bio-based textile reinforced composites increased strongly in the past decade. According to the Nova Institute, the market share of bio-based composites in Europe will rise from an annual volume of over 400,000 tonnes in 2017 to more than 800,000 tonnes by 2027 (Composites Europe). However, bio-based composites still have limitations and barriers compared to traditional (e.g., carbon-based) composites. The large variety within the fibers is a challenging barrier in developing high performing composites. Composites for high performance structural applications contain carbon fibers, since bio-based fibers do not have the required properties. The challenge remains in producing carbon fibers of renewable resources but maintaining the strength of conventional carbon fiber.

Opportunities in developing new bio-based textile and composite products arise; however, the next challenges involve also the design of bio-based sustainable products which can be recycled many times. There is a need for bio-based polymers with high stiffness, high impact, and high durability without impairing recyclability. The BIO4SELF project tackles these drawbacks by developing bio-based self-reinforced composites, based on two polylactic acid grades: one is used to compose the matrix, the second grade to form the high stiffness reinforcing fibers (BIO4SELF, 2019).

10.5 Conclusion

Coated textiles and textile reinforced composites are being used in a wide range of applications. New regulations and the societal concern triggered the search for new products nontoxic for the environment, but there are also the concerns about the depletion of fossil resources. By developing bio-based textiles and composites, the dependency of fossil resources is minimized. In this chapter the implementation of bio-based materials (textile structures and polymer matrices) in textile reinforced composites and textile coatings is discussed. The properties that can be realized with the bio-based technologies and some applications are discussed.

Acknowledgments

The authors gratefully acknowledge the financial support of Flanders innovation and Entrepreneurship (grant agreements IWT120273, IWT145006, HBC.2016.0295 and HBC.2016.0038) and Interreg. Research work is partly financed within the Interreg V program France-Wallonia-Flanders (http://www.interreg-fwvl.eu/nl), a cross-border collaboration program with financial support of the European Fund for Regional Development, and cofinanced by the province West Flanders and the Walloon Region and partly within the Interreg V program Vlaanderen-Nederland, the transregional cooperation program with the financial aid of the European Fund for Regional Development.

References

ABV-Algemeen Belgisch Vlasverbond, February 22, 2019. http://www.vlasverbond.be/EN/the-belgian-flax-industry.

Alongi, J., Carosio, F., Malucelli, G., 2012. Layer by layer complex architectures based on ammonium polyphosphate, chitosan and silica on polyester-cotton blends: flammability and combustion behaviour. Cellulose 19, 1041–1050.

Alongi, J., Carletto, R., Di Blasio, A., Cuttica, F., Carosio, F., Bosco, F., Malucelli, G., 2013. Intrinsic intumescent-like flame retardant properties of DNA-treated cotton fabrics. Carbohydr. Polym. 96, 296–304.

Alzate-Sanchez, D., Smith, B., Alsbaiee, A., Hinestroza, J., Dichter, W., 2016. Cotton fabric functionalized with a β-cyclodextrin polymer captures organic pollutants from contaminated air and water. Chem. Mater. 28, 8340–8346.

Anacarda, February 25, 2019. CARDAMINE. http://www.anacarda.com/.

Anon, March 24, 2002. Laminating and Coating: Flexible Future. Textile Month.

AOC, February 25, 2019. AOC | Sustainability. https://www.aoc-resins.com/sustainability-leadership.html.

Aquafil, February 15, 2019. https://www.aquafil.com/sustainability/econyl/.

Archroma, February 14, 2019. http://www.bpt.archroma.com/earthcolorsbyarchroma/.

Ariadurai, S., 2012. Bio-composites: current status and future trends. In: 5th International Technical Textiles Conference.

Arkema, February 25, 2019. Bio-Based Products. Arkema.com. https://www.arkema.com/en/innovation/bio-based-products/.

Ashland, February 25, 2019. Ashland | Envirez™ Unsaturated Polyester Resins. https://www.ashland.com/industries/energy/renewable/envirez-unsaturated-polyester-resins.

Barkoula, N., Garkhail, S., Peijs, T., 2010. Biodegradable composites based on flax/polyhydroxybutyrate and its copolymer with hydroxyvalerate. Ind. Crops Prod. 31, 34–42.

Benega, M., Raja, R., Blake, J., 2017. A preliminary evaluation of bio-based epoxy resin hardeners for maritime application. Proc. Eng. 200, 186–192.

Bensadoun, F., Verpoest, I., Baets, J., Müssig, J., Graupner, N., Davies, P., Gomina, M., Kervoelen, A., Baley, C., 2017. Impregnated fibre bundle test for natural fibres used in composites. J. Reinf. Plast. Compos. 36, 942–957.

Bertomeu Perelló, D., Garcia-Sanoguera, D., Fenollar Gimeno, O., Boronat, T., Balart Gimeno, R., 2012. Use of eco-friendly epoxy resins from renewable resources as potential substitutes of petrochemical epoxy resins for ambient cured composites with flax reinforcements. Polym. Compos. 33, 683–692.

Bhaskara-Amrit, U., Agrawal, P., Warmoeskerken, M., 2011. Applications of β-cyclodextrins in textiles. Autex Res. J. 11, 94–101.

Bhatt, P., Rani, A., 2013. Textile dyeing and printing industry: an environmental hazard. Asian Dyer 10, 51–54.

Bio-based News, 2018. Good Market Developments for Bio-Based Composites. http://news.bio-based.eu/good-market-developments-for-bio-based-composites/. February 22, 2019.

BIO-ON, February 25, 2019. Bio-on | What. http://www.bio-on.it/what.php?lin=inglese.

BIO4SELF, February 21, 2019. https://www.bio4self.eu/.

BIOMER, PHB, February 25, 2019. http://www.biomer.de/IndexE.html.

Biotex. Biotex Flax. https://compositesevolution.com/biotex/. February 26, 2019.

Bittner, N., 2017. BIO World Congress on Industrial Biotechnology. https://www.bio.org/sites/default/files/1030AM-%20Natalie%20Bittner.pdf.

Bledzki, A., Mamun, A., Lucka-Gabor, M., Gutowski, V., 2008. The effects of acetylation on properties of flax fibre and its polypropylene composites. Express Polym. Lett. 2, 413–422.

Bleys, G., July 2015. Technology watch: plasticizers. Retrieved from: http://puipatents.com.

Bos, H., Van Den Oever, M., Peters, O., 2002. Tensile and compressive properties of flax fibres for natural fibre reinforced composites. J. Mater. Sci. 37, 1683–1692.

Bosco, F., Carletto, R., Alongi, J., Marmo, L., Di Blasio, A., Malucelli, G., 2013. Thermal stability and flame resistance of cotton fabrics treated with whey proteins. Carbohydr. Polym. 94, 372–377.

Brighton, J., Zhu, J., Zhu, H., Abhyankar, H., Avril, C., Immonen, K., 2015. Novel hybrid flax reinforced supersap composites in automotive applications. Fibers 3, 76–89.

Business Wire, January 31, 2019. https://www.businesswire.com/news/home/20160510006018/en/Global-Bio-based-Polyurethane-Market-2016-2020.

BYK Additives & Instruments, February 11, 2019. https://www.byk.com/en/additives/additives-by-name/ceraflour-1000.php.

Caillol, S., 2018. Cardanol: a promising building block for biobased polymers and additives. Curr. Opin. Green Sustain. Chem. 14, 26–32.

Campaner, P., D'Amico, D., Longo, L., Stifani, C., Tarzia, A., 2010. Innovative Bio-Based Composites for Automotive Applications. JEC Magazine #56.

Carbonell-Verdu, A., Garcia-Sanoguera, D., Jordá-Vilaplana, A., Sanchez-Nacher, L., Balart, R., 2016. A new biobased plasticizer for poly(vinyl chloride) based on epoxidized cottonseed oil. J. Appl. Polym. Sci. 133 (27), 1–10.

Cardolite, February 25, 2019. Homepage – cardolite corporation – chemistry for tomorrow. https://www.cardolite.com/.

Chaudhary, B., Nguyen, B., Zamanskiy, 2015. Dialkyl furan-2,5-dicarboxylates, epoxidized fatty acid esters and their mixtures as bio-based plasticizers for poly(vinylchloride). J. Appl. Polym. Sci. 132, 42382.

Chen, J., Wang, Y., Huang, J., Li, K., Nie, X., Jiang, J., 2018. Epoxidized castor oil-based diglycidyl-phthalate plasticizer: synthesis and thermal stabilizing effect on pol(vinyl chloride). J. Appl. Polym. Sci. 135, 47142.

Cheng, X., Ma, K., Li, R., Ren, X., Huan, T., 2014. Antimicrobial coating of modified chitosan onto cotton fabrics. Appl. Surf. Sci. 309, 138–143.

Coatings World, January 30, 2019. https://www.coatingsworld.com/issues/2018-06-01/view_europe-reports/europe-moves-toward-bio-based-renewable-resources.

Conway, R., 2016. Chapter 8: Coating of Textiles, Handbook of Technical Textiles. Woodhead Publishing, pp. 211–229.

Corbion, February 25, 2019. Total Corbion PLA. https://www.corbion.com/bioplastics/total-corbion-pla.

Costes, L., Laoutid, F., Dumazert, L., Lopez-cuesta, J.-M., Brohez, S., Delvosalle, C., Dubois, P., 2015. Metallic phytates as efficient bio-based phosphorous flame retardant additives for poly(lactic acid). Polym. Degrad. Stabil. 119, 217–227.

Costes, L., Laoutid, F., Aguedo, M., Richel, A., Brohez, S., Delvosalle, C., Dubois, P., 2016. Phosphorus and nitrogen derivatization as efficient route for improvement of lignin flame retardant action in PLA. Eur. Polym. J. 84, 652–667.

Covestro (a), January 30, 2019. https://www.coatings.covestro.com/en/Products/Impranil.

Covestro (b), January 30, 2019. https://www.coatings.covestro.com/en/Products/Desmodur/ProductListDesmodur/201508271411/Desmodur-eco-N-7300.

Danisco, February 14, 2019. http://www.danisco-softnsafe.com/fileadmin/user_upload/softnsafe/documents/Brochures/brochure-softnsafe.pdf.

De Smet, D., Weydts, D., Vanneste, M., 2015. Chapter 1: Environmentally Friendly Fabric Finishes, Sustainable Apparel Production, Processing and Recycling. Woodhead Publishing, pp. 1–33.

de Weyenberg, I., Ivens, J., De Coster, A., Kino, B., Baetens, E., Verpoest, I., 2003. Influence of processing and chemical treatment of flax fibres on their composites. Compos. Sci. Technol. 63, 1241–1246.

Dehabadi, V., Buschmann, H.-J., Gutmann, J., 2014. A novel approach for fixation of ß-cyclcodextrin on cotton fabrics. J. Inclusion Phenom. Macrocycl. Chem. 79, 459–464.

Demedts, B., 2018. De blauwe keten. Retrieved from: https://www.centexbel.be/en/projects/de-blauwe-keten.

Di Landro, L., Janszen, G., 2014. Composites with hemp reinforcement and bio-based epoxy matrix. Compos. B Eng. 67, 220–226.

DSM, February 25, 2019. EcoPaXX® – PA410: the Bio-Based, High-Performance Polyamide | DSM Engineering Plastics. https://www.dsm.com/markets/engineering-plastics/en/products/ecopaxx.html.

Dumas, L., Bonnaud, L., Olivier, M., Poorteman, M., Dubois, P., 2014. High performance benzoxazine/CNT nanohybrid network – an easy and scalable way to combine attractive properties. Eur. Polym. J. 58, 218–225.

Dumas, L., Bonnaud, L., Olivier, M., Poorteman, M., Dubois, P., 2016. High performance bio-based benzoxazine networks from resorcinol and hydroquinone. Eur. Polym. J. 75, 486–494.

Eadie, L., Gosh, T., 2011. Biomimicry in textiles: past, present and potential. An overview. J. R. Soc. Interface 8, 761–775.

ECHA, January 29, 2019. https://echa.europa.eu/nl/-/restriction-proposal-on-diisocyanates-and-several-authorisation-applications-agreed-by-rac-and-seac.

Econyl, February 15, 2019. In: https://www.econyl.com/the-process/.

Ekşi, S., Genel, K., 2017. Comparison of mechanical properties of unidirectional and woven carbon, glass and aramid fiber reinforced epoxy composites. Acta Phys. Pol. A 132, 879–882.

Elanchezhian, C., Ramnath, B., Ramakrishnan, G., Rajendrakumar, M., Naveenkumar, V., Saravanakumar, M., 2018. Review on mechanical properties of natural fiber composites. Mater. Today: Proc. 5, 1785–1790.

Entropy Resins, February 25, 2019. Epoxy Resins Glue System | Bio Resin | Epoxy Adhesive. https://entropyresins.com/.

Eriksson, P., Jakobsson, E., Fredriksson, A., 2001. Brominated flame retardants: a novel class of developmental neurotoxicants in our environment? Environ. Health Perspect. 109, 903–908.

European Plasticisers, February, 12, 2019. https://www.plasticisers.org/regulation/reach/.

EUROSTAT, January 29, 2019. https://ec.europa.eu/eurostat/statistics-explained/index.php/Waste_statistics.

Eve, S., Jernot, J., Gomina, M., Breard, J., Charlet, K., 2009. Tensile deformation of a flax fiber. Proc. Eng. 1, 233–236.

Evonik, February 25, 2019. VESTAMID®—A History of the Evolution of High-Performance Polyamides. https://www.vestamid.com/product/vestamid/en/products-services/VESTAMID-terra/.

Farah, S., Anderson, D., Langer, R., 2016. Physical and mechanical properties of PLA, and their functions in widespread applications — a comprehensive review. Adv. Drug Deliv. Rev. 107, 367–392.

Farrington, D., Lunt, J., Davies, S., Blackburn, R., 2005. Poly(lactic acid) fibers. Biodegrad. Sustain. Fibres 191–220.

Feldmann, M., Bledzki, A., 2014. Bio-based polyamides reinforced with cellulosic fibres – processing and properties. Compos. Sci. Technol. 100, 113–120.

Ferry, L., Dorez, G., Taguet, A., Otazaghine, B., Lopez-Cuesta, J., 2014. Chemical modification of lignin by phosphorus molecules to improve the fire behavior of polybutylene succinate. Polym. Degrad. Stabil. 113, 135–143.

Fu, X., Shen, Y., Jiang, X., Huang, D., Yan, Y., 2011. Chitosan derivatives with dual-antibacterial functional groups for antimicrobial finishing of cotton fabrics. Carbohydr. Polym. 85, 221–227.

Fung, W., 2002. Chapter 1: General Survey, Coated and Laminated Textiles. Woodhead Publishing and CRC Press, pp. 1–23.

Gao, Y., Cranston, R., 2008. Recent advances in antimicrobial treatments of textiles. Textil. Res. J. 78, 60–72.

Gardetti, M., Muthu, S., 2015. Handb. Sustain. Luxury Text. Fash. 1, 1–211.

Giannis, S., Arnold, E., Hoydonckx, H., Weager, B., Martin, R., 2008. Development of high performance bio-composite based on furan bio resin for vehicle panel. In: 13th European Conference on Composite Materials, pp. 1–10.

Glawe, A., Reuscher, R., Koppe, R., Kolbusch, T., 2003. Hot melt application for functional compounds on technical textiles. J. Ind. Text. 33, 85–92.

Goutianos, S., Van der Schueren, L., Beauson, J., 2019. Failure mechanisms in unidirectional self-reinforced biobased composites based on high stiffness PLA fibres. Compos. Appl. Sci. Manuf. 117, 169–179.

Graupner, N., Müssig, J., 2011. A comparison of the mechanical characteristics of kenaf and lyocell fibre reinforced poly(lactic acid) (PLA) and poly(3-hydroxybutyrate) (PHB) composites. Compos. Appl. Sci. Manuf. 42, 2010–2019.

Greco, A., Brunetti, D., Renna, G., Mele, G., Maffezzoli, A., 2010. Plasticizer for poly(vinyl chloride) form cardanol as a renewable resource material. Polym. Degrad. Stabil. 95, 2169–2174.

Greco, A., Ferrari, F., Maffezzoli, A., 2016. Effect of epoxidation yield of cardanol derivative on the plasticization and durability of soft PVC. Polym. Degrad. Stabil. 134, 220–226.

Grigoriu, A.-M., Luca, C., Grigoriu, A., 2008. Cyclodextrins applications in the textile industry. Cellul. Chem. Technol. 42, 103–112.

Guo, J., Xiang, H.-X., Wang, Q., Guan, F.-C., 2010. Research Progress of Hot-Melt Adhesives. China Adhes, 2010-07.

Haag, K., Padovani, J., Fita, S., Trouvé, J.-P., Pineau, C., Hawkins, S., De Jong, H., Deyholos, M., Chabbert, B., Müssig, J., Beaugrand, J., 2017. Influence of flax fibre variety and year-to-year variability on composite properties. Ind. Crops Prod. 98, 1–9.

Hahn, C., Keul, H., Möller, M., 2012. Hydroxyl functional PUs and polyesters: synthesis, properties and potential biomedical application. Polym. Int. 61, 1048–1060.

Horrocks, A., 2018. Chapter 9: Smart Flame Retardant Textile Coatings and Laminates, Smart Textile Coatings and Laminates. Elsevier, pp. 205–236.

Hosney, H., Nadiem, B., Ashour, I., Mustafa, I., El-Shibiny, A., 2018. Epoxidized vegetable oil and bio-based materials as PVC plasticizer. J. Appl. Polym. Sci. 135, 46270.

Hu, R., Lim, J., 2007. Fabrication and mechanical properties of completely biodegradable hemp fiber reinforced polylactic acid composites. J. Compos. Mater. 41, 1655–1669.

Huntsman, February 4, 2019. http://www.huntsman.com/textile_effects/a/Solutions/Product%20Highlights/Chemicals/Zelan%20R3.

IMCD, January 30, 2019. https://www.imcdus.com/unirez/.

Jia, P., Zhang, M., Hu, L., Feng, G., Bo, C., Zhou, Y., 2015. Synthesis and application of environmental castor oil based polyol ester plasticizers for poly(vinyl chloride). ACS Sustain. Chem. Eng. 3, 2187–2193.

Jia, P., Zhang, M., Hu, L., Zhou, Y., 2016a. Green plasticizers derived from soybean oil for poly(vinyl chloride) as a renewable resource material. Korean J. Chem. Eng. 33, 1080–1087.

Jia, P., Zhang, M., Hu, L., Zhou, Y., 2016b. A novel biobased plasticizer prepared form palm oil and its plasticizing effect on poly(vinyl chloride). Pol. J. Chem. Technol. 18, 9–14.

Jia, P., Xia, H., Tang, K., Zhou, Y., 2018. Plasticizers derived from biomass resources: a short review. Polymers 10, 1303.

Jimenez, M., Guin, T., Bellayer, S., Dupretz, R., Bourbigot, S., Grunlan, J., 2016. Micro-intumescent mechanism of flame-retardant water-based chitosan-ammonium polyphosphate multilayer nanocoating on cotton fabric. J. Appl. Polym. Sci. 43783, 1−12.

Joseph, K., Sabu, T., Pavithran, C., 1996. Effect of chemical treatment on the tensile properties of short sisal fibre-reinforced polyethylene composites. Polymer 37, 5139−5149.

Joshi, M., Butola, B., 2013. Chapter 14: application technologies for coating, lamination and finishing of technical textiles. In: Advances in the Dyeing and Finishing of Technical Textiles. Woodhead Publishing, pp. 355−411.

Joshi, M., Wazed, A., Purwar, R., 2009. Ecofriendly antimicrobial finishing of textiles using bioactive agents-based on natural products. Indian J. Fiber Text. Res. 34, 295−304.

Jungbunzlauer, February 13, 2019. https://www.jungbunzlauer.com/en/products/specialties.html.

Kaneka, February 25, 2019. Kaneka Belgium − Kaneka Biodegradable Polymer PHBH™. http://www.kaneka.be/new-business/kaneka-biodegradable-polymer-phbh.

Keul, H., Mommer, S., Möller, M., 2013. Poly(amide urethane)s with functional/reactive side groups. Eur. Polym. J. 49, 853−864.

Ku, H., Wang, H., Pattarachaiyakoop, N., Trada, M., 2011. A review on the tensile properties of natural fiber reinforced polymer composites. Compos. B Eng. 42, 856−873.

Lepelaar, M., Hoogendoorn, A., Blok, R., Teuffel, P., 2016. Bio-based composite pedestrian bridge. Part 2: materials and production process. In: Proceedings of IASS Annual Symposia, IASS 2016 Tokyo Symposium: Spatial Structures in the 21st Century − New Approaches, Materials & Construction Methods, pp. 1−10 (10).

Li, X., Tabil, L., Panigrahi, S., 2007. Chemical treatments of natural fiber for use in natural fiber-reinforced composites: a review. J. Polym. Environ. 15, 25−33.

Li, X., Nie, X., Chen, J., Wang, Y., 2017. Preparation of epoxidized cardanol butyl ether as a novel renewable plasticizer and its application for poly (vinyl chloride). Polym. Int. 66, 443.

LIBRE − Lignin Based Carbon Fibres For Composites, February 26, 2019. http://libre2020.eu/.

Lis, M., Carmona, O., Carmona, C., Bezerra, F., 2018. Inclusion complexes of citronella oil with β-cyclodextrin for controlled release in biofunctional textiles. Polymers 10, 1324.

Lowell Center for Sustainable Production at the University of Massachusetts Lowell, 2011. Pthalates and Their Alternatives: Health and Environmental Concerns. Retreived from: http://ec.europa.eu/environment/aarhus/pdf/35/Annex_11_report_from_Lowell_Center.pdf.

Lubrizol, January 30, 2019. https://www.lubrizol.com/Engineered-Polymers/Products/Pearlbond-TPU/Pearlbond-ECO-TPU.

Lunder, S., Sharp, R., 2003. Record High Levels of Toxic Fire Retardants Found in Breast Milk From American Mothers. https://www.ewg.org/research/mothers-milk-0. (Accessed 4 February 2019).

Malpani, S., 2013. Antibacterial treatment on cotton fabric from neem oil, aloe vera & tulsi. Int. J. Adv. Res. Sci. Eng. 2, 35−43.

Mamish, A., 1990. Co-extrusion coating of woven fabrics. J. Coat. Fabr. 20, 108.

Mansfield, R., 2003. Combining nonwovens by lamination and other methods. Text. World 153, 22−25.

Martin, N., Mouret, N., Davies, P., Baley, C., 2013. Influence of the degree of retting of flax fibers on the tensile properties of single fibers and short fiber/polypropylene composites. Ind. Crops Prod. 49, 755−767.

Mehta, B., Kathalewar, M., Mantri, J., 2014. Biobased co-plasticizer for PVC in addition with epoxidised soyabean oil. Res. Rev. J. Eng. Technol. 3, 20−30.

Michelena, A., Graham-Jones, J., Summersclaes, J., Hall, W., 2005. Chapter: eco-friendly flax fibre/epoxy resin/composite system for surfboard production. In: Natural Fibres: Advances in Science and Technology Towards Industrial Applications, RILEM Bookseries, vol. 12. Springer, pp. 267–277.

Milbrandt, A., Booth, S., 2016. Carbon Fiber from Biomass. https://www.nrel.gov/docs/fy16osti/66386.pdf. February 26, 2019.

Mishra, S., Mohanty, A., Drzal, L., Misra, M., Parija, S., Nayak, S., Tripathy, S., 2003. Studies on mechanical performance of biofibre/glass reinforced polyester hybrid composites. Compos. Sci. Technol. 63, 1377–1385.

Mitsui Chemicals, January 30, 2019. https://www.mitsuichem.com/en/service/packaging/coatings/stabio/index.htm.

Mohammed, L., Ansari, M., Pua, G., Jawaid, M., Islam, M., 2015. A review on natural fiber reinforced polymer composite and its applications. Int. J. Polym. Sci. 1–15. article ID 243947.

Mohanty, A., Khan, M., Hinrichsen, G., 2000. Influence of chemical surface modification on the properties of biodegradable jute fabrics – polyester amide composites. Compos. Appl. Sci. Manuf. 31, 143–150.

Mohanty, A., Misra, M., Drzal, L., 2001. Surface modifications of natural fibers and performance of the resulting biocomposites: an overview. Compos. Interface 8, 313–343.

Mohanty, S., Nayak, S., Verma, S., Tripathy, S., 2004. Effect of MAPP as a coupling agent on the performance of jute–PP composites. J. Reinf. Plast. Compos. 23, 625–637.

Murariu, M., Dubois, P., 2016. PLA composites: from production to properties. Adv. Drug Deliv. Rev. 107, 17–46.

Nakajima, H., Dijkstra, P., Loos, K., 2017. The recent developments in biobased polymers toward general and engineering applications: polymers that are upgraded from biodegradable polymers, analogous to petroleum-derived polymers, and newly developed. Polymers 9 (523), 1–26.

NatureWorks, February 25, 2019. NatureWorks | Home. https://www.natureworksllc.com/.

Nishino, T., Hirao, K., Kotera, M., Nakamae, K., Inagaki, H., 2003. Kenaf reinforced biodegradable composite. Compos. Sci. Technol. 63, 1281–1286.

Noreen, A., Zia, K., Zuber, M., Tabasum, S., Zahoor, A., 2016. Bio-based polyurethane: an efficient and environment friendly coating systems: a review. Prog. Org. Coating 91, 25–32.

Nostro, A., Rocarro, A., Bisignano, G., Marino, A., Cannatelli, M., Pizzimenti, F., Cioni, P., Procopio, F., Blanco, A., 2007. Effects of oregano, carvacrol and thymol on *Staphylococcus aureus* and *Staphylococcus epidermidis* biofilms. J. Med. Microbiol. 56, 519–523.

Ochi, S., 2008. Mechanical properties of kenaf fibers and kenaf/PLA composites. Mech. Mater. 40, 446–452.

Oksman, K., Skrifvars, M., Selin, J., 2003. Natural fibres as reinforcement in polylactic acid (PLA) composites. Compos. Sci. Technol. 63, 1317–1324.

Oleon, February 13, 2019. http://www.oleon.com/product/derivatives/acetylated-mono-and-diglycerides.

Opwis, K., Gutmann, J., 2011. Surface modification of textile materials with hydrophobins. Textil. Res. J. 81, 1594–1602.

OrganoClick, January 30, 2019. http://www.organoclick.com/site/assets/files/1583/organoclick_oc_biobinder_folder.pdf.

OrganoClick, February 4, 2019. http://organoclick.com/products/performance-textiles-nonwoven/organotex/.

Palacios, O., Rincon, P., Corriou, J.-P., Pardo, M., Fonteix, C., 2014. Low-molecular-weight glycerol esters as plasticizers for pol(vinyl chloride). J. Vinyl Addit. Technol. 20, 65–71.

Pickering, K., Efendy, M.A., Le, T., 2016. A review of recent developments in natural fibre composites and their mechanical performance. Compos. Appl. Sci. Manuf. 83, 98–112.

Pil, L., Bensadoun, F., Pariset, J., Verpoest, I., 2016. Why are designers fascinated by flax and hemp fibre composites? Compos. Appl. Sci. Manuf. 83, 193–205.

Pisupati, A., Park, C.-H., Goethals, F., Willaert, L., Bonnaud, L., 2019, BIOCOMPAL – Résultats. https://www.biocompal.eu/Resultats/, February 25, 2019.

Prömpers, G., Keul, H., Höcker, H., 2005. PUs with pendant hydroxy groups: polycondensation of *D*-mannitol-1,2:5,6-dicarbonate with diamines. Des. Monomers Polym. 8, 547–569.

Prömpers, G., Keul, H., Höcker, H., 2006. PUs with pendant hydroxy groups: polycondensation of 1,6-Bis-*O*-phenoxycarbonyl-2,3:4,5-di-*O*-isopropylidene-galactitol and 1,6-Di-*O*-phenoxycarbonyl-galactitol with diamines. Green Chem. 8, 467–478.

Rahaman, M., Ismail, A., Mustafa, A., 2007. A review of heat treatment on polyacrylonitrile fiber. Polym. Degrad. Stabil. 92, 1421–1432.

Rahmani, H., Najafi, S., Ashori, A., 2014. Mechanical performance of epoxy/carbon fiber laminated composites. J. Reinf. Plast. Compos. 33, 733–740.

Ramon, E., Sguazzo, C., Moreira, P., 2018. A review of recent research on bio-based epoxy systems for engineering applications and potentialities in the aviation sector. Aerospace 5, 110.

Reichhold, February 25, 2019. ENVIROLITE®. http://www.reichhold.com/en/composites-products.aspx?cat=Brands&pid=171.

Rubia Natural Colours, February 14, 2019. https://www.rubia-nc.com/.

Rudolf Group, February 12, 2019. https://www.rudolf.de/en/products/textile-auxiliaries/finishing/silicone-softening-agents/.

Sachan, K., Kapoor, V., 2007. Optimization of extraction and dyeing conditions for traditional turmeric dye. Indian J. Tradit. Knowl. 6, 270–278.

Saling, P., Kicherer, A., Dittrich-Krämer, B., Wittlinger, R., Zombik, W., Schmidt, I., Schrott, W., Schmidt, S., 2002. Eco-efficiency analysis by BASF: the method. Int. J. Life Cycle Assess. 7, 203–218.

Savvashe, P., Kadam, P., Mhaske, S., 2015. Ester-amide based on ricinoleic acid as novel primary plasticizer for poly(vinyl chloride). J. Appl. Polym. Sci. 132, 41913.

Schoeller Technologies, February 4, 2019. https://www.schoeller-textiles.com/en/technologies/ecorepel-bio.

Schuster, J., Govignon, Q., Bickerton, S., 2014. Processability of biobased thermoset resins and flax fibres reinforcements using vacuum assisted resin transfer moulding. Open J. Compos. Mater. 4, 1–11.

Shahidi, S., Aslan, N., Ghoranneviss, M., Korachi, M., 2014. Effect of thymol on the antibacterial efficiency of plasma-treated cotton fabric. Cellulose 21, 1933–1943.

Shim, E., 2010. Chapter 2: Coating and laminating processes and techniques for textiles. In: Smart Textile Coatings and Laminates. Woodhead Publishing, pp. 10–41.

Shim, E., 2018. Chapter 2: Coating and laminating processes and techniques for textiles. In: Smart Textile Coatings and Laminates. Elsevier, pp. 11–46.

Shin-Etsu, February 26, 2019. http://www.shinetsusilicone-global.com/.

Sicomin, February 25, 2019. GreenPoxy® Bio Resins Facilitate Sustainable Production Practices for ZAG™ Skis. http://www.sicomin.com/news/greenpoxy-and-zag-skis.

Smith, W., 2018. Chapter 1: Commentary/overview of textile coating and lamination. In: Smart Textile Coatings and Laminates. Elsevier, pp. 3–10.

Sood, M., Dwivedi, G., 2018. Effect of fiber treatment on flexural properties of natural fiber reinforced composites: a review. Egypt. J. Pet. 27, 775–783.

The matrix: CompositesWorld, February 20, 2019. https://www.compositesworld.com/articles/the-matrix.

TransFurans Chemicals, February 25, 2019. Furfural Chemicals. https://www.polyfurfurylalcohol.com/transfuranschemicals.

Trejo-Machin, A., Puchot, L., Verge, P., 2019. Chapter 4: Design and Synthesis of Bio-Based Benzoxazines. Paint Coatings Industry, IntechOpen, pp. 53–69.

Ubaghs, L., Fricke, N., Keul, H., Höcker, H., 2004. PUs with pendant hydroxyl groups. Synthesis and characterization. Macromol. Rapid Commun. 25, 517–521.

U.S. Environmental Protection Agency, February 12, 2019. https://www.epa.gov/sites/production/files/2015-09/documents/phthalates_actionplan_revised_2012-03-14.pdf.

Varga, C., Miskolczi, N., Bartha, L., Lipóczi, G., 2010. Improving the mechanical properties of glass-fibre-reinforced polyester composites by modification of fibre surface. Mater. Des. 31, 185–193.

Vencorex chemicals, January 30, 2019. https://www.vencorex.com/product/tolonate-x-flo-100/.

Venkateshwaran, N., Elaya Perumal, A., Arunsundaranayagam, D., 2013. Fiber surface treatment and its effect on mechanical and visco-elastic behaviour of banana/epoxy composite. Mater. Des. 47, 151–159.

Vercet, February 14, 2019. https://vercet.natureworksllc.com/~/media/Files/Vercet/Products/NatureWorks_Vercet-Polymeric-Resins_pdf.pdf.

Walentowska, J., Foksowicz-Flaczyk, J., 2013. Thyme essential oil for antimicrobial protection of natural textiles. Int. Biodeterior. Biodegrad. 84, 407–411.

Whiteman, R., 1993. On-line gauging, control and benefits for the coating/laminating process. J. Coat. Fabr. 23, 87–104.

Wicks, Z., Jones, F., Pappas, S., 1999. Chapter 1: Introduction to coatings. In: Organic Coatings: Science and Technology. Wiley, pp. 1–5.

Woodruff, F. Exciting prospect for PUR compounds. Int. Dyer 187, 34.

World Bank, January 29, 2019. http://www.worldbank.org/en/topic/urbandevelopment/brief/solid-waste-management.

Xie, Y., Hill, C., Xiao, Z., Militz, H., Mai, C., 2010. Silane coupling agents used for natural fiber/polymer composites: a review. Compos. Appl. Sci. Manuf. 41, 806–819.

Zeng, X., Mooney, S., Sturrock, C., 2015. Assessing the effect of fibre extraction processes on the strength of flax fibre reinforcement. Compos. Appl. Sci. Manuf. 70, 1–7.

Zhang, Y., Li, Y., Thakur, V., Wang, L., Gu, J., Gao, Z., Fan, B., WU, Q., Kessler, M., 2018. Bio-based reactive diluents as sustainable replacements for styrene in MAESO resin. RSC Adv. 8, 13780–13788.

Zhou, N., Yao, L., Liang, Y., Yu, B., Ye, M., Shan, Z., Qiu, Y., 2013. Improvement of mechanical properties of ramie/poly (lactic acid) (PLA) laminated composites using a cyclic load pre-treatment method. Ind. Crops Prod. 45, 94–99.

Zhou, Z., Weiren, B., Youbo, D., Jinming, D., 2015. Preparation and characterisation of cyclo-triphosphazenekeratin/viscose fibers. Fibers Polym. 16, 550–564.

Zickler, D., 1978. Coating industrial fabrics by means of the melt roll process. J. Coat. Fabr. 8, 121.

Further reading

Singh, R., Srivastava, S., 2017. A critical review on extraction of natural dyes from leaves. Int. J. Home Sci. 3, 100–103.

Index

Note: 'Page numbers followed by "f" indicate figures, "t" indicates tables'.

D

E

F

G

T

CPSIA information can be obtained
at www.ICGtesting.com
Printed in the USA
LVHW080957310120
645445LV00015B/383

9 780128 184936